The Rejection of
CONTINENTAL DRIFT

The Rejection of CONTINENTAL DRIFT

Theory and Method in
American Earth Science

NAOMI ORESKES

New York   Oxford
Oxford University Press
1999

Oxford University Press

Oxford New York
Athens Auckland Bangkok Bogotá Buenos Aires Calcutta
Cape Town Chennai Dar es Salaam Delhi Florence Hong Kong Istanbul
Karachi Kuala Lumpur Madrid Melbourne Mexico City Mumbai
Nairobi Paris São Paulo Singapore Taipei Tokyo Toronto Warsaw

and associated companies in
Berlin Ibadan

Copyright © 1999 by Oxford University Press, Inc.

Published by Oxford University Press, Inc.
198 Madison Avenue, New York, New York 10016

Oxford is a registered trademark of Oxford University Press

All rights reserved. No part of this publication may be reproduced,
stored in a retrieval system, or transmitted, in any form or by any means,
electronic, mechanical, photocopying, recording, or otherwise,
without the prior permission of Oxford University Press.

Library of Congress Cataloging-in-Publication Data
Oreskes, Naomi.
The rejection of continental drift : theory and method
in American earth science / Naomi Oreskes.
p. cm.
Includes bibliographical references and index.
ISBN 0-19-511732-8; ISBN 0-19-511733-6 (pbk)
1. Continental drift.
2. Geology—United States—History—20th century.
I. Title.
QE511.5.O74 1999
551.1'36—dc21   98-4161

1 3 5 7 9 8 6 4 2
Printed in the United States of America
on acid-free paper

To Shlomo Flötzgebirge, my faithful companion,
and to K. B., who sees connections that no one else notices
and always keeps me on track.

PREFACE

This book began in 1978 when I first studied geology at Imperial College in London. I had completed two years as a geology major at a leading U.S. university and counted myself lucky to have chosen a field of science heady in the wake of revolutionary upheaval: geologists around the globe were reinterpreting old data and long-standing problems in the new light of plate tectonics. It seemed a good time to be an aspiring young earth scientist. Imagine my surprise—and dismay—to discover in England that the radically new idea of plate tectonics had been proposed more than half a century before by a German geophysicist, Alfred Wegener, and widely promoted in the United Kingdom by the leading British geologist of his era, Arthur Holmes. The revolution that had been described by my professors in the United States as the radical revelation of a dramatically new vision of the earth was viewed by many of my professors in England as the pleasing confirmation of a long-suspected notion. Whereas my textbooks in the United States had proclaimed the explanatory power of the new ruling theory, Dorothy Rayner, the doyenne of British Stratigraphy, dryly instructed in her text, *Stratigraphy of the British Isles*, "our stratigraphy and history has certainly been illuminated by the current hypothesis, but so far the light shed is somewhat uneven." And although I had only just learned of these ideas two years before, my English flatmate could pull out the dog-eared copy of Arthur Holmes's 1945 textbook she had read in elementary school.

The seeds of an intellectual inquiry were sown. For some years they lay dormant while I lived the life of a field geologist, although the ground in which they lay was being heavily fertilized. Working as a professional geologist in Australia, I learned— often from mildly indignant colleagues—not only that many Australian geologists knew about and believed in the idea of continental drift in the 1940s and 1950s but also that in several instances they were ridiculed at international meetings or on visits to the United States by rude and arrogant Americans. I also learned that other theories of crustal mobility, including the expanding earth hypothesis, had been advocated and in some cases were still being advocated by Australian geologists. One Australian who was receptive to the idea of an expanding earth was my employer, the director of exploration of Australia's third-largest mining company, who periodically circulated papers on this topic among his scientific staff. It was evident that the recent history of earth science was much more complex, much more nationalistic, and much more *interesting* than my professors and textbooks—or my readings in the philosophy of science—had ever suggested.

In the early 1980s, I returned to the United States to pursue graduate studies in geology and again encountered a conundrum. My English training and Australian experience had inculcated in me an inductive methodology, in which scientific problems originated in the observation of geological phenomena in the field, but many of my American professors disdained inductive science and what they pejoratively dismissed as "outcrop" geology. They encouraged me to pursue a deductive strategy and to rely primarily on the tools of laboratory analysis. This was particularly true of younger professors and those who had achieved a high level of professional recognition. The issue was not one of theoretical belief but of methodological commitment. My American and British professors promoted contradictory and ultimately incompatible views about the right way to generate scientific knowledge. Two strands began to merge: divergent visions of the recent history of earth science and divergent methodological commitments. Was there some relation between the two? So began the active portion of the inquiry represented by this book.

My debts are thus spread over several continents. My research advisors at Stanford University, Peter Galison and Marco T. Einaudi, encouraged me to pursue the questions raised in this book while still engaging in scientific research. Neither of these men has ever allowed his thinking to be constrained by the historically contingent boundaries of academic disciplines, and for this I am deeply grateful. Among the Stanford faculty, past and present, I am also indebted to Nancy Cartwright, who profoundly influenced my thinking; to Dennis Bird, John Bredehoeft, John Dupree, Jane Maienschein, James O'Neil, Tjeerd van Andel, and Norton Wise; and to fellow graduate students David Magnus, Carey Peabody, Barbara Bekken, Lisa Echevarria Benatar, Peter Mitchell, Hilary and Jon Olson, Nicolas Rose, and Allan Rubin. In Australia I am indebted to Roy Woodall, formerly Director of Exploration of Western Mining Company, and to colleagues at Western Mining and BP Minerals who shared historical anecdotes; at Imperial College, London, to John Knill, Paul Garrod, Angus Moore, Mike Rosenbaum, Ernie Rutter, Richard Sibson, and the late Janet Watson, all of whom inspired me in important ways; and at Dartmouth College to Claudia Henrion, Richard Kremer, and the late, great Chuck Drake, who I dearly wish had lived to read this book.

Throughout this project I have benefited from the intellectual generosity and moral support of many colleagues: Duncan Agnew, Ron Doel, Robert Dott Jr., Mott Greene, David Kaiser, Homer LeGrand, Phil Pauley, Ron Rainger, Martin Rudwick, and Kenneth Taylor commented on the manuscript; Allan Allwardt, Richard Creath, Henry Frankel, Eli Gerson, Carl-Henry Geschwind, Bruce Hevly, Rachel Laudan, Leo Laporte, Chandra Mukerji, Robert Smith, and Don White provided feedback and information; Helen Wright Greuter and Finley Wright gave me access to their father's letters; Michele Aldrich sent historical tips. Among librarians and archivists, I am indebted to Deborah Day of the Scripps Institution of Oceanography, Charlotte Dirksen and Henry

Lowood at Stanford, Barbara DeFelice and the late Susan George at Dartmouth College, Ronald Brashear at the Huntington Library, Susan Vasquez and Shaun Hardy at the Carnegie Institution of Washington, and the staff of the Yale University Libraries. In Ireland, I thank Patrick Wyse Jackson and the staff of the Trinity College Archives; in South Africa, Leonie Twentyman-Jones and Jill Gribble at the University of Cape Town, the family of Alexander du Toit in Johannesburg, and Mr. Gerry Levin of the Geological Society of South Africa, all of whom provided extremely generous help. Thanks are also due to my research assistants Mary Ann Marcinciewicz, Drew Tenenholz, Jennifer Wehner, Virginia Terry, and Laurie Norris; to Karen Endicott, who proofread the entire text; to the staff of the Dartmouth College Day Care Center and the University Plaza Nursery School; and to Joyce Berry and Cynthia Garver at Oxford University Press.

I have had generous financial support from the National Science Foundation (SBE-9222597 and EAR-9357888), the National Endowment for the Humanities (Fellowships for University Teachers), the Mellon Foundation New Directions Fellowship (Stanford University), the William and Mary Ritter Memorial Fellowship of Scripps Institution of Oceanography, and the Burke Research Initiation Award Program at Dartmouth College. For continuous intellectual support and stimulation, I am grateful to my parents, Susan and Irwin Oreskes; my siblings Michael, Daniel, and Rebecca Oreskes; my husband, Kenneth Belitz, and, above all, my children, Hannah and Clara Belitz, who I hope will never stop asking "why?"

New York N. O.
March 1998

CONTENTS

Introduction: The Instability of Scientific Truth   3

I Not the Mechanism
   1 Two Visions of the Earth   9
   2 The Collapse of Thermal Contraction   21
   3 To Reconcile Historical Geology with Isostasy: Continental Drift   54
   4 Drift Mechanisms in the 1920s   81

II Theory and Method
   5 From Fact to Theory   123
   6 The Short Step Backward   157
   7 Uniformitarianism and Unity   178

III A Revolution in Acceptance
   8 Direct and Indirect Evidence   223
   9 An Evidentiary and Epistemic Shift   262
  10 The Depersonalization of Geology   288

Epilogue: Utility and Truth   313
Notes   319
Bibliography   371
Index   404

The Rejection of
CONTINENTAL DRIFT

Evidence is hard to come by, it is largely circumstantial, and there is never enough of it.
—J. Hoover Mackin

We only know causes by their effects.
—William Whewell

Introduction

The Instability of Scientific Truth

Scientists are interested in truth. They want to know how the world really is, and they want to use that knowledge to do things in the world. In the earth sciences, this has meant developing methods of observation to determine the shape, structure, and history of the earth and designing instruments to measure, record, predict, and interpret the earth's physical and chemical processes and properties. The resulting knowledge may be used to find mineral deposits, energy resources, or underground water; to delineate areas of earthquake and volcanic hazard; to isolate radioactive and toxic wastes; or to make inferences and predictions about the earth's past and future climate. The past century has produced a prodigious amount of factual knowledge about the earth, and prodigious demands are now being placed on that knowledge.

The history of science demonstrates, however, that the scientific truths of yesterday are often viewed as misconceptions, and, conversely, that ideas rejected in the past may now be considered true. History is littered with the discarded beliefs of yesteryear, and the present is populated by epistemic resurrections. This realization leads to the central problem of the history and philosophy of science: How are we to evaluate contemporary science's claims to truth given the perishability of past scientific knowledge? This question is of considerable philosophic interest and of practical import as well. If the truths of today are the falsehoods of tomorrow, what does this say about the nature of scientific truth? And if our knowledge is perishable and incomplete, how can we warrant its use in sensitive social and political decision-making?[1]

For many, the success of science is its own best defense. From jet flight to the smallpox vaccine, from CD players to desktop scanners, contemporary life is permeated by technology enabled by scientific insight. We benefit daily from the liberating effects of petroleum found with the aid of geological knowledge, microchips manufactured with the aid of physical knowledge, materials synthesized with the aid of chemical knowledge. Our view of life—and death—is conditioned by the results of scientific research and the capabilities of technological control. Our moral and political judgments are colored by what science tells us is natural. Nearly every aspect of our lives has been affected by scientific knowledge or technical innovation facilitated by that knowledge. But science and technology have also brought us to the edge of an environmental abyss with its familiar litany of silent crises: global deforestation, ozone depletion, greenhouse effects, groundwater contamination, nuclear annihilation. Most of us stand as mute witnesses to this leviathan: we know surprisingly little about the process of scientific research that has brought us this confusing largesse.

Why do we know so little about how scientific knowledge is generated? Throughout the first half of this century, philosophers and historians of science were concerned primarily with the progression of scientific theories and rarely with the processes by which those theories came to be.[2] The mechanics of scientific research—the choice of questions and methods, the construction and application of instruments, the resolution of contradictions between overlapping data sets—traditionally have received scant attention. The reason was deliberate: analytic philosophers were interested in the logic of scientific theories and their demonstration, not in their genealogy. Viewing science as a rational enterprise, philosophers such as Hans Reichenbach explicitly denied the relevance of the methods by which scientific ideas came to fruition. According to Riechenbach, the philosopher of science was "not much interested in the . . . processes which lead to scientific discoveries. . . . He is interested not in the context of discovery, but in the context of justification."[3]

Reichenbach's ideal has been radically challenged in the last several decades. Most contemporary historians and sociologists of science now believe that the social context of science—the context of discovery—is *more* important that the context of justification.[4] Any theory can be rationally reconstructed to sound logical, even inevitable, in retrospect, but this tells us little about how scientific knowledge actually develops.[5] Scientists are not free agents, historians and sociologists have argued, and the social context of their work not only delimits their options but may even determine the content of their knowledge.[6] And if all knowledge is socially constructed, then objectivity is a chimera. This radical claim strikes at the heart of scientists' beliefs about their enterprise.

Not surprisingly, scientists (and many others) resist such a view and argue that these historians have missed the point. Scientists are interested in truth, and since scientific knowledge works, it must be more or less correct. Human factors sometimes "get in the way" of objective knowledge, but the point of science is to resist such influences. If scientists permit social pressures to distort their perceptions of the physical world, then they are failing to live up to the scientific ideal. Of course, scientists do fail, because they are human, but in most cases it scarcely matters because bad science gets weeded out by the collective scrutiny of the scientific community and—imperfect or not—the surviving knowledge *works*.[7] To attack science simply because it exists within human culture is pettiness, at best. At worst, some believe, it is to attack the very foundations of rationalism.

Historians would argue that it is scientists who have missed the point. *All* science is socially structured, both the good and the bad, and so is the peer review system that adjudicates between them. But can historians prove this point? Historical case studies can illustrate how the development of a particular idea—including our best science—

reflects the constraints of historical situations, and in recent years historians have produced many such studies.[8] But in many such contextualized histories of science, social context is a kind of miasma that pervades scientific thinking in an intangible and ultimately inexplicable manner.[9] The evidence for the role of social forces in the production of scientific knowledge is almost always circumstantial. If scientists are consciously struggling to generate knowledge independent of time and place and historians are claiming that this is an illusory or even meretricious goal, then either scientists are self-deluded or historians are disingenuous.[10] Can these positions be reconciled? Can knowledge be both contingent and transcendent?

Part of the answer may lie in the realm of scientific methods and the process of scientific research. Until recently, few historians, philosophers, or scientists focused their concern on the epistemic status of scientific process.[11] The scientific method, always in the singular, has been taken as monolithic and unproblematic—a textbook cliché, the one sure thing we all know about science. But is there such a thing as *the* scientific method? The answer is clearly no. From the past two decades of historical scholarship, one insight has emerged unequivocally: the methods of science are complex, variegated, and often local. Throughout history scientists have drawn on a wide variety of epistemic commitments and beliefs, linguistic and conceptual metaphors, and material and cognitive resources, all of which have changed with time and varied in space. At different times or in different social contexts, scientists have preferred either inductive or deductive modes of reasoning and argumentation, experimental or theoretical approaches to problem-solving, laboratory- or field-based methodologies. Various attempts have been made to prove the superiority and permanently establish the hegemony of particular methodological approaches, but most of these attempts have failed. The diversity of scientific methodology persists. And whatever methods scientists have chosen, it is through these choices that representations of the natural world have been forged from the infinite sea of sensory perceptions. Scientific practices are the tools with which scientists link the phenomenological world and their representations of that world. Perhaps they provide the link between science and the human world as well.

The Problem of Continental Drift

The story of continental drift illustrates how choices about methods constrained possibilities for scientific truth. In the early part of the twentieth century, a number of scientists—most notably Alfred Wegener—proposed that the relative positions of the continents of the earth were not fixed. Among the leading North American scientists of the 1920s and 1930s—members of the U.S. National Academy of Sciences, presidents of scientific societies, professors at distinguished universities, and men for whom we have medals named today—continental drift was widely discussed and almost uniformly rejected, not merely as unproved, but as wrong, incorrect, physically impossible, even pernicious. American scientists were much more hostile to the idea than their European counterparts; some even labeled the theory *unscientific*.[12] Yet forty years later, the basic central idea of continental drift—that the continents are not fixed but move horizontally over the face of the earth—was established as scientific fact. Why did distinguished scientists adamantly reject as false a claim now universally accepted as true?[13]

The standard answer, to be found in most geology textbooks and many works on the history of science, is that continental drift was rejected for lack of an adequate causal mechanism. Because scientists could not explain *why* continents moved, they concluded that they *could not move*. But this is an example of a rational reconstruction in the history of science. Because moving continents are now accepted as fact, scientists have tended to assume that if the idea was rejected by earlier scientists, then those scientists must have had a good reason to reject it. Perhaps there was not enough evidence. Perhaps the arguments were inadequately articulated. Perhaps the people who proposed the ideas were not well-known, or published their ideas in obscure places. When scientists and historians in the 1960s and 1970s looked back at Alfred Wegener's work, they found an obvious deficit in that the mechanism he proposed was not the same as that accepted today. Indeed, it was patently implausible by today's standards. The implausibility of Wegener's mechanism was taken as the obvious explanation of why his theory was rejected.

But this explanation is anachronistic. What matters in historical argument is not what seems plausible to us, looking backwards, but what was said and done at the time the events took place. What matters is what was plausible to *them*. Viewed this way, the standard account is demonstrably false. Part I of this book shows how Wegener's mechanism was not considered hopelessly implausible at the time it was first proposed; it drew on a large body of work in geodesy and physical geology that pointed to flow in a plastic substrate beneath the crust. Wegener argued that this flow, widely accepted in the 1920s as a demonstrated phenomenon, could help to account for continental drift. Broadly construed, Wegener's argument holds today. But more important, Wegener's proposed mechanism was just the opening round in a series of discussions and debates about the mechanism of continental drift. By the early 1930s, a number of mechanisms had been proposed for continental drift, including the one that is generally accepted as the driving force of plate tectonics: convection currents in the earth's asthenosphere.

So why was continental drift rejected? The thesis of this book is that American earth scientists rejected the theory of continental drift not because there was no evidence to support it (there was ample), nor because the scientists who supported it were cranks (they were not), but because the theory, as widely interpreted, violated deeply held methodological beliefs and valued forms of scientific practice. The idea of the motion of continents, the empirical evidence for it, and the mechanical explanation of it developed by Arthur Holmes have all been corroborated by contemporary earth science.[14] But to accept these ideas in the 1920s or early 1930s would have forced American geologists to abandon many fundamental aspects of the *way they did science*. This they were not willing to do.

The conclusion of this book, therefore, is that *science is not about belief; it is about how belief gets formulated*. At any given moment, only a finite set of knowledge satisfies the reigning criteria for the formulation of scientific belief, and only *this* knowledge is eligible as truth. But the discriminating criteria are historically contingent; over time and across communities, they shift, they evolve, they are overthrown, they transmute. The changing criteria for the formulation of belief provide the pathways through which cultural context delimits the boundaries of scientific knowledge.

I

NOT THE MECHANISM

Joly and Holmes have a beautiful theory,
and I believe it will be epoch-making.
—*Chester Longwell, January 1926*

1

Two Visions of the Earth

Plate tectonics is the unifying theory of modern geology. This theory, which holds that the major features of the earth's surface are created by horizontal motions of the continents, has been hailed as the geological equivalent of the "theory of the Bohr atom in its simplicity, its elegance, and its ability to explain a wide range of observation," in the words of A. Cox.[1] Developed in the mid-1960s, plate tectonics rapidly took hold, so that by 1971, Gass, Smith, and Wilson could say in their introductory textbook in geology:

> During the last decade, there has been a revolution in earth sciences ... which has led to the wide acceptance that continents drift about the face of the earth and that the sea-floor spreads, continually being created and destroyed. Finally in the last two to three years, it has culminated in an all-embracing theory known as "plate tectonics." The success of plate tectonics theory is not only that it explains the geophysical evidence, but that it also presents a framework within which geological data, painstakingly accumulated by land-bound geologists over the past two centuries, can be fitted. Furthermore, it has taken the earth sciences to the stage where they can not only explain what has happened in the past, and is happening at the present time, but can also predict what will happen in the future.[2]

Today moving continents are a scientific fact. But some forty years before the advent of the theory of plate tectonics, a very similar theory, initially known as the "displacement hypothesis," was proposed and rejected by the geological fraternity. In 1912, a German meteorologist and geophysicist, Alfred Wegener, proposed that the continents of the earth were mobile; in the decade that followed he developed this idea into a full-fledged theory of tectonics that was widely discussed and debated and came to be known as the theory of continental drift. To a modern geologist, raised in the school of plate tectonics, Wegener's book, *The Origin of Continents and Oceans*, appears an impressive and prescient document that contains many of the essential features of plate tectonic theory.[3] His ideas were substantiated by an impressive volume of data culled from diverse branches of earth sciences. Yet Wegener's theory was rejected by most of his contemporaries. In the British Commonwealth and in continental Europe, he gained a minority following; in the United States his ideas were resoundingly rejected and in some quarters ridiculed. Thirty years after the last edition of Wegener's book was published, however, a major revolution in earth sciences occurred that incorporated many of the essential features of Wegener's theory. In light of the considerable overlap between Wegener's concepts and a modern theory that commands virtually universal support, why was the theory of continental drift rejected?[4] Why did American geologists in particular react so negatively to the idea of moving continents? To answer these questions, we must first ask where the theory of continental drift came from and what it sought to replace.

Tectonics in the Nineteenth Century

One of the outstanding problems of early-twentieth-century geology was the origin of mountains. Throughout the eighteenth and nineteenth centuries, geologists had focused interest on mountain ranges, for their economic importance as physiographic boundaries and sources of mineral resources, for their scientific importance as repositories of large-scale exposures of the earth's internal structure, and for the psychic impression made by their size and beauty. In the nineteenth century, causal theories of mountain-building became a primary focus of geological concern in both the United States and Europe.[5] Theorists such as Eduard Suess in Austria and James Dana in the United States proposed that mountains formed through compressive stresses generated by a gradual thermal contraction of the whole earth. But these two men—and the colleagues they influenced—held very different views of contraction.

Suess's Collapsing Earth

Eduard Suess (1831–1914) was a field geologist and professor of geology at the University of Vienna (figure 1.1) who dedicated his career to unraveling the structure of the Swiss Alps. From 1883 to 1904, Suess published his magnum opus, a four-volume treatise entitled *Das Antlitz der Erde*, in which he articulated the theory of thermal contraction as the earth's mechanism of mountain-building. (These volumes were published in English between 1904 and 1909 as *The Face of the Earth*, translated by the English geologist Hertha Sollas.) Suess likened the process of terrestrial contraction to the wrinkling of the skin of a desiccating apple and suggested that mountains resulted from a wrinkling of the earth's crust to accommodate diminishing surface area. The earth's contraction, he proclaimed, allowed us to witness "the breaking up of the terrestrial globe."[6]

According to Suess's theory, the features of the earth were explained in terms of vertical motions of its crust. Initially, the entire earth was covered by a continuous continental crust. As the earth cooled and began to shrink, portions of the outer crust collapsed and began to form the ocean floor, causing the earth to become differentiated into continents and oceans. With continued cooling, the remaining uplifted portions were undermined and became unstable, and they collapsed to form the next generation of ocean basins. What had formerly been ocean became dry land. With

Figure 1.1. Eduard Suess. (From Zittel 1901, p. 513.)

each additional increment of cooling, what had been continent became ocean basin and what had been ocean became continent.

The continual transformation of oceans into continents and vice versa explained a number of puzzling geological phenomena, such as the presence of marine deposits on dry land and the numerous alternating episodes of marine and terrestrial conditions recorded in stratigraphic successions. Perhaps most important, the theory explained the well-known similarity of fossil assemblages in widely separate continents: animals and plants had migrated and been dispersed across now sunken continents. Darwin had explained the divergence of plant and animal species as the result of natural selection in isolated communities under contrasting environmental conditions, but paleontologists had found the *same* fossil forms in widely separated continents with radically different biological and climatic environments. Paleontologists concluded that these areas must once have been contiguous and must later have broken apart. Suess's world of gradually decreasing environmental continuity could account for gradually increasing faunal diversity.

This evidence from historical geology was a primary motivation for most paleontologists and stratigraphers to accept the concept of "Gondwanaland"—named after the Gondwana system of India—the term used by Suess to describe the giant supercontinent that had once united much or all of the globe.[7] Sunken portions were referred to as "land bridges" across which ancient species had traveled (although the broad swatches of sunken continents were more like platforms than bridges). But Suess's theory also had consequences for structural geologists: it suggested that mountains would occur across the globe—on all continents and all parts of continents—a prediction borne out by the widespread occurrence of mountains throughout Europe and nearby parts of Asia and North Africa. The collapse of continental blocks was also linked to the origin of igneous intrusions and volcanism, as molten rock escaped from the earth's interior along radial cracks formed during periods of collapse.

Contraction theory was a unifying account of global progression that explained both the history of life and the history of the planet on which life evolved. Writing in

1901, Karl Zittel, professor of geology at Munich and president of the Bavarian Royal Academy of Sciences, triumphantly (and to his mind generously) declared:

> In the hands of one of the most accomplished of foreign geologists and one of the strictest logicians of any age, crust-tectonics [has] been elevated into a new inductive philosophy of earth-configuration.... A work like that of Suess, so cosmopolitan in its standpoint, reminds all workers of their community of aim, rouses each one from the particular to the general, and brings him back with renewed vigor and mental insight to the particular. The time was ripe for an effort to establish systematic clearness in the acquired abundance of detail and to seek for comprehensive laws and principles.

Quoting Marcel Bertrand, pioneer in determinative mineralogy and petrology, Zittel continued, "The creation of a science, like that of a world, demands more than a single day; but when our successors write the history of our science, I am convinced that they will say that the work of Suess marks the end of the first day, *when there was light*."[8]

The European Tradition of Secular Cooling

Suess's work was the culmination of a long-established tradition of interpreting earth history in terms of secular cooling, an idea promulgated in the early nineteenth century by geologists such as Léonce Elie de Beaumont and Henry De la Beche. In 1834, De la Beche had written:

> If we suppose with M. Élie de Beaumont, that the state of our globe is such that, in a given time, the temperature of the interior is lowered by a much greater quantity than on its surface, the solid crust would break up to accommodate itself to the internal mass; almost imperceptibly when time and the mass of the earth are taken into account, but by considerable dislocations according to our general ideas on such subjects.[9]

The first director of the British Geological Survey and founder of the Royal School of Mines, Henry De la Beche (1796–1855) was a principal proponent of the thermal contraction theory in Great Britain, which stood in opposition to Charles Lyell's uniformitarianism, which subsequently has been better known. Lyell argued that past processes must be interpreted in terms of presently observable and incremental causes, even when the empirical evidence on the face of it might suggest otherwise. Thus large gaps in the rock record—unconformities—could and should be explained by slow uplift and erosion, rather than by a sudden event that stripped away the missing strata. In contrast, De la Beche argued that there was strong empirical evidence and good theoretical justification for episodic geological upheaval, progressively decreasing in intensity through geological time.

De la Beche found support for his progressionist views among many geologists who have been labeled in retrospect "catastrophists," such as Adam Sedgwick, who objected to the *a priori* and rigid character of Lyell's philosophy, and William Conybeare, who found steady-state uniformitarianism inadequate to account for many aspects of the geological record, including the obvious power of past orogenic activity displayed in mountain belts.[10] In 1831, De la Beche arranged for publication of part of Elie de Beaumont's theory in the British *Philosophical Magazine,* and his own *Manual of Geology* (1831) and *Researches in Theoretical Geology* (1834) drew heavily on his French colleague's work. Paralleling the French geologist's arguments, De la

Beche proposed a series of periodic, moderate upheavals in earth history to account for the earth's distinct mountain chains.

De la Beche's argument for greater intensity of force at some intervals during geological history was subsequently seconded by William Whewell in his famous account of "The Two Antagonist Doctrines of Geology."[11] Using an empiricist argument against Lyell's "pseudo-empirical" uniformitarianism, Whewell wrote:

> It must be granted at once, to the advocate of . . . geological uniformity, that we are not arbitrarily to assume the existence of catastrophes. The degree of uniformity and continuity with which terramotive forces have acted, must be collected, not from any gratuitous hypothesis, but from the facts of the case. We must suppose the causes which have produced geological phenomena, to have been as similar to existing causes, and as dissimilar, as the effects teach us. . . . But when Mr. Lyell goes further, and considers it a merit in a course of geological speculation that it *rejects* any difference between the intensity of existing and past causes, we conceive that he errs no less than those he censures. . . . We are in danger of error, if we seek for slow causes and shun violent agencies further than the facts naturally direct us, no less than if we were parsimonious of time and prodigal of violence. *Time*, inexhaustible and ever accumulating his efficacy, can undoubtedly do much for the theorist in geology; but *Force*, whose limits we cannot measure, and whose nature we cannot fathom, is also a power never to be slighted: and to call in the one to protect us from the other, is equally presumptuous, to whichever of the two our superstition leans.[12]

Despite Whewell's warnings about the excesses of Lyellianism, De la Beche was overpowered by the more eloquent and more adamant Lyell (although thermal contraction would be taken up again later in the century by British geophysicists Osmond Fisher and George Darwin).[13] In contrast, Léonce Elie de Beaumont (1798–1874) became one of the most influential figures in French science of the mid-nineteenth century. A founder of the Société Géologique de France, Elie de Beaumont was inspired by his teacher at L'Ecole des Mines, René Just Haüy, to look for mathematical regularity in geological phenomena. Professor of physics and mineralogy, Haüy pioneered the science of crystallography and is credited with formulating the mathematical law of rational intercepts to explain crystal morphology and symmetry.[14] As a field geologist with the Corps des Mines, responsible for producing the first geological map of France and delineating the contained mineral resources, Elie de Beaumont sought to find in macroscopic geological phenomena the geometric regularity his teacher had found within microscopic phenomena.

The geometric analysis of geological strata was well established in mines, where tracing of dipping and intersecting strata was essential for exploiting ore horizons and coal seams. But Elie de Beaumont wanted to extend this form of analysis to larger scale geological features, including the largest of all—mountain belts.[15] One key empirical observation was that the orientations of folded structures within mountain belts were not random: deformed structures of similar age tended to be oriented along similar trends, and the observed regularities commonly extended over large geographical areas. Systems of mineral veins within mountain chains characteristically followed a preferred orientation as well. Elie de Beaumont concluded that the orientation of a mountain chain was related to the age of its formation and that the forces involved were global in scale: all of the world's mountains could be connected along a series of great circles that defined a pentagonal network, or *réseau pentagonal*, along which global contraction occurred. Building on the ideas of his predecessor, Georges Cuvier (1769–1832), who had proposed periodic organic upheaval to explain the extinction of fossil species, Elie de Beaumont proposed periodic physical upheaval, caused by gradual secular cooling and thermal contraction of the earth.[16]

Elie de Beaumont's notion of secular cooling was drawn from a consensus forged by several of the leading scientists of seventeenth- and eighteenth-century Europe. Descartes, Leibniz, and Buffon had all supposed that the earth had formed by cooling from a molten mass. Using this assumption, Buffon (1707–1788) produced one of the earliest estimates of the age of the earth—approximately 70,000 years.[17] In 1796, the origin of the solar system by condensation from a hot gaseous cloud was developed theoretically by Laplace, in what came to be known as the Kant–Laplace nebular hypothesis, and the theory was widely accepted by European scientists in the nineteenth century.[18]

If the earth had formed by consolidation from a primordial nebular gas, this had obvious consequences for its structure and subsequent history. One consequence was that the earth was probably still cooling. This seemed to be supported by the available empirical evidence: it was well known that mines were hot, and the temperatures in mines greatly increased with depth. Volcanic phenomena also pointed to a hot, if not necessarily molten, interior. In 1837, Joseph Fourier applied his new method of mathematical analysis to the theory of secular cooling and concluded that the shape of the earth and its thermal gradient proved that it must have formed at high temperature and have been still in the process of cooling from this original hot state.[19] By the mid-nineteenth century, the hot origin of the earth and its subsequent conductive cooling was a well-established scientific principle, one that could be drawn upon in developing theories of the earth's geological history and evolution. Cuvier's and Elie de Beaumont's *révolutions* were part of the normal evolution of the earth's crust. Periodic upheaval was the consequence of continuous secular cooling and resulted in lateral pressure along zones of weakness within the earth. The geological consequence was the formation of mountain belts.[20]

Suess's *Das Antlitz der Erde* was thus the culmination both of one man's life's work and of a long-standing and distinguished tradition in continental European science bound to a dynamic, changing earth. But while contraction theory in Europe was viewed in opposition to Lyell's steady-state uniformitarianism, the situation in the United States was different. There, James Dana had developed an entirely different version of contraction in which oceans and continents were permanent features of the earth's crust.

Dana's Stable Continents and Permanent Oceans

James Dwight Dana (1813–1895) was a professor of geology at Yale University in the latter half of the nineteenth century, the son-in-law of Benjamin Silliman, America's first professor of geology and founder of the *American Journal of Science* (figure 1.2).[21] As a student under Silliman at Yale, Dana focused his early work on the chemistry and properties of minerals. In 1837, at the age of 24, he published his *System of Mineralogy*, a book that remains the basis of modern mineralogy texts.[22] The following year, Dana joined the Wilkes expedition—America's first federally financed scientific expedition. After his return, he married his mentor's daughter and published exten-

Figure 1.2. James Dwight Dana. (Reproduced with permission of Yale University Art Gallery, Bequest of Edward Dana Salisbury, B.A., 1870.)

sively on the results of his observations of geology, volcanology, and paleontology in the Pacific region. In 1850, Dana succeeded his father-in-law as Silliman Professor of Geology at Yale, where he continued his work on the results of the Wilkes expedition. Following in his father-in-law's footsteps, Dana served as editor of the *American Journal of Science* from 1846 to 1890, and he passed the position on to his own son, Edward S. Dana.

In addition to his work on the Wilkes expedition, Dana published some 250 scientific articles on a wide range of subjects, including mineralogy, petrology, and dynamical geology, and a number of popular and pedagogical texts. His *Textbook of Geology* (1863) was widely used in burgeoning geology courses around the United States; this book, along with his position as editor of the *American Journal of Science*, helped to establish Dana as one of the most prominent academic geologists of late-nineteenth-century America.[23]

Dana's interests were both systematic and synthetic, and in the mid-1840s he began work on a theory of the origins of the earth's major physiographic features. His theory rested on three legs: his detailed knowledge of mineralogy, his experience and observations on the Wilkes expedition, and his reading about the properties of the lunar surface. Astronomical observations indicated that the lunar surface was heterogeneous, cratered and mountainous, and perhaps not unlike the surface of the earth.[24] Presuming that lunar craters were similar to the volcanic craters he had seen in the Pacific Islands, Dana theorized that they had formed from similar materials and by similar processes. Dana connected this supposition to the mineralogical observation that different minerals had different melting temperatures and proposed that the lunar topography had formed when the surface of the moon first became solid in an inhomogeneous manner, just as "a melted globe of lead or iron ... when cooling unequally, becomes depressed by contraction on the side which cools last."[25] The lunar craters were the areas of lowest melting temperature; the lunar highlands, the areas of highest temperatures. By analogy, Dana suggested that the continents and oceans of the earth were primordial features formed when the earth solidified from a molten globe.

> The areas of the surface constituting the continents were first free from eruptive fires. These portions cooled first, and consequently the contraction in progress affected most the other parts. The great depressions occupied by the oceans thus began; and for a long period afterward, continued deepening by slow, though it may have been unequal, progress.[26]

In Dana's view, the oceans were permanent, primordial features that continued to be the site of thermal contraction for a long period of geological history; Charles Darwin's study of the subsidence of coral islands in the Pacific Ocean suggested that oceanic sinking still continued. Dana explained the presence of marine deposits on the continents as the product of a former era when "the depth [of former oceans] would be too shallow to contain the seas; and consequently the whole land would be under water."[27] The occurrence of mountain ranges on the edges of continents Dana attributed to a "lateral pressure" induced by "the greater subsidence of the oceanic parts relative to the continental."[28] As the whole earth contracted, its rocks would be squeezed, and the squeezing would be greatest on the continental margins, which were the sites of greatest differential stress.[29] Local vertical uplift might occur in regions subjected to the expansive force of ascendant magma, but the net direction of crustal movement was down (figure 1.3).

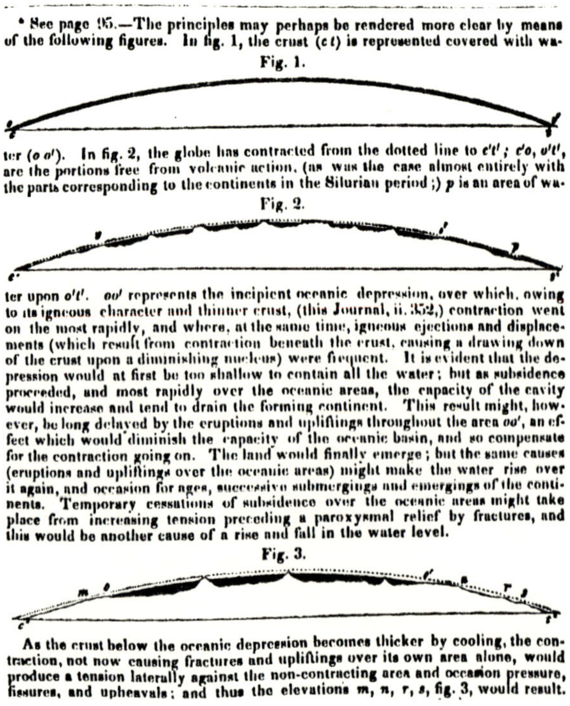

Figure 1.3. Dana's representation of contraction theory. In Dana's model of the earth, the oceans and continents formed by differential thermal contraction early in earth history. (From Dana 1847c.)

* See page 95.—The principles may perhaps be rendered more clear by means of the following figures. In fig. 1, the crust (c t) is represented covered with water (o o').

Fig. 1.

In fig. 2, the globe has contracted from the dotted line to c't', c'o, o't', are the portions free from volcanic action, (as was the case almost entirely with the parts corresponding to the continents in the Silurian period ;) p is an area of water upon o't'. oo' represents the incipient oceanic depression, over which, owing to its igneous character and thinner crust, (this Journal, ii. 352,) contraction went on the most rapidly, and where, at the same time, igneous ejections and displacements (which result from contraction beneath the crust, causing a drawing down of the crust upon a diminishing nucleus) were frequent. It is evident that the depression would at first be too shallow to contain all the water; but as subsidence proceeded, and most rapidly over the oceanic areas, the capacity of the cavity would increase and tend to drain the forming continent. This result might, however, be long delayed by the eruptions and upliftings throughout the area oo', an effect which would diminish the capacity of the oceanic basin, and so compensate for the contraction going on. The land would finally emerge; but the same causes (eruptions and upliftings over the oceanic areas) might make the water rise over it again, and occasion for ages, successive submergings and emergings of the continents. Temporary cessations of subsidence over the oceanic areas might take place from increasing tension preceding a paroxysmal relief by fractures, and this would be another cause of a rise and fall in the water level.

Fig. 2.

Fig. 3.

As the crust below the oceanic depression becomes thicker by cooling, the contraction, not now causing fractures and upliftings over its own area alone, would produce a tension laterally against the non-contracting area and occasion pressure, fissures, and upheavals; and thus the elevations m, n, r, s, fig. 3, would result.

In contrast to Suess's work, a consequence of Dana's theory was that mountains would only form on the edges of continents, and this was borne out by the Appalachians, the principal mountain chain of American experience. Dana's theory also suggested that the highest mountains should correlate with the greatest depth of adjacent ocean floor, and this, too, seemed to be confirmed by his observations of the Andes on the Wilkes expedition. Because continents and oceans were primordial and not interchangeable features, the theory implied that the continents would be composed of different material from the ocean floor; this seemed to be confirmed by what little was known about the ocean floor from dredging and from the study of volcanic islands. On the other hand, unlike Suess's later account, Dana's theory had no explanation for why fossil assemblages should be so alike in distant places.

Because continents and oceans were not interchangeable in Dana's version of contraction, the concept came to be known as "permanence theory." Although clearly a type of contraction theory, resting as it did on the premise of global cooling, Dana's work was perceived by his American colleagues as standing in opposition to the "contraction" hypothesis articulated by Suess. Although starting from the same premise, Suess and Dana had come to very different conclusions about the present state of the globe. Permanence theory became widely accepted in the United States; in the early twentieth century, the permanence of ocean basins was considered by most geologists to be a fact. In 1910, Bailey Willis, a well-known professor of geology at Stanford University, wrote emphatically in the journal *Science* that *"the great ocean basins are permanent features of the earth's surface and they have existed, where they are now, with moderate changes of outline, since the waters first gathered."*[30]

In some ways "permanence" theory was a misnomer, because Dana believed that the earth was still contracting and therefore still changing. But how was it changing? Dana had still to address the central question of tectonics—the origins of mountains. To do this, he had to incorporate, and to some extent subsume, the work of his foremost adversary, James Hall Jr.

Permanent Oceans and the Geosynclinal Theory

The principal mountain belt of North American experience was the Appalachians, and any American theory of mountain-building would have to account for its geology. Many of the details of Appalachian geology had been worked out by two of the United States' most famous early geologists, brothers Henry Darwin Rogers and William B. Rogers, who mapped large portions of the Appalachians in the 1830s and 1840s.[31] One of the principal conclusions of their work was that the asymmetric pattern of deformation and overthrusting evident throughout the Appalachians indicated a compression from east to west—suggesting large-scale tangential movement of the type advocated by De la Beche and Elie de Beaumont as the consequence of global contraction. Rather than interpreting these asymmetric folds as the result of thermal contraction, however, the Rogers brothers attributed them to the disturbing effects of thermal expansion of molten material and gases beneath the crust.

> We suppose the strata of [a folded] region to have been subjected to excessive upward tension, arising from the expansion of the molten matter and gaseous vapors, the tension relieved by linear fissures, through which much elastic vapor escaped, the sudden release of pressure adjacent to the lines of fracture, producing violent

pulsations on the surface of the liquid below. This oscillating movement of the [underlying] fluid mass would communicate a series of temporary flexures to the overlying crust and those flexures would be rendered permanent . . . by the intrusion of molten matter. If, during this oscillation, we conceive the whole heaving tract to have been shoved (or floated) bodily forward in the direction of the advancing waves, the union of the tangential with the vertical wave-like movement will explain the peculiar steepening of the front side of each flexure, while a repetition of similar operations would occasion the folding under, or inversion, visible in the more compressed districts.[32]

The Rogers' complex interpretive framework was not widely accepted by their contemporaries, but it did alert people to the "stupendous mechanical problems" involved in interpreting the complex fold-thrust belts of Appalachian geology.[33] It also motivated James Hall and James Dana to consider these problems in greater detail.

James Hall (1811–1898), State Paleontologist of New York, was the first president of the Geological Society of America (1889), a charter member of the U.S. National Academy of Sciences (1863), and an early president of the American Association for the Advancement of Science.[34] Hall emphatically rejected the Rogers' dynamic interpretation as unsupported by empirical evidence. However, he was greatly impressed by their documentation of the thick sequences of sedimentary rock that had been deformed to create the Appalachian chain and set out to explain this important observation. Borrowing an idea first proposed by John Herschel, Hall proposed the concept that Dana later named a "geosyncline": an area of subsidence and large-scale sedimentary accumulation adjacent to the continents.

Hall suggested that large accumulations of sediments along continental margins contributed to the depression of that continental margin in a kind of positive-feedback loop that resulted in the gradual buildup of a huge sedimentary pile. Sediments, then, accumulated along continental margins; the weight of these sediments depressed the crust along those margins; and this permitted more sediments to accumulate. Eventually, the pile would become so thick that the sedimentary material would be heated and compressed, and, by some mechanism not exactly explicated, it would be uplifted and deformed into a marginal mountain range. Thus the origin of mountains was causally linked to the accumulation of sediments along continental margins, and any apparent periodicity of mountain-building was a consequence of the time required for sedimentary accumulation.[35]

Dana famously criticized Hall's hypothesis as a theory of the elevation of mountains in which "the elevation of mountains is left out"—because the actual mechanism of uplift remained unexplained—and sought to replace Hall's interpretation of the Appalachians with his own.[36] He argued that thick sedimentary piles were not the *cause* of depressions along continental margins but the *result* of them. The Rogers' pattern of asymmetric folding from east to west was clear evidence of lateral compression generated along the American continental margin by the differential thermal contraction of the Atlantic basin and the adjacent North American continent. The zone of sedimentary accumulation formed in response to stresses caused by this differential contraction (not the weight of the sediments themselves); its subsequent deformation was the consequence of the same effect. Although Hall vigorously opposed Dana's theory as excessively speculative, Dana's subsumation of Hall's work was highly effective, and later American colleagues came to see their views as complementary, referring to the "geosynclinal theory of Hall and Dana."[37]

The details of geosyncline theory would be debated well into the twentieth century, but agreement had been forged on a number of counts. Thick piles of sediments accumulated on continental margins where they were altered by heat and pressure — that much was clear — and somehow they became deformed and uplifted into mountain ranges along those margins. Dana's theory gave an account — albeit incomplete — of why this was so: differential contraction produced lateral pressure along the continental margins, which alternately resulted in downfolds and sedimentary accumulation, as well as uplift and deformation of the same. Thus the tallest mountains occurred next to the deepest oceans, and nearly all mountain chains occurred along continental margins.

Dana's account also explained the occurrence of multiple episodes of deformation within a given mountain chain. Continual contraction resulted in repeated episodes of downwarping, sedimentary accumulation, compression, and uplift, and the result was the gradual growth of the continents through accretion of successive mountain belts. Dana's was a unifying theory, bringing together the best of American field work with the long-standing European tradition of interpretation based on the premise of secular cooling.

Two Continents, Two (or More) Theories

At the end of the nineteenth century, geology was an international science, and geologists frequently traveled abroad to visit colleagues and to see sites of interest. Nevertheless, European and American geologists found themselves subscribing to incompatible views of earth evolution. From the same starting point — the secular cooling of the earth — two different pictures emerged. In the European view, the earth was in a state of continual flux with complete interchangeability of its parts. Ocean basins could be elevated into continents, continents could collapse to form ocean basins, and change occurred across the globe. In the American view, the basic outlines of the earth had been set at the beginning of geological time and had not changed fundamentally since then. Continents were always continents, oceans were always oceans, and change was confined to discrete zones at the interface between them. The two theories also differentially weighed the available facts. The American perspective emphasized the physical properties of minerals, the contrasting compositions of continental rocks and the ocean floor, and the asymmetry of folding in the Appalachians. The European view emphasized the biogeographical patterns, the stratigraphic evidence of interchangeability of land and sea, and the diverse patterns of folding in European and African mountain belts.

In Great Britain — which had little in the way of mountains to explain — neither theory was entirely accepted. A tension persisted between the uniformitarian and progressionist perspectives, in which field-oriented geologists leaned mostly toward the former and mathematically oriented physical geologists held to the latter. On the face of it, geologists who accepted the uniformitarian view might have been inclined toward Suess, whose earth history was less obviously directional than Dana's. But the *tone* of Suess's theory was cataclysmic — with its collapsing continents and "breaking up of the terrestrial globe." In contrast, Dana's theory was explicitly progressionist—Dana believed that the earth's heat was running down and the globe would ultimately decay to darkness—but the tone of his theory was gradualistic, if not nearly steady-state.[38] The major patterns of the earth had been established at the beginning of geological time, and the subsequent changes were local and gradual. For progressionists, Dana's theory was sufficiently progressive, but for uniformitarians, it was easily interpreted as portraying an essentially steady-state earth.

Thus, by the start of the twentieth century, two different versions of earth history were being handed to the next generation: Europeans inherited a vision of a constantly shifting earth—with an unsteady history and an uncertain future—whereas Americans inherited a vision of a much more stable place.

2

The Collapse of Thermal Contraction

In 1901, Karl Zittel, president of the Bavarian Royal Academy of Sciences, declared that "Suess has secured almost general recognition for the contraction theory" of mountain-building.[1] This was wishful thinking. Suess's *Das Antlitz der Erde* was indeed an influential work, but by the time Suess finished the final volume (1904), the thermal contraction theory was under serious attack. Problems were evident from three different but equally important quarters.

Horizontal—Not Vertical—Displacement in the Swiss Alps

The most obvious problem for contraction theory arose from field studies of mountains themselves. As early as the 1840s, it had been recognized that the Swiss Alps contained large slabs of rock that appeared to have been transported laterally over enormous distances.[2] These slabs consisted of nearly flat-lying rocks that might be construed as undisplaced, except that they lay on top of younger rocks. In the late nineteenth century, several prominent geologists, most notably Albert Heim (1849–1937), undertook extensive field work in the Alps to attempt to resolve their structure. Heim's detailed field work, beautiful maps, and elegant prose convinced geological colleagues that the Alpine strata had been displaced horizontally over enormous distances. In some cases, the rocks had been accordioned so tightly that layers that previously extended horizontally for hundreds of kilometers were now reduced to distances of a few kilometers. But in even more startling cases, the rocks were scarcely folded at all, as if huge slabs of rocks had been simply lifted up from one area of the crust and laid down in another.

Heim interpreted the slabs of displaced rock in his own Glarus district as a huge double fold with missing lower limbs, but in 1884 the French geologist Marcel Bertrand (1847–1907) argued that these displacements were not folds but faults. Large segments of the Alps were the result of huge faults that had thrust strata from south to north, over and on top of younger rocks.[3] August Rothpletz (1853–1918), an Austrian geologist, realized that the Alpine thrust faults were similar to those that had been earlier described by the Rogers brothers in the Appalachians.[4] By the late 1880s, thrust faults had been mapped in detail in North America, Scotland, and Scandinavia.

In many cases the thrust faults appeared to have formed along the bottom of huge recumbent folds—folds whose axial planes were horizontal, rather than vertical, implying that they had been formed by compression on one side only (figure 2.1). Apparently, these folds had been subjected to so much lateral compression that they had rolled over on their sides and dislocated from the underlying strata. By the end of the century, the existence of such nappes, as they came to be called, was considered established, and in 1903 Pierre Termier (1859–1930), formerly a student of Bertrand, used the theory of nappes to produce the first unified geological interpretation of the European Alps.[5]

Although considerable argument ensued over the mechanism of nappe formation, most Alpinists accepted the reality of their existence. The most widely cited account of how they might form came from Heim. On the basis of detailed microscopic examination, Heim demonstrated that fossil molluscs in rocks from the basal portions of the Glarus nappes had been flattened and stretched up to ten times their original lengths. Heim suggested that during recrystallization the rocks had behaved in a plastic manner, thus accounting for the stretched rather than broken fossils, allowing the

Figure 2.1. The formation of nappes. (A) Most geological folds have near-vertical axial planes, with strata sloping outward on both sides, implying bilateral compression. (B) However, if compression is asymmetrical, the fold will incline away from the direction of maximum stress and the dip of the strata will be steeper on one side of the fold than on the other. (C) If the asymmetry of compression is great enough, the fold will "roll over" onto its side; in this case, the axial plane and the strata will lie in a nearly horizontal position—this is a nappe. Note that the sequence of strata on the upper limb is the same as if the rocks had not been folded at all, but on the lower limb the order is reversed. (Illustration by Melody Brown.)

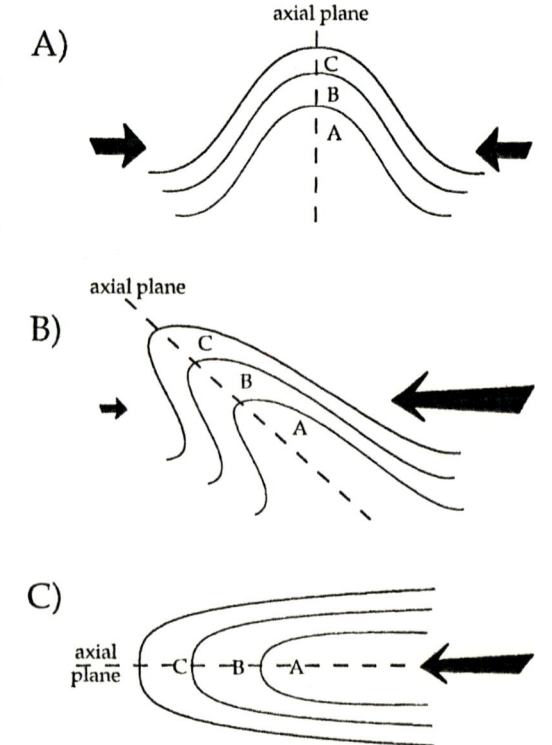

rocks to flow rather than crumple under pressure. In his classic work on the geology of the Alps, *Untersuchungen über den Mechanismus der Gebirgsbildung*, published in 1878, Heim postulated an upper "zone of fracture" in the crust underlain by a lower "zone of flow." But regardless of the exact mechanism, the geological evidence of lateral displacement was clear. So were the theoretical implications: to account for these huge displacements by terrestrial shrinkage would have required an impossibly large primordial earth.[6]

On Heim's original double-fold theory, the strata of the Glarus district were interpreted to have been compressed to 50 percent of their original length; now it appeared that they had been reduced to 20 percent or less. Such extreme shortening could not be a consequence of terrestrial contraction. Contrary to conventional geological wisdom, there was simply no way to shrink the earth enough to generate this much compression. Some other mechanism was required. Dana's version of contraction theory could perhaps withstand this attack: his theory limited mountains to the edges of ocean basins, and extreme shortening in focused regions was perhaps plausible; it need not be extrapolated to the entire earth. But Suess's version of global generation of mountains by contraction appeared to be completely at odds with these data, gathered in his own geological backyard. It was becoming evident that mountains were not caused by vertical movements of the crust, as contraction theory would have it, but by focused horizontal shortening. And this was as true in the Appalachians as it was in the Alps. Speaking on the occasion of his presidential address to the Geological Society of America in 1893, J. William Dawson declared, "[Both] the Alps and the Appalchians are mountains of crumpling, showing evidence of enormous lateral pressure proceeding from the adjoining sea basins, and to this, it is now almost universally admitted, their elevation must in great part be due."[7] Vertical uplift was a side effect of horizontal compression, not the reverse.

The Idea of Isostasy

A second major problem arose from the rapidly expanding fields of geodesy and geophysics. While Heim and his colleagues were mapping the details of the Swiss Alps, workers with the Great Trigonometrical Survey of India were undertaking extensive geodetic measurements to produce accurate maps of British colonial holdings.[8] In the early 1850s, Colonel (later Sir) George Everest, the Surveyor-General of India, discovered a discrepancy in the measured distance between two geodetic stations there, Kaliana and Kalianpur, 370 miles apart. When measured on the basis of triangulation, the latitude difference was five seconds greater than when computed on the basis of astronomical observation. Everest thought that the difference might be due to the gravitational attraction of the Himalayas on surveyors' plumb bobs and enlisted John Pratt, a Cambridge-trained mathematician and the Archdeacon of Calcutta, to look more closely at the problem. Pratt computed the expected deflection based on the observable mass of the Himalayas and discovered that the discrepancy was actually *less* than it should have been: it was as if part of the mass of the Himalayas were somehow missing. When Pratt pursued the question, he found that the converse was true too: in coastal regions, plumb bobs were deflected toward the oceans, implying that the force of gravity over the oceans was not as low as expected considering the low density of water (figure 2.2).[9]

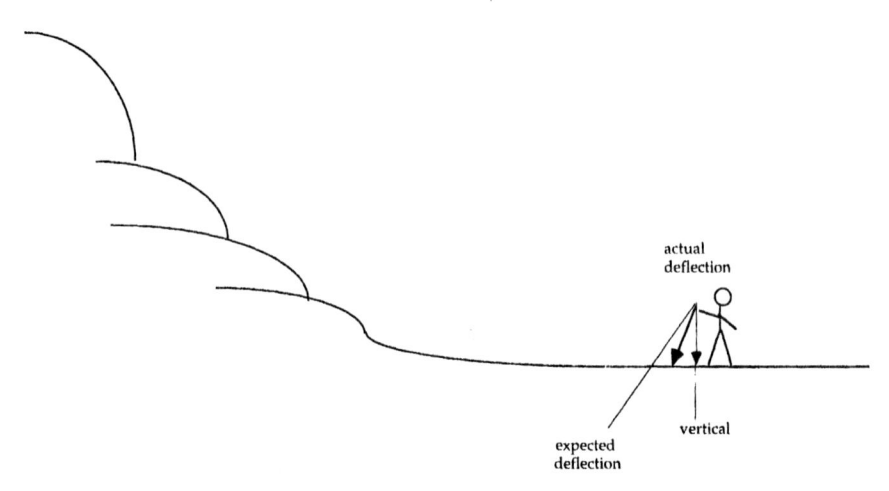

Figure 2.2. The detection of isostatic compensation. Adjacent to mountain ranges, a surveyor expects the plumb bob to be deflected by the gravitational attraction of the adjacent rock masses. However, surveyors in the Himalayas found the plumb bob deflected far less than expected. This implied that the surface mass excess of the Himalayas was somehow compensated by a subterranean mass deficit. The gravitational attraction of the Himalayas was first computed by John Henry Pratt in 1855.

Pratt published these results in the *Philosophical Transactions of the Royal Society* in 1855, where the paper was reviewed by George Biddell Airy, the Astronomer Royal of the United Kingdom. In a follow-up paper, Airy suggested that the discrepancy could be explained if the observed surficial mass of the Himalayas was gravitationally compensated by a subterranean mass deficit. Compensation could be achieved if a low-density crust, overlying a denser substrate, were thickened under high mountains, allowing the continents to float in the heavier substrate like icebergs at sea (figure 2.3).[10] But Pratt later put forward an alternative explanation: that compensation was achieved by underground differences in density that make up for above-ground topography. The crust, in this view, has constant thickness but varies markedly in density; compensation occurs because rocks comprising mountain belts are less dense than those comprising low lands and ocean deeps. "The density of the crust beneath the mountains," he wrote, "must be less than that below the plains, and still less than that below the ocean beds."[11] In a view rather similar to Dana's, Pratt supposed that these density differentials were produced by differential radial contraction during the early history of the globe: the areas that had contracted the most were the densest; those that had contracted the least were the lightest.[12]

One observational problem had engendered two different solutions. The observed phenomena could be explained by variations in either the thickness or the density of the crust. For Airy, mountains were like icebergs, supported by invisible roots beneath, and the size of the roots was proportional to the height of the mountains. For Pratt, mountains were like dough that was well risen—flatlands were a leaden loaf—and there was no need for roots below them. In either model, there would be some level within the earth where the total overlying mass was the same; this came to be known as the depth of compensation, referring to the point at which the extra mass above was

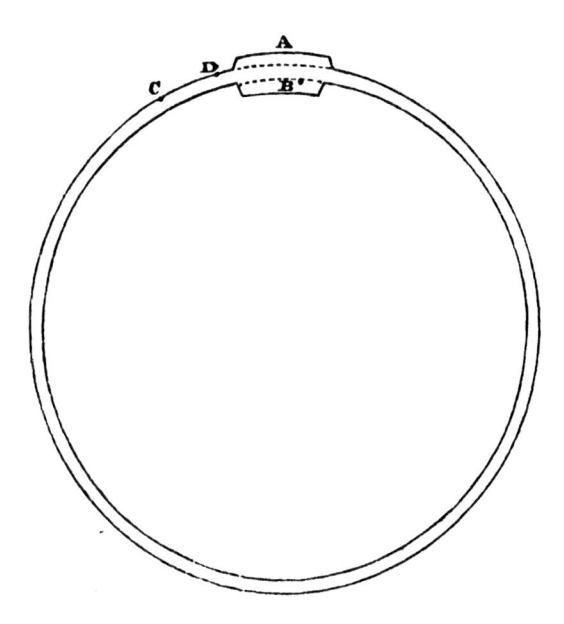

Figure 2.3. Airy's roots of mountains hypothesis. George Biddell Airy reviewed Pratt's calculations and suggested that the missing gravitational effect could be explained if the low-density earth crust floated in a higher density liquid substrate and if mountainous areas were supported by hidden "roots" that compensated for the extra mass above. Point A represents a continental block, B is the root below it. C and D are points on the oceanic crust, where no compensation is required. (From G. B. Airy 1855.)

compensated, or balanced out, by a lack of mass below. In Airy's model, the depth of compensation varied according to the depth of the roots; in Pratt's model the depth of compensation was everywhere the same. Thus the two models became known as the "roots of mountains" and the "uniform depth of compensation" hypotheses, respectively, or the Airy and Pratt models for short.[13] In either model, at some unknown depth, the weight of the overlying rocks would be the same everywhere, a condition that the American geologist Clarence Dutton named *isostasy*—meaning equal standing. Isostasy implied that the continents and oceans were balanced in a delicate equilibrium. Collectively, these ideas came to be known as the *theory* of isostasy—essentially a restatement of Archimedes' principle applied to the structure of the earth. To maintain the earth's surface features in a condition of hydrostatic equilibrium, elevated areas such as mountain ranges had to be compensated by a mass deficit below them, and lowlands had to be compensated by a mass surfeit.

Isostasy as a Theory of the Earth: Osmond Fisher

The phenomenon of isostatic compensation was discovered in the utilitarian context of geodetic mapping and surveying, but geologists soon considered its theoretical implications. Among these were the Reverend Osmond Fisher (1817–1914), the rector at the parish of Harlton, near Cambridge, a Fellow of the Geological Society of London and a close friend of Cambridge geologist Adam Sedgwick. Fisher was one of a number of British scientists in the late nineteenth century who were attempting to quantify questions pertaining to the structure and history of the earth.[14] In a series of papers published in the 1870s, Fisher attempted to mathematicize the concept of contraction and thereby demonstrate its sufficiency as an explanation for the earth's surface features (figure 2.4). Instead, he proved the reverse: mathematical analysis showed that thermal contraction was incapable of causing observed differences in ele-

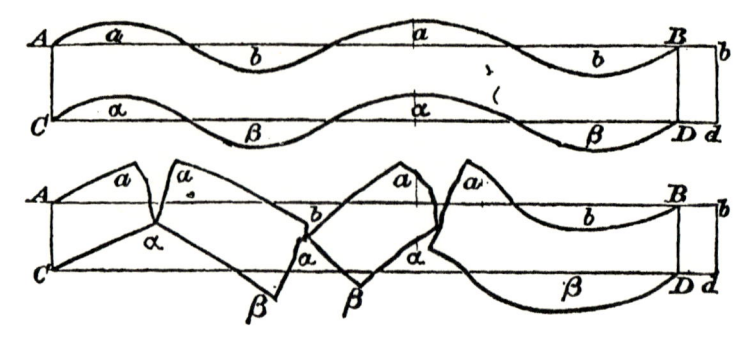

Figure 2.4. Fisher's geometric representation of the problem of terrestrial contraction. In the mid-nineteenth century, Osmond Fisher attempted to interpret the theory of earth contraction mathematically to demonstrate its sufficiency as a mechanism of surface deformation. His goal was to find "some simple laws which must govern the disturbed strata in spite of the confusion which appears to reign among them." (From Fisher 1881, p. 46.)

vation around the globe. "To my excessive surprise," he declared, "the result showed the utter inadequacy of the contraction hypothesis."[15]

The assumption of a solid earth was premised on the arguments of William Thomson (later Lord Kelvin) that the earth's internal pressure would render it solid.[16] Therefore contraction had to be evaluated in the context of conductive cooling of a solid sphere. Upon failure of the model of a solid earth, Fisher worked with ideas of a formerly liquid earth, a partially liquid earth losing water vapor, and a cooling crust resting on a thin fluid layer, but none of these modifications appeared adequate to save the hypothesis.[17] In 1881, Fisher laid out his refutation of the contraction hypothesis in *Physics of the Earth's Crust*—the first book in English dedicated to what would come to be called theoretical geophysics.[18] Whether solid, partially liquid, previously liquid, or partially gaseous, the earth simply could not contract sufficiently to do the work required of it. At best, contraction would produce elevation differentials of eight to nine hundred feet.[19]

Fisher was sensitive to the assumptions involved in his calculations and the uncertainties they introduced, and he tried to differentiate between constraints that were measurable—or could be calculated from other measurements, such as the mean density of the earth—and those that were purely guesswork. Ideally, the model should be built as far as possible on the former. The problem was that even "known" constraints could often "be satisfied in more ways than one."[20] An example was the constraint of rigidity. Kelvin had forcefully argued that the earth must be not only solid throughout but also fairly rigid.[21] His argument was based on the tides: if the earth were mostly fluid with only a thin solid crust, then it would respond to the tidal force in a similar manner as the surface waters, and there would be no tides, or only very small ones. That the oceans ebbed and flowed, while the earth beneath stayed still, was proof for him of a solid earth.[22] Fisher responded that the tides proved only that the earth was *mostly* solid. If the crust were sufficiently thick and rigid to resist the tidal force, the possibility remained of a fluid or plastic layer beneath it. Fisher pointed out that if the crust were solid due to its low temperature and the core were solid due to its high pressure, then at some intermediate level there could be a crossover zone where temperatures were high enough to cause melting but pressures were low enough to sustain a liquid.[23] Kelvin's objections notwithstanding, the earth might yet contain an internal fluid layer somewhere beneath the crust, and Fisher calculated that it could occur at a depth of twenty to twenty-five miles.

At this point in his analysis, Fisher turned from mathematical and physical considerations to geological evidence. Did geology support the notion of a fluid substrate? Yes—and more—it *required* it. The most conspicuous feature of mountain belts, Fisher argued, was the widepread evidence of lateral compression: this recognition had motivated the contraction theory in the first place. Yet, how could rocks be transported laterally without being crushed altogether? The answer lay with Fisher's fluid substratum. "The shifting of the crust towards a mountain range," he wrote, "which is testified by the corrugation of the rocks of which it is formed, requires a more or less fluid substratum to admit of it."[24] Rocks could move laterally if the substrate beneath them were yielding. "*What is required to explain the phenomenon is a liquid, or at least a plastic substratum for the crust to rest on, which will allow it to undergo a certain amount of lateral shift,*" Fisher wrote emphatically. A fluid substrate was also required to explain the accumulation of thick piles of sediment in coastal regions: the solid crust could sink into the fluid substrate, thereby making room for further accumulation of surface sediment. "The sinking of areas such as deltas, and other regions of deposition, demands [that] the crust . . . be in a condition of approximate hydrostatical equilibrium, such that any considerable addition of load will cause any region to sink, or any considerable amount denuded off an area will cause it to rise." Once the notion of a fluid substratum was accepted, Fisher concluded, "many of the facts [of geology] are more easily explained." A fluid substrate was not only a deductive consequence of physical principles but also inductively "necessary for the explanation of the phenomena of geology."[25]

The idea of a fluid substrate was also necessary for Airy's version of isostatic compensation, and Fisher presented Airy's model and the Himalayan data on which it was based as confirmatory evidence for his views. A fluid substrate would permit the continents to float in just the manner required by the roots of mountains hypothesis. Although Fisher quoted extensively from Pratt's analysis of the Himalayan data, he disagreed with Pratt's argument that isostatic compensation "may be produced in an infinite variety of ways and therefore, without data, it is useless to speculate regarding the arrangement of matter which actually exists in the solid parts below."[26] Although the effect could indeed be produced in a variety of ways, Fisher argued that there *were* relevant data—the data of observational geology—and that only the Airy explanation fitted them.

His reasoning here was quite specific:

> [Because] the crust . . . must accommodate itself to compression by being crushed together and thickened in place, . . . the very important consequence follows that elevations above the datum level will be accomodated by depressions beneath. The anticlinals will not be filled with fluid from below, but will be the upper portions of double bulges, which will dip into the fluid below, as well as rise into the air above. . . . It is analogous to the case of a broken-up area of ice, re-frozen and floating upon the water. The thickened parts which stand higher above the general surface also project deeper into the liquid below.[27]

Roots of mountains were the inevitable result of lateral compression of a rigid crust floating above a partially fluid substrate. In a world without compression, the Pratt model might hold, but that was not the world that geology revealed.

This conclusion led Fisher back to the question that had motivated Dana and Hall and virtually all geologists interested in the large scale features of the crust: What causes lateral compression? Pursuing the iceberg analogy, Fisher argued that the roots of mountains would be larger than their exposed portions. Denudation of the exposed tops would lead to gradual uplift of the hidden roots. Concurrently, sediments, carried by streams draining the mountains, would be deposited in adjacent coastal areas, and the result would be a net transfer of material from mountains to coastal plains. But, because mountains are not entirely free-floating, but are attached to their adjacent oceanic crust, the crust would be stressed along the boundary between the sinking ocean and the floating continent—the same boundary where Hall had placed geosynclines. Consequently, the crust must fissure, and magmas (either already liquid or liquified by pressure release) would escape from the fluid substrate and intrude into the crust along the lines of fissures, thus explaining the occurrence of volcanoes and ocean islands near continental margins. The ocean islands, having no roots themselves, would tend to sink (which, it was well known from the work of Charles Darwin, they did), while the igneous intrusions on the continents would compress the zone around them to produce mountain belts. Meanwhile, the gradual melting of the mountain roots would provide a mechanism for further generation of igneous rocks.

Although Fisher credited his ideas to Airy's insight, his vision of the earth was a blend of the Pratt and Airy models. The Airy model assumed uniform density for the entire crust, but Fisher held out for an oceanic crust that was denser than the continental crust. Otherwise, he argued, there was no way to explain why water pools in ocean basins, rather than being spread uniformly across the earth. This blended view of isostasy implied that the earth was fundamentally heterogeneous in a manner inconsistent with the theory of the interchangeability of continents and oceans, and it left Fisher with the unresolved question of what gave rise to the continents and oceans in the first instance. Fisher addressed this issue in a paper in *Nature* published in 1882. He was particularly troubled by the enormous size of the Pacific basin, which covers nearly half the globe. Concluding that such a "remarkable circumstance" warrants a remarkable cause, he lent his support to to the fissiparturition theory of his Cambridge colleague George H. Darwin (son of Charles), namely, that the moon had formed by separation from the earth early in planetary history.

In 1879, Darwin *fils* had proposed that the primordial earth, partially consolidated and covered by a thin crust, and subject to large tidal forces as it rotated rapidly, had split, and the moon had gradually receded from the slightly larger earth. Darwin called this process fissiparturition—birth through fission; Fisher extended Darwin's idea to argue that the scar left where the moon had broken off became the Pacific basin. Magmas welled up into the fractured region, producing a secondary crust of somewhat greater density than the original. The fragmented primary crust became the continents, preserving a "rude parallelism" of the coastlines; the secondary crust became the ocean basins. The major features of the earth were relics of this primordial event. The overall result was a geological model consistent with Airy isostasy: relics of low density primordial continental crust floating in a younger, denser oceanic substrate.[28]

Fisher's theory of volcanic compression did not have a profound impact on his colleagues, but his refutation of contraction and insights about isostasy did. Forty years later, John Joly, Professor of Geology at Trinity College, Dublin, and a Fellow of the Royal Society of London, concluded that Fisher wrote with a "perfect vision of what is involved in this theory [of isostasy] and of the essential simplicity and adequacy of the condition as a physical basis of the structure of the earth."[29] Writing more immediately in the United States, Clarence Dutton proclaimed that Fisher "has rendered most effectual service in utterly destroying the [contraction] hypothesis." Anyone who read Fisher's book, Dutton argued, would see that the contraction hypothesis was "nothing but a delusion and a snare, and that the quicker it is thrown aside and abandoned the better it will be for geological science."[30] But having dispensed with the contraction hypothesis, what was one to replace it with? Dutton was working on an answer of his own.

Isostasy as a Theory of the Earth: Clarence Dutton

One of the first Americans to challenge contraction theory was Clarence Edward Dutton (1841–1912), who is credited with coining the term *isostasy* and bringing the subject to the attention of American geologists.[31] Known in retrospect as one of the first American exponents of a quantitative approach to geological problems, Dutton was known in his own time as both a theoretician and an outstanding field geologist. In a review in *Nature* in 1880 by A. Geikie, Dutton's most famous work, his *Report on the Geology of the High Plateaus of Utah*, was praised as "one of the very best of the many admirable contributions to geology which have recently been made by the official surveys of the United States."[32]

After graduating from Yale University in 1860, C. E. Dutton joined the United States Army to fight in the American Civil War and then served as an officer in Army Ordnance. In 1867, Dutton joined G. K. Gilbert and John Wesley Powell in the *United States Geographical and Geological Survey of the Rocky Mountain Region*, one of the precursors of the U.S. Geological Survey (subsequently founded in 1879). Powell had already become famous for his dramatic explorations of the Colorado Plateau and its Grand Canyon, the "last completely blank area on the country's map."[33]

Dutton was deeply impressed by the spectacular geology of the Western regions, much of which did not fit standard American views based on Appalachian geology. His reports described abundant evidence of active and recent vulcanism, hot springs which indicated water at depth, and, above all, the Colorado Plateau and its Grand Canyon, which revealed the surprising power of erosional processes and the possibility of large-scale uplift without compressive deformation. Dutton was particularly impressed by the much greater prominence of igneous rocks in the Sierra Nevada than in the Appalachians, and the former gradually replaced the latter for him as the archetypical mountain belt.

Beginning in the early 1870s, Dutton delivered a series of papers developing isostasy in lieu of contraction as a framework for approaching large-scale geological problems. His most pointed attack, "A criticism upon the contractional hypothesis," was delivered in J. D. Dana's own journal, the *American Journal of Science*.[34] While accepting the reality of secular cooling, Dutton provided a devastating critique of its ability to affect the earth's surface processes. As it was for Fisher, Dutton's starting point was

Kelvin's analysis not only that the earth must be solid but also that the bulk of its cooling must necessarily take place in a relatively thin outer layer. Below some finite depth, the increase of temperature with depth must be virtually nil, and contraction equally nil. Given actual measurements of geothermal gradients taken from mines, and measurements of the thermal conductivity of rocks recently completed by Kelvin and J. D. Forbes, one could calculate how old the earth was: approximately 100 million years.[35]

Dutton favorably summarized Kelvin's argument but added the crucial caveat that the accuracy of the calculations depended critically on one's estimates of thermal conductivity and near-surface geothermal gradient. Or, as he sagely understated it, "The only ground of controversy must be the values to be assigned to the constants [in Fourier's theorem]." Dutton repeated the calculations using a range of plausible values for these constants and came to a rather different result from Kelvin's: the earth was anywhere between 98 million and 2.5 billion years old. But for Dutton the age of the earth was not the central issue. The central issue was that, given any reasonable geological estimate of thermal conductivity and geothermal gradient, secular cooling was an ineffectual mechanism of surface change. Significant contraction was limited to a relatively thin outer crust. "The unavoidable deduction from [Fourier's] theorem," he wrote, " is that the greatest possible contraction due to secular cooling is insufficient in amount to account for the phenomena attributed to it by the contractional hypothesis."[36]

Fisher had linked his theory to data collected in government geodetic surveys of the Himalayas, and Dutton linked his ideas to physical evidence from geological surveys in the American West. For example, Fourier's theorem required that secular cooling be necessarily greater in early periods of earth history, but this was inconsistent with the Rocky Mountains and the Sierra Nevada, both of which showed evidence of considerable disturbance since the Cretaceous period. Contraction theory also implied that orogeny should be a general feature of the globe, but, as Dutton noted, even the older Appalachians consist of a highly focused zone of folding adjacent to a huge undeformed plain. Why were surface disturbances restricted to narrow zones? Why did North America have great mountain chains on either coast, with a great, yawning, flat divide between them? Returning to geodetic evidence, Dutton pointed out that if, by some as yet unexplained mechanism, the earth contracted only along narrow zones of weakness, then how did it maintain an ellipsoid of revolution? "It is here that the analogy of the withered apple fails," Dutton wrote decisively. "If [the earth] is corrugated irregularly by shrinkage, it fails to preserve its original figure; and conversely, if it preserves its original figure, it must be corrugated uniformly."[37] One could not have it both ways. Contraction theory was inconsistent with both theoretical geophysics and observational geology. Dutton set to work on an alternative account of the processes that cause surface geological change.

American geologists on the East Coast had traditionally viewed mountains primarily as folded sediments, but Dutton's experience in the West led him to view mountains as igneous rocks intruding into sedimentary sequences. He set aside the Appalachians as an orogenic archetype and focused instead on the Rockies and the Sierra Nevada. Generalizing from them, he wondered if magmas were not the driving force of orogeny. If they were, then the key question was not so much the origin of mountains but that of magmas. Dutton's thought moved to the role of heating and melting in forming mountain belts and changing the physical properties of their substrates. Rocks that were heated near their own melting points would necessarily become plastic and, upon melting, would increase in volume and decrease in density. Low-density molten rock would work its way toward the surface, disturbing the adjacent and superincumbent strata on its way. Left behind would be a plastic zone in which isostatic adjustment could occur. But what force would heat and melt rocks in the first place? The answer was the pressure of isostatic adjustment in response to erosion and sedimentation.

One point of agreement among virtually all North American geologists was that the thick geological strata of the Appalachians were mainly deposited in shallow waters. Therefore, as James Hall had argued, they must have subsided as they were deposited. But if they subsided, then they must have displaced the matter beneath them; "And what becomes of the displaced matter?" Dutton asked rhetorically.[38] Borrowing the notion of a plastic substrate introduced by Babbage and Herschel earlier in the century and developed further by Fisher, Dutton answered that the displaced material moved laterally, away from the zone of sedimentation and toward the uplifted zone of erosion. Sedimentary loading presses on the subjacent plastic zone and causes lateral displacement toward adjacent, less heavily weighted areas, in the process generating physical changes in the substrate. The result is igneous intrusion in uplifted areas and folding of the sinking sediments in the area from which substrate has been removed (figure 2.5). In a widely cited address to the Philosophical Society of Washington in 1889, Dutton closely paralleled Fisher's argument:

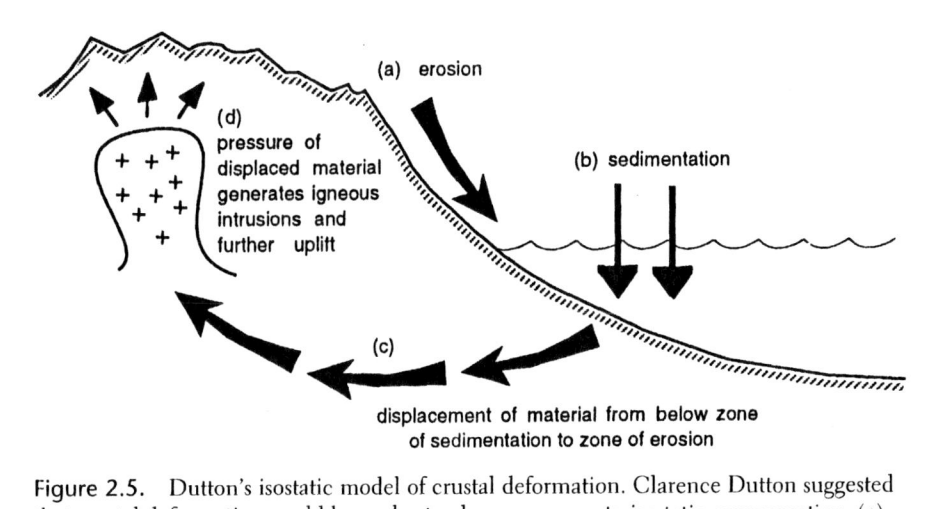

Figure 2.5. Dutton's isostatic model of crustal deformation. Clarence Dutton suggested that crustal deformation could be understood as a response to isostatic compensation. (a) The uplifted portions of the continents are eroded, and material is transported to coastal regions. (b) The weight of the deposited sediments causes subsidence along the continental margins, which causes displacement of material at depth. (c) The displaced material moves laterally and results in (d) igneous intrusions and further uplift of the continent, thus renewing the cycle. Dutton's idea draws freely on the work of Herschel, Dana, and Hall. (Illustration by Melody Brown.)

> A little reflection must satisfy us that the secular erosion of the land and the deposit[ion] of sediment along the shore lines constitute a continuous disturbance of isostasy. The land is ever impoverished of material—is continuously unloaded; and the littoral is as continuously loaded up. The resultant forces of gravitation tend to elevate the eroded land and to depress the littoral to their respective isostatic levels.[39]

Erosion and sedimentation start the process and isostatic adjustment continues it in a self-reinforcing way. The direction of isostatic readjustment must always be from areas of sedimentation to areas of denudation, that is, from coastlines toward the interiors of the continents. The occurrence of coastal mountain belts with regionally consistent deformation trends is thus explained: "This gives us a force of the precise kind that is wanted to explain the origin of systematic plications [i.e., folding, by] viscous flow of the loaded littoral [area] inward toward the unloaded continent."[40] The Alps were dislocated from south to north, the Appalachians from east to west, and the Sierra Nevada from west to east, because these were the directions of isostatic adjustment in the substrate. The timing and episodic nature of deformation was explained as well: "These plications, according to the isostatic theory, are the results of the disturbance of isostasy, and follow immediately upon that disturbance . . . and cease with it." In short, isostasy gave an explanation for the localization, timing, and directional patterns of fold belts across the globe.

Dutton's view of surface processes was the same as Fisher's, but his mechanism for generating surface change was different. Whereas Fisher had looked to a mechanical stress in the crust, Dutton emphasized a disturbance of the substrate. He concluded:

> Thus the general theory here proposed gives an explanation of the origin of plications. It gives us a force acting in the direction required, in the manner required, at the times and places required, and one which has the intensity and amount required and no more. The contractional theory gives us a force having neither direction nor determinate mode of action, nor definitive epoch of action. It gives us a force acting with a far greater intensity than we require, but with far less quantity. To provide a place for its action it must have recourse to an arbitrary postulate assuming for no independent reason the existence of areas of weakness in [the] crust which would have no raison d'être except that they are necessary for the salvation of the hypothesis.[41]

Isostasy, Dutton argued, could be geology's unifying theory.

But an obvious problem was whether the forces of isostatic imbalance were sufficient to overcome the earth's resistance to flow. Dutton realized that the notion of a flowing earth contradicted both scientific and common intuitions that the earth was "hard as rock," but he pointed to the flow of glaciers as a reasonable parallel: in pieces, ice is hard and immobile, yet the great ice sheets of Greenland were known to be moving.

> The forces called into play to carry the glacier along horizontally do not seem to differ greatly in intensity or amount from the described forces, and the rigidity of the ice itself may not exceed the mean rigidity of the rock masses beneath the littoral [zone]. . . . Whether these forces shall become kinetic and produce actual movement or flow will depend, first, upon their intensity; second upon the rigidity of the earth by which such movement is resisted.

The forces, he concluded, were sufficient when considered over geological time; the rigidity, he admitted, could not be assigned a definite value, but—risking circular reasoning—one had to conclude that the observed evidence of actual movement and deformation suggested that rocks at depth were far less rigid than at the earth's surface. Lacking independent confirmation, one had to accept the observed geological evidence on face value: rocks deform, and therefore they cannot be always rigid. In an obvious response to Kelvin, Dutton concluded that theoretical physics could only be taken so far.

> The geologic changes which have taken place may be regarded as experiments conducted by Nature herself on a vast scale, and from her experiments we may by suitable working hypotheses draw provisional conclusions: . . . the mean rigidity of the subterranean masses [must] be far less than that of ordinary surface rocks. . . . Pure physics alone would not have enabled us to reach such a conclusion, for the equations employ constants of unknown value. But geologic inquiry may, and I believe does, furnish us with narrow limits within which those values must be taken.[42]

Isostasy in America: G. K. Gilbert

Dutton's work brought the problem of isostasy to the attention of other American geologists, including his survey colleague Grove Karl Gilbert (1843–1918). Like Dutton, Gilbert's early experience came on a federal survey, in this case the *United States Geographical Surveys West of the 100th Meridian* (1871–1873). Officially led by a Lieutenant G. M. Wheeler, the Wheeler Survey came to be known among geologists as the Gilbert Survey because of Gilbert's pioneering descriptions of the structure of the Basin and Range Province and the dry beds of what was still optimistically called Lake Bonneville. In 1874, Gilbert joined the Powell Survey and became famous for his detailed descriptions of the geology and geography of the Henry Mountains (which Powell named after the Smithsonian Institution's Joseph Henry), the last unexplored mountain range in the contiguous United States. On the founding of the U.S. Geological Survey in 1879, Gilbert became one of its six senior geologists and served as its chief geologist from 1888 to 1892. A founding member of the Geological Society of America, Gilbert was elected president of that organization in 1892 and again in 1909, the only person to have twice held this position.[43]

Like Dutton, Gilbert documented evidence of orogenic processes distinct from those inferred from the Appalachians and the Alps. In the Basin and Range, Gilbert mapped structures that suggested the role of normal faulting and tilting in response to crustal tension rather than compression in generating uplift. In the Henry Mountains, Gilbert emphasized the role of erosion and sedimentation in controlling their shape and structure. At Lake Bonneville, Gilbert demonstrated that the old lake shorelines had been extensively uplifted without compressive deformation, and he suggested that they had experienced isostatic rebound in response to removal of the weight of water when the lake dried out.[44]

In his presidential address to the *Geological Society of America* in 1892, Gilbert considered the evidence for isostasy as a motive force in geological change.[45] Deep-sea soundings of the ocean basins, recently completed on the *Challenger* and *Blake* expeditions, proved the differential structure of continents and oceans (figure 2.6). What, then, holds up the continents? Gilbert argued that there were two possible answers: rigidity and isostasy. Either the continents were sufficiently strong to support their own weight, or they were less dense than the surrounding ocean basins and floated

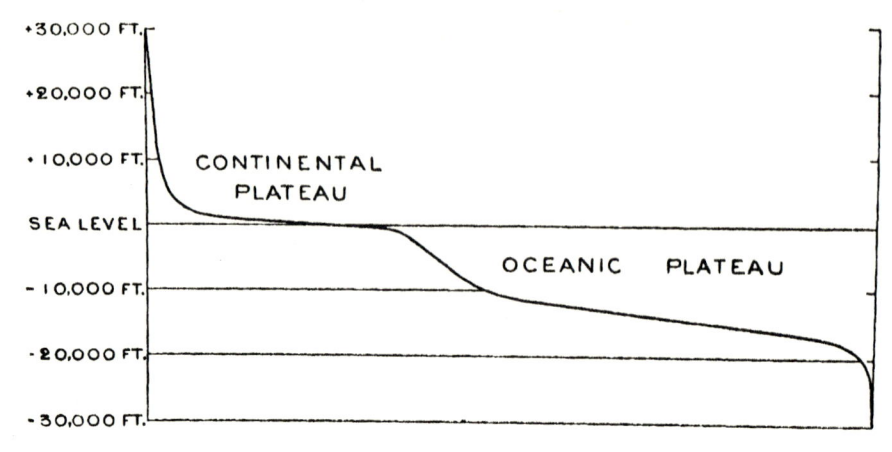

Figure 2.6. G. K. Gilbert's illustration of the structure of continents and oceans. Gilbert argued that the differential elevation of continents and oceans was one of the basic problems of physical geology and suggested that some form of isostasy was required to hold up the continents. (From Gilbert 1893, p. 180). Reproduced with permission of the publisher, the Geological Society of America, Boulder, Colorado, U.S.A. Copyright © 1893, Geological Society of America.

according to the theory of isostasy. Gilbert argued for the latter interpretation on the basis of a simple physical calculation. Imagine a base level, located at the bottom of the ocean deeps but extended beneath the continents as well. Then consider a column of water above the oceanic base and an equivalent column of rock above the continental base. Because the continental column is taller than the oceanic column, and rocks are denser than water, the pressure under the continental column must be greater than under the column of water. The differential pressure can be calculated: Gilbert proffered a figure of approximately 12,000 pounds per square inch. Yet evidence from laboratory experiments and quarries suggested that 12,000 psi was more than enough to crush rocks completely. Even allowing for the compensatory effects of confining pressure at depth, it was extremely unlikely that the rocks at the base of the continents were strong enough to withstand such enormous pressures. Yet the continents did not collapse. Something else must hold them up, and isostasy provided an answer: the continents were not so much strong as they were buoyant.

Yet isostasy was no panacea. It remained to explain what forces caused the relative movements of land and sea. For if there was one fact that all geologists agreed upon, it was that "the geologic history of every district of the land includes alternate submergence and emergence from the sea." But if the continents were floating and could not sink, then how to account for this fact? Contra Dutton and Fisher, isostasy alone could not explain it, because "the transfer of masses by erosion and sedimentation . . . is essentially conservative." Continents remain continents, and oceans remain oceans. Gilbert's provisional conclusion was that processes of both "terrestrial degradation" and "isostatic restoration" took place, and that somehow this permitted the oceans periodically to transgress the lands.[46] Precisely how remained unanswered.

This uncertainty left an opening for defenders of secular contraction. As the end of the century approached, a number of American geologists tried to reconcile the opposing camps of contraction and isostasy by supposing that the former was a general force that disturbed the earth's equilibrium, whereas the latter was a local force that tended to re-establish it. Robert Woodward, a vice-president of the American Association for the Advancement of Science and later the second president of the Carnegie Institution of Washington, saw a toehold in the relative rates at which each process operated. Commenting on Dutton's 1889 paper at the Philosophical Society of Washington, he demurred:

> While not unmindful of the difficulties of the contraction hypothesis, [it remains] an essential basis for the hypothesis of isostasy. The process of isostasy tends at a relatively rapid rate towards equilibrium; it ought apparently to run down in a comparatively brief geologic age. The process of contraction goes on at a relatively slow rate and is continually opposing the equilibrium to which isostasy leads. Both processes tend to produce crumpling along lines of weakness, and though that of isostasy may have been the more effective of the two, it appears to require secular contraction for its maintenance.[47]

For Woodward—ever an astute politician—isostasy was not so much a replacement for contraction as a necessary adjunct to it. Isostasy restored the balance that contraction periodically disturbed.

Bailey Willis (1857–1949), a geologist with the U.S. Geological Survey and later Professor of Geology at Stanford University, sagely reasoned that if the principal objection to the contraction theory was that the force it provided was insufficient to account for the observed effects, this was hardly a disqualifying complaint, for it could be applied to isostasy as well. Indeed, an apparent lack of adequate causal power could be applied to *all* global theories and thus did nothing to differentiate them. "To every hypothesis brought forward to account for the folding of stratified rocks," he wrote, "there is one objection made by its opponents: the cause is not quantitatively equal to the task required of it. For argument's sake, admitting for each and every one that the criticism is sound, I do not understand that it disposes of any [hypotheses] which are based on good inferences from observed facts."[48]

A Fact or a Theory?

Fisher and Dutton had presented two variations on the theme of isostasy as an explanation of the major surface features of the earth, and Gilbert lent support to crucial aspects of their proposal. But was there any evidence that the idea was actually true? Was isostasy a fact, or just speculation? As Bailey Willis's criticism indicated, for many American geologists it was but one of several hypotheses that might be developed to account for the large-scale physiography of the earth. And with two competing models of the same phenomenon—Airy and Pratt—it was evident that they could not both be right. For Fisher, it was virtually axiomatic that the Airy model was geologically correct; geologists "would hardly be satisfied with [Pratt's] explanation," he wrote, given the overwhelming evidence of lateral compression in orogenic processes. Although the Colorado Plateau suggested vertical uplift without lateral compression, most mountain ranges consisted of rocks that had been pushed horizontally. "I have shown," he reminded his readers, "in my 'Physics of the Earth's Crust,' that, if the substratum is plastic, this process would involve the production of the kind of downward protuberance that Airy postulates."[49]

But Fisher underestimated the impact that the Colorado Plateau had on American thinking. Dutton *did* subscribe to the Pratt model, referring repeatedly in his papers to the work of "Archdeacon Pratt"; indeed, he went so far as to claim that plumb-bob observations demonstrated that "the elevated regions and continents are of lighter matter and the depressed regions and ocean basins of denser matter [as suggested by] Archdeacon Pratt," despite Airy's demonstration that his hypothesis could equally well account for the observed geodetic effects.[50] Likewise, Gilbert, in his presidential address to the Geological Society of America in 1893, listed the nature and cause of the density differences between the continents and oceans as " the second problem of the continents"—that is, second only to the question of their ultimate origin.[51] Perhaps Americans preferred Pratt's theory for its consistency with Dana's model, which, like Pratt's version of isostasy, was based on density differences between the oceanic and continental crusts. Whatever the reasoning, for Gilbert and many other American geologists, regional density differences were not a hypothesis but a fact in need of explanation.

In 1886, Fisher had attempted to prove the theory of isostasy by showing how it could account for gravity variations in the Himalayan region.[52] Using the assumption of isostatic compensation, Fisher calculated a predicted value for gravity at each of twenty-one geodetic stations in India and compared these with observed values, and with equivalent predictions made without the assumption of isostatic compensation. The results clearly favored isostasy. The match between predicted and observed values was far better with the assumption of isostasy than without it. But Fisher's demonstration appears to have had little impact—perhaps because the data were too few, or perhaps because of the circularity of using data to confirm a theory that had been developed to explain them. In any event, the match was still not exact.

Gravity data were clearly critical to the interpretation of isostasy: if mountains were compensated, then their observed mass would not contribute to an increase in the local acceleration of gravity. In short, one could use gravity data to test the theory of isostasy. However, in the United States, there were as yet no systematic data available on the variation of gravity. When preliminary data became available, they led Gilbert to question his earlier convictions about isostasy. In his 1893 presidential address, Gilbert had advocated isostasy, but two years later he began to differentiate the problem of the overall elevation of the continents from the local elevation of mountains. The former might be explained by isostasy, but the latter perhaps could not. Gravity data, he wrote, "tend to show that the earth is able to bear on its surface greater loads than American geologists, myself included, have been disposed to admit."[53] By 1889, Gilbert had rejected isostasy as an account of topographic features within the continents; these were held up by the strength of the crust: "Mountains, mountain ranges, and valleys of magnitudes equivalent to mountains exist generally in virtue of the rigidity of the earth's crust; continents, continental plateaus, and oceanic basins exist in virtue of isostatic equilibrium in a crust heterogeneous as to density."[54] Gilbert's distinction was lost on many geologists, who were left uncertain, if not entirely confused.[55] Isostasy was a powerful idea, but without clear confirmatory data, perhaps it was only that. The situation changed in the early 1900s, when work by John Hayford and William Bowie provided dramatic confirmation of isostasy.

Elevating Theory into Fact

In the United States, gravity fell in the domain of the U.S. Coast and Geodetic Survey. Founded under Thomas Jefferson in 1807, the Coast Survey, as it was then called, was the first and arguably the dominant American scientific agency of the nineteenth century. But it had been conceived of as temporary: it was to map the coasts and then disband. As it persisted nevertheless and even grew in the late nineteenth century, there was substantial pressure for the Survey to justify its continued existence. The 1886 Allison Commission Hearings into the role and organization of science in the federal government (named for the Alabama senator who chaired the proceedings) eventually confirmed the value of the agencies it investigated, but not without first demonstrating that governmental support was predicated on clear evidence of utility.[56]

In 1867, the Coast Survey's powerful director, Alexander Dallas Bache, died, and the helm passed to Harvard mathematician Benjamin Peirce. Peirce turned the Survey's attention increasingly inland, and in 1878 the name of the body was changed to the U.S. Coast and Geodetic Survey. One of Peirce's accomplishments was the completion of a transcontinental triangulation arc connecting the Survey's Atlantic and Pacific coastal work—the geodetic equivalent of the golden spike. By the 1890s, fueled by the need for maps of the new states and territories, the task of transcontinental triangulation had expanded into the development of a primary geodetic network for the whole of the United States. This demand led in turn to a heightened concern with the figure of the earth.

To make accurate topographic maps on a regional scale, one needed a precise model for the shape or "figure" of the earth. This could be achieved either directly by geodetic measurement—by measuring the variation of the length of a degree of latitude along a meridian—or indirectly by measuring the acceleration of gravity as revealed by the period of a pendulum. Peirce's son, Charles Saunders Peirce, had pioneered pendulum measurements in the United States, and in 1894 the Survey, now under the direction of W. C. Mendenhall, undertook a program of transcontinental gravity measurements. Mendenhall, like his predecessors Bache and Peirce, was motivated by both theory and utility, and his program of gravity measurement was designed to "give light not only on geodetic problems but also on problems of peculiar interest to geologists."[57] To assist in the geological interpretation, Mendenhall enlisted the U.S. Geological Survey's G. K. Gilbert.

Gilbert wanted to use gravity as a test of isostasy. If topography were fully compensated, it should have no effect on the value of gravity. Conversely, if topography did affect gravity, then it was not fully compensated. Therefore, if one assumed that isostasy applied, there should be little or no "residual," or extra, gravity after correcting for topographic effects. The problem was how, exactly, to correct for topography. Gilbert made the assumption that the Great Plains of America, being flat and undeformed, must be close to isostatic equilibrium. Therefore the average value of gravity over this interior plain could be used as a reference standard. Each of the remaining stations could be thought to lie on a hypothetical "mean" plane whose elevation was equivalent to the average of the 30 mile radius around the station. Then, if one assumed that topography was broadly compensated, one could detect local variations by subtracting the effect of a hypothetical "plate of rock as thick as the space between the two plains" (figure 2.7). Gilbert estimated the average elevations of his mean planes from available topographic maps and the average rock densities by a detailed study of the geology of the districts surrounding each station. He called his correction the "adjustment to the mean plain."[58] The difference between the adjusted gravity at the station of interest and the average value over the Great Plains was the residual gravity and hence a measure of the degree of isostatic disequilibrium at that station. If isostasy applied, there should be little or no residual gravity after this adjustment.

Although most of Gilbert's residuals were small, significant positive residuals occurred at the two Colorado Rocky Mountain stations—Pike's Peak and Gunnison—equivalent to the gravitational effect of a rock layer 2200 to 2300 feet thick. Strikingly, this value was just about the average elevation difference between the Rockies as a whole and the western edge of the Great Plains. Gilbert concluded that the Rocky Mountains were not isostatically compensated but did in fact represent an excess load on the earth's crust. "The evident suggestion," Gilbert wrote, "is that the whole Rocky Mountain Plateau, regarded as a prominence of a broader plateau, is sustained by the rigidity of the interior." Similar residuals were found at Yellowstone, where three stations indicated "gravitational excesses of 1500, 1800, and 2300 rock-feet."[59] Isostasy was not as complete as Gilbert had previously imagined. It could account for the overall elevation difference between continents and oceans, but not for the local variations within continents.

Gilbert's counterpart at the Coast and Geodetic Survey, G. R. Putnam, took a different approach to data reduction and came to a different conclusion. Rather than assuming any particular reference standard—that is, without assuming that any part

h_s = height of station
h_r = average elevation of region
m_s = excess mass in slab of rock between station and sea level

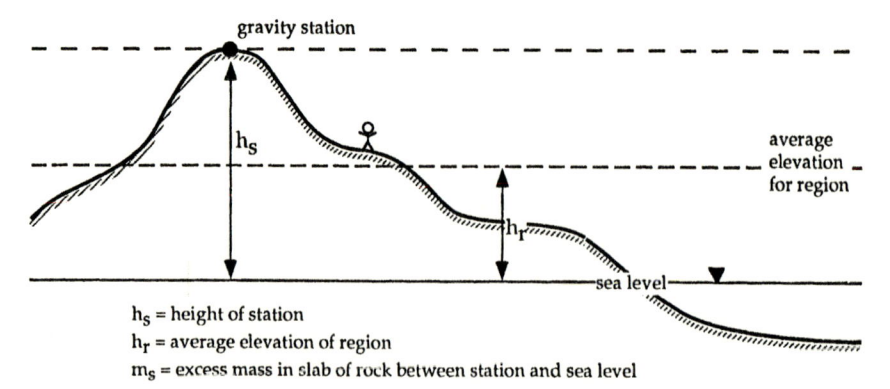

Figure 2.7. Gilbert's reduction of isostatic data by adjustment to the mean plain. Gilbert attempted to interpret early gravity data by assuming that topography was broadly compensated. If this was true, then one could detect local variations by treating the gravity station as if it lay on an elevated plain above a background "mean plain" and subtracting the effect of a hypothetical "plate of rock as thick as the space between the two plains." (From Gilbert 1859b). (Illustration by Melody Brown.)

of the crust was more or less equilibrated—Putnam corrected the measured values of gravity at the fifty-nine available stations and compared these with those predicted on the basis of the earth's spheroidal shape.[60] A positive anomaly meant that the corrected values were greater than expected; a negative anomaly meant that they were less than expected. Putnam used three different methods to correct the observed gravity values. The *Bouguer* correction assumed an uncompensated crust, and subtracted the attractive effect of the mass of rock between the station and sea level. The *free-air correction* assumed a completely compensated crust and corrected only for the height of the station above sea level—as if the pendulum were floating in "free air"—without regard to the mass of rock below it. The *average elevation* or *Faye* reduction was similar to Gilbert's adjustment to the mean plain, in that it corrected for the effect of a slab of rock beneath the station based on the average elevation of the district, but without regard to regional variations in density.[61]

Putnam found that the largest anomalies were produced using the Bouguer reduction, thus "show[ing] the incorrectness of the hypothesis of uncompensated topography above sea level."[62] In areas of moderate topography, the anomalies were smallest when only the free-air correction was applied; making a Bouguer correction increased rather than decreased the anomalies. This suggested that isostasy did apply; the Bouguer correction was handling a problem that the earth had already disposed of. However, in regions of extreme topography, the best results were produced with the average elevation correction. This might have been viewed as support for Gilbert's conclusion that compensation of large topographic features was incomplete, except that Putnam's residuals were far smaller than Gilbert's. For example, over Pike's Peak—one of the stations that led Gilbert to conclude that the Rocky Mountains were an excess load on the crust—Putnam's average elevation reduction produced a value of 0.006 dynes, less than a tenth of Gilbert's value of 0.078. This was equivalent to an excess load of only two hundred feet—and was well within the range of observational error and analytical uncertainty. Contra Gilbert, Putnam concluded that the results "confirmed the validity of the equilibrium, or isostatic, theory of the condition of the earth's crust."[63] Using different methods of data reduction, Putnam and Gilbert had come to opposite interpretations. Isostasy appeared to have reached an impasse.

Isostasy from the Figure of the Earth

Gravity was an interesting but essentially peripheral activity for the Coast and Geodetic Survey; its chief obligation at the turn of the century was to complete a primary network of triangulation for the United States. As John Hayford (1868–1925), Inspector of Geodetic Work and Chief of the Computing Division of the Survey, put it, the primary business of the Survey was to obtain from "field observations, the best available lengths, azimuths, and positions of lines and points fixed by triangulation, and in furnishing these values to persons who are to use them as controls for surveys."[64] Thus the Survey established the United States Standard Datum (renamed the North American Datum when united with Canadian and Mexican work in 1913) as the basis for all future surveys. It also collected the world's largest data set on deflections of the vertical.

By this is meant the difference between vertical as defined by a surveyor's plumb bob and the normal to the surface of the earth's ellipsoid. On a perfect sphere, the direction defined by a plumb bob is coincident with the normal to the surface of the sphere. But on the earth (or any ellipsoid), the vertical is not coincident with the normal to the surface, because of polar flattening. (The result is that the distance between two points is different if measured by triangulation than if calculated on the basis of latitude—the problem encountered by Sir George Everest.) The deflection of the vertical is the difference between the plumb bob (vertical) and the normal to the ellipsoid. However, there is no direct way to measure the normal to the ellipsoid. Instead, geodetic practice is to infer the direction of this normal at one point relative to another by measuring the distance between the two points and using an assumed ellipsoid (or figure of the earth). Then one can compare this with the angle between the verticals defined by the difference in the angle between the vertical and a fixed star at each point. The difference is the deflection of the vertical (figure 2.8).

By the early 1900s, the Survey had collected data on the deflection of the vertical at more than 507 geodetic stations connected by continuous triangulation across the United States. Previous workers had generally ignored small deflections of the vertical as inconsequential and treated large deflections as surveying errors that had to be discarded.[65] Hayford realized that these "errors" were information: one could use them to obtain an improved value for the figure of the earth. Indeed, Robert Woodward had explicitly made this point some fifteen years before, in an address to the American Association for the Advancement of Science. "The deviations [of the plumb bob]," he wrote, "are the exact numerical expression of our ignorance" regarding the shape of the earth.[66] In 1904, Hayford began the project of computing the figure of the earth based on deflections of the vertical in the United States.

Hayford began by compiling the raw data—a list of comparisons of positions from astronomy and triangulation. However, he realized that the computed deflections had to be corrected for the effect of local topography—the problem that had confronted Pratt half a century before. If the earth were a perfect ellipsoid with uniform mass distribution, the deflection of the vertical would be solely the result of the figure of the earth. In practice, the deflections are also affected by surface topography and near-surface heterogeneity. The traditional method for calculating topographic deflections was to get a good local map, calculate the volume of adjacent hills and valleys, and subtract their gravitational effect. But Hayford realized that, at least in principle, one needed to account for topographic effects "at all distances and in every direction from the station so as to include, if necessary, the whole surface of the earth. If such a computation were successfully made with absolute accuracy, the resulting computed deflection would be . . . the [true] topographic deflection at the station." Needless to say, to do this for even one station would be a laborious task; to do so for more than 500 was almost unimaginable. Yet, without accurate topographic corrections, Hayford's data would be useless. Thus he concluded,

> Very early in the investigation, it was realized that it would be necessary to compute the topographic deflection for each station, and that the computation to serve its full purpose must extend to a great distance from the station. It was also realized that to make such computations by any method known to have been used heretofore would have been impossible on account of the great expenditure of time and money involved. It was necessary, therefore, to devise some new method of computation, or to modify old methods, so as to make these computations feasible.[67]

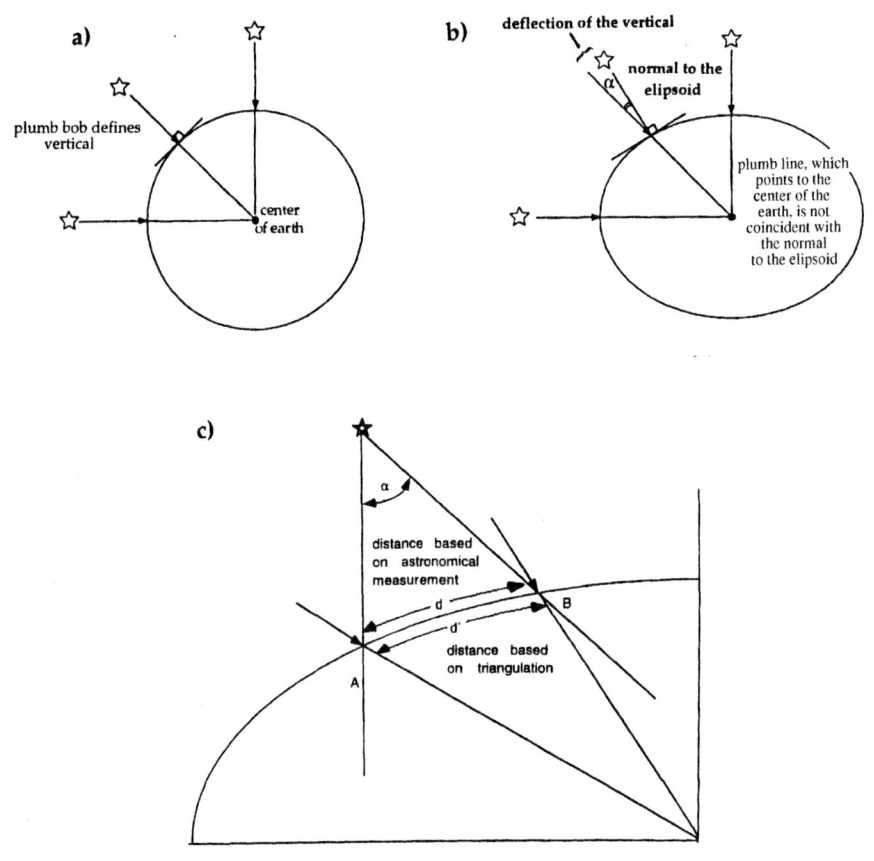

Figure 2.8. Figure of the earth based on the deflection of the vertical. (a) On a perfect sphere, the direction of vertical as defined by a plumb bob is coincident with the normal to the surface of the sphere. (b) On an ellipsoid such as the earth, the vertical is not coincident with the normal to the surface, because of the effect of polar flattening. The deflection of the vertical is the difference between the plumb bob (vertical) and the normal to the ellipsoid. (c) Because there is no independent way to measure the normal to the ellipsoid, the deflection of the vertical must be determined indirectly. This can be done by measuring the difference in the angle to a fixed star at two points and comparing this angle with what it should be based on how far apart the two points are. If the earth were a perfect ellipsoid with uniform mass distribution, the deflection of the vertical would be solely the result of the figure of the earth. In practice, the deflections are also affected by surface topography and near-surface mass distributions. (Illustrations by Melody Brown.)

Hayford's innovation was to create a series of transparent celluloid templates, composed of concentric circles and radial lines, that could be overlain on the relevant maps with the geodetic station at the center of the template. If the templates were drawn to an appropriate scale, the sectors of the template would be small enough so that all or most of each sector would fall within a single contour interval, and the average elevation could be estimated at a glance (figure 2.9). Thus, the key to Hayford's program was the use of finely divided templates—on a variety of scales appropriate for

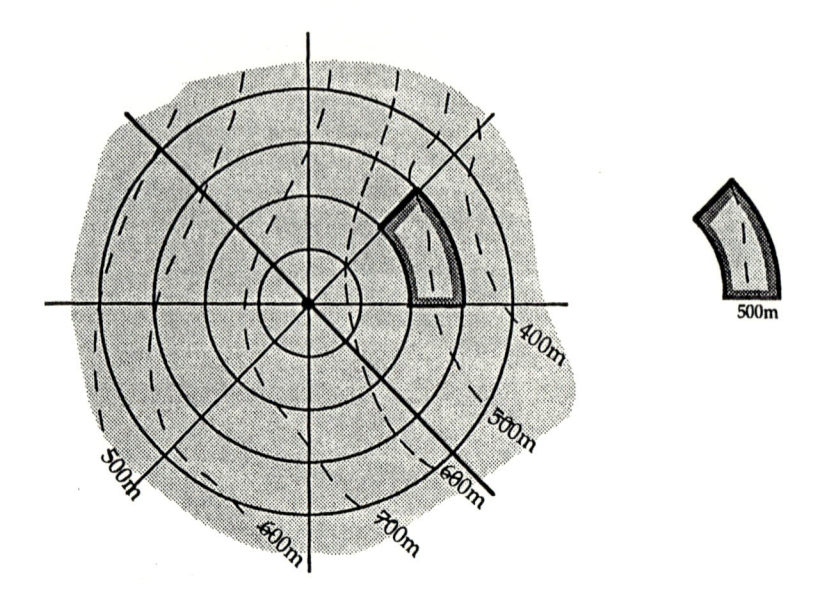

Figure 2.9. Hayford's template method for calculation of topographic mass. Because the deflection of the vertical is also affected by topographic features, Hayford sought a simple method to calculate the mass of surface topography. He began by creating a series of templates of various sizes that could be overlain on topographic maps with the center point at the location of the geodetic station. Each template was divided into sectors, and the idea was to choose a sufficiently fine template so that each of its sectors was crossed by only one contour interval on the underlying topographic map. The value of that contour could then be read and taken as the average elevation for that sector. (In the sector highlighted, the average elevation is 500 meters.) The point was to make the estimation of elevation as quickly as possible so as to account for the largest possible area surrounding the geodetic station. (Illustration by Melody Brown.)

estimating topography at a variety of map scales—and the use of a large number of employees to do the voluminous estimations. For the latter, he relied on the Survey's "computers"—modestly paid, generally female workers.[68] In cases where sectors comprised more than one contour interval, results were reestimated by a second computer and, if the two estimates did not agree, "the computers revised their work together with extra care, subdividing into smaller compartments, if necessary, until they had agreed within the specified limits."[69] The relevant topographic information was compiled into tables in which the elevations were multiplied by an appropriate constant to get their equivalent topographic mass. For the calculation of distant topographic effects, Hayford developed special templates that accounted for the distortions inherent in large-scale map projections.[70] Using this technique, it took one computer, on average, nine and a half hours to calculate the topographic deflections for each geodetic station. When the topographic deflections were calculated, they would be subtracted from the observed deflections to get that portion of the deflection caused only by errors in the assumed value of the oblateness of the earth (figure 2.10).

Hayford's computers mass-produced topographic corrections, completing calculations for 507 geodetic stations at distances up to 4126 km from each station. Five years

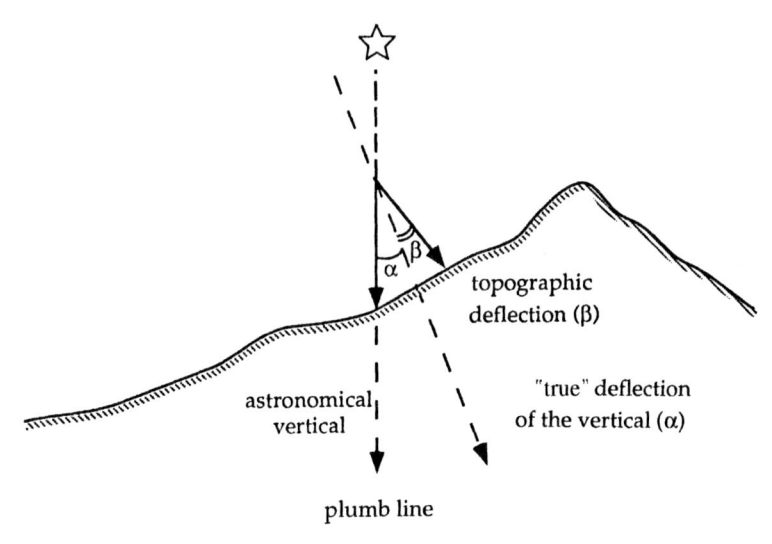

Figure 2.10. Hayford's corrections for the effect of surface topography. Having estimated the average elevation of an area surrounding a geodetic station (figure 2.9), Hayford attempted to account for surface topographic features by subtracting their predicted gravitational effect from the deflection of the vertical determined from the ellipsoid (figure 2.8). In principle, the deflection of the vertical should be increased by the effect of surface topographic features relative to the deflection based solely on the figure of the earth. Therefore, Hayford subtracted the computed topographic effects (β) from the total deflections ($\alpha + \beta$). However, when he did this, he found that the topographic corrections were too big. In many cases, the predicted topographic effects were *larger* than the total deflection of the vertical. Thus Hayford concluded, following the earlier results of Pratt, that something must be compensating for the effect of surface mass. In his words, "the logical conclusion from . . . the fact . . . that the computed topographic deflections are much larger than the observed deflections of the vertical, is that some influence must be in operation which produces [a] counterbalancing of the deflections produced by the topography." (From Hayford 1909, p. 65.) (Illustration by Melody Brown.)

later, the results of this tremendous labor—all performed as a by-product of necessary survey operations—were documented in a 176-page publication of the U.S. Government Printing Office. The results were the same as suggested by Pratt's simple calculations more than half a century before: the observed deflections of the vertical were much less than those predicted on the basis of topographic effects. Hayford's topographic corrections were too big—or, conversely, the observed deflections were too small. The expected topographic effects were simply not there. With the same results as Pratt, Hayford came to the same conclusion: Something must be compensating for the effect of surface mass. "The logical conclusion," he wrote, "from . . . the fact . . . that the computed topographic deflections are much larger than the observed deflections of the vertical, is that some influence must be in operation which produces [a] counterbalancing of the deflections produced by the topography."[71]

The key point for Hayford was how remarkably consistent these results were. In over 500 stations, encompassing mountains, plains, and coastal stations—and probably exceeding the total number of such measurements available in the rest of the world virtually every one produced the same result. The effects were not to be explained away by local irregularities but demanded a general principle or process. Isostasy provided one. Hayford made the connection with Pratt's data and his explanation explicit:

> There must be some general law of distribution of subsurface densities which fixes a relation between subsurface densities and the surface elevations such as to bring about [a] balancing of deflections produced by the topography on one hand against deflections produced by variation in subsurface densities on the other hand. The theory of isostasy postulates precisely such a relation between subsurface densities and surface elevations.[72]

What Pratt had suggested on the basis of a few score measurements, Hayford had now demonstrated on the basis of five hundred, analyzed and documented in painstaking detail. The principle of isostasy could account for the systematic discrepancies between observed topography and observed deflections. Hayford therefore set to recalculating the topographic deflections using the assumption of isostatic compensation. But how, exactly, was this to be done?

By making two simplifying assumptions, Hayford rendered this awkward problem tractable. Following the Pratt model, he assumed that the required density deficiencies were uniformly distributed from the surface down to a fixed (but unknown) depth: the depth of compensation. He further assumed that, in a region whose mean elevation is h_i, the excess mass above sea level, ρh_i, where ρ is the mean surface density, is compensated by the reduced mass below sea level. Thus $\rho h_i = -\Delta\rho_i h_c$, where $\Delta\rho_i$ is the "compensating defect of density" below the surface and h_c is the depth of compensation (figure 2.11). Because ρ is approximately known from measurements of rocks (Hayford used a value of 2.67 gm/cc), and h_i may be read off a topographic map, this leaves two unknowns, $\Delta\rho_i$ and h_c. By choosing an arbitrary value for h_c, one could solve for $\Delta\rho_i$, compute the deflection, and end up with a set of the residual errors. Hayford then repeated this process using a series of different values for the depth of compensation and then plotting the mean square of the residuals versus the several values of h_c. The value that minimized the residuals he took to be the best estimate for the actual depth of isostatic compensation: this turned out to be 113.7 km (figure 2.12). Finally, he subtracted the minimized residuals from the observed deflections to obtain a corrected value for the deflection of the vertical at each station.

As in the first part of the study, the calculations involved were, in Hayford's own words, "difficult, extensive, and unusual," but their value was demonstrated in two respects.[73] First, Hayford was able to calculate new values for the equatorial and polar dimensions of the earth. Both were notably greater than previous estimates, but this was consistent with theoretical expectation: ignoring the deflection of the vertical—as previous workers had done—would cause distances to be systematically underestimated. Second, and perhaps more important, the new values were in agreement with results of the latest gravity measurements as reported by F. R. Helmert, Director of the International Geodetic Association and widely considered to be the world's foremost geodesist. Helmert had recalculated the earth's polar flattening based on gravity measurements worldwide, and his value, expressed as the reciprocal of the ratio of polar flattening, was the same within errors as Hayford's: 298.3 ± 0.7 versus 297.8 ± 0.9. Not only did this support Hayford's estimate per se but also it vindicated his approach. Hayford wrote: "The close agreement of the value from this investigation with Helmert's

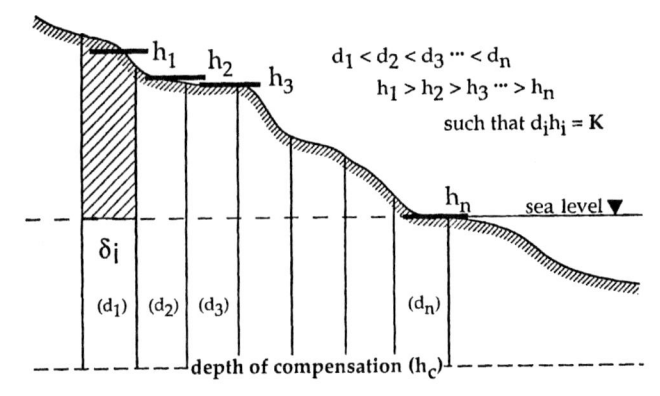

Figure 2.11. Hayford's model of isostatic compensation. Following the Pratt model of isostatic compensation, Hayford assumed that the required density deficiencies were uniformly distributed from the surface down to a fixed (but unknown) depth: the depth of compensation. He also assumed that, in a region whose mean elevation is h_i, the excess mass above sea level (ρh_i, where ρ is the mean surface density) is compensated by the reduced mass below sea level. Thus, $\rho h_i = -\Delta\rho_i h_c$, where $\Delta\rho_i$ is the "compensating defect of density" below the surface and h_c is the depth of compensation. Because ρ is approximately known from measurements of rocks (Hayford used a value of 2.67 gm/cc), and h_i may be read off a topographic map, this leaves two unknowns, $\Delta\rho_i$ and h_c. Thus, if one could estimate the depth of compensation, one could solve for the density differences and recalculate the topographic corrections. The problem was how to estimate the depth of compensation. For this, Hayford employed an iterative method, illustrated in figure 2.12. (Ilustration by Melody Brown.)

value is significant as showing what accuracy can be secured with the geodetic observations of a single country when treated by the method pursued in this investigation."[74] Americans, working in America, had achieved results as good as or better than Europeans in their global empires. Hayford presented his work at the 1909 meeting of the International Geodetic Association, and within a few years geodesists around the globe were using his method. In 1924, the "Hayford spheroid" was officially adopted by the International Geodetic and Geophysical Union as the reference standard for the figure of the earth.[75]

The success of Hayford's method appeared to validate his assumption of isostatic compensation. Without this assumption, the observed phenomena were inexplicable; with it, they seemed to make sense. However, Hayford did not claim that isostasy had been proven but simply that geodetic evidence required an explanation along these lines. As he put it in 1909, "it is certain that the assumption that the condition called isostasy exists is a much closer approximation to the truth than the assumption that it does not exist."[76]

From Triangulation to Gravity

Hayford's next step was to reevaluate gravity in light of isostasy. In 1909, he began an expanded program of gravity measurement, working in conjunction with his assistant chief of computing and geodetic work, William Bowie.[77] Later that year, Hayford

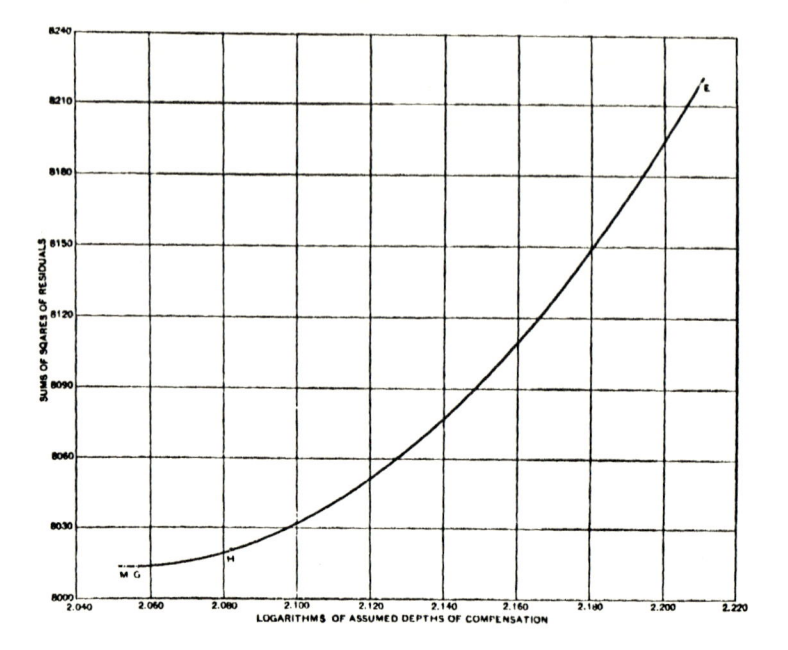

Figure 2.12. Hayford's calculation of the depth of compensation. Hayford calculated the depth of compensation by choosing the value that minimized the residuals of the topographic deflections. He did this following an iterative method. First, he chose an arbitrary value for h_c and used it to solve for $\Delta\rho_i$. The computed values of $\Delta\rho_i$ were used to produce a new set of estimates for the topographic corrections. Then he subtracted these new values from the original deflections to end up with a set of residual errors, and repeated this process using a series of different values for the depth of compensation. Then he plotted the mean square of the residuals vs. the several values of h_c. The value that minimized the residuals he took to be the best estimate for the actual depth of isostatic compensation: 113.7 km. Finally, he subtracted the minimized residuals from the observed deflections to obtain a corrected value for the deflection of the vertical at each station. (From Hayford 1909, p. 145.)

accepted the position of director of the College of Engineering at Northwestern University, but the work was continued by Bowie, now elevated to Hayford's former position. Between 1909 and 1912, Hayford and Bowie completed the gravity measurements and data reductions for 105 new gravity stations — 16 abroad and the rest in the United States.

Bowie and Hayford applied the same technique to the reduction of gravity data that Hayford had applied to the deflections of the verticals. Survey computers once again calculated the topographic effects surrounding each station, but this time the work included the topography for the whole of the surface of the earth. (Each set of calculations now took seventeen hours per station.) Then Hayford and Bowie compared the gravity anomalies using an isostatic correction with those produced using Bouguer and free-air corrections. The isostatic correction reduced the anomalies nearly to naught. By assuming isostasy, one could, with a high degree of accuracy, predict the value of gravity across the United States. It appeared that isostasy was ubiquitous and, in Bowie's words, "practically perfect."[78] The Bouguer corrections could be ignored, as could the complications of local geology that Gilbert had tried to account for—unnecessarily as it now appeared. Hayford and Bowie's "isostatic reduction" thus dramatically streamlined the calculations needed to process gravity data: in just over ten years, Bowie had more than doubled the number of gravity stations in the United States.[79]

The gravity data were in every way consistent with the deflections of the verticals and supported the assumption of isostatic compensation of topographic features. Thus Bowie wrote:

> The isostatic compensation of topographic features is so complete that the deflections of the vertical are only approximately 10% of what they would be if the land masses were excess loads on the earths's crust[, and] the difference between the observed and the computed values for the intensity of gravity are on an average not more than 15% as large . . . as they would be on the theory that the earth's crust is rigid. . . . What the geodetic observations have done is really to measure indirectly the differences in weight of the masses in the various unit columns of the Earth's crust.[80]

Isostasy, in short, made sense of geodesists' empirical observations.

Hayford had not exactly claimed to have proved isostasy in his early published work, but now he did. Writing in the journal *Science* in 1911, Hayford began by stating, "Within the past ten years, geodetic observations have furnished positive proof that a close approximation to the condition called isostasy exists in the earth and comparatively near its surface." Bowie was even more assertive. From 1912 onward, Bowie published widely and repeatedly referred to his work, with Hayford's, as the "establishment," "demonstration," and "proof" of isostasy.[81] Bowie was no doubt partial, but his colleagues generally seconded his self-appraisal. In one of his last published papers, G. K. Gilbert reversed his position on isostasy one last time, as a consequence of the work by Hayford and Bowie, and wrote, "The success of the hypothesis [of isostasy] in reducing anomalies of the vertical and anomalies of gravity is held—properly, as I think—to show that isostatic adjustment in the earth's crust is nearly perfect. Starting with geodetic and topographic data . . . Hayford and Bowie have demonstrated isostasy."[82] George Becker, speaking on the occasion of his Presidential Address to the Geological Society of America in 1914, cited Helmert, who now called isostasy the "Pratt–Hayford hypothesis, in recognition of the importance of Hayford's work in establishing the actuality of isostasy."[83]

Colleagues on the other side of the Atlantic were also convinced. In a discussion at the Royal Geographical Society in London in 1923, the Cambridge geophysicist Dr. (later Sir) Harold Jeffreys claimed that "the work of the United States Coast and Geodetic Survey in putting the theory of isostasy on its present basis is one of the outstanding scientific achievements of our time."[84] One year later, John Joly, in his Halley Address to the Royal Society, went further: "Isostasy must, today, be regarded as *fact* which has been tested by many observations. . . . Hayford's work in America (later Hayford's and Bowie's) may be regarded as having decisively elevated theory into fact."[85]

Isostasy Refutes Collapsing Continents

Isostasy's implications for geological theory now had to be addressed. In his 1909 treatise, Hayford had allowed himself a rather long discussion of isostasy, which he justified because "there are many other theories of geology and geophysics which are so intimately related to this theory, that they must stand, or fall, or be greatly modified, according to its fate."[86] And so it was. Isostasy directly contradicted Suessian contraction. Because it required the continents to be fundamentally different, either in structure or in composition, from the ocean floors, isostasy was irreconcilable with the interchangeability of continents and oceans postulated by Suess's theory. And if continents and oceans were *not* interchangeable, then they were, by implication, permanent—as Dana had argued some sixty years before. As Bailey Willis put it in 1910, a year after Hayford's first monograph, "There is a relation between the intensity of gravity and the relative altitude of a continental or oceanic plateau, which proves that the plateaus have assumed different altitudes according to the densities of the subjacent material. The transformation of a continent into an ocean basin, or *vice versa*, would require, therefore, a change in density of an enormous volume of material, and there is neither evidence nor explanation of such a change."[87] Or, as Willis would repeat as a mantra in later years, "Once a continent, always a continent; once an ocean, always an ocean." The confirmation of isostasy was the death knell for Suess's hypothesis of contraction. Suess had lived to witness not the collapse of the earth, but the collapse of his own theory.

And isostasy appeared to bolster Dana's version of contraction. Dana had held that the continents were fundamentally different from the oceans and therefore might represent primordial, and permanent, heterogeneities, a view that appeared to be fully compatible with isostasy. Indeed, Pratt had attributed density differences to differential radial contraction, precisely as had Dana. In his 1871 treatise, *The Figure of the Earth*, Pratt had written: "Supposing that the earth was once in a fluid state, as it became solid it contracted unequally, leaving mountains where it contracted least and ocean beds where it contracted most. . . . As contraction went on in subsequent ages, no doubt the partial sinking of these huge masses might . . . cause corresponding though subordinate upheavals in neighbouring parts of the crust."[88] This was what Dana had independently proposed on the basis of the properties of minerals and on the basis of his analogy with the features of the moon. Density differentials between continents and oceans were inherent in Dana's version of contraction. Isostasy might thus have been construed as definitive support for the American version of contraction, but for one thing: the discovery of radiogenic heat, which undermined the very premise of a cooling earth.

The Discovery of Radiogenic Heat

In 1903, Marie Curie shared the Nobel prize with her husband, Pierre, and their colleague Henri Becquerel for the discovery of radioactivity. Becquerel had first observed emanations of energetic rays from salts of uranium in 1896; soon thereafter, the Curies demonstrated that these emanations were a nuclear phenomenon attributable to a previously unknown chemical element, which they named radium. Following on the Curies' work, Ernest Rutherford and Frederick Soddy demonstrated that radium was chemically unstable—most likely formed by the transmutation of uranium—and that prodigious quantities of energy were released during this transmutation.[89] The immediate result of these discoveries, besides capturing the attention and imagination of the world at large, was to settle geology's long-standing debate with Lord Kelvin over the age of the earth.

Since the 1860s, Kelvin had systematically attacked geological estimates of the age of the earth and, in consequence, Lyell's principle of uniformitarianism. Calculations based on physical principles demonstrated, he argued, that the earth could not be more than a few tens of millions of years old and the sun not more than a few hundred million. If a uniformitarian interpretation of the rock record suggested otherwise, then so much the worse for that notion. As the years had passed, Kelvin's estimates had become increasingly miserly, decreasing from the 100 million years that Dutton had quoted—and quite a few geologists had been willing to accept—to a preferred value of only 20 million. In the words of Sir Archibald Geikie, president of the British Association for the Advancement of Science in 1892 (the year of Kelvin's elevation to the peerage), Kelvin had placed geologists "in the plight of King Lear when his bodyguard of one hundred knights was cut down." Like Goneril, Lear's faithless daughter, the "inexorable physicist [had] remorselessly struck slice after slice for his allowance of geological time."[90]

Although Kelvin's pronouncements were at odds with the intuitions of nearly all professional geologists, few of them felt able to challenge him on his own terms. One exception was T. C. Chamberlin, the American geologist who near the turn of the century developed the theory of solid earth accretion.[91] In a pair of papers published in 1899, Chamberlin pointed out that Kelvin's determinations of the ages of the sun and earth were premised on two questionable assumptions: one, that both had originated as liquids, and two, that neither had been subject to additional heat inputs during the course of geological history. With respect to the first assumption, Chamberlin pointedly stated that it was precisely "an *assumption* . . . which [is] still quite as much on trial as the chronological inferences of the biologist or geologist."[92] With respect to the second, Chamberlin prophetically reasoned that the discovery of an additional heat source would render Kelvin's argument moot.

> Is present knowledge relative to the behavior of matter under such extraordinary conditions as obtain in the interior of the sun sufficiently exhaustive as to warrant the assertion that no unrecognized sources of heat reside there? What the internal constitution of atoms may be is yet an open question. It is not improbable that they are complex organizations and the seat of enormous energies. Certainly, no careful chemist would affirm either that the atoms are really elementary or that there may not be locked up in them energies of the first order of magnitude.[93]

And so it was. The discovery of radiogenic heat made clear that Kelvin's arguments rested on a faulty foundation. The implications for geology were plain. As a former student of Kelvin's put it bluntly to him, "If the sun or the earth contains even a small amount of radioactive matter, all physical calculations of age & heat are overturned, & the old ratios of the geologists are restored (or may be)."[94] Or, as John Joly, Professor of Geology at Trinity College, Dublin, put it with barely concealed satisfaction, "the estimates derived from physical speculations are now subject to modification in just the direction which geological data required."[95]

The modifications were soon in hand. In 1905, R. J. Strutt, later the fourth Baron Rayleigh, used the helium content of a radium bromide salt to produce a new estimate for the age of the earth of 2.4 billion years—two orders of magnitude greater than Kelvin's preferred estimate. Meanwhile, Strutt's results were independently confirmed in the United States by Bertram Boltwood at Yale University. Having concluded that lead was the end product of uranium decay, Boltwood used the ratio of these two elements to date a series of rock samples: the results ranged from 410 million to 2.2 billion years old.[96] The uranium–lead technique was refined by Arthur Holmes, a student of Strutt's at Imperial College, and in 1913 Holmes produced geology's first absolute time scale.[97]

The debate over the age of the earth received enormous popular attention, and the vindication of observational geology over mathematical physics has remained a parable for professional geologists to this day.[98] But of broader significance for geologists than the precise age of the earth—which was still not fully resolved—was the prospect of radiogenic heat as a geological force. If radium released huge amounts of energy during its transmutations, then this energy was potentially available as an agent of geological change. Geologists had argued with Kelvin over the results of his calculations, and Fisher and Dutton had argued that contraction was ineffectual, but virtually everyone had accepted the premise that the earth was a cooling body. Now even that premise was open to dispute. If radium was a ubiquitous element and radiogenic heat widely distributed, then the earth might not be cooling down at all. It might be in a steady state or perhaps even heating up. One could no longer assume that global contraction occurred at all—with or without the requisite force to build mountains.

One of the first to suggest and pursue this line of thought was John Joly (1857–1933). In a letter to *Nature* in 1903, Joly suggested that the earth's geothermal gradient might be wholly or in part controlled by radiogenic heat.[99] His suspicion was confirmed in part by Strutt's researches, in which he demonstrated the presence of radium in a variety of rocks, minerals, and soils. But Joly went further, by showing the actual effect of radioactivity in one of the most common rock-forming minerals, biotite mica. In a series of papers published between 1907 and 1917, Joly demonstrated that pleochroic haloes in biotite were caused by radiation damage. These haloes—dark circles found ubiquitously in biotite and occasionally in other minerals—consisted of small dark spots or concentric rings, in some cases surrounding a tiny visible grain of zircon or apatite. Strutt had shown that zircon and apatite were minerals of higher than average radioactivity; this suggested to Joly that the haloes were zones of radiation damage induced by emissions from the mineral grains. Just as radiation could darken photographic plates, Joly supposed that it could darken sections of minerals. To test his idea, he exposed samples of mica to tiny specks of radium and found that identical haloes could be induced by such exposure (figure 2.13). Furthermore, the sizes of the haloes were proportional to the amount of exposure, and they were consistent with what would be predicted based on Rutherford's recent studies of the depth of penetration of alpha emissions.[100]

Joly had demonstrated that radioactivity was not a freakish effect confined to uranium ores: it was pervasive in everyday rocks and minerals. Yet, at the same time, he had also demonstrated "the sporadic and unequal distribution of radium."[101] If radium were selectively concentrated in particular minerals, then it might also be concentrated in particular areas of the globe and could be a force for focused geological effects. It was, as Joly put it, proof of radioactive energy as a *vera causa* (true cause) of geological change.[102] Meanwhile, as Joly considered the implications of his discoveries, Marie Curie successfully isolated pure radium and found that its heat production was approximately 100 calories per gram per hour (a value Rutherford later refined to 132).[103] It was now clear that radioactive materials were a prodigious source of energy for geo-

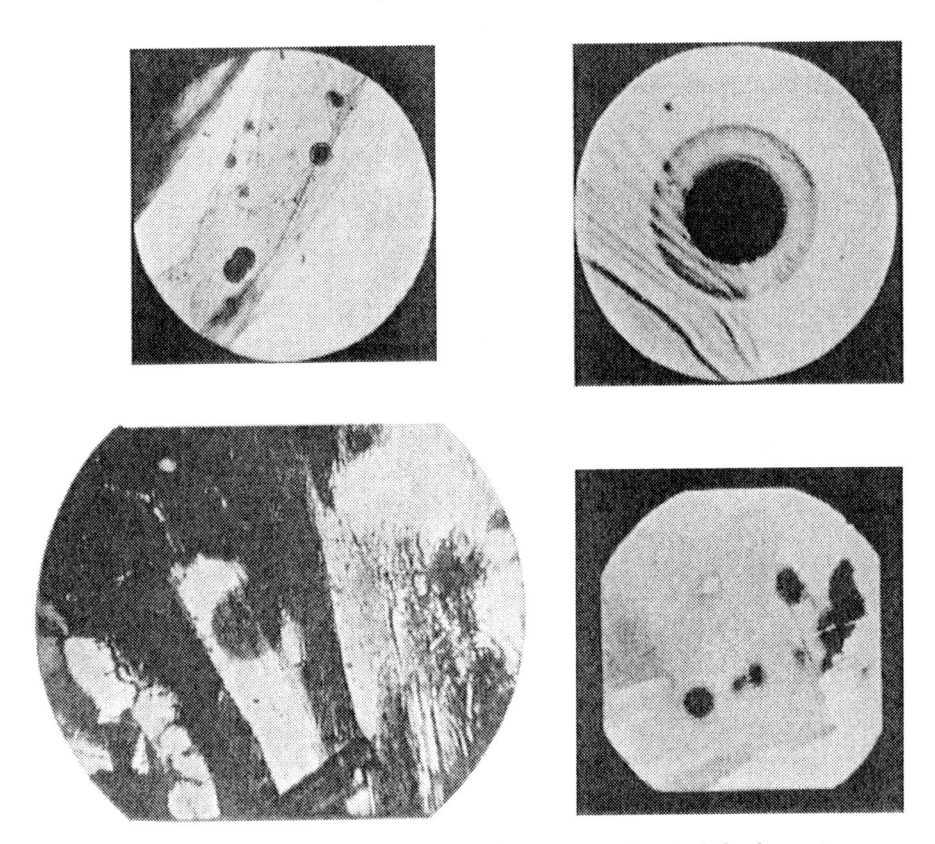

Figure 2.13. Joly's illustration of pleochroic haloes in mica. On the left, photomicrographs of natural pleochroic haloes in micas, visible in thin section. On the right, synthetic haloes produced by exposing micas to a radium source. (Each halo is about one millimeter in diameter.) The similarity of the two strongly suggests that pleochroic haloes in nature are the result of radiation damage from trace amounts of radioactive elements in rocks. The prevalence of these haloes in rocks therefore indicated that radioactivity was ubiquitous and that it could have geological-scale effects. (From Joly 1909, p. 67; also see 1915, p. 228a.)

logical processes. As Bailey Willis put it some years later, "Thanks to Madame Curie, the inexhaustible energies of the atoms of the globe ... are potentially available to geological speculation."[104] Atomic energy could power not only the world but also geological theory.

Epistemological Affinities

If field mapping in the Swiss Alps caused geologists to question the adequacy of contraction theory, isostasy and radioactivity conclusively proved its inadequacy. By demonstrating large-scale horizontal displacement of intact slabs of rock, field mapping first called into question the adequacy of the contraction hypothesis. Isostasy then killed Suess's version of contraction by refuting its proposed mechanism, while radioactivity undermined both Suess's and Dana's versions by removing their driving force.[105] By refuting contraction theory, field mapping re-opened the question of the origin of compressive forces in mountain belts. By refuting the hypothesis of sunken continents, the confirmation of isotasy left geology with no account of the basic facts of historical geology and paleontology. By refuting the hypothesis of a cooling earth, the discovery of radiogenic heat had left geology with no driving force for orographic change.

Three distinct research traditions—field mapping, geodetic surveying, and atomic physics—focused on different problems, using different methods, and pursued by different individuals, had led to the same conclusion: contraction was not the answer to geologists' questions. Mapping provided observational evidence derived from geologists' own traditional methods and long-standing concerns; geodesy provided quantitative data from a utilitarian tradition motivated by economic and imperial aspirations; and atomic physics provided theoretical constraints from an independent scientific discipline, albeit one to which geologists had close ties. The best geological ideas of the nineteenth century had been refuted, and geologists found themselves with no viable account of their most basic and agreed-upon observational phenomena.

Suess's light had been reduced to heat. And yet the facts of geology remained. Mountains existed on every continent, but still no one knew how they formed. In the Alps, whole stratigraphic sections had been moved kilometers horizontally, yet somehow the sequences themselves remained almost undisturbed. Across the globe, there was undisputed evidence of repeated oscillations of sea level, and the fossil record—given increased prominence in light of Darwin's theory of evolution through natural selection—continued to support the now seemingly discredited idea that the continents had previously been connected. As Joly put it in 1909, "That former continents of any antiquity or magnitude are not hidden beneath the waves [now] seems certain, [but this] argument, of course, does not touch on the former existence of those causeways which biological history requires as the highways of organic migration."[106]

What Dana had written in 1873 was true yet again, that "the great oscillations of the earth's surface . . . still remain facts, and present apparently a greater difficulty than ever."[107] When geologists began to develop new explanations, they found that their problem was not too few "facts," but too many. Various theoretical ideas were available that could explain some of the available data, but no one theory seemed able to explain them all. In response to this situation, different accounts were developed by different scientists, but a common element was that—for the most part—geologists gave most weight to their own locally available evidence. British scientists were heavily influenced by the Trigonometrical Survey data; Americans emphasized field results from geological and geodetic surveys of the Western states; and those who had backgrounds in physics, such as Joly and Holmes, were moved by the significance of radioactive heat.

Historian Homer LeGrand has called this "localism"—Henry Frankel has referred to it as "provincialism"—when geologists are most influenced by data found close to them in time and space and within their own disciplinary specialties.[108] LeGrand has suggested that field geologists, who worked in specific geographic localities on which they focused their attention, were particularly prone to such localism. But the "local" data were not always *physically* proximate, as the case of the British Trigonometric Survey shows, nor were they necessarily field based, as in the influence of discoveries in radioactivity on the work of John Joly. Often, localism was nationalistic or disciplinary. Thus we might better call this phenomenon epistemological chauvinism, or perhaps less pejoratively, epistemological *affinity*. Geologists and geophysicists at the end of the nineteenth century responded to the difficulties in their discipline by relying on and emphasizing the information that was geographically, nationally, and/or disciplinarily closest to themselves. They did not necessarily ignore other information, but they did weigh the available evidence differentially, based on their affinities. And while these affinities expressed themselves epistemologically—in terms of differential weightings of evidence—their sources were largely social.

3

To Reconcile Historical Geology with Isostasy

The Theory of Continental Drift

Alfred Wegener (1880–1930) first presented his theory of continental displacement in 1912, at a meeting of the Geological Association of Frankfurt. In a paper entitled "The geophysical basis of the evolution of the large-scale features of the earth's crust (continents and oceans)," Wegener proposed that the continents of the earth slowly drift through the ocean basins, from time to time crashing into one another and then breaking apart again. In 1915, he developed this idea into the first edition of his now-famous monograph, *Die Entstehung der Kontinente und Ozeane*, and a second edition was published in 1920. The work came to the attention of American geologists when a third edition, published in 1922, was translated into English, with a foreword, by John W. Evans, the president of the Geological Society of London and a fellow of the Royal Society, in 1924 as *The Origin of Continents and Oceans*. A fourth and final edition appeared in 1929, the year before Wegener died on an expedition across Greenland.[1] In addition to the various editions of his book, Wegener published his ideas in the leading German geological journal, *Geologische Rundschau*, and he had an abstract read on his behalf in the United States at a conference dedicated to the topic, sponsored by the American Association of Petroleum Geologists, in 1926.

The Origin of Continents and Oceans was widely reviewed in English-language journals, including *Nature*, *Science*, and the *Geological Magazine*. Although a number of other geologists had proposed ideas of continental mobility, including the Americans Frank Bursey Taylor, Howard Baker, and W. H. Pickering, Wegener's treatment was by far the best developed and most extensively researched.[2]

Wegener argued that the continents are composed of less dense material than the ocean basins, and that the density difference between them permitted the continents to float in hydrostatic equilibrium within the denser oceanic substrate. These floating continents can move through the substrate because it behaves over geological time as a highly viscous fluid. The major geological features of the earth, he suggested — mountain chains, rift valleys, oceanic island arcs—were caused by the horizontal motions and interactions of the continents. Originally, the entire earth was covered by a thin, continuous continental crust, which gradually broke apart and thickened via crumpling of the moving pieces; but during the Mesozoic era, the major continental pieces were reunited in a supercontinent called Gondwanaland.

On one level, there was nothing new in this theory. Half of it was a restatement of isostasy, while the other half was mostly a restatement of Suess's insight of continental contiguity. Wegener had borrowed the idea of a mobile substrate from isostasy and both the idea and the name of Gondwanaland from Suess. But whereas other workers viewed isostasy and Gondwanaland as incompatible, Wegener saw a way to reconcile them. Instead of sinking Gondwanaland away, as Suess had tried to do, Wegener proposed to break it apart sideways. In Wegener's view, Gondwanaland was still with us, albeit in pieces.

Wegener was explicit that his purpose was to effect a grand synthesis between geology and geophysics and to reunite the divergent lines of thinking in earth science just as continental drift had reunited the broken pieces of Gondwanaland.[3] Thus he began his theoretical presentation with the following observation:

> It is a strange fact, characteristic of the incomplete state of our present knowledge, that totally opposing conclusions are drawn about prehistoric conditions on our planet, depending on whether the problem is approached from the biological or the geophysical viewpoint. Paleontologists as well as zoo- and phytogeographers have come again and again to the conclusion that the majority of those continents which are now separated by broad stretches of ocean must have had land bridges [connecting them] in prehistoric times and that across these bridges undisturbed interchange of terrestrial fauna and flora took place. The paleontologist deduces this from the occurrence of numerous identical species that are known to have lived in many different places, while it appears inconceivable that they should have originated simultaneously but independently in these areas. . . .
>
> It is probably not an exaggeration to say that if we do not accept the idea of such former land connections, the whole evolution of life on earth and the affinities of present-day organisms occurring even on widely separated continents must remain an insoluble riddle.[4]

Geologists, in short, had a problem. Paleontological data demanded prior land connections, and sunken continents were the obvious explanation. Thus, Wegener wrote, "even today [contraction theory] possesses a degree of attractiveness, with its bold simplicity of concept and wide diversity of application."[5] But contraction theory had been refuted.

Wegener summarized the arguments from Alpine mapping and the discovery of radiogenic heat, but the most important refutation, he argued, came from isostasy. Isostasy was incompatible with sunken continents. Americans had recognized this and therefore tended toward the view developed by James Dana—now expounded by Bailey Willis—that continents and oceans were permanent features of the earth's crust. This would be a logical position, Wegener noted, but for the paleontological data. Permanence theory gave no account of the overwhelming evidence of fossil homology. Conversely, many European geologists respected the paleontological evidence and clung to land bridges in disregard of the American proof of isostasy. Here were the horns of a dilemma.

> Isostasy theory proves the impossibility of regarding present-day ocean floors as sunken continents, ... [and therefore] "permanence theory" appears to be a logical conclusion from our geophysical knowledge, disregarding, of course ... the distribution of organisms. So we have the strange spectacle of two quite contradictory theories of the prehistoric configuration of the earth being held simultaneously—in Europe, an almost universal adherence to the idea of former land bridges, in America to the theory of the permanence of ocean basins and continental blocks.[6]

Geologists were divided along both disciplinary and national lines. If one could reconcile historical geology with isostasy, one would reconcile European and American geologists as well.

Wegener attempted a reconciliation by proposing a novel solution: the present continents *were* permanent, in that they were the only ones that had ever been, but they had moved *horizontally*, rather than vertically, relative to one another. In short, both sides were right in what they affirmed but wrong in what they denied.

> This is the starting point of displacement or drift theory. The basic "obvious" supposition common to both the land-bridge and permanence theory—that the relative position of the continents, disregarding their variable shallow-water cover, has never altered—must be wrong. The continents must have shifted. South America must have lain alongside Africa and formed a unified block which was split in two in the Cretaceous; the two parts must then have become increasingly separated over a period of millions of years like pieces of a cracked ice floe in water. ... If drift theory is taken as [our] basis, we can satisfy all the legitimate requirements of the land-bridge theory and of permanence theory. This now amounts to saying that there were land connections, but formed by contact between blocks now separated, not by intermediate continents which later sank; there is permanence, but of the area of ocean and area of continent as a whole, but not of individual oceans or continents.[7]

Wegener's goal was nothing less than a grand unifying theory—a reconciliation of two seemingly irreconcilable positions. Like Kant in his attempted reconciliation of rationalism and empiricism, Wegener concluded that both sides made a common mistake: the shared assumption that continents were fixed in their horizontal positions. Let the continents move horizontally, Alfred Wegener argued, and many of the largest problems of earth history would disappear.

The Evidence of Drift

Wegener considered paleontology to provide the most important evidence of drift, perhaps because the data were already established and accepted in a different context. In the 1850s, the British zoologist Philip Sclater had noticed that the island of Madagascar possessed almost none of the common African animals such as monkeys, giraffes, and lions but hosted numerous species of lemurs, an animal common to India. Indeed, some of the Madagascarian lemurs were nearly identical with Indian types. To explain this peculiar phenomenon, Sclater had postulated that a sunken continent, Lemuria, formerly connected Madagascar with India. The advent of Darwin's theory of evolution in the 1860s lent credence to Sclater's idea: the lemurs were too similar to have evolved independently. Subsequent work by paleontologists such as Herbert von Ihering in Germany and Charles Schuchert in the United States had established many other examples of faunal and floral similarities across the oceans. By the 1920s, former connections had been postulated between Australia and India, Africa and Brazil, Madagascar and India, and Europe and North America.[8]

Wegener summarized the work of various paleontologists who had demonstrated global resemblance of fossil forms. In addition to Sclater's lemurs, there was evidence from the Glossopteris seed fern, the Mesosauridae reptiles, the lumbricidae earthworms, and many other organisms.[9] The earthworms' distribution seemed to be particularly significant, because earthworms can neither swim nor fly, nor do they have resilient seeds or a dormant life-cycle or free-floating larval stage, which could permit passive distribution. Furthermore, and most tellingly, in several cases the distributions of species clung to the coastlines of either side of an ocean yet failed to extend very far into their respective continents. This strongly suggested a fragmented habitat: if the continents had not moved, the required connections would have been extremely long and crossed several climatic zones; it seemed unlikely that a species would persist, morphologically unchanged, over such a distance, only to stop its migration precisely at the other end of a land bridge. Likewise Wegener noted that Australian marsupials were almost identical with their South American counterparts—even to the point of having the same parasites—yet were totally unlike their near neighbors in Asia. "One cannot deny," he concluded, "that, in this case, our view . . . does justice to the peculiar nature of the Australian animal kingdom in a way quite different from that of the theory of sunken continents, which is geophysically impossible, anyway."[10]

The faunal evidence alone, Wegener argued, would be enough to compel an explanation along the lines of drift, but it was also supported by direct geological data. The most obvious was the "jigsaw-puzzle" fit of the continents. Indeed, in his introduction, Wegener claimed that the idea of continental drift first came to him "when considering a map of the world, under the direct impression produced by the congruence of the coastlines on either side of the Atlantic."[11] The morphological match of South Africa with South America and of Europe with North America was too close to be coincidental. Wegener's explanation was that they had been contiguous but had broken apart in geologically recent time. Minor discrepancies in their reconstruction could be attributed to "tearing" and rotation during rupture, or to uncertainties regarding the outlines of the continental shelves. After all, it was the edges of the continental shelves that had to match, not the visible coastlines, which were to some extent artifacts of sea level and erosion.

But the jigsaw-puzzle fit of the continents had been noted many times before.[12] More important than the congruence of the coastlines was the conformity of the rocks along them. Throughout the geological literature, there were accounts of stratigraphic sequences and structural patterns that were nearly identical on opposite sides of oceans. If one reunited North America and Europe, as Wegener proposed, the Caledonian and Appalachian fold belts matched, as did the southern extent of Pleistocene glacial moraines. In South America and Africa, there were nearly identical stratigraphic sequences on either sides of the South Atlantic. But, above all were the Gondwana beds of India, from which Suess had taken the name of his great continent. This interbedded sequence of sedimentary and volcanic rocks contained economically important coal deposits and was nearly identical to the equally famous Karroo sequence of southern Africa.

Taken individually, any one of these matches might be dismissed as coincidence, but the "totality of these points of correspondence constitutes an almost incontrovertible proof," Wegener held.[13] Not only did the coastlines fit, but the rocks along them fit too. This proved that continental breakup had occurred *not* in the early history of the earth, as Darwin and Fisher held, but during the entire course of geological time.

> Of crucial importance here is the fact that, although the blocks must be rejoined on the basis of other features—their outlines, especially—the conjunction brings the continuation of each formation on the farther side into perfect contact with the end of the formation on the near side. It is just as if we were to refit the torn pieces of a newspaper by matching their edges and then check whether the lines of print run smoothly across. If they do, there is nothing left but to conclude that the pieces were in fact joined in this way. If only one line was available for the test, we would still have found a high probability for the accuracy of fit, but if we have n lines, this probability is raised to the nth power.[14]

Devotees of jigsaw puzzles fit pieces together by shape, and when the picture comes together, they know for sure that they have got it right.

Wegener's professional specialty was meteorology and geophysics—he was the author of an important and widely used textbook, *Thermodynamik der Atmosphäre*—and the geological evidence closest to his own expertise was paleoclimatic.[15] Across the globe, geological indicators such as glacial deposits, coal seams, coral reefs, and evaporites showed that the earth's climate had been subject to repeated and in some cases severe fluctuation. Fossil assemblages in Europe recorded considerably warmer conditions than at present; glacial deposits in the Permian of South Africa and Australia indicated previously much colder climates. Some of North America's most prominent geologists—Louis Agassiz, G. K. Gilbert, T. C. Chamberlin—had devoted major parts of their careers to the documentation and explanation of glacial deposits and landforms.[16] Although some nineteenth century geologists had interpreted prior warm conditions as evidence of continuous secular cooling, by the twentieth century this view had been abandoned, and geologists agreed that both paleontological and lithological criteria demonstrated repeated climatic fluctuation. In the United States, Bailey Willis had described the cause of climatic fluctuations as one of "the most obscure" problems of paleogeography, and concluded that there must be "some general cause, so subtle that it has yet eluded distinct recognition."[17]

Wegener proposed drift as that general cause. As a meterologist, he knew that there were many possible causes of climate change; he also knew that the Pleistocene glaciations of North America and Europe had occurred too recently to be accounted for by continental drift. Thus he allowed that "drift is only one of many causes of variation in climate, and not even the most important for the most recent periods."[18] Nevertheless, viewed alongside the paleontological and geological data, paleoclimatic data helped to constitute a coherent body of evidence. Furthermore, insofar as latitude is the *predominant* variable affecting modern climate, latitudinal drift was a parsimonious explanation of ancient change. At minimum, climate data provided corroboration of the theory. Pushed to the maximum, they demonstrated the explanatory power of the theory: while the paleontological and lithological data could be used to determine the *timing* of episodes of continental amalgamation and fragmentation, the paleoclimatic data constrained their *location*. By examining when fossil and stratigraphic assemblages were coherent and when they diverged, one could determine the timing of drift events; by examining the corresponding paleoclimatic data, one could estimate the latitude of continental blocks before and during breakup. In short, continental drift would permit the complete reconstruction of earth history in a manner yet to be achieved by any other theory (figure 3.1).

The Consequences of Drift

Wegener took pains throughout his book to point out where his theory had testable consequences. For example, whereas Dana and Suess had been at odds over the loci of mountain-building, drift could account for mountains both on the edges of continents and within them: Mountains would *form* only on the edges of continents (consistent with Dana) but could be found *within* them wherever two blocks had collided and amalgamated (consistent with Suess). Consequently, mountains deep within continents should be older than those on continental margins, and in general this seemed to be true. The theory also implied that fossil and stratigraphic similarities should end relatively abruptly as compared with the length of geological time during which they cohered, and this too appeared to be true. But the most important prediction of the theory—indeed the only true *prediction*, in the temporal sense—was that if continents had moved in the past, they would likely be moving still. Wegener assumed that the "majority of scientists will probably consider this to be the most precise and reliable test of the theory."[19] One could quarrel about the details of paleontological and stratigraphic correlations—and Wegener's critics did—but if one could show that the continents were moving at present, there would be nothing left to dispute. All the rest would be commentary.

In the fourth edition of his *Origin of Continents and Oceans*, the geodetic evidence—a relatively minor point in previous editions—became Wegener's first line of argument. Given the difficulty of testing most geological theories, which aimed to explain the inaccessible events of the distant past, the testability of drift theory was a distinct strength:

> Compared with all other theories of similarly wide scope, drift theory has the great advantage that it can be tested by accurate astronomical position finding. If continental displacement was operative for so long a time, it is probable that the process is still continuing, and it is just a question of whether the rate of movements is enough to be revealed by our astronomical measurements in a reasonable period of time.[20]

Wegener believed that such movements had been revealed, albeit inadvertently. Longitudinal measurements on a series of expeditions to East Greenland, in 1823, 1870, and 1907, had apparently revealed an active westward drift. Wegener had been a member of the last of these expeditions, sponsored by the Danish government to explore the inland ice cap; he was the expedition meteorologist. (This was not Wegener's first adventure: in 1906, he and his brother Kurt set a world record of fifty-two and a half hours in a balloon on a flight across Germany to test navigational equipment.)[21] Wegener calculated the following rates of displacement of Greenland vis-à-vis the European continent according to the expedition measurements:

1823–1870   420 meters (9 meters/year)
1870–1907   1190 meters (32 meters/year)

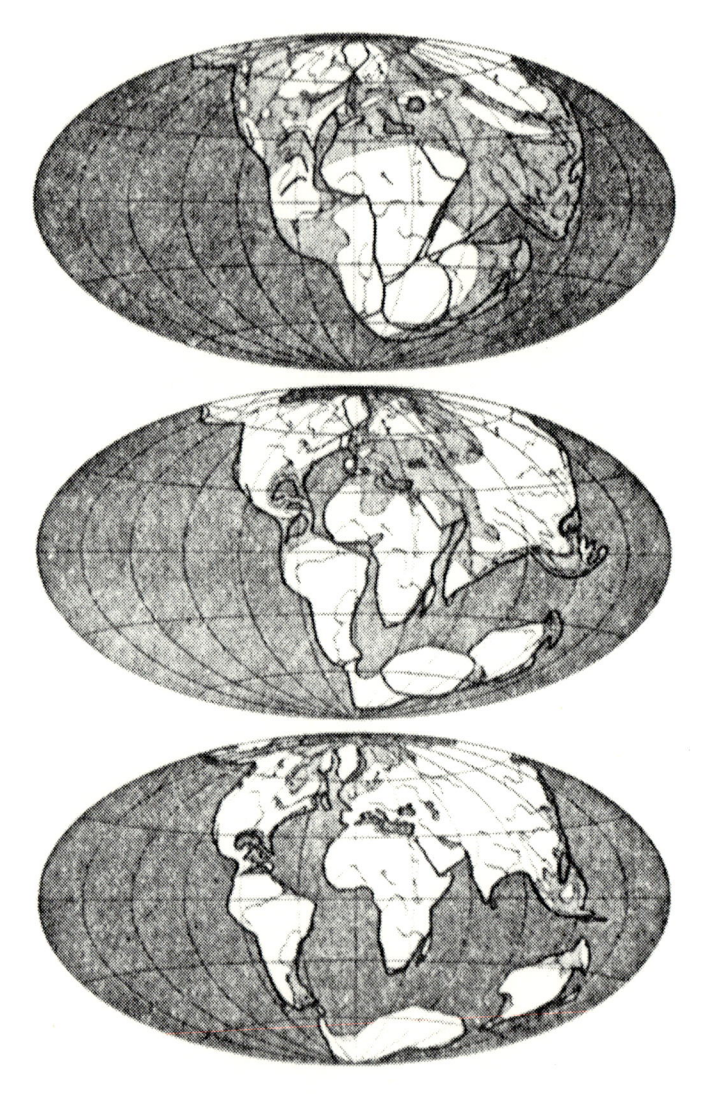

Figure 3.1. Wegener's reconstruction of the world. In this diagram, Wegener illustrates how the present configuration of continents and oceans arose from the breakup of Pangea from Carboniferous (top) through Eocene (middle) to Quaternary times (bottom). The stippled areas represent land masses that were under shallow-water cover; present-day rivers are shown to aid identification. Wegener's point was that, given the assumption of drift, geological evidence could be used to determine the timing and position of the continents at various epochs—that is, to reconstruct earth history. (From Wegener [1929] 1966, p. 18.) Reproduced with permission of Dover Publications, Inc.

In an appendix to the fourth and final edition of his book, Wegener cited evidence of a displacement of North America relative to Europe of between 0.3 and 0.6 meters per year, on the basis of radio wave transmission times, and a displacement of Madagascar relative to Europe of 60 to 70 meters per year. Wegener compared these values with estimates of the rate of drift deduced from continental reconstructions and concluded that they were roughly compatible (figure 3.2).

The Greenland data were based on lunar measurements of longitude, which were known to be imprecise and subject to systematic error, and the determinations from the 1823 and 1870 expeditions were not made in exactly the same place. This led to considerable uncertainty regarding the accuracy of these measurements. Therefore, Wegener had enlisted the help of J. P. Koch, the 1907 expedition cartographer, to calculate whether the observed differences could be attributed to measurement or location error. Koch concluded that "the sources of error, whether taken individually or all together, are insufficient to explain the [observed] difference."[22] Since that time, the Danish Geodetic Institute had experimented with longitude measurements on the basis of radio telegraphy; comparing these with the earlier measurements suggested an average drift of Greenland of 20 meters per year.

Wegener summarized these measurements with a diagram showing that the calculated drift was much greater than the measurement error and therefore must be at least qualitatively correct (figure 3.3). He wrote in response to criticisms of the earlier editions of his book:

> While it is true that all these comparisons suffer wholly or partly from the fact that they are based on lunars and *may* therefore contain systematic errors that cannot be checked, this accumulation of similar results which do not stand in opposition to any others makes it highly improbable that it is all just a matter of an unfortunate combination of extreme errors of observation.[23]

Even if the Greenland data were wholly in error, independent evidence of drift had been detected in the United States. Walter Lambert, a geodesist at the U.S. Coast and

Area	Relative drift to date (km)	Period since separation, approx. (10^6 years)	Displacement per year, average (m)
Sabine Island–Bear Island	1070	0.05–0.1	21–11
Cape Farewell–Scotland	1780	0.05–0.1	36–18
Iceland–Norway	920	0.05–0.1	18–9
Newfoundland–Ireland	2410	2–4	1.2–0.6
Buenos Aires–Cape Town	6220	30	0.2
Madagascar–Africa	890	0.1	9
India–southern Africa	5550	20	0.3
Tasmania–Wilkes Land	2890	10	0.3

Figure 3.2. Wegener's chart of continental displacements. Wegener used longitude measurements from Danish expeditions to Greenland to estimate rates of continental drift. By modern standards his estimates were too high by 1 to 3 orders of magnitude. (From Wegener [1929] 1966, p. 26.) Reproduced with permission of Dover Publications, Inc.

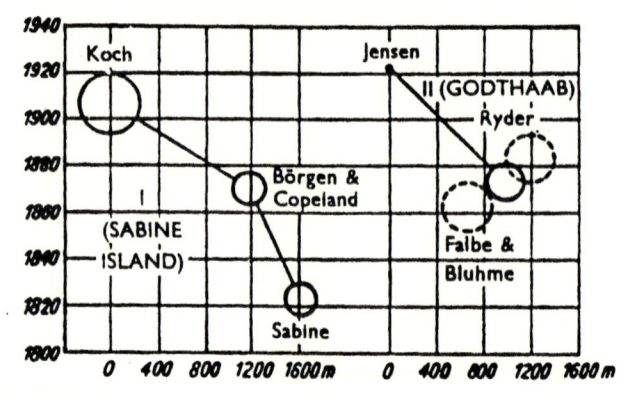

Figure 3.3. Wegener's calculation of the displacement of Greenland. Wegener plotted the displacement of Greenland as calculated from longitude measurements on a series of Danish expeditions. The implied displacement rates were very large, and Wegener acknowledged that the lunar method of longitude reckoning was poor, but he argued that even if the quantitative evidence was in error, the qualitative results pointed consistently in the direction of a westward displacement. (From Wegener, [1929] 1966, p. 28.) Reproduced with permission of Dover Publications, Inc.

Geodetic Survey, had reported evidence of latitude change in geodetic stations in California. Wegener quoted Lambert with satisfaction: "A systematic study of such anomalies would be highly desirable."[24] Wegener agreed, and in 1930 he returned to Greenland to confirm the longitudinal drift. But while attempting to cross the inland ice cap, Alfred Wegener died. The guess of those who found him was that he died in his tent, probably of a heart attack, in early November 1930.[25] But the debate about his theory lived on.

The Rejection of Drift: Lack of a Causal Mechanism?

The theory of continental drift presented a possible solution to the old geological problem of the origin of mountains at a time when existing explanations were seriously in doubt. Unlike contraction theory, drift could resolve the seemingly conflicting data of geophysics and paleontology and unify a wide range of observations from stratigraphy and paleoclimatology. Moreover, drift did not rely on the dubious postulate of a cooling earth, nor on any particular interpretation of planetary genesis. The evidence for it had been collected and documented by leading geologists and paleontologists from diverse localities across the globe, and there was even limited evidence, although admittedly inconclusive, that a present-day drift might actually be occurring.[26] In short, Wegener had both evidence and explanation. Yet the notion of moving continents was rejected by the geological community of the 1920s and 1930s.

The usual explanation for the rejection of continental drift is that Wegener lacked an adequate causal mechanism. This view was presented to a generation of students by Frank Press and Raymond Siever in their widely used 1974 textbook, *Earth*, one of the first major texts written in the context of plate tectonics. The problem with continental drift, they said, was that "proponents could not come up with a plausible driving force."[27] A similar view was presented by the geophysicist Seiya Uyeda in his 1978 book on the development of the theory of plate tectonics. "What was the driving mechanism of drift?" Uyeda asked. "An explanation of *effect* was meaningless if the *cause* could not be identified."[28]

But *is* an explanation of effect meaningless without an explanation of cause? More to the point, was it meaningless to geologists in the early twentieth century? Historical evidence suggests not. As the geologist-historian Anthony Hallam has pointed out, many empirical geological phenomena have been accepted before their causes were known.[29] The ice ages (whose causes are still being debated today), the Alpine overthrusts (whose existence helped stimulate the drift debate), and geomagnetic polarity reversals (which were critical to the establishment of plate tectonics) are three important examples; there are many others.[30] The late Marshall Kay, Professor of Geology at Columbia University, went so far as to describe the operating premise of geologists as "anything that has happened, can."[31]

There was a long tradition in geology of accepting the reality of phenomena without requiring causal accounts of them. As historian Martin Rudwick has pointed out, the issue of causation can be logically separated from the issue of occurrence, and many geologists found it helpful to make just such a separation: "Many geologists wisely concentrated on establishing that certain events *had* occurred, recognizing that their causation was a logically separate question."[32] A historically significant example is found in the tension between James Hall and James Dana over geosynclines. Dana criticized Hall's work as a theory of the origin of mountains "with the origin of mountains left out" because of its lack of a causal mechanism, but Hall emphatically defended his position as based on an appropriate demarcation between "justified inference" (that the crust *must* have subsided gradually to permit the accumulation of thousands of feet of shallow water sediment) and unwarranted "speculation" (an explanation of how those sediments got transformed into a mountain range).[33] The former was based on *evidence*, the latter transcended the observable.[34] Hall's distinction may have been self-serving—some historians have suggested that he was rather poor at creative speculation—but the historical point is that it was possible to accept geosynclines without knowing the mechanism of uplift, and geologists did accept them, because there was evidence for them.

Nor was Hall the only one to erect linguistic demarcations along these lines.[35] European geologists in the eighteenth century used the term *geognosy* to describe knowledge of phenomena, *geology* to describe systems of interpretation.[36] The great German mineralogist Abraham Gottlob Werner, professor at the Bergakademie in Freiburg, Saxony, promoted *geognosy* as the knowledge of rocks, minerals, and ores.[37] His idea of geognosy as "certain knowledge of the solid framework" of the earth, and his close relation to the Saxon mining industry, suggest that diffidence about causes might have origins in geology's utilitarian roots: one could exploit and develop mineral resources using Werner's system of classification yet remain aloof from any particular interpretation of them.[38] Or perhaps the desire to separate evidence from explanation arose from the historical nature of geology: before one could explain the events of geological history, one first had to accept that there was a geological history—and not everyone did.[39] Whatever the reasons, the existence and use of these complementary terms reflected a belief that it was both possible and desirable to decouple phenomenological knowledge from theoretical interpretation. Although the appropriate placement of the boundary between "geognosy" and "geology" was a matter of recurrent dispute, this terminological demarcation persisted into the early twentieth century.[40]

The Alpine overthrusts brought home the point in the early twentieth century. The evidence for their existence was overwhelming, yet the mechanism by which huge sheets of rocks could be moved hundreds of kilometers was unfathomable. John Joly concluded that the very fact that geologists *had* accepted them, solely on the basis of phenomenological evidence and despite a complete lack of causal explanation, was the best possible proof of their *reality*:

> Many felt unequal to the acceptance of structural features involving a folding of the earth-crust in laps which lay for scores of miles from country to country, and the carriage of mountainous materials from the south of the Alps to the north, leaving them finally as Alpine ranges of ancient sediments reposing on foundations of more recent date. The historian of the subject will have to relate how some who finally were most active in advancing the new views were at first opposed to them. In the change of conviction of these eminent geologists we have the strongest proof of the convincing nature of the observations and the reality of the tectonic features upon which the recent views are founded.[41]

Arthur Holmes suggested that unimaginable things were possible in geology because of the stupendous amounts of time available. In his 1913 book *The Age of the Earth*, in which he explained precisely how stupendous they were, he quoted from Herodotus: "If one is sufficiently lavish with time, everything possible happens."[42] As Marshall Kay suggested, what was possible was not to be assumed a priori but deduced from the geological record of what had happened.

Given this history, the absence of a causal explanation of drift can hardly be the singular cause of its rejection. To accept the existence of continental drift without knowing its cause would have been well within the mainstream of geological tradition.[43] In the mid-1920s, Chester Longwell, a professor of geology at Yale University, suggested precisely this option. At a 1926 symposium on continental drift sponsored by the American Association of Petroleum Geologists (AAPG), Longwell addressed the mechanism issue and concluded that geophysical arguments, although "generally unfavorable to the displacement hypothesis," were in the end "not conclusive." The interior of the earth and the forces affecting it were not well enough known. The problem of drift would have to be decided on the strength of the geological evidence of whether or not it had actually occcurred—and this was the task now before geologists. "Geologists alone can determine whether the geophysical forces have had 'in geologic history an appreciable influence on the position and configuration of our continents,'" Longwell said, quoting from American geodesist Walter Lambert. "Let us therefore re-examine some of our own evidence, to see whether it is compelling. If it is," he concluded, "then physical geologists should be content to accept the *fact* of displacement, and leave the explanation to the future."[44]

Longwell's prescription came true when plate tectonics was established. As Ursula Marvin explained in 1973, in the wake of the new theory, "Today, we stand rather close to where Wegener stood. The clear evidence of plate motions may be seen at every hand [although] no adequate driving mechanism has been agreed upon."[45] Since then, scientists have increasingly acknowledged this paradox in their own account: If continental drift were rejected for lack of a causal mechanism, why was plate tectonics not rejected for the same reason? Claude Allegre, a well-known geochemist, recently made this point: "No doubt [Wegener's] willingness to present explanatory theories without convincing mechanisms contributed greatly to the rejection of his theory. Paradoxically, global tectonics has been accepted today without apparent underpinnings of a causal theory." Allegre notes that scientists who worked on explanatory models of plate tectonics did so ex post facto: "None of the scientists who developed the concepts of sea-floor spreading and plate tectonics attempted to develop a theory explaining the primary causes of drift *at the same time*. It was only *after* acceptance of the phenomenological theory that the scientific community began to research its causes. In no way did this prevent mobility theory from prevailing."[46] Even today, the driving force of plate tectonics is poorly understood. In the words of one leading geophysicist: "Although the theory of plate tectonics is well established, the engine that drives the motions of the lithospheric plates continues to defy easy analysis."[47] Or more pointedly: "Plate tectonics . . . describes the motions of the plates, but not the forces driving those motions."[48] The situation was summarized by geologist Vincent Saull, writing in 1986 in a leading geological journal: "Popular acceptance of plate tectonics is not based on knowing the mechanism that moves the plates."[49]

Dynamics versus Kinematics

A number of geologists and historians, including Stephen Jay Gould, have presented a slightly different interpretation of the mechanism issue. Gould has written that the problem with drift was not that it lacked an adequate driving force but that it lacked an adequate mechanical account. In his words, drift "was dismissed because no one had devised a physical mechanism that would permit the continents to plow through an apparently solid oceanic floor."[50] Rachel Laudan refined this argument by introducing the terminology of kinematics and dynamics—that is, the how and the why, respectively, of crustal motion. She has argued that the acceptance of horizontal crustal motion occurred in the 1960s when an acceptable kinematics was developed, even though the dynamics remained (and remains) an unsolved problem. Wegener's failure, she suggested, was inevitable, because "the problem with drift was not that there was no *known* mechanism or cause, but that any *conceivable* mechanism would conflict with physical theory."[51] But Laudan and Gould are wrong. There *were* conceivable mechanisms—mechanisms that did not require continents to plow through a rigid oceanic floor—and they were grounded in isostasy.

In *The Origin of Continents and Oceans*, Wegener argued that the substrate in which the continents reside could behave, over geological time, in a fluid manner, just as glass can flow if subjected to small stresses over sufficient time. This notion was inherent in the theory of isostasy, which, as Wegener noted, was "nothing more than hydrostatic equilibrium according to Archimedes' principle, whereby the weight of the immersed body is equal to that of the fluid displaced." Plainly, continents could float in hydrostatic equilibrium if and only if the substrate in which they were imbedded behaved as a fluid. And if the substrate *were* fluid, then the continents could, at least in principle, move through it. Calling the substrate fluid was merely another way of stating the theory of isostasy. However, as Wegener noted, "the introduction of a special word [isostasy] for this state of the earth's crust has some point because the liquid in which the crust is immersed apparently has a very high viscosity, one which is hard to imagine."[52] Indeed, the kinematic state of the substrate *was* hard to imagine and had been a topic of debate for some time. The net result of this debate, however, was that moving continents *were* conceivable. To understand why, we need to look more closely at the history of thinking about the earth's interior.

Fathoming the Interior of the Earth

What is the interior of the earth made of? Is it solid or liquid? These are among the most fundamental questions in the earth sciences, questions that geologists and geophysicists have long tried to answer. Geologists worried particularly about the zone immediately below the crust, the zone from which igneous rocks and surface dislocations were presumed to originate. In the early twentieth century, when Alfred Wegener proposed continental drift, most of them believed that the best answer to the question—solid or liquid?—was neither. The crustal substrate appeared to be something in between.

Volcanoes and hot springs made it seem obvious to many eighteenth-century observers that the interior of the earth must be partly or wholly liquid—the view implicit in versions of contraction theory that supposed the collapse of a shrinking crust into a molten substrate.[53] But strong arguments against a fluid interior derived from physical astronomy. In the 1840s, William Hopkins, who called himself a "physical geologist," suggested that the precession and nutation of the earth's axis was inconsistent with a fluid Earth: the solid crust must be close to 1000 miles thick to account for the earth's rigid behavior. Lord Kelvin, as we have seen, argued that the existence of the tides proved conclusively that the earth was solid (and therefore its age could be determined from its cooling history), because if it were not solid, it would deform along with its surface waters and there would be no tides. From this line of reasoning came his famous pronouncements that the earth as a whole was more rigid than a globe of solid glass, and probably more rigid than a globe of solid steel.[54]

Those who leaned toward mathematical analysis and physics-based accounts found such arguments compelling. Writing in the early 1870s, the American geologist Joseph LeConte concluded:

> The foregoing objections have induced many of the best geologists to believe that the earth . . . may be regarded as *substantially solid*. . . . While much of the prestige of great names, such as Humboldt, Von Buch, Elie de Beaumont, &c. are on one side, most of the substantial argument is on the other. I feel convinced that the whole theory of igneous agencies—which is little less than *the whole foundation of theoretic geology*—must be reconstructed on the basis of a solid earth.[55]

But if a liquid earth was inconsistent with geophysics, there were problems with a solid earth as well. However well supported by mathematics, the theory of the solid earth did not have a ready account of many geological processes. The most important of these, LeConte acknowledged, were the "phenomena of movement of the earth's crust . . . especially those great and wide-spread oscillations which . . . have left their impress in the general unconformability of the strata." These oscillations, LeConte confessed, "I cannot explain."[56]

The oscillations of the earth's crust were a central focus of the advocates of isostasy. The problem of reconciling Kelvin's physical insights with those of observational geology and geodesy had led Osmond Fisher to argue that the earth must be at least partially fluid. "If there be granted a layer of fluid matter of a thickness even very small compared to the radius of the earth," Fisher wrote in response to Thomson, "even that will suffice us."[57] Fisher hailed from the same school as Hopkins and Kelvin, and like them he applied mathematical and physical principles to geological problems. But unlike them, he was sensitive to the observational evidence of geologists and recognized that a wholly solid earth was impossible to reconcile with isostasy. The compensation of topographic features required a fluid substrate beneath the crust. As Fisher put it emphatically, *"What is required to explain the phenomenon [of isostasy] is a liquid, or at least a plastic substratum for the crust to rest on."*[58]

Fisher's preference was for a fully liquid substrate, but his qualification "or at least plastic" was deference to the physicists' arguments and followed George Biddell Airy's original suggestions. In his classic 1855 paper, Airy had emphasized that the "fluidity" entailed by isostasy did not necessitate liquidity. "[The] fluidity of the earth may be very imperfect," Airy had suggested; "it may be mere viscidity; it may even be little more than that degree of yielding which (as is well known to miners) shows itself by changes in the floors of subterranean chambers."[59] A slowly yielding substrate could permit isostatic adjustment over geological time frames yet behave like a solid with respect to thermal conductivity and short-duration astronomical processes.[60]

Kelvin, however, was as unyielding as he imagined the earth to be, and Fisher became increasingly adamant that the physicist's analysis was simply wrong. Any mathematical treatment that "proved" the complete solidity and rigidity of the earth was clearly missing some essential element. Fisher wrote:

> With respect to the yielding of the crust, I think we cannot but lament that mathematical physicists seem to ignore the phenomena on which our science founds its conclusions, and, instead of seeking for an admissible hypothesis, the outcome of which, when submitted to calculation, might agree with the facts of geology, they assume one which is suited to the exigencies of some powerful method of analysis, and having obtained their result, on the strength of it bid bewildered geologists to disbelieve the evidence of their senses. Such appears to most of us the conclusion that the earth is excessively rigid from its centre to its surface. For we know that down to a quite recent period it has yielded freely to pressure.[61]

Fisher's thesis of a yielding substrate was seconded by American advocates of isostasy. In the widely cited 1889 paper in which he coined the term *isostasy*, Clarence Dutton noted that isostasy necessarily implied the transfer of substratal material from areas of sedimentation to areas of denudation. "The result," he concluded, "is a true viscous flow" in the substrate.[62] Dutton allowed that this notion challenged the idea, common in the culture, of the earth as the very model of stability, but used the analogy, earlier made by Fisher and later invoked by Wegener, of glacial ice. Viewed instantaneously, both rocks and ice are rigid. Yet we know that ice can flow, and so perhaps could rocks if subjected to small but sustained stress:

> There may be in this proposition some degree of violence to a certain mental prejudice against the idea that the rock-ribbed earth, to which all our notions of stability

and immovableness are attached, can be made to flow. It may assist our efforts if we reflect upon the motion of the great ice sheet which covers Greenland. . . . The forces called into play to carry the glacier along horizontally do not seem to differ greatly in intensity or amount from [those required by isostasy], and the rigidity of the ice itself may not exceed the mean rigidity of the rock masses beneath the littoral [zone].[63]

G. K. Gilbert likewise concluded that the theory of isostasy required a mobile substrate. In his well-known discussion of the beds of Lake Bonneville, Gilbert closed by saying that "the hypothesis of the hydrostatic restoration of equilibrium by the underflow of heavy earth-matter is the only explanation which explains . . . the observed facts."[64]

In the year that Dutton coined the term isostasy, the American astronomer and geodesist Robert Woodward, later director of the Carnegie Institution of Washington, summarized the state of scientific knowledge regarding the earth's interior. In an address to the American Association for the Advancement of Science entitled "The Mathematical Theories of the Earth," Woodward allowed that the problem of the "liquidity or solidity of the interior, and the rigidity of the earth's mass as a whole" was a "vexed question." Nevertheless, he maintained,

> The conclusion seems unavoidable that at no great depth the pressure is sufficient to break down the structural characteristics of all known substances, and hence to produce viscous flow. . . . Purely observational evidence . . . of a highly affirmative kind in support of this conclusion is afforded by . . . experiments on the flow of solids and by the abundant proofs in geology of the plastic movement and viscous flow of rocks."[65]

Woodward's belief in a mobile substrate was a consequence of his own geodetic work and had been stated explicitly in his memoir "On the Form and Position of Sea Level" published by the U.S. Geological Survey the previous year. In this seminal work (which articulated the problem that John Hayford would later solve) Woodward concluded that deviations of the geoid from the reference ellipsoid were due to insufficient recognition by geodesists of the effects of isostatic adjustment in a plastic substrate. Woodward wrote:

> It seems probable that these forces producing these deviations [in the form of the geoid] have their seat in a comparatively thin terrestrial crust resting on a fluid or plastic substratum (or nucleus), or that such was the antecedent condition of the earth, and that our failure to perceive the relations of the crust to the substratum is the chief obstacle to improvement [of geodetic results].[66]

Although Woodward was convinced that the substrate was plastic, he acknowledged that the issue was "still lingering on the battle fields of scientific opinion." Revisiting a metaphor that Kelvin had used self-referentially—Dryden's vain king who "thrice slew the slain"—Woodward concluded that the battle over the interior constitution of the earth would yet be "fought o'er again":

> Many of the "thrice slain" combatants in these contests would fain risk being slain again; and whether our foundation be solid or liquid, or to speak more precisely, whether the earth may not be at once plastic under the action of long continued forces and highly rigid under the periodic forces of short period, it is pretty certain that some years must elapse before the arguments will be convincing to all concerned.[67]

Woodward's comments were more prescient than he imagined; the debate was to continue not merely for several more years, but for several decades.

Isostatic Undertow in a Plastic Substrate

Among those who again took up the question of the nature of the earth's interior was the colossus of American geology, T. C. Chamberlin (1843–1928). Professor of Geology at the University of Chicago and founding editor of the *Journal of Geology*, Chamberlin was perhaps the single most influential person in the history of American geology. Chamberlin was not an advocate of a fluid earth—indeed, he was in the process of formulating a new theory of the earth on the basis of solid accretion—but he nevertheless allowed that the theory of isostasy and the gravity work of Putnam and Gilbert provided "both theoretical and observational grounds for the belief [in] the slow fluency or quasi-fluency of ... rocks."[68] In his and R. Salisbury's massive three-volume work of 1909, *A College Text-book of Geology*, Chamberlin followed Gilbert's original argument for the isostatic basis of continental elevation and postulated an isostatic creep at the base of the continents:

> The continental platforms now stand forth above the ocean bottoms as great plateaus [and] their weight develops a pressure on their basal parts that may be safely estimated at 15,000 to 20,000 pounds to the square inch. This pressure *tends to cause the platforms to spread by lateral creeping*. This is opposed by the strength of the rock, and by the hydrostatic pressure of the oceans against the sides of the platforms. If this last be estimated at 5000 pounds per square inch at the bottom, there remains an unbalanced pressure of 10,000 to 15,000 pounds per square inch. [A] pressure of this magnitude constantly exerted for a prolonged period, might cause some spreading of the great continental platforms. ... Whether it is really very effective or not, it is at least a constant agency *tending* to produce lateral creep.[69]

According to Chamberlin, isostatic creep would result in a gradual lowering of the continents and a concomitant rise in the ocean bottom and surface of the sea; continental elevations would be restored during episodes of diastrophism. The net result would be a geological history characterized by long periods of continental relaxation interrupted by short periods of rejuvenation—precisely as indicated by the rock record. But Chamberlin was not sure how effective isostatic creep really was. On the one hand, it seemed to be of limited extent, because if "the continental rock-masses spread with anything like the facility that continental ice-masses do, the land would probably have been completely submerged between each of the great periods of [continental] deformation." This was evidently not the case. On the other hand, normal faults provided evidence of tensional deformation of the continent, which might be interpreted as evidence of relaxation.

> Normal faulting may therefore finally prove to be an evidence of glacier-like, but extremely slow, creep of the continental masses ... Cracking and spreading of this kind is what might be expected if the deeper parts [of the earth] have crept laterally under pressure, just as similar cracks open and widen on the surface of asphalt when it slowly spreads under its own weight. Without affirming that this is the explanation of crevicing, or that it is evidence of continental spreading, the mutual fitness of the two is worthy of consideration.[70]

Like many American scientists, Chamberlin was more cautious in his published claims than in his private correspondence. In a letter to Bailey Willis in 1894, discussing Willis's widely read *Mechanics of Appalachian Structure* (1893), Chamberlin declared unequivocally, "flowage [of the subcrust] under pressure is an unquestioned fact."[71] Meanwhile, as Chamberlin articulated his notion of isostatic creep, J. F. Hayford published the first of his two seminal memoirs documenting the geodetic evidence of isostasy. Hayford rarely spoke to the geological implications of his work, but he inferred the necessity of a mobile substrate for the maintenance of isostatic equilibrium. In an address to the Washington Academy of Sciences in 1906, he revived Dutton's idea that the expansion of crustal columns in areas of erosion reduced the pressure at the bases of those columns, while loading in areas of sedimentation increased the pressure there. The resulting pressure gradient caused "a great slow undertow from the ocean areas towards the continents, and a tendency to outward creeping at the surface from the continents towards the oceans."[72] The evidence of tangential stress and horizontal compression in mountain belts could be thus explained: "The great undertow toward the continents is attached to the surface strata by continuous material and tends to carry them inward" (figure 3.4). Commenting on Hayford's presentation, Dutton, now the grand old man of isostasy, was pleased. He especially liked Hayford's relatively shallow depth of compensation, because it "permits us to infer a larger amount of horizontal displacement in the underlying masses than if it were much deeper."[73]

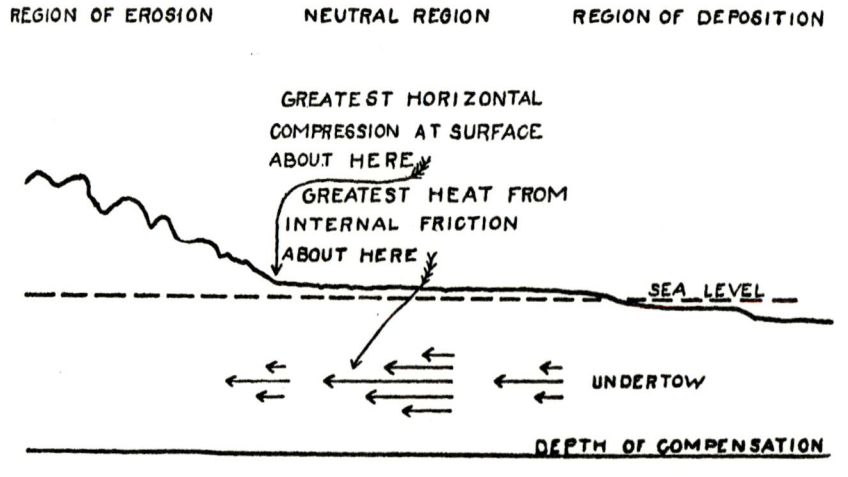

Figure 3.4. Isostatic undertow. The existence of isostasy had been taken by Hayford and others to imply that some form of plastic flow must occur beneath the earth's crust. In order to maintain isostatic equilibrium, material had to flow from regions of loading (sedimentation) to regions of unloading (erosion); this was the same argument earlier made by Dutton. Thus Hayford suggested the idea of an isostatic "undertow." (From Hayford 1911, p. 205.) Wegener extended this line of reasoning to argue that plastic flow beneath the crust could be a cause of continental drift.

A Consensus Emerges: A Mobile Substrate

In 1914, with the confirmation of isostasy receiving widespread attention in American scientific circles, G. K. Gilbert summarized the geological evidence of a mobile substrate:

> There is much geologic evidence of mobility somewhere below the surface. Part of this evidence is volcanic. The continuous or secular relations of pressure, temperature, and density in the subterranean region from which liquid rocks arise at intervals may be assumed to be such that moderate change of condition either induces liquefaction or else so lowers the density of rock already liquid as to render it eruptible; and such a balancing of conditions implies some sort of mobility. Other evidence is diastrophic. In some regions, such as the Appalachians, overthrusts and folds testify to the great reduction in the horizontal extent of rocks near the surface. . . . If the subjacent portion of the nucleus had been correspondingly forced into narrower space there would have resulted an enormous mountain range, but the actual uprising was of moderate amount. Plausible explanations of the phenomena necessarily include horizontal movements of the upper rocks without corresponding movements of the nucleus and thereby imply mobility in an intervening layer.[74]

Like Dutton and Woodward, Gilbert was sensitive to the difficulty of conceptualizing the properties of the "mobile" layer and reconciling it with astronomical evidence, and he cautioned that it was "not necessary to suppose that the degree of mobility is that of a liquid at the surface, [nor is it] necessary to think of the degree of mobility as uniform, either from place to place or from time to time. Its place variation would naturally be coordinate with that of rock types, and its time variation coordinate with epochs of elevation and subsidence."[75] Regardless of how one imagined it, a mobile substrate was geologically evident and consistent with geodetic results. Turning Hayford's reasoning around, Gilbert argued that the geodetic data could be viewed as confirmation of ideas required on geologic grounds:

> The conception thus engendered, of a relatively mobile layer separating a less mobile layer above from a nearly immobile nucleus, appears to me in full accord with the evidence which geodesy affords of isostatic adjustment. The geodetic "depth of compensation" agrees with such suggestions as to the position of the horizon of maximum mobility as might be afforded by the volcanic and diastrophic phenomena. The existence of a horizon of mobility accords with the inference of approximate perfection of isostatic adjustment.[76]

According to one's preference, one could view the mobile substrate as an inductive inference from geology or as a deductive consequence of geodesy. Either way, the result was the same. By the early 1900s, the concept of a viscous fluid layer was so well established in the American geological literature that geologists began to conduct compression experiments to determine its depth (figure 3.5).[77]

For those physically oriented earth scientists who still resisted the notion, perhaps because they had trouble imagining how a quasi-fluid earth might work, George Becker reiterated that a mobile substrate "must be accepted as demonstrated by the geodesists." Thinking that physicists might thus better understand the idea, he used their favorite metaphor, a heat engine, in his 1915 Presidential Address to the Geological Society of America:

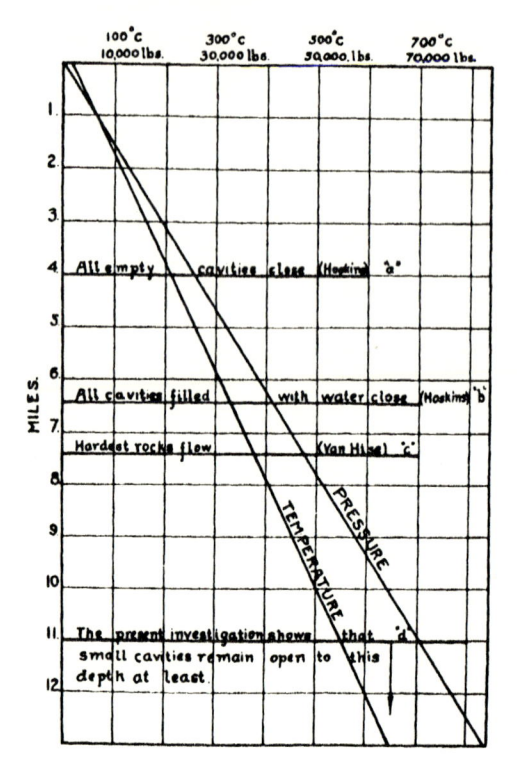

Figure 3.5. Zone of plastic flow in rocks. Based on the arguments of Hayford, Dutton, Chamberlin, and others, a number of scientists in the early twentieth century attempted to calculate the depth of the zone of plastic flow in rocks. Among Americans, the most widely cited of these calculations were those of Charles van Hise, Professor of Geology and later president of the University of Wisconsin. Van Hise's calculations, which suggested a zone of plastic flow at a depth of 7–8 miles, stimulated Frank Dawson Adams, a petrologist at the Carnegie Institution of Washington, to undertake some of the earliest rock compression experiments. Adams concluded that Van Hise's estimates were too shallow, because small cavities in rocks remained open in his experiments to pressures equivalent to depths of eleven miles. Therefore, the actual depth of the zone of flow must be greater than twelve miles. (From Adams 1912, p. 99.)

> As soon as the oceans came into existence and erosion began, the superficial transfer of material combined with a deep-seated solid flow in the nature of an undertow would establish a system analogous to incipient convective circulation in a mass of hyperviscous fluid. Such an unequally heated globe reduces to a species of heat engine.

By linking isostasy to the problem of terrestrial heat, "the general features of the dynamical system resulting in isostasy thus become intelligible."[78] The idea of moving continents was perhaps not as great a conceptual leap as might otherwise appear. Chamberlin, Dutton, Hayford, and others had written explicitly of "lateral creep" and continental "spreading"; the unknown issue was the *scale* of these effects, and whether they operated in a cyclical manner, as Chamberlin seemed to suggest, or whether they could actually produce a net lateral motion, as Hayford might be interpreted to imply.[79]

How Wide a Consensus?

It might be argued that geologists who were not *au fait* with the latest geodetic results might not have accepted this work, nor perhaps even known about it. Perhaps they pursued local and regional stratigraphy and structural geology in blissful ignorance of geodetic work. The prominence of the advocates of isostasy makes this suggestion unlikely. Gilbert and Chamberlin were among the most renowned of American earth scientists: both were recognized for major contributions to the scientific literature and both had served as president of the Geological Society of America. By the early 1920s, William Bowie—now chief of the Geodesy Section of the U.S. Coast and Geodetic Survey and president of the Geodesy Section of the International Union for Geodesy and Geophysics—was publishing widely on his and Hayford's results in specialist and generalist publications. This work was furthered on the West Coast by University of California Berkeley professor Andrew Lawson, who promoted the idea of subcrustal flow in interpreting the geology and seismic activity of California.[80] In 1921, the Geological Society of America devoted its annual symposium to isostasy, and the papers delivered there were published in its *Bulletin* the following year. American geologists in the early twentieth century were beginning to divide into disciplinary subspecialties and may not have been as catholic in their readings as previously, but it is unlikely that any geologist active on the national level could have been ignorant of this work.

Even if one were ignorant, or perhaps just skeptical, of geodetic progress, one would still not have avoided the necessity of a mobile substrate, because it was inherent in geosynclines—a theory of which no North American geologist was ignorant. The accumulation of sediments in slowly subsiding geosynclines necessarily required the yielding of the underlying crust. James Hall had acknowledged his intellectual debts to Babbage and Herschel—two of the earliest advocates of a mobile substrate—writing that he had "necessarily incorporated the general philosophic views so long ago clearly set forth by Babbage [and] Herschel . . . since these had early been fixed in my mind as a part of the elements and principles of geological science."[81] Indeed, U.S. geologists in the 1890s interpreted Hall's theory as an early step in the development of isostasy, retrospectively crediting his insight as "one of the most important contributions ever made to the theory of isostasy."[82] Even advocates of a solid earth, such as LeConte, acknowledged that geosyncline theory required local plasticity or fluidity; LeConte attributed this fluidity to the heating of buried sediments, which could produce local "pastiness" or partial melting of the rocks.[83]

Dutton had explicitly linked isostasy to geosynclines, knowing how close the latter were to American hearts. If one could show that isostasy was entailed by the existence of geosynclines, this would make it virtually axiomatic for North Americans. Dutton reminded his readers of this:

> The attention of the early Appalachian geologists was called, as soon as they had acquired a fair knowledge of their field, to the surprising fact that the Paleozoic strata in that wonderful belt, though tens of thousands of feet in thickness, were all deposited in comparatively shallow water. The Paleozoic beds of the Appalachian region have a thickness ranging from 15,000 to over 30,000 feet, yet they abound in proofs

that when they were deposited, their surfaces were the bottom of a shallow sea whose depth could not probably have exceeded a few hundred feet. No conclusion is left us but that sinking went on *pari passu* with the accumulation of the strata.... It seems little doubtful that these subsidences of accumulated deposits ... are, in the main, results of gravitation restoring the isostasy which has been disturbed by ... sedimentation.[84]

Writing in 1914, the Yale geologist Joseph Barrell agreed that geosyncline theory presupposes a "delicate" isostatic balance.[85] If one accepted geosynclines, and virtually all Americans did, then one implicitly accepted isostasy.[86]

Empirical Proof: The Fennoscandian Uplift

A mobile substrate was inherent in the theories of isostasy and geosynclines, but Wegener argued that there was compelling phenomenological evidence as well: the Fennoscandian uplift. It was, he argued, a natural experiment in which the evidence for plasticity had been laid out before geologists' eyes. The earliest evidence for the perceptible uplift of Scandinavia was discussed by Charles Lyell in his famous *Principles of Geology*:

> In the year 1807, Von Buch, after returning from a tour in Scandinavia, announced his conviction, "that the whole country [sic] was slowly and insensibly rising." ... He was led to these conclusions principally by information obtained from the inhabitants and pilots, and in part by the occurrence of marine shells of recent species, which he had found at several points on the coast of Norway above the level of the sea.[87]

Lyell outlined in some detail the occurrences of raised beaches with fossils characteristic of the modern Baltic Sea but now occurring well above sea level, and the phenomenon of the *skär*—a fringe of small rocky islands, sometimes referred to as the "outer coastline" of Scandinavia—which, much to the consternation of local navigators, appeared to be rising as much as one or two feet per century:

> To a stranger, indeed, who revisits it after an interval of many years, its general aspect remains the same; but the inhabitant finds that he can no longer penetrate with his boat through channels where he formerly passed, and he can tell of countless other changes in the height and breadth of isolated rocks, now exposed, but once only seen through clear water.

Numerous other examples of visible topographic changes were attested to by "the memory of living witnesses on several parts of the coast."[88]

Following Lyell's description, the British geologist T. F. Jamieson suggested that the observed uplift was the effect of postglacial rebound.[89] Jamieson noted that the area of uplift coincided almost exactly with the estimated limits of Pleistocene glacial ice, and he proposed that the weight of glacial ice had formerly depressed the underlying crust, which was now recoiling in response to its removal. Osmond Fisher recognized in Jamieson's suggestion an empirical confirmation of isostasy. Writing in the *Geological Magazine* in 1882, Fisher pointed out that "to any one who believes in a yielding condition of the earth's crust ... the depression of a tract of country upon its being loaded with any adventitious accumulation of matter appears an inevitable consequence." Depression in response to load, and uplift on its removal, was a deductive consequence of isostasy. Conversely, taking the Fennoscandian uplift as an empirical fact, it showed that "the crust, irregular though its surface be, nevertheless has a form appropriate to its equilibrium, and that there must be a plastic, if not a fluid, substratum to allow of the observed changes of level, when the load is freshly distributed."[90]

The Fennoscandian uplift was discussed in detail by Suess in volume two of his *Face of the Earth*, as the tales of "inhabitants and pilots" were being corroborated by detailed survey work. In 1888, the Scandinavian scientist Gerard De Geer published a detailed map of the distribution of uplift in central Scandinavia. Seeing a parallel between his own work and that of G. K. Gilbert at Lake Bonneville, De Geer concluded that the Fennoscandian uplift was in fact the result of isostatic rebound. A few years later, R. Witting mapped the change in the level of high water marks along the Baltic Sea since the start of record keeping and showed that uplift was occurring at the staggering rate of 2–10 mm per year (figure 3.6). Furthermore, the pattern of uplift was coherent, increasing steadily toward the northernmost reaches of the Baltic Sea. Subsequent investigation of nineteenth-century benchmarks confirmed these results, and preliminary gravity work found negative anomalies over the areas of uplift, as expected if isostatic adjustment were as yet incomplete.[91]

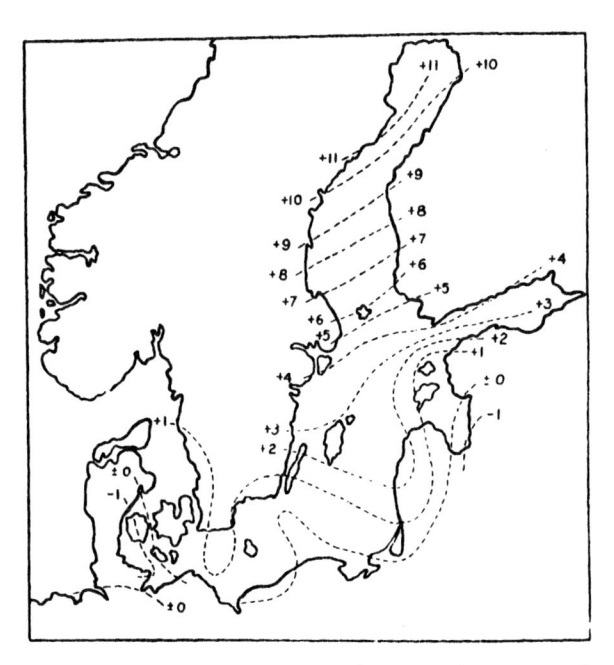

Figure 3.6. The uplift of Scandinavia. In the early twentieth century, a Scandinavian scientist, R. Witting, following the earlier work of De Geer, mapped the change in the level of high-water marks along the Baltic Sea since the start of record keeping and showed that uplift was occurring at the staggering rate of 2–10 mm per year. The resulting contour map, illustrated here, confirmed the long-standing popular belief that Scandinavia was rising. (The contour labels represent millimeters of uplift.) Furthermore, the pattern of uplift was coherent, increasing steadily toward the northernmost reaches of the Baltic Sea, suggesting that uplift was the result of isostatic rebound after the removal of Pleistocene glacial ice. (From Daly 1940, p. 314, after DeGeer 1888.)

Still further confirmation came from the work of the great Norwegian geologist and explorer Fritjof Nansen. In 1922, Nansen published the first of two important studies of the "strandflat" of Norway—a low platform along the western and northwestern coastlines, consisting of thousands of low islands, many barely above sea level. Nansen showed that the strandflat actually contained three distinct levels—two above sea level and one submerged—and attributed them to isostatic rebound during previous interglacial periods: the levels of the strandflat reflected the former positions of sea level. "The land has been ... raised by isostatic movement to the new level of equilibrium after the disappearance of the ice cap of each glacial period."[92] In short, it was clear that since the end of the Pleistocene glacial era, a large portion of Norway, Sweden, and Finland had become gradually elevated relative to sea level. The elevation was greatest where the ice had been thickest and rebound was continuing to this day. But, as Fisher had long since pointed out, this explanation could only work if the substrate behaves in a fluid manner, moving around the oscillating continent. As Wegener put it, "vertical movements of such a large portion of the crust obviously sets up flow in the substrate, so that displaced material is carried outwards"[93] (figure 3.7). It was as if nature had performed the experiment. The Fennoscandian uplift was positive proof of a mobile zone beneath the crust.

If Isostasy, Then Why Not Drift?

By the time Wegener's *Origin of Continents and Oceans* was translated into English in 1924, a rough consensus had emerged among geologists in the United States and Europe that the earth contained a mobile layer beneath the crust that provided both the seat of isostatic adjustment and the kinematic explanation for surficial horizontal dislocations. By imagining the mobile zone as either a plastic solid or a highly viscous fluid, geologists reconciled their empirical observations with the theoretical constraints

Figure 3.7. Mechanism of a mobile substrate. Wegener argued that the Fennso-scandian uplift was a "natural experiment" that demonstrated the existence of a mobile substrate. (**A**) Depression of the continent under load of glacial ice can only occur if the substrate is sufficiently mobile to flow out of the way. (**B**) Rebound occurs after removal of the glacial load when the substrate flows back under the continent.

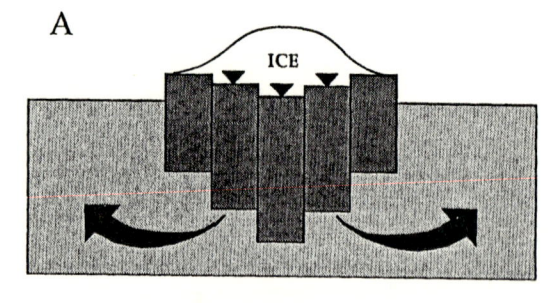

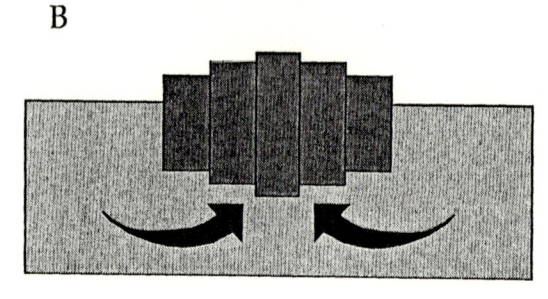

provided by physics and astronomy. Continents were not viewed as rigid, fixed bodies, but more like large rafts. Their vertical motions were accepted as fact, and from these vertical motions, Hayford and others had argued, horizontal dislocations arose as a consequence.[94]

Wegener built his theory on this consensus but challenged its last aspect. Rather than viewing horizontal displacement as a side effect of vertical oscillation, he viewed it as a fundamental process in its own right. If continents could move vertically through the substrate, he argued, then at least in principle they could move horizontally as well. Thus the novel point in Wegener's argument was not the idea of horizontal mobility per se but the scale and extent of that mobility. Given a mobile substrate, large-scale horizontal motions were at least plausible. In his words:

> The theory of isostasy . . . provide[s] a direct criterion for deciding the question of whether continents can drift horizontally, [because] the whole isostasy theory depends on the idea that the crustal underlayer has a certain fluidity. But if this is so and the continental blocks really do float on a fluid, even though a very viscous one, there is clearly no reason why their movements should only occur vertically and not also horizontally.[95]

Was Wegener correct in his reasoning? Conventional wisdom among geologists and geophysicists today is that Wegener's model was fundamentally flawed (and fundamentally different from present understanding) because he had the continents plowing *through* the rigid ocean floor. That is, contra Wegener, vertical and horizontal movements are not conceptually interchangeable. In the case of vertical motions, continents dislodge the underlying plastic substrate, whereas in Wegener's model of horizontal motion, they have to dislodge the adjacent rigid oceanic crust. In short, Wegener's model required the rigid continental crust to plow through equally rigid oceanic crust, and this simply could not be done, mobile substrate or no. As philosopher Michael Ruse has put it, "a solid lump of rock stuck in a solid lump of rock simply cannot move. That is that. Hence because it violated the geological game-rules, Wegener's hypothesis was rejected."[96]

But this objection is anachronistic, ignoring both the larger context of the discussion and Wegener's own response to it. Wegener's argument hinged on the belief that the ocean floor was more like the crustal substrate than the continental blocks — or, to use the terminology of the day, it was simatic (rich in silicon and magnesium) rather than sialic (rich in silicon and aluminum). This was not a particularly controversial view: it had long been suggested by evidence from ocean dredging and the basaltic composition of most oceanic islands. In the 1920s, it was further supported by studies of gravity and magnetism — which were at the forefront of geophysical research — and these suggested that the rocks of the ocean floor were denser and more ferruginous than those on the continents; most probably, therefore, they were basalt. And if the ocean floor *was* primarily composed of basalt, then the true isostatic situation combined elements of both the Pratt and the Airy model: the continents had deep roots *and* the ocean basins were composed of denser material than the continents. If so, then the continents plowed mostly through plastic substrate and needed only to dislodge a thin veneer of crust at the top.

In the third edition of *The Origin of Continents and Oceans*, Wegener expressed this view implicitly in his illustration of the structure of the earth's crust (figure 3.8);

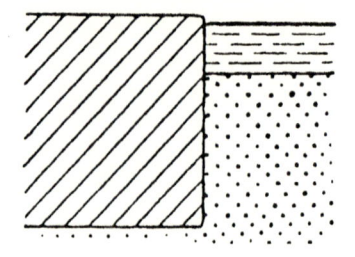

Figure 3.8. Wegener's interpretation of the crustal structure. Wegener's interpretation was a blend of Pratt and Airy isostasy, invoking both density and thickness differences between the continental and oceanic crusts. This was critical for his argument, because it placed the forward continental edge against only a very thin oceanic crust and thus allowed the continent to move. (From Wegener 1924, p. 31.) Reproduced by permission of Methuen Publishers.

in the fourth edition he said so explicitly, no doubt in response to criticism on this issue. "The correct interpretation [of isostasy] may be found in an amalgamation of both [Pratt and Airy] concepts. In the case of mountain ranges, we have to do basically with thickening of the light continental crust, in Airy's sense; but when we consider the transition from continental block to ocean floor, it is a matter of difference in type of material, in Pratt's sense."[97] Drawing on the gravity work of Helmert, Wegener estimated that the thickness of the oceanic crust was no more than 5 km (a figure confirmed by modern evidence) (figure 3.9). If the continental blocks were, on average, 100 kilometers thick, then the obstruction faced by their advancing edges was 95 percent plastic substrate and only 5 percent rigid or semirigid oceanic crust. The obstacle to drifting continents was much less than appeared at first sight.

How significant were these 5 km of oceanic crust? Very little direct evidence was available to answer the question. Wegener therefore approached it indirectly, draw-

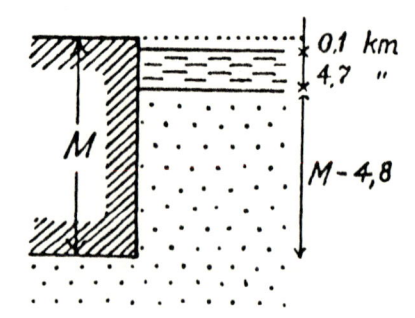

Figure 3.9. The thicknesses of the oceanic and continental crusts. Wegener calculated the thicknesses of the two types of crust on the basis of recent gravity measurements and concluded that the thickness of the oceanic crust was no more than five kilometers (a figure in line with modern values). If the continental blocks were, on average, 100 kilometers thick, then the obstruction faced by their advancing edges was 95 percent plastic substrate and only 5 percent rigid or semirigid oceanic crust. Therefore, he argued, the obstacle to drifting continents was much less than at first sight appeared. (From Wegener 1924, p. 37.) Reproduced by permission of Methuen Publishers.

ing again on isostasy. Responding in the third edition of *The Origin* to the complaint that the advance of a drifting continent should produce visible deformation of the ocean floor, Wegener pointed out that isostasy precluded the formation of significant elevation in the dense material of the ocean floor. Deformation had to be taken up, therefore, by "the material [of the ocean floor being] squeezed out below or to the sides, just as water between two approaching icebergs."[98] The apparent absence of deformation of the ocean floor was the predictable result of its material properties, because ocean floors were both denser and weaker than the continental rafts that plowed through them.[99]

Wegener also responded, in both the third and the fourth edition, to the complaint that seismic evidence pointed to a rigid earth. From the rate of propagation of earthquake waves, seismologists had calculated the viscosity of the earth's interior and defended Kelvin's conclusion that the earth as a whole was more rigid than steel. But Wegener rightly pointed out that these calculations were essentially irrelevant to his arguments: many materials behave in a rigid manner in response to short, sharp blows, but the same materials behave in a plastic manner when the applied pressures are small, steady, and slow. The propagation of seismic waves was a response of earth materials to short-duration disturbances; the problem of continental drift involved incremental deformation over geological times. Indeed, it would be fine if the earth *were* as rigid as steel, Wegener pointed out, because even steel "loses its rigidity . . . and becomes plastic" if subjected to sufficient pressure over a sufficient period of time.[100]

To drive home the point, Wegener made an analogy, now famous, with sealing wax. The effect of temperature on the material properties of such wax was well known; the not-so-well-known effect of time was similar. "If a stick of sealing wax be thrown on the floor," Wegener wrote, "it breaks into splintery fragments; but if it be left supported only at two points, a bending can be noticed after some weeks; and after some months the unsupported portions will be hanging practically vertical." Indeed, Wegener argued, one could use geophysical calculations of viscosity to determine a time equivalence of deformation. "If the sima has a viscosity coefficient 10,000 times as large as sealing wax, then a month for the sealing wax is equivalent to a thousand years for the sima."[101] And geologists had many thousands of years to work with. Wegener thus concluded, like Darwin and Lyell before him, that the key to understanding earth history was the element of geological time," insufficiently appreciated in previous literature, but . . . of the greatest importance in geophysics."[102]

A Plausible Kinematic Account

Today, geologists have a different kinematic model for the motions of the crust. According to the theory of plate tectonics, both continental and oceanic crusts migrate decoupled from the plastic substrate below. In the context of present knowledge, Wegener's model is wrong, even ridiculous. But in the context of the geological and geophysical knowledge of the 1920s, it was not ridiculous. It was consistent with the available understanding of terrestrial kinematic properties. It was supported by a wealth of geological evidence, a conceptual context of explaining crustal movements by reference to a yielding substrate, and a time-honored disciplinary tradition of accounting for monumental geological events by the incremental accumulation of small effects. A mobile substrate proved the possibility of mobile continents, and with the discovery of radioactivity, geological time was without doubt in full supply. Wegener had a plausible kinematic account. What he lacked was a dynamic explanation.

Were there geological forces capable of moving continents? This was the key question. Continents can move, Wegener argued, "provided only that there are forces in existence which tend to displace continents, and that these forces last for geological epochs." In the case of vertical displacements, the forces were known: the weight and subsequent removal of superincumbent loads of sediment, water, or ice. But whence an equivalent horizontal power? This was the problem geology faced, and so Wegener concluded with perfect geological logic: "That these forces do actually exist is proved by the orogenic compressions."[103] Continental drift had happened, therefore it could. The problem was to find out how. In the mid-1920s, some of the world's most prominent geologists took up the question of the driving force of continental drift.

4

Drift Mechanisms in the 1920s

The final chapter of the third edition of *The Origin of Continents and Oceans* was devoted to the dynamic causes of drift, and Wegener's tone in these final fifteen pages was decidedly more tentative than in the rest.[1] Frankly acknowledging the huge uncertainties surrounding this issue, he proceeded on the basis of a phenomenological argument. Mountains, Wegener pointed out, are not randomly distributed: they are concentrated on the western and equatorial margins of continents. The Andes and Rockies, for example, trace the western margins of North and South America; the Alps and the Himalayas follow a latitudinal trend on their equatorial sides of Europe and Asia. If mountains are the result of compression on the leading edges of drifting continents, then the overall direction of continental drift must be westward and equatorial. Continental displacements are not random, as the English word *drift* might imply, but coherent.

This coherence had been the inspiration for an earlier version of drift proposed by the American geologist Frank Bursley Taylor (1860–1938). A geologist in the Glacial Division of the U.S. Geological Survey under T. C. Chamberlin, Taylor was primarily known for his work on the Pleistocene geology of the Great Lakes region.[2] But his knowledge extended beyond regional studies: as a special student at Harvard, he had studied geology and astronomy; as a survey geologist under the influence of Chamberlin and G. K. Gilbert, he had published a number of articles on theoretical problems. One of these was an 1898 pamphlet outlining a theory of the origin of the moon by planetary capture; in 1903, Taylor developed his theoretical ideas more fully in a privately published book. Turning the Darwin–Fisher fissiparturition hypothesis on its head, Taylor proposed that the moon had not come from the earth but had been captured by it after the close approach of a comet. Once caught, the tidal effect of the moon increased the speed of the earth's rotation and pulled the continents away from the poles toward the equator.[3] In 1910, Taylor pursued the geological implications of this idea in an article in the *Bulletin of the Geological Society of America* entitled "Bearing of the Tertiary Mountain Belt on the Origin of the Earth's Plan."[4] Taylor argued that the morphology of the Alps and the Himalayas implied a recent great compression in the near-equatorial regions. To explain this compression, he proposed that two large crustal sheets had migrated from the north and south poles analogously with the advance of continental ice sheets recorded in the Great Lakes district (figure 4.1). Similar migrations might explain other crustal features including the mid-Atlantic Ridge.[5]

In later years, American geologists would often refer to drift as the Taylor–Wegener hypothesis, but in 1910 Taylor's ideas garnered little attention, which lends credence to Wegener's later claim that he developed his ideas independently. Wegener did credit Taylor, however, with the observation of equatorially oriented migration.[6] In the third edition of *The Origin*, Wegener agreed that movement from the poles toward the equator was "manifest in Eurasia in the arrangement of the great girdle of Tertiary folding of the Himalayas and the Alps," but he rejected Taylor's mechanism and proposed instead that continental migration was driven by tidal effects arising from the earth's oblate shape.[7] Following the work of Hungarian physicist Roland (Lorand) von Eötvös (1848–1919), Wegener argued that there was a potential for continents to migrate toward the earth's center of gravity.

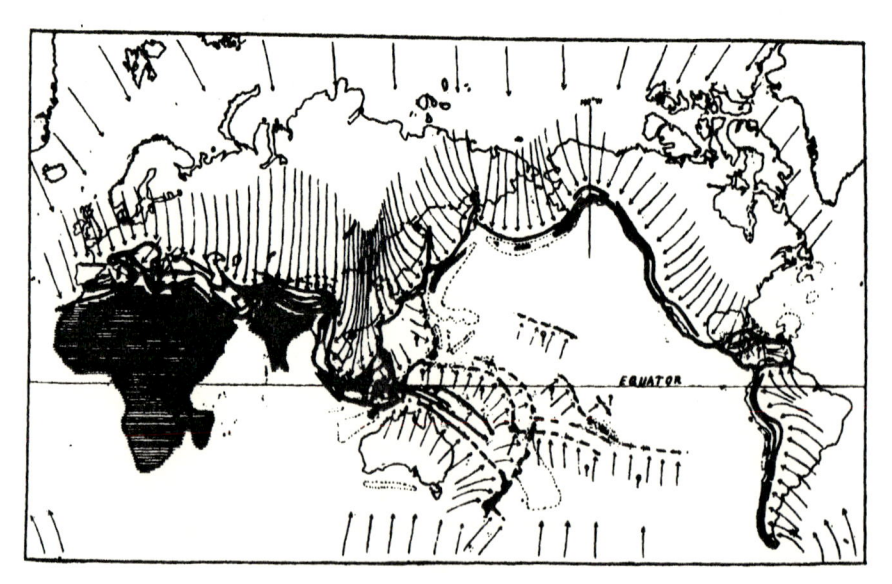

Figure 4.1. F. B. Taylor's model of continental drift. Taylor proposed that the Tertiary mountain belts of the world—the Alps, Himalayas, Indonesian archipelago, Rockies, and Andes—could all be explained by a general shifting of the continents toward the equator. His arrows indicate the inferred direction of movement; their size is proportional to the inferred magnitude of movement. The darkened areas mark the locations of oceanic deeps, whose origin Taylor attributed to crustal compression and buckling. (From Taylor 1910. Reproduced with permission of the publisher, the Geological Society of America, Boulder, Colorado, U.S.A. Copyright © 1910, Geological Society of America.)

Because the earth is an oblate spheroid, its equatorial bulge exerts a gravitational attraction on the rest of the globe—the Eötvös force. The existence of this force was broadly acknowledged, but did it have any measurable effect? On a completely rigid earth, the answer would almost surely be no. But given a plastic substrate, continents might be slowly—imperceptibly—pulled toward the equator. Excess mass in the equatorial bulge would pull the rest of the earth away from the poles, so the term pole-flight force was coined. The magnitude of this force was a function of the concentration of mass at the equator; its efficacy in moving continents depended on the viscosity of the substrate. The former was known from Eötvös's work, but the latter could only be guessed at from phenomenological arguments.

The westward migration of the continents was even more evident, given the dispositions of the North American Rockies and the South American Andes, but harder to explain. Wegener suggested the role of tidal retardation: the slowing down of the earth's rotation and consequent lengthening of the day caused by the frictional effect of the tides. Perhaps this set up strains within the solid earth? Given the earth's eastward direction of rotation, there would be a small drag, particularly felt on the earth's outer surfaces, which might "lead to a slow, sliding movement of the whole crust, or of the individual continental blocks." As with the Eötvös force, the efficacy of tidal retardation in causing crustal motions depended on the strength and material properties of the subcrust. Given a rigid substrate, there would be no effect; given a plastic substrate, there might be an "impulse to a slight progressive displacement of the crust ... which accumulates from day to day by an amount which is certainly very slight, and so not at all apparent in the daily measurements, but which can nevertheless lead to considerable displacements in the course of millions of years."[8] The effect might be significant if it were sustained over geological time. This was uniformitarianism applied to the question of crustal motion: the incremental effect of small forces was enough not only to wear down mountains but also to move them. "The question," Wegener finally concluded, "as to the forces which have caused and now cause the continental displacements is still too much in a state of flux to permit of a complete answer satisfactory in every detail."[9] Their relative efficacy remained to be determined, but conceivable forces did exist.

How Large a Force Is Needed to Move Continents?

In arguing for the reality of continental drift and outlining possible causes of it, Wegener was well within the norms of North American geology. In 1915, when *The Origin of Continents and Oceans* was first published in Germany, seismologist Harry O. Wood structured an argument about crustal displacement along precisely the same lines. Writing in the *Bulletin of the Seismological Society of America*, Wood proposed isostatic adjustment in response to sedimentation as a driving force of earthquakes. Just as the continents responded to the removal of glacial ice, he argued, so they must respond to the addition or subtraction of sedimentary loads, and one response might be faulting. But the test of this or any speculative hypothesis, he held, was "(a) that it depends upon true tendencies, however sufficient or insufficient these may be quantitatively; and (b) that it offers explanation, in matters of detail and sense, of the phenomena observed."[10] Wood continued: "If the action described is real, it would continue uninterruptedly for long intervals of time." Thus barely detectable forces could effect obvious geological change, as they had in the Fennoscandian uplift. But how did one know if a tendency were "true"? The reality of the Fennoscandian uplift was indisputable, and it helped to cement the widespread acceptance of isostasy. Yet the reality of continental drift was precisely what was disputed, and most American geologists considered that the "tendencies" that Wegener was proposing to account for it were woefully inadequate. But why?

By all accounts, the Eötvös effect was tiny. Eötvös himself was famous for developing a delicate, double-armed torsion balance for determining minute variations in gravitational force. With this highly refined instrument, he had measured the rotation of the earth, the horizontal pull of a mountain, and the loss in weight of a body traveling eastward: the raison d'être of the Eötvös balance was to measure tiny gravitational effects. At best, as Wegener acknowledged in the third edition of *The Origin*, the force available to pull continents toward the equator was no more than a two- or three-millionth part of the earth's total gravitational field. But what did such a number mean? Was it sufficient to move mountains given enough time?

In the proceedings of the 1926 American Association of Petroleum Geologists symposium on continental drift, its convenor, W. J. A. M. van Waterschoot van der Gracht, reminded his colleagues that no one knew the size of the force required. The "lack of sufficient explanation for the mechanism" of drift was, he allowed, the most serious objection to the theory. But the reality of Alpine thrust sheets was once doubted on the same grounds. Perhaps geologists should learn from their own history. "The facts have since proved beyond any doubt that these [thrust] sheets exist," van der Gracht argued pointedly, "not only in the Alps, but universally. Still their detailed mechanism, their 'possibility,' remains almost as much a riddle as it was then. The possibility has only been demonstrated by fact, not explained." Here, then, was the demarcation between phenomenon and causation applied to continental drift: one could accept their reality on the basis of geological evidence while reserving judgment on the question of causation. In fact, this was precisely what Emile Argand proposed to do. "Argand," van der Gracht recounted, "mindful of the Alps . . . refuses to explain, although he warmly advocates, drift." But van der Gracht considered Argand's position too extreme. "We should at least try to see whether or not [drift] is utterly impossible." The difficulty, as van der Gracht cogently observed, was that the question of possibility depended upon "conditions of matter in the depths of the earth, which we can neither observe directly, nor reproduce in our laboratories."[11] Geologists, like the continents, appeared to be stuck. Geodesists, however, considered that the condition of matter in the interior of the earth *was* known: it was plastic, and isostasy had proven it. Hence the discussion turned back once again to forces, and the American who first analyzed the possible forces was Walter Lambert.

Walter Lambert was known primarily as a mathematician, but he had been involved in the Survey's measurements of latitude changes associated with Californian seismicity. In 1921, he published a paper in the *American Journal of Science* analyzing the Eötvös force. Bodies do not fall to the geographical center of a body, he pointed out; they fall toward its center of mass. The earth's gravitational center is its equatorial bulge, so a frictionless ball on the surface must roll toward the equator (figure 4.2). The equatorially oriented component of gravity along a terrestrial meridian turned out to be one eleven-millionth of the force of gravity—an admittedly small force. But then Lambert calculated the effect this force would have on his hypothetical ball: if

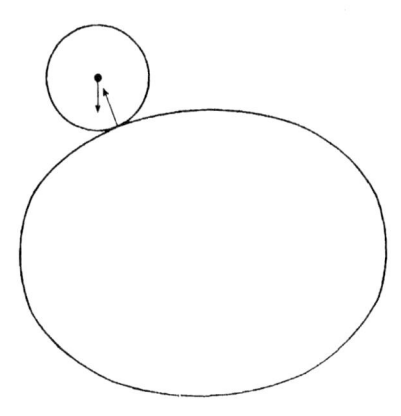

Figure 4.2. Walter Lambert's calculation of the "pole flight force." Lambert, a mathematician at the U.S. Coast and Geodetic Survey, calculated that the gravitational attraction of the earth's equatorial bulge would cause a free sphere on the surface to roll toward the equator. Here, he illustrates the counterintuitive result that the direction of the force of gravity acting on a sphere is not perpendicular to the tangent to its surface. (From Lambert 1921.) Reproduced with permission of the American Journal of Science.

the ball were 1 km in diameter and started its journey at a latitude of 30°, it would reach the equator in just under fourteen and a half days at a speed of 4.3 meters per second.[12]

Could this force move continents? Lambert preferred to view the problem the other way around. "A well-known geologist," he wrote, "told me that he had been convinced by geologic evidence that a part or a whole of each continental block has shifted its position in past time by moving towards the equator.... He had been led to think along these lines by two articles [Taylor 1910; Wegener 1912], published quite independently of each other, in which the authors evolved the hypothesis of continental creep in explanation of mountain building. If we accept this idea, the inevitable question is: What forces caused this motion of the continental masses?"[13] So Lambert applied his analysis of the free-floating ball to continents floating in a viscous substrate. If the substrate were a liquid, even a highly viscous one, then continents *would* move, he argued, because "a liquid, no matter how viscous, will give way before a force, no matter how small, provided sufficient time be allowed.... The peculiarities of ... gravity will give us minute forces, as we have seen, and the geologists will doubtless allow us eons of time for the action of the forces."[14] Lambert thus endorsed the argument for continental drift; indeed, he might seem to have solved it. The pole-flight force was real, and there was ample time for it to work. But there was a stumbling block: Did the substrate really behave as a viscous liquid?

> The viscosity of the liquid [substrate] may be of a different nature from that postulated by the classical theory, so that the force acting might have to exceed a certain limited amount before the liquid would give way before it, no matter how long the small force in question might act. The question of viscosity is a troublesome one, for the classical theory does not adequately explain observed facts and our present knowledge does not allow us to be very dogmatic. The equatorward

force is present, but whether it has had in geologic history an appreciable influence on the position and configuration of our continents is a question for geologists to determine.[15]

Lambert was responding to the arguments of Dr. (later Sir) Harold Jeffreys (1891–1989), who was adamant against treating the substrate as a viscous liquid. Heir to the tradition of mathematical geophysics at Cambridge developed by Hopkins and Kelvin—and soon to become Plumian Professor of Astronomy (succeeding Sir Arthur Eddington)—Jeffreys paralleled his colleagues' arguments for a solid and rigid earth: that the geophysical behavior of the earth, particularly the propagation of seismic waves at depth, precluded a fluid interior. Seismic shear waves do not travel through liquids, no matter how viscous, yet the new science of seismology was showing that they did travel through the region below the crust. In fact, they seemed to travel *faster* there. Geodesists might demand a plastic substrate, but seismologists were demanding a solid one. For the latter, no small force would be sufficient. Thus Jeffreys concluded, discussing tidal retardation at a meeting on continental drift in 1922, that "the rotational force which could be invoked to explain the movements of the continents was very small and quite insufficient to produce the crumpling up of the Pacific Ranges."[16]

In the 1920s, Jeffreys was also arguing for contraction; at the time of his death in 1989, he was one of the few prominent holdouts, and perhaps the most prominent holdout, against plate tectonics.[17] Many geophysicists have assumed that Jeffreys's objections were a key reason for the rejection of continental drift.[18] In the 1930s and 1940s, Sir Jeffreys became internationally renowned for his work on seismic wave travel times and his demonstration of the existence of the earth's liquid outer core. His book, *The Earth: Its Origin, History, and Physical Consitution*, first published in 1924 and still in print in 1976, has been described as "immortal."[19] Were his objections a critical factor in the rejection of continental drift?

In fact, Jeffreys' arguments proved quite insufficient to move most geologists. At the time of Lambert's paper, Jeffreys was only thirty years old—*The Earth* had not yet reached parturition, much less immortality—and British geologists indicated little inclination to accept Jeffreys's pronouncements if geological data warranted otherwise. Having lived to see Lord Kelvin disinherited, few felt bound to accept the demands of his heirs.[20] Lambert pointedly rejected Jeffreys's arguments as too dogmatic; the nature of the substrate was simply not known.[21] Arthur Holmes complained that Jeffreys refused to discuss ideas frankly, or "is unpleasant in manner if he does, though we are old friends."[22]

Bailey Willis, whose opposition to continental drift might have led him to rely on Jeffreys's support, considered the young man an interloper who, together with British mathematical physicist J. H. Jeans, had borrowed shamelessly—if not stolen—from Chamberlin and his co-worker F. R. Moulton. Willis considered Jeffreys and Jeans little short of enemies for their "gross discourtesy or worse" toward Chamberlin, and for the generally dismissive manner with which they treated American theoretical work, not the least Willis's own.[23] Writing to his wife, Margaret, after visiting Moulton in Chicago, just after Chamberlin's death in 1928, Willis wrote that "Moulton . . . is full of fight on the Jeffries[sic]–Jeans plagiarisms and says that they will do well to meet the situation honestly, for he is prepared to show up the weaknesses of their mathematics as well as their morals, if they continue to ignore Chamberlin and seek to wear his robe."[24] Willis was thus primed to dislike Jeffreys when he met him in Cambridge four months later:

> Jeffries [sic] is not at all the man I had expected to meet: not tall, but short, not a lean greyhound but a fattish, workaday chap; not dignified and well mannered but awkward, schoolboyish; not frankly affirmative and cock-sure, but silently self-satisfied, not an honest fighter, but a shifty opponent. He did not once look me in the eye. He made no answer to the openings . . . I gave him to discuss controversial topics. "He's a great mathematician" [people say]. Well, I guess the devil is also, but he, if he proved the false to be true, would do it charmingly and like a gentleman.[25]

Willis encapsulated his opinion in a letter to Reginald Daly: "Beware of the mathematician."[26]

In the conclusion to the fourth and final edition of *The Origin of Continents and Oceans*, Wegener sought to distance himself from dynamic arguments by reminding his readers that it was normal in the course of science to outline a phenomenon first and explain it later: "The determination and proof of relative continental displacements . . . have proceeded empirically . . . without making any assumptions about the origin of these processes." Invoking his reading of the history of science, Wegener continued, "The formulation of the laws of falling bodies and of the planetary orbits was first determined purely inductively, by observation; only then did Newton appear and show how to derive these laws deductively from the one formula of universal gravitation. This is the normal scientific procedure, repeated time and again. The Newton of drift theory has not yet appeared."[27] Wegener not altogether unreasonably hoped to gain acceptance for his theory by decoupling the fact of drift from its cause. Certainly, other scientists had made this argument in other situations, not least of all the "mathematicians" themselves when confronted with vexing questions like the origin of the sun's internal energy.[28] And had not Newton himself famously declined to speculate on the cause of gravity? Yet Wegener's situation was better than he imagined: by the late 1920s, several prospective Newtons had appeared. One of them was Reginald A. Daly (1871–1957), Professor of Geology at Harvard University.

Gravity Sliding: The Work of Reginald Daly

Frank Taylor and Walter Lambert were respected members of the rank and file of North American earth science, but Reginald Daly was one of its leaders (figure 4.3). Raised in Ontario, Canada, the youngest of nine children, Daly was born to a family of tea merchants of Irish descent. After graduating from high school, he undertook general studies at the University of Toronto, where he earned a B.A. degree in 1891 and then a B.S. degree while serving as a mathematics instructor in 1892. Encouraged by his geology professor, A. P. Coleman, Daly decided to pursue graduate studies. He enrolled at Harvard University in the United States and obtained an M.A. (1893) and a Ph.D. (1896) in geology. Then on a traveling fellowship from Harvard, he spent two years in postdoctoral studies in Heidelberg and Paris. During this time he attended the Seventh International Geological Congress in Russia and traveled widely. His European experience exposed him to recent advances in laboratory techniques and to the forefront of geological theory. But he lacked extensive field experience. Thus after a two-year stint lecturing at Harvard, Daly took a position as a field geologist for the

Figure 4.3. Portrait of Reginald A. Daly. (From Biographical Memoirs 24, opposite p. 31. Courtesy of the National Academy of Sciences.)

Canadian International Boundary Commission. From 1901 to 1907, he worked mapping scant outcrop in thick forests in a 400 x 10 mile wide swath along the Canadian–U.S. border. This exceedingly arduous work (from which he collected over 1,500 samples and studied 960 thin sections) exposed him to a wide range of geological problems, many of which he revisited throughout his professional life. It also led to a lifelong conviction that the foundation of practical geology was theory and that the foundation of geological theory was fieldwork.[29]

After completing his work on the Boundary Commission, Daly became a professor of physical geology at the Massachusetts Institute of Technology. In 1912, he succeeded William Morris Davis as the Sturgis Hooper Professor of Geology at Harvard, where he remained for the rest of his career. Catholic in his interests, Daly worked on the origins of igneous rocks and ore deposits, the deformation of mountain belts, the causes of glaciation and isostatic responses to it, the formation of submarine canyons, and the strength and structure of the crust. He was one of the first to look systematically at igneous rock compositions, and he shared this interest with a field assistant, Norman Levi Bowen, whose work later superseded his own. (It was Daly who convinced Bowen that his study of the evolution of igneous rocks — perhaps the most influential book ever written in American geology — should begin with basalt.[30]) His interest in the properties of earth materials helped him forge a close friendship with Harvard physicist and philosopher Percy W. Bridgman, upon whom Daly prevailed to include rocks and minerals in his study of the behavior of materials under extreme pressure. Like many geologists, Daly traveled extensively, in his case focusing particularly on the northeastern portions of the United States and Canada, Scandinavia, and the Pacific Islands. Also like many geologists, he had a talented and supportive wife—née Louise Haskell Porter—a Radcliffe graduate who often traveled with him and always typed and edited his manuscripts; many colleagues considered her help essential to his productivity, which ultimately totaled seven books and nearly 150 articles.[31]

Daly's work was considered distinctive for his consistent efforts to find quantitative data relevant to qualitative geological problems encountered in the field. He was also known for his commitment to causal interpretations of geological phenomena. Chester Longwell, a Yale colleague, described him as one who "strikes boldly into the no-man's land of obscure or hidden causes. To him, the study of geology is not a mere recital of orthodox facts and principles, but a search for the meaning of things."[32] Daly would have liked Longwell's characterization; in the Harvard brochure describing geology as a field of concentration, Daly emphasized its interpretive aspects. "The supreme faculty of the human brain is imagination," he wrote, "the ability to see things as they are and not as they appear to be. Fact is the raw material on which imagination works. Knowledge is only the beginning of wisdom; imagination is power." Geology demanded that one look beyond surface phenomena to imagine their hidden causes; this is what made it both exciting and didactic. "The chief aim of Harvard College," he continued, "is not so much to gain knowledge of this or that subject as it is to gain the power to understand. Growth in understanding means the strengthening of the imagination. Hence geology in college is a cultural study of the first rank."[33]

Although the role of imagination was debated—sometimes testily—in American geological circles (see chapter 5), Daly was rewarded richly for the products of his own. A member of the U.S. National Academy of Sciences, he received numerous honorary degrees and awards, including the Penrose Medal of the Geological Society of America and the Bowie Medal of the American Geophysical Union, and he served as president of the Geological Society of America in 1932. In a lavish biographical memoir, Francis Birch, a Harvard colleague, described Daly as possessed of a "personality of exceptional force and single-mindedness." But in this memoir, published in 1960—just a few years before the development of plate tectonics theory—Birch made no mention of the issue that commanded Daly's attention from the 1920s onward: continental drift.

Speculating on Causes

Daly was the geologist who stimulated Lambert to calculate the strength of the pole-flight force.[34] As Lambert indicated, Daly had been interested in continental drift since reading Taylor's and Wegener's early articles. Fluent in French and German, Daly read widely and was one of the few North Americans who had read Wegener's original 1912 piece. In January 1921, Daly replied to the letter Lambert had sent along with the manuscript of his paper. Given that drift was a highly speculative idea, Lambert asked Daly's advice on temper and tone. Daly suggested he say that, given the evidence outlined by Taylor and Wegener, "the inquiring geologist naturally is looking for forces to account for the inferred motion of the continental masses." Daly outlined four explanatory options.

Guess No. 1, as Daly put it, was tidal retardation. This was the force that Jeffreys dismissed, but Daly saw a loophole in the geophysicist's argument. He explained: "Jeffreys . . . suggests that oceanic tides give friction enough to explain the observed change [in the earth's rate of rotation] and that 2/3 of the total dissipation of energy involved may be found in Bering Sea alone." That is, tidal retardation was mainly taken up by the oceans, and the frictional effect therefore depended on the surface area of the seas. If shallow seas were more extensive in the geological past—which they were during warm periods when the polar ice caps melted and raised global sea levels—then the frictional effect would have been greater. Continental drift might thus be intermittent—greater in the past, less today—and linked to glaciation, geology's great agent of surficial change. Guess No. 2 was gravitational instability arising from isostatic adjustment below geosynclines. At present, isostasy was said to be "nearly perfect," but Daly wondered if it were "not possible to think that the present is a special stage in the earth's history when strains have been largely relieved during the Tertiary orogenic revolution?" If so, then isostatic underflow might have been greater in the past and tangential stresses accordingly stronger. Whereas Wegener had tried to work within a uniformitarian framework, focusing on forces that were operating at present, albeit perhaps ineffectually, Daly was considering the option that the present was not necessarily the key to the past—or at least not the whole key. Perhaps conditions were different in the past in ways that made small forces more efficacious.

Guess No. 3 was the gravitational effect of the earth's triaxial shape, and Guess No. 4 was gravitational stress arising from the overthrusting of one crustal segment over another during orogeny.[35] "Is there anything in these speculations that might warrant the idea that the solid-elastic crust, floating on highly viscous liquid, might be torn to pieces and the 'floes' pulled around the curve of the earth?" Daly asked rhetorically. He was receptive to the theory, perhaps even provisionally entertaining it, but he doubted that the question could be resolved one way or the other, at least not by geologists:

> All these guesses are bound to hang in midair until Bridgman or someone else tells us what to expect re rigidity, viscosity, and strength in depth. This is now so clear that it seems almost hopeless to make much progress along these lines of thought. Yet I sketch them here, in the belief that some of them you can already show to be quantitatively n[o] g[ood]. That would be a gain![36]

Lambert honored Daly's request: he rejected the effect of the earth's ellipsoid shape and, perhaps inadvertently, focused Daly's attention on the gravitational effect of subsidence and sedimentation in geosynclines. He wrote:

> I do not think that the existence of such an uncompensated excess of mass as is implied in Helmert's result (for the earth's triaxial ellipsoid) would help much to explain a continental creep towards the Pacific Ocean. The introduction into the earth of this excess of matter would of course set up stresses in the earth, but I do not see why these stresses should affect the continents very differently from the way in which they would affect the crust immediately adjacent to the continents. In other words, I do not see why there should be any great differential motion of continent and its surroundings. Of course there is the effect of the elevation of a floating con-

tinent. . . . Bulge the globe somewhat, or deform it as you like . . . but the relative configuration of the continents is very little changed. To explain the creep, forces are needed that act on the elevated portions of the crust of the continents in a different way from that in which they act on the more depressed portions of the crust.[37]

Lambert encouraged Daly to think locally rather than globally: large-scale astronomical and geodetic considerations would affect the whole earth more or less equally. What was needed was a force that differentially affected continents and oceans. Sedimentary processes did: the ocean floor subsides in geosynclines under the effect of sedimentary loads. Continents rebound in response to denudation. Daly's attention thus was focused on this speculation, but before he had a chance to do more, he embarked on a trip that convinced him that progress on the question was not only possible but also essential. South Africa convinced Daly of the reality of continental drift.

An American among Drifters

In early 1922, Daly left for Africa with Frederick E. Wright (1877–1953), a petrologist at the Geophysical Laboratory of the Carnegie Institution of Washington.[38] In that year, when the third edition of *The Origin of Continents and Oceans* was first reviewed in English, the Carnegie Institution provided funds for a major field excursion organized in conjunction with Harvard University, the Shaler Memorial Expedition. The goal of the journey was to examine the phenomenal igneous intrusions of southern and eastern Africa with an eye toward understanding their implications for the origin of igneous rocks. The nine-month journey, which lasted from January to September and covered a total of 33,250 miles, took Daly and Wright across Africa from south to north.[39] For Wright, nearly every aspect of the trip was a geological and social revelation—from witnessing labor unrest in the Transvaal to eating his first nectarine—but for Daly one thing stood out above all else: the Bushveld Igneous Complex.[40] The world's largest single formation consisting entirely of igneous rocks, the Bushveld was (and is) the world's largest source of chromium and platinum and a major source of copper, nickel, and gold. Beyond its economic significance, it was unique in the world as an example of a layered mafic (rich in iron and magnesium) intrusion.

Most of the igneous intrusions that Daly had studied were of modest size and relatively homogeneous composition, but the Bushveld Complex was different. It was layered, consisting of repeated sequences of distinctive composition, and it was huge. Extending for more than two hundred miles across the African desert, just to describe it was a monumental task, much less to explain how it had formed. Its bulk chemistry appeared to be that of basalt, but individual layers ranged from highly mafic to highly silicic (rich in silicon and aluminum), with the densest layers at the base and the lightest at the top. The obvious implication was that the layers had formed by progressive crystallization—a phenomenon just then being studied by Daly's student, Norman Bowen—but under the influence of gravity. Some of the layers had features such as cross-bedding and ripple marks—normally found in sedimentary formations—as if mineral grains had settled out of a vast magma sea.

Daly and Wright were stunned by what they saw—it was entirely unlike anything they had seen before—and their insights greatly stimulated their South African colleagues. In his classic memoir on the Complex, South African geologist A. L. Hall wrote that "the year 1922 is a landmark [in the history of geological research in South

Africa], for it saw the visit of the Shaler Memorial Expedition."[41] If the Bushveld was the tangible highlight of the trip, however, the intangible one was talking about continental drift with G. A. F. Molengraaf and Alexander du Toit.

Professor of Geology at Delft, the Netherlands, Molengraaf (1860–1942) had joined Wright en route, sailing with him from Europe to Cape Town, and he guided the Americans on the South African portion of their trip.[42] A former State Geologist of the Transvaal, Molengraaf was now the world's leading expert on the geology of the Dutch East Indies, and he had declared in favor of drift—at least in general terms, if not in all of Wegener's details—for its evident applicability to the distinctive geology of that region.[43] He also saw the applicability of drift to the interpretation of African geology. Like Daly, Molengraaf was a geologist of catholic interests and copious knowledge. His early work in the Transvaal (1897–1898) included mapping large portions of the Permian Karroo system. Molengraaf's attention had been particularly captured by the Dwyka conglomerate, the basal unit of the Karroo over much of the Transvaal; Molengraaf accepted the then somewhat controversial view that the Dwyka was a glacial deposit, and he became a vocal advocate of a Permian glacial event in the Southern Hemisphere.[44]

Molengraaf subsequently worked for more than a decade in the Dutch East Indies and published extensively. This experience convinced him of the importance of submarine information for interpreting tectonic history—a radical position when scant submarine geology was known worldwide. He suggested that the extreme submarine topography of the Indonesian archipelago, and extensive subsidence as evidenced by thick coral reefs, were caused by large-magnitude crustal motions. Considered in conjunction with adjacent on-land overthrusts, the evidence pointed to large-scale regional compression and down-warping. In a paper published in 1916, Molengraaf first suggested that besides including "the generally accepted orogenetic and epeirogenetic movements," this motion might also include "horizontal movements of continental blocks such as Wegener assumes in his bold hypothesis about the origins of the continents and oceans."[45] By the early 1920s, Molengraaf had become an advocate of drift, concluding that the origins of the Indonesian archipelago could be found in the convergence of two large slabs of continental crust. After a lecture by Molengraaf at the Royal Geographical Society in London in 1920, J. W. Evans summarized:

> The surface of the earth is like an ice-floe composed of great solid blocks which are always crumbling against one another. These blocks as a whole seem almost immune from any change, but in between them are very energetic movements. It was just one of the intervening regions [Indonesia] which Prof. Molengraaf has been telling us about to-night. To the north is one of the solid floating masses; to the south another, and between them is an area subject to great lateral pressure.[46]

Two years later, Molengraaf was in South Africa discussing these ideas with Daly and Wright as they camped out under the southern lights. Playing his second was their South African host, Alexander du Toit.[47]

Alexander Logie du Toit (1878–1948) was South Africa's leading field geologist, perhaps the best field geologist in the world (figure 4.4). As a geologist for the Irrigation Department in Pretoria, he had already published thirty-two papers and two books on South African geology. More important, he was the world's leading expert on the Karroo formation, said to be virtually identical to equivalent-aged se-

Figure 4.4. Portrait of Alexander Logie du Toit, from a commemorative postage stamp issued by the Republic of South Africa. Reproduced with permission of the South African Post Office, Ltd.

quences in India, Australia, and South America. This was the claim that had originally motivated Suess's idea of Gondwanaland; now, it supported Wegener's hypothesis of continental drift. Du Toit spent several weeks in February and March with the field party, where he and the Americans became fast friends during their days in the field, with their "innumerable discussions, practically endless,"[48] and their nights camping amidst the "baboons and the zodiacal lights."[49] On the 10th of April the foursome traveled together to Johannesburg where Molengraaf lectured on Wegener's theory.[50]

A Mechanism for Drift

If Daly went to South Africa pondering drift, he came back accepting it. Immediately upon returning to the United States, he published an article proposing a mechanism for it. He began by inverting the questions he had been posing in the field about the Bushveld Complex. Rather than ask what kind of crustal processes formed the complex, he asked what the complex revealed about crustal processes. His answer was this: continental drift could be explicated in terms of the structure and properties of a layered earth. In a paper entitled "The earth's crust and its stability," published in the *American Journal of Science* in 1923, Daly began by pointing out a fundamental flaw in seismologists' arguments for a solid, rigid earth: seismologists conflated rigidity with solidity. Unlike Bailey Willis, Daly took Harold Jeffreys's arguments seriously, but like Wegener and Lambert, he concluded that they were incomplete. Seismic studies supported the *rigidity* of crustal substrate, but this did not necessarily imply its *solidity*. Glass, Daly pointed out, appears at room temperature to be a brittle solid, but it flows under pressure and over time because it is really a frozen liquid. In terms of material properties, the relevant distinction was not between liquid and solid—as geologists and geophysicists generally assumed—but between *crystalline* and *noncrystalline* materials. If the substrate were noncrystalline, like glass, it could be rigid in response to short-term effects, such as the propagation of seismic waves, but plastic in response to long-term effects, such as isostatic adjustment—precisely what was needed.[51] Perhaps the substrate was in fact glass.

Glass forms from volcanoes on the earth's surface when liquid magma is quenched, but could it form in the earth's interior? Daly argued that it could, and would, because of the countervailing effects of temperature and pressure. Given available estimates of geothermal gradients, the temperature in the earth at a depth of 40 km should be about 1200° C, more than sufficient to melt basalt. If the bulk composition of the substrate were basaltic, like the bulk composition of the Bushveld, it would, ceteris paribus, be molten at the base of the crust. But other things were not equal, because of the effect of pressure on material properties. Daly knew from Bridgman's work that temperature and pressure worked opposite effects on material properties: increased temperature softened and eventually melted solids, but increased pressure hardened and eventually solidified melts. If decreased temperature solidifies melts to produce volcanic glass at the earth's exterior, increased pressure might do the same in the earth's interior.[52] The countervailing effects of temperature and pressure would maintain a basaltic melt in a rigid or semirigid state at depth, thus reconciling the contradictory demands of seismologists and geodesists.

Daly accepted Jeffreys's conclusions, based on discontinuities in the transmission of seismic waves, that the earth was divided into three principal zones: a crust extending to a depth of 35–40 km, where seismic velocities notably increased, a middle zone (today called the mantle) ending at a depth of approximately 2500 km, and a core. But here he parted company with the young Englishman. Jeffreys assumed that both the crust and the mantle were rigid, because both propagated seismic waves, but Daly argued that the middle zone could be at best only semirigid, at least in upper portions.[53] Borrowing terminology developed by Yale geologist Joseph Barrell, Daly referred to the semirigid portion of the substrate as the *asthenosphere*—the zone without strength—and this, he argued, is where the difficulties of geodesy and geology would be resolved:

> Admitting the . . . effects of pressure, one has little difficulty in imagining the asthenospheric glass to have [the properties] meeting the requirements of a proper theory of tidal, nutational, and seismic strains. At the same time, the subcrustal shell permits isostatic undertow, with additional yielding in the deeper interior, so that a nearly hydrostatic condition of the earth is preserved in spite of the transfer of masses on the surface. In other words, the substratum yields almost as if it were a viscous liquid, and practically quite like a plastic solid with a very low limit of elasticity.[54]

The middle zone had the properties to satisfy all interested parties.

There was, however, a potentially fatal flaw in Daly's theory. If the oceanic crust were basaltic, as most geologists believed, and the substrate beneath it were a liquid version of the same, then the former would be denser than the latter—a physically unstable situation. The cold, dense, oceanic crust should sink into the hotter, lighter substrate. Daly argued that this apparent flaw was actually a virtue, because it helped to explain the kinematics of continental drift. The crust *was* fundamentally unstable. That was the point. The basaltic portions were held aloft by virtue of their internal strength and their continuity with the lighter continental crust, but if once they were ruptured, they would begin to sink, dragging along the attached continents:

> Under ordinary conditions . . . the strength of the crust is ample to prevent foundering. . . . On the other hand, if tangential compression causes the suboceanic crust

to be locally depressed, ruptured, and underthrust, the downward pull due to differential density becomes increasingly important. The immersed part of the crust tends to pull down with it the horizontal part to which it is solidly attached.[55]

Daly called his theory gravity sliding; today geologists would call it a "slab-pull" model of subduction: that crustal plates may be dragged along by the weight of a descending slab as it sinks back into the earth's interior.[56] "Is not this cause of crustal instability significant in connection with the hypothesis of extensive horizontal movement of continental blocks?" he asked.[57] Crediting Taylor with the first articulation of the idea (and with "greater tact" than Wegener), Daly threw his support behind continental drift, suitably reinterpreted.

Like Wegener, who drew on his own experience in Greenland, Daly drew on his own experience with igneous rocks. Wegener had conceptualized drift by analogy to icebergs; Daly, whose speciality was igneous rocks, conceptualized drift by analogy with lava flows. An ordinary lava flow, he pointed out, "after solidification of its upper part, continues to move, the thick upper elastic layer sliding on the deeper, hotter, more purely viscous part of the flow. Downstream the solid layer is seen to be folded, overfolded, and broken into thrust blocks; upstream this layer has been torn apart."[58] So it was with the continents. Daly's analogy allowed one to visualize the kinematics of crustal sliding by likening it to an observable geological process, while his glassy substrate accounted for how it could happen.

Following his discussions with Lambert, Daly outlined four possible forces for initiating rupture: the gravitational effects of differential elevations arising from the weight of the ocean, tidal retardation, isostatic adjustment to erosion and sedimentation, and differential thermal contraction in response to heterogeneous distribution of radioactive elements. "A highly trained mathematical physicist," Daly concluded, was needed to attack the problem in its quantitative aspects. Geologists, meanwhile, should work harder at the phenomenology:

> To ask [the physicist's] sympathy with the sliding hypothesis and his help in its proper discussion is one of [my] main purposes. . . . Small as may be the immediate success in the search for the force or forces which could tear a continent to pieces, it seems eminently wise to retain as a working hypothesis the daring yet majestic idea of Taylor and Wegener. In Europe, its discussion is lively and many geologists, as well as geophysicists, are convinced that the hypothesis of continental displacement on the large scale should not be summarily rejected. As yet American geologists have not done their share in developing the possibilities and probabilities of the case. A hypothesis which explains so many details and major features of the earth's plan merits their best thought in an exceptionally difficult field of investigation.[59]

Daly did his share to develop the possibilities and probabilities of drift in his book *Our Mobile Earth*, published in 1926. If there were any doubt about his views, it was removed by his epigram recalling Galileo: *E pur si muove!* The debate over drift was now becoming noisy, Daly wrote, and "many geologists have found [the] idea bizarre, shocking; yet an increasing number of specialists in the problem are already convinced that it must be seriously entertained as the true basis for a sound theory of mountain-building." Reproving his colleagues, he concluded, "every educated person cannot fail to be interested in this revolutionary conception."[60]

A focus of the book was the causes of crustal warping—the issue raised by Lambert—and Daly focused on the effect of sedimentary accumulation in geosynclines, drawing on his knowledge of the Appalachians. Mountains are composed primarily of sedimentary rocks, originally deposited in a marine environment. How were these sediments incorporated into continental mountain ranges? This was the fundamental unresolved problem of geosynclines—the origin of mountains that James Hall had left out—and Daly saw an opportunity to resolve this long-standing anomaly and simultaneously explain the initiation of continental drift. He argued as follows: The accumulation of sediments in geosynclines necessarily involved crustal subsidence—this was widely agreed. Down-warping in geosynclines must cause melting as the base of the crust sinks into hotter portions of the earth.[61] The result, then, should be a thin, weak crust below the geosyncline, which continues to sink and melt as sediments are piled upon it (figure 4.5). Meanwhile isostatic undertow away from the area of sedimentation generates uplift in the adjacent continental block. Prolonged sedimentation and associated isostatic adjustment would eventually produce a significant elevation difference between the continental block and its adjacent sedimentary depression. Ultimately, the weakened crust would rupture and the elevated continent would slide down under the force of gravity, folding the adjacent sediments in its wake (figure 4.6). Repeated over time, this mechanism could explain both the migrations of the continents, and the source of deformation long missing from geosynclinal theory.[62]

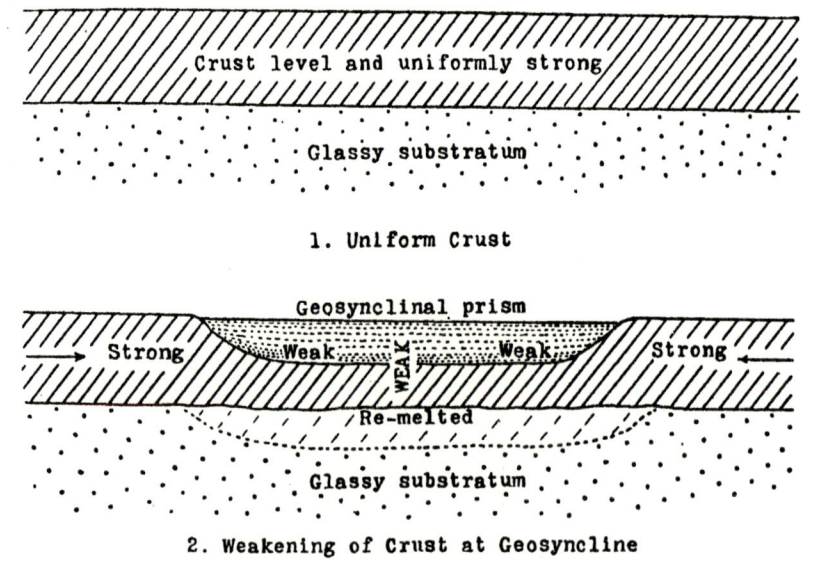

Figure 4.5. Daly's model of crustal weakening below geosynclines. Daly argued that the accumulation of sediments in geosynclines pushed the underlying crust into hotter regions, where it would begin to melt. The resulting thinned crust would be weaker than continental crust elsewhere and therefore susceptible to rupture. (Reprinted with the permission of Scribner, a Division of Simon and Schuster from *Our Mobile Earth*, by Reginald A. Daly. Copyright © 1926 by Charles Scribner's Sons, renewed 1954 by Reginald A. Daly; p. 205.)

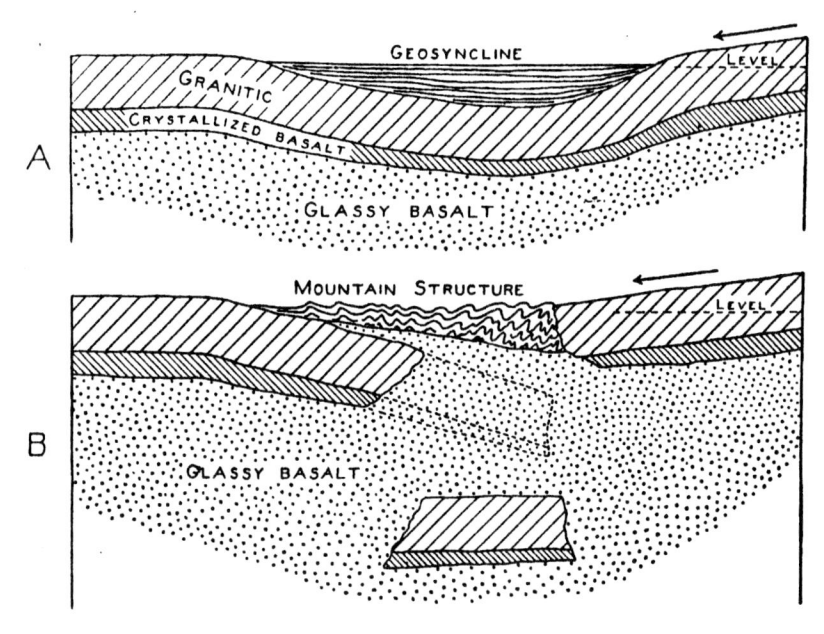

[Figure: two cross-section diagrams labeled A (showing GEOSYNCLINE with GRANITIC, CRYSTALLIZED BASALT, and GLASSY BASALT layers) and B (showing MOUNTAIN STRUCTURE with GLASSY BASALT below)]

Figure 4.6. Daly's model of continental drift by gravity sliding. Once the crust was weakened beneath the geosyncline, Daly argued, it would be susceptible to gravitational instabilities generated by the depression of the crust under the geosyncline. The crust would rupture, and the elevated regions would slide laterally under the force of gravity, crumpling the adjacent sedimentary rocks. Daly also included a mechanism of crustal consumption: once ruptured, portions of the cold dense crust would sink into the hot substrate beneath. (Reprinted with the permission of Scribner, a Division of Simon and Schuster from *Our Mobile Earth*, by Reginald A. Daly. Copyright © 1926 by Charles Scribner's Sons, renewed 1954 by Reginald A. Daly; p. 269.)

Answering the Objections to Drift

Daly's model answered several substantive objections to Wegener's version of continental drift. It answered the dynamic problem by invoking the full force of gravity rather than a minor side effect. It answered the kinematic objection by having the continents slide along the top of the substrate rather than plowing through it; the substrate did not need to be completely devoid of strength, only weak enough to be pliable. The model also answered the objection, raised by Bailey Willis, that if the oceanic floor were yielding, as Wegener argued, then it would deform, rather than the continents, so drift could not be a mechanism of *continental* deformation.[63] In Daly's model, deformation was taken up by the folding and incorporation of sediments from the geosynclinal prism into the continental masses—a phenomenon no one disputed (figure 4.7).[64]

By making the connection to geosynclines, Daly linked drift to a keystone concept in American theoretical geology. Yet, elsewhere Daly's model parted company with American theoretical views, which could have been cause for dispute. In advocating a layered earth, Daly was committing himself to a molten protoearth, whose layers had formed by density separation as in the Bushveld Complex. Such a view was

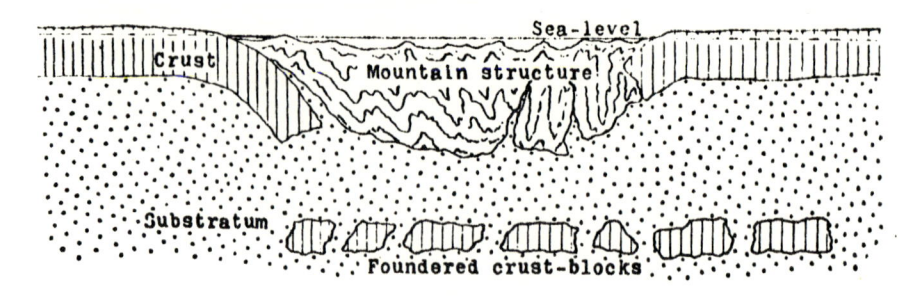

```
                                    Sea-level
┌─────────┐                                          ┌─────────┐
│  Crust  │      Mountain structure                  │         │
└─────────┘                                          └─────────┘
    Substratum
              Foundered crust-blocks

1. Mountain belt is low just after folding
```

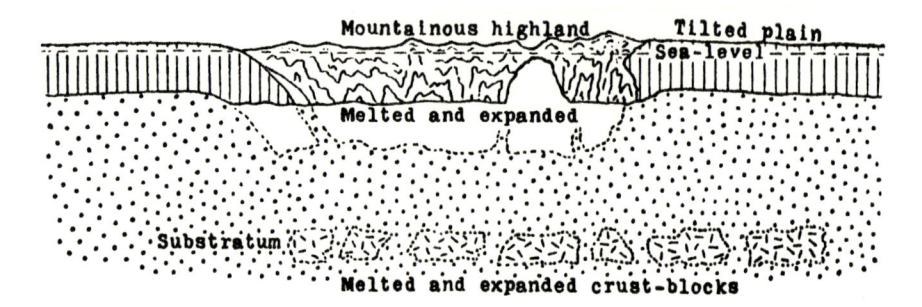

```
              Mountainous highland    Tilted plain
                                        Sea-level
              Melted and expanded

    Substratum
              Melted and expanded crust-blocks

2. Later upheaval of mountain-belt because of
   thermal expansion of mountain roots and of
   foundered crust-blocks
```

Figure 4.7. Daly argued that his model accounted for both continental drift and mountain-building. The crumpled sedimentary rocks would be uplifted due to forces associated with the down-warping and local melting of the subjacent crustal slab. This mechanism, he suggested, was the principal cause of orogeny and crustal uplift. Note the lack of vertical exaggeration—Daly emphasized that even lofty mountains were relatively minor disturbances of the overall structure of the crust. (Reprinted with the permission of Scribner, a Division of Simon and Schuster from *Our Mobile Earth*, by Reginald A. Daly. Copyright © 1926 by Charles Scribner's Sons, renewed 1954 by Reginald A. Daly; p. 276.)

consistent with the long-standing views of many British and European theorists but at odds with T. C. Chamberlin's theory of solid accretion. Rollin T. Chamberlin, son of T. C., made this a general argument against continental drift: that it was a "foot-loose" theory, "detached and free-floating," rather than fitting in with preexisting knowledge like the planetesimal hypothesis, "an integral part of geological philosophy."[65] Rollin was of course partial; the reality was that American geologists were already abandoning the planetesimal hypothesis.[66] Bailey Willis—ever loyal to Chamberlin—acknowledged this unhappily in a letter to his wife in September 1928 after a visit from Arthur Day, head of the Carnegie Institution's Geophysical Laboratory. "We touched on Chamberlin," Willis wrote to Margaret, "and Day said he gathered that geologists were tacitly agreed not to criticize his work while he lived, but then there would be a 'flood of criticism' when he was gone. I asked for specification and got one: that Chamberlin postulates a homogeneous earth, whereas it is clearly heterogeneous."[67]

Others might have felt that Daly went too far in his speculations, but Chester Longwell defended this foray into hidden causes, linked as it was to Daly's undisputed geological expertise:

> Conservative geologists may feel that [*Our Mobile Earth*] would be better if the hypothetical discussions in the last chapters had been omitted. However, this material is very properly included, as it is the product of the author's serious thinking through many years. Daly is one of the few American geologists who have been strongly attracted by the doctrine of continental displacement. As the original proponents of this hypothesis have not suggested any force that appears adequate to do the work of moving continents and making mountains, Daly devotes his attention chiefly to this phase of the problem. The conception of continental *sliding* on the flanks of great bulges or domes is proposed in place of the continental *drifting* favored by Wegener. . . . [His] hypothesis [has the benefit that it] recognizes the ordinary principles of mechanics, and . . . makes use of older helpful conceptions [that] Wegener discards . . . as . . . outworn tool[s].[68]

Daly had tried to show that continental drift could be linked to facts and theories that American geologists already knew or believed, and to some extent he succeeded. Charles Schuchert, for example, while arguing against drift at the 1926 American Association of Petroleum Geologists (AAPG) meeting, nevertheless acknowledged that "when one turns to the Alps and is told by the best of authorities that their present width of some 150 miles was originally 500 and perhaps 625, which means that their southern limit has moved from south to north, he begins to remember the statement of Galileo in regard to the earth: 'And yet it does move.'"[69] Schuchert, of course, had not been reading Galileo; he had been reading Daly.

Ultimately Daly's theory foundered on the problem it was meant to address: it was still not clear if the proposed forces were sufficient to make the crust rupture. The problem of forces had not been resolved, just displaced. Longwell thus concluded his AAPG presentation:

> Daly, aware of the work involved and the magnitude of forces required, seeks to substitute *sliding* for *drifting*, assuming that broad domes or bulges form at the earth's surface, and on the flanks of these domes the continental masses slide downward, moving over hot basaltic glass as over a lubricated floor. This is an interesting speculation; but Daly has not yet demonstrated that the doming can actually occur on the scale he assumes, or that the [thermal] gradient will be sufficient to give the desired effect. At present, therefore, his suggestion must be regarded as purely speculative.[70]

Daly, a North American, had addressed the problem of drift from the perspective of isostasy and geosynclines, signal contributions of North American geology. John Joly, an Irishman, was meanwhile addressing it from the perspective of radioactivity, one of the great developments of *début de siecle* European science.

Periodic Fusion: The Work of John Joly

Born in Holywood, Kings County, Ireland, John Joly (1857–1933) was the third son of a family descended from European nobility (figure 4.8). His father, the Reverend John Plunkett Joly, traced his lineage from fifteenth-century French counselors to the

Figure 4.8. Portrait of John Joly. (From *Obituary Notices of the Royal Society of London* 3, 1934.) Reproduced with permission of the Royal Society.

King; his mother, Julia Anna Maria Georgina, Comtesse de Lussi, hailed from a German-Italian family ennobled by Frederick the Great. (Among Joly's bequests to Trinity College was a relic said to be a piece of the true cross, given to his maternal great-grandfather by Pope Pius VI.) Joly was a man of extraordinary intellectual versatility whose life's work included contributions to physics, geology, mineralogy, medicine, marine navigation, and the advancement of Irish higher education. Among other things, he invented a technique used in World War I for determining the altitude of a star when the horizon is obscured, pioneered the technology of color photography, and published two volumes of original verse.[71] His geological sonnets evinced his profound sense of process and time; in "Oldhamia," inspired by the brachiopod of same name, the poet speaks to the fossil as witness to the ages and to the evanescence of human existence:

> Is nothing left? Have all things passed thee by?
>   The stars are not thy stars. The aged hills
>   Are changed and bowed beneath the ills
> Of ice and rain, of river and of sky;
> The sea that riseth now in agony
>   Is not thy sea. The stormy voice that fills
>   This gloom with man's remotest sorrow shrills
> The memr'y of thy lost futurity.
>   We—promise of the Ages? Lift thine eyes.
> And gazing on these tendrils intertwined
> For aeons in the shadows, recognize
> In hope and joy, in heaven-seeking mind,
> In faith, in love, in reason's potent spell
> The visitants that bid a world farewell.[72]

As an undergraduate at Trinity College, Dublin, Joly studied physics, chemistry, mineralogy, engineering and modern literature, taking first class honors in the last of these. He continued to write poetry throughout his life, but his vocation was science. He served as secretary of the Dublin University *Experimental Science Association*, for which he wrote a paper estimating the height of the Krakatoa eruption based on eyewitness accounts, and throughout his undergraduate years and beyond he kept detailed diaries of experiments and theoretical ideas. His diary from 1881, entitled "Notes on Work Done, Proposed, and Scientific Speculation," includes precise sketches of experimental designs intermingled with notes on geological questions, sketches on chemical theories, and critiques of the previous Sunday's sermon. By 1882, a philosophy of science emphasizing the role of experimental and laboratory data in answering geological questions was evident. Discussing an explanation provided by Dana for the origin of certain features in slates, Joly wrote critically, "[Is this] not to be verified by experiment? . . . Anything, otherwise, suspicious."[73] While Daly looked to the field to answer his theoretical questions, John Joly looked to the laboratory.

Upon graduating from Trinity in 1882, Joly was appointed lecturer of engineering at an annual salary of £100. (A compulsive compiler of data, he recorded his yearly salary thoughout his life: £500 when he became chair of geology; £800 when he retired at the age of 69.) During the next decade, he made significant contributions to physics, particularly in designing equipment to measure the physical properties of materials. Inventions from this period include the meldometer, a device for measuring the melting temperatures of metals and minerals by reference to the thermal expansion of a thin ribbon of platinum attached to the sample, a hydrostatic balance for measuring the specific gravity of small samples of materials, and the steam calorimeter, with which he became the first person to measure the specific heats of gases at constant volume.[74] For this, he was elected to the Royal Society of Dublin (later serving as its president), and in 1891 he was offered the position of assistant professor of experimental philosophy at Trinity. In 1892, he was inducted into the Royal Society of London at the exceptionally young age of thirty-four; in later years he won virtually every major British scientific award, including the Royal (1910) and Boyle (1911) Medals of the Royal Society, and the Murchison Medal (1923) of the Geological Society.[75]

Ironically, Joly's move to the Trinity physics department coincided with developments that pushed his research toward geology. Arthur Rambaut, a colleague at the Dunsink astronomical observatory in Dublin, asked Joly to apply his inventive talents to designing an improved shutter for stellar photography. Joly obliged and soon developed a general interest in photography, which led to his invention of the first-ever technique for producing color photographs from a single plate. Prior to this invention, color images had been created as optical effects, projected on a wall or a screen, by superimposing three separate images taken with red, green, and blue filters. Joly designed a method to create an image from a single glass plate by etching it with red, green, and blue lines. The "Joly process," patented in 1894, was superseded by chemical methods, but not before Joly had used it to produce the first color images of minerals in thin section.[76] In 1897, partly on the basis of this work, he was offered the Trinity Professorship of Geology and Mineralogy, which he held for the remainder of his life.

In his application for the professorship, Joly proposed to devote his energies to research "along the lines where Physics and Geology . . . meet."[77] These lines had recently been battle fronts; Joly sought rapprochement. Keenly aware of the rancorous dispute between physicists and geologists over the age of the earth, Joly sought an independent estimate of it based on chemical or physical principles. Uniformitarianism, he argued in a paper published in 1899, should be a method, not a doctrine. "Properly restricted, [it] defines the only scientific attitude open to the geologist in dealing with the past and the future. Rightly defined this doctrine is no other than that held and lived up to by every scientific man."[78] Scientists should build their arguments on observable evidence, no more or less. The task of the geologist was not to argue for unlimited stores of time but to use the evidence of geology to estimate that time. Joly thus suggested the following idea, uniting the uniformitarian principle with the data of geochemistry: if the salts of the oceans come from the weathering of rocks, and if the mean rate of weathering has been constant through geological time, then one could determine the age of the earth from the quantity of salt in the sea. Measure the concentration of sodium in rivers today, multiply this by the volume of water supplied by rivers to the oceans each year to get the annual rate of salt transfer, and then divide the total salt content of the sea by the annual supply rate to get the age of the ocean. The result, Joly found, was approximately 90 million years.[79]

In retrospect, Joly's estimate was far too low, but it was considerably greater than the 20 to 40 million year figure Kelvin had finally demanded that geologists accept. Perhaps more important, Joly's geological colleagues relished the way he fought back on Kelvin's own quantitative terms. Thus it was not surprising that Joly became the first person to recognize the implications of radiogenic heat for Kelvin's calculations of the age of the earth. Radiogenic heat explained why Kelvin's numbers were too low; the rapprochement between physics and geology appeared to be at hand. In 1903, in perhaps the most ironic statement in the history of geology, Joly wrote with satisfaction: "the hundred million years which the doctrine of uniformity requires, may, in fact, yet gladly be accepted by the physicist."[80] In fact, the earth would turn out to be much older than even this.

Radioactivity as a Vera Causa of Tectonics

It took a decade for scientists to demonstrate—and a few years more for Joly to accept—that the earth was billions of years old.[81] Meanwhile, Joly published his work on pleochroic haloes and showed that the size and intensity of natural haloes are directly proportional to the stratigraphical age of the rocks that contain them, thus providing the first empirical evidence that the decay of radioactive elements had been constant throughout geological time and might be used to determine the absolute ages of rocks. In 1913, working with Rutherford, Joly refuted his own earlier work and dated rocks of the Devonian period at approximately 400 million years.[82] During this period, he also corresponded with nearly every major scientist working in radioactivity—Rutherford, Frederick Soddy, Marie and Pierre Curie, Paul Langevin, Bertram Boltwood—and considered various means by which radioactivity might affect geological processes, and vice versa. He wrote to Rutherford in 1920 to ask whether relativistic effects might somehow alter the radioactive clocks in rocks, and Rutherford replied tartly: "Life is too short to worry unduly about relativity."[83]

Joly was perhaps hoping that radioactive decay rates were not constant: if that were the case, then his own 90-million-year estimate for the age of the earth might yet be saved. But in the end, Joly not only embraced radioactivity but also took it one step further. Rutherford had shown that a by-product of radioactive decay was the release of significant amounts of heat, and Joly began to ponder the geological significance of this. For if radioactive elements were everywhere, then radiogenic heat also was everywhere, and it might be a significant energy source affecting geological processes. Writing in his private notebook—perhaps to convince himself—Joly proclaimed radioactivity a revolutionary discovery whose implications were yet to be fully realized:

> The old belief in the immutability of matter appears to be dying hard. Some of our most valued leaders still seek some explanation of the facts of radioactivity which will leave the old view [of matter] untouched. Efforts of this kind are unhealthy and [impede] the cause of scientific advance. [But] we must remember that the subject of the mutability of matter had fallen into disrepute, and had come to be associated with horoscopes and the philosopher's stone.[84]

Displaying an extreme empathy with the objects of his study, as well as his intense poetic sensibility, Joly continued. "Could we only see with our eyes what goes on 'round a speck of radium assuredly we would be impressed. Let us fancy the atom of matter is visible to us in all its beauty and order. . . . Tens of thousands of atoms respond to its energy; quiver, sway, and shed some part of their circling systems of electrons . . . "[85] The remainder of Joly's scientific career was dedicated to developing this idea.

In 1909, Joly published *Radioactivity and Geology*, in which he proposed radiogenic energy as a *vera causa* of tectonic processes:

> If . . . the ubiquity of uranium be conceded, then the influence of radioactive energy on the surface dynamics of the globe becomes a matter of first importance. The efficacy of uranium as an almost eternal source of thermal energy seems to be unquestionable. . . . We need submit the uranium ore to neither chemical nor physical processes. It is sufficient to take it from the rocks, and place it in the calorimeter, when the constant outflow of heat will be apparent. . . . With these fundamental facts before us, why should we regard as premature their presentation as a *vera causa* in geological dynamics?[86]

Lyell and Kelvin had wrongly devolved geological interpretation toward two extremes: a complete steady state on the one hand and an inexorable progression toward heat death on the other. Both views were incompatible with the geological evidence. The history of the earth was cyclic. Mountain-building occurred as discrete, topographically rejuvenating events punctuating longer periods of denudation: the Caledonians and Appalachians formed in the early Paleozoic, the Rockies in the late Mesozoic, the Alps and Himalayas in the late Cenozoic. But what caused this rejuvenation, absent terrestrial contraction? Joly's answer was radiogenic heat, which would build up over geological time until it was sufficient to melt rocks, generating magmas that would disrupt the overlying crust. But the eruption of these magmas, and the crustal disturbances they caused, also dissipated the accumulated heat. And so the cycle would begin again. The driving force behind the earth's discontinuous history of mountain-building was the cyclic buildup and dissipation of radiogenic heat. This became the informing principle of Joly's theoretical views.

Joly introduced the idea of convection as a driving force behind tectonic processes—not as understood today, but as an indirect effect of the transfer of heat-bearing radioactive materials. Joly's research had confirmed the earlier results of R. J. Strutt that granites were more radioactive than basalts and that the continental crust was therefore more radioactive than the oceanic. Erosion of continental materials thus resulted in a net transfer of energy—radiogenic heat—from the continents to geosynclines. "In the deposition of sediments—their uplifting into mountain chains—their subsequent removal and their re-deposition elsewhere—there is a convective movement of radioactive materials on the surface of the earth."[87] Uranium-bearing continental materials carried *energy*.

Joly suggested that the huge thicknesses of uranium-rich sediment in geosynclines led to the buildup of radiogenic heat, which in turn caused melting in the lower, hotter parts of the pile.[88] The melted rocks became the igneous cores of mountain belts, while the adjacent, now-plastic substrate provided the rheological conditions for detachment and over-thrusting. Joly discussed the Alps specifically, noting, as had van der Gracht, that the lack of a clear mechanical explanation had greatly troubled some geologists. Others, however, had "contented themselves by recording the facts without advancing any explantory hypothesis beyond that embodied in the incontestable statement that such phenomena must be referred to the effects of tangential forces acting in the earth's crust." Neither position now seemed necessary to Joly. "It would appear that the explanation of the phenomena of recumbent folds and their *déferlement* is to be obtained directly from the temperature conditions prevailing throughout the stressed pile of rocks."[89] Heat was not only the cause of orogenesis but also the explanation of the correlation between sedimentary accumulation and uplift: the heat generated within the sedimentary pile drove its deformation.[90] In essence, sedimentary sequences sowed the seeds of their own destruction. And this explained why geosynclinal wedges were so thick: the pile had to be thick enough to generate heat enough to impart plasticity to the underlying portions. "Sedimentation, from this point of view," he concluded, "is a convection of energy."[91]

Joly's cyclic vision was an optimistic one. Not condemned by the laws of thermodynamics to a frigid death by dissipation, the world possessed its own internal potential for renewal and rejuvenation. Small forces could effect huge change. It was even possible to imagine earlier epochs of now-obliterated geological history:

> Wonderful possibilities are brought before us. Peaceful cooling may await the earth or catastrophic heating may lead in a new era of life. Our geological age may have been preceded by other ages, every trace of which has perished in the regeneration which has heralded our own. . . . The planets may now be in varying phases of such great events. And when a star appears in the heavens where before we knew of none, may not this be a manifestation of the power of the infinitely little over the infinitely great—the unending flow of energy from the unstable atoms wrecking the stability of the world?[92]

Radioactive heat had enabled geologists to recover the geological time scale; perhaps it would also now enable them to explain the causes of large-scale geological change.

From Cyclicity to Continental Drift

Deformation caused by rising geothermal gradients in areas of sedimentation was not a new idea—it had been used by Babbage, Herschel, Hall, and Hayford—but Joly added a specific and focused source of heat.[93] And like his predecessors, Joly worked from the assumption that the basic configuration of the continents was not at issue. Dislocations were something that occurred along the *margins* of continents, as evidenced by the fact that "over 94% of all recorded [earthquake] shocks lie in the geosynclinal belts" and that the "girdle of volcanoes around the Pacific and the earthquake belt coincide" (figure 4.9).[94] Yet even as Joly assumed the constancy of continental configuration, his imaginative mind was toying with the alternative. For although he was well-versed in the isostatic arguments against sunken continents, he was equally versed in the biological arguments for ancient continental connections.[95] Thus, while introducing the idea of erosion and deposition as a kind of forced convection, Joly also raised the specter of true convective heat transfer within the earth's fluid interior.

The concentration of radioactive materials in the crust proved, Joly argued, that convective overturn had occurred during the early history of the earth. Most geophysicists and geochemists accepted this: the surface concentration of uranium could be accounted for if, and really only if, the earth had differentiated during an early molten phase. The highly radioactive materials, being hottest and therefore least dense, rose to the surface "just as a region of heated water beneath the ocean must slowly rise to the surface." The intriguing question was whether similar episodes had occurred in later geological history and might be implicated in crustal motions. Joly thought so:

> Such convective actions may have taken place principally in the early stages of terrestrial heating, [but] there is no improbability that such movements, acting with extreme slowness over geological time, and requiring long periods for the accumulation of radioactive heat, might [also] affect the already formed crust of later geological times. Some of the great lava flows of comparatively recent date may be associated with such actions.[96]

Joly was referring to the famous Deccan traps of India, of late Cretaceous age, covering more than 200,000 square miles, and the Tertiary Antrim basalts of Giant's Causeway in his own Ireland. Convection could explain the huge outpourings of basalt that occurred around the globe at discrete times in geological history.

Joly treated these ideas in passing in his early work, but he returned to them forcefully in the 1920s.[97] In his notes for the prestigious Halley lecture, delivered at Oxford University in 1924, he focused on flood basalts as direct evidence of an active, mobile earth.

> We perceive that the picture presented to us is not that of a slowly cooling world dying into quiescence.... We see no such decadences ... We know the mountains have been but recently uplifted: that great floods of molten rock have in recent ages risen from beneath and overwhelmed vast territories. We know that the continents and islands are still restless after scores of millions of years of thermal wastage. No! The slow exhaustion of primitive heat has not been the history of our planet. Our world is not decrepit by reason of advancing years. Rather, we should consider it as rejoicing in the gift of perpetual youth. Endlessly rejuvenated, its history begins afresh with each great revolution.[98]

Like Wegener, Joly situated the problem of tectonics firmly in the framework of isostasy, which he called the "clue to the fundamental architecture of the earth."[99] How could one reconcile isostasy—continents balanced in a delicate equilibrium—with the massive geological evidence of crustal dislocations and the energetic implications of radiogenic heat? Wegener had focused on the contradiction between the geodetic

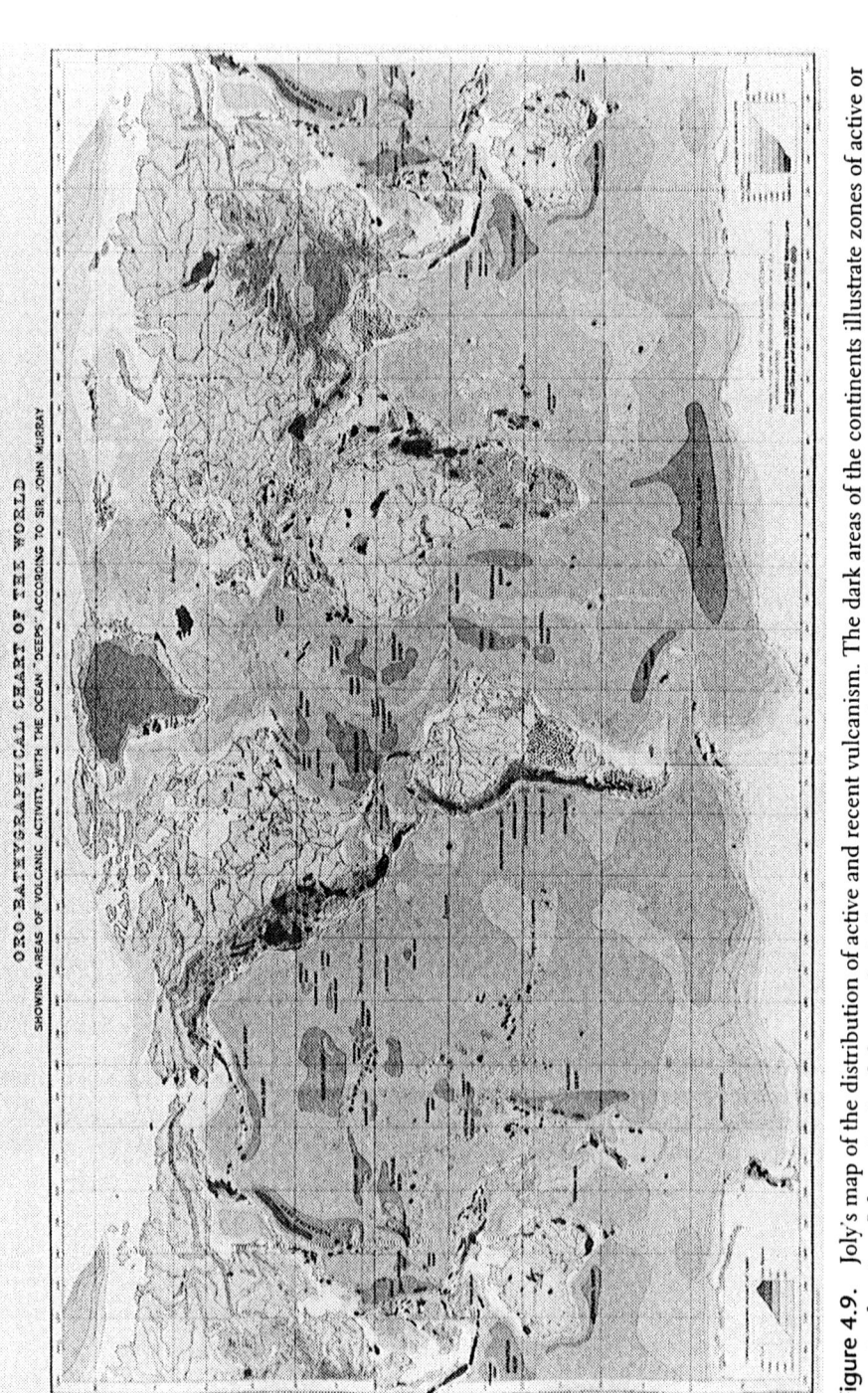

Figure 4.9. Joly's map of the distribution of active and recent vulcanism. The dark areas of the continents illustrate zones of active or recent vulcanism. Joly maintained that geological activity was not randomly or uniformly distributed but confined to well-defined and relatively narrow zones, usually along the margins of continents. (From Joly 1925.)

refutation of sunken continents and the paleontological demand for prior land connections; Joly focused on reconciling isostatic equilibrium with the instability implied by radiogenic heat. The conclusion Joly came to was conceptually similar to Wegener's (although independent of paleontological arguments): isostasy maintains the continents as continents, but radiogenic heat provides the energy that intermittently disturbs them.

In *The Surface-History of the Earth*, published in 1925, Joly expanded his lecture into a full-fledged tectonic theory. Its essence was the idea of thermal cycles: periodic fusion and solidification of the subcrust as a consequence of the buildup and dissipation of radiogenic heat. As heat built up, the subcrust beneath the continents would become first plastic and then molten, finally resulting in convective overturn, the generation of flood basalts and other volcanic eruptions, and large-scale dislocations of the crust. Joly called these periods "revolutions" and identified them with the major known orogenic episodes: the Caledonian, the Appalachian/Hercynian, the Laramide, and the Alpine[100] (figure 4.10). As the geological record showed, each revolution would be followed by a period of relative quiescence, during which radiogenic heat would begin slowly to accumulate again. "Events of a paroxysmal nature would very probably have occurred, a period of convective heat loss being followed by comparative quiescence and crusting of the surface. This would entail a rise of temperature within and an eruptive phase would again prevail." The key point, which Joly emphasized repeatedly in his public lectures and private notes, was that thermal cycles accounted

	Barrell and Schuchert (1924)	Haug (1921)	Marr (1898)	Sonder (1924)	De Lapparent (1900)
Recent					
Pleistocene					
Pliocene					
Miocene	Cascadian Revolution	Alpine and Dinaric	'Continental'	'Emergence'	Great Alpine Uplift.
Oligocene					
Eocene					Uplift of Pyrenees and Apennines.
Cretaceous	Laramide Revolution				
Jurassic					
Triassic					
Permian	Appalachian Revolution	Hercynian	'Continental'	'Emergence'	Hercynian Chains formed.
Carboniferous					
Devonian					
Silurian	Caledonian disturbances	Caledonian	'Continental'	'Emergence'	Caledonian Chains formed.
Ordovician					
Cambrian					
Keweenawan } Huronian	Killarney Revolution	Huronian	'Continental'	'Emergence' (?)	Huronian Chains formed.
Timiskamian	Algoman Revolution				
Loganian	Laurentian Revolution				Continents outlined.

Figure 4.10. Joly's Table of Revolutions. (From Joly 1925, p. 133.)

for indisputable geological evidence of periodic energetic rejuvenation right up to the present. "We are not quite done with [this process] even today. Witness the impressive lava floods of western North America. The Deccan Traps of India mark a more remote outbreak, the Keweenawan floods one yet more remote."[101] Contra Kelvin, the history of the earth was not directional in any simple sense.

Flood basalts also provided the link to isostasy: they were samples of the high-density material in which the granitic continents floated. Geodesy proved that the substrate had to be something denser than granite, and flood basalts showed what that something was: "The evidence is of the sort that would convince the occupants of a ship, when water flowed in [from] a casual leak, that their vessel must be floating in water."[102] Ships leak water; the earth leaked basalt. The implication was obvious.

The idea of molten basalt beneath the continents was inconsistent with seismic evidence, but just as Daly saw a solution to this apparent dilemma—that the basalts were rigid but not solid—so did Joly: his theory postulated a periodically and locally, rather than continuously and uniformly, molten substrate. Geophysicsts assumed that the present evidence of solidity was proof that the substrate had always been solid, but this, Joly argued, was uniformitarianism of the most primitive kind. It was ironic indeed that the followers of Kelvin, who so famously dismissed geologists for accepting uniformitarianism as an act of faith, should now fall into the same trap. The state of the substrate throughout geological history was a question that had to be answered, not a known fact. Its present condition was but one constraint among many. Joly posited: "The simplest conclusion we can come to—and the most probable in that it involves the minimal thermal change—is that at the present time the basaltic [substrate] is in the viscous-solid state very near its melting point. In the past it was in a state of fluidity [also] near its melting point."[103] A substrate nearing its melting point would be first elastic and then plastic, until sufficient heat rendered it liquid. Intermittent melting was proven by the temporally sporadic distribution of flood basalts; for the rest of geological history, "the assumption of temperatures generally prevailing in the outer crust, beyond what are sufficient to confer plasticity towards the secular force responsible for secular movements, is gratuitous."[104]

Besides the mistake of assuming temporal invariance in the condition of the substrate, geophysicists also mistakenly assumed spatial homogeneity. In fact, Joly pointed out, the geophysical evidence of solidity was "not at variance . . . with the existence of molten matter beneath large areas, even of continental magnitude, provided such areas are mutually isolated."[105] Seismic waves could travel through a partially molten substrate provided that the remaining solid portions were interconnected. The substrate could be rigid in most places and for most of geological time, as seismologists would have it, yet still be plastic or even liquid at the times and in the places where geologists and geodesists needed it to be. And this was precisely what geological evidence required, because orogenic and volcanic activity were not uniformly distributed around the globe but focused in discrete zones. One could account for the geophysical evidence of a solid substrate by the temporal and physical localism of crustal dislocations.

But so what if the substrate were periodically molten? How would this account for the creation of mountains? Joly suggested that stresses, arising in part from the volume increases associated with melting of the substrate and in part from melting at the base of the crust during thermal peaks, would separate the continents from the oceans and generate the forces that caused crustal deformation along continental margins:

> We expect the development of rifts in the stretched surface crust, [above zones of heating and magma generation] . . . where special differential stresses arise [that] tend to part the continents from the basaltic floor of the ocean. Through such rifts the melted magma will be forced at high pressure. It will flood out over the sea-floor where rifts have developed therein and great dikes have been formed. Such lava will soon re-solidify under the cooling effects of the ocean. . . . Where the lava flows out along the continental margin we find such resulting surface features as we see to-day along the coasts of North-Western Europe—the basaltic coastal features of Northern Ireland, Western Scotland, &c. Again, sometimes these fractures will release the imprisoned magma upon the land, as happened in the out-flows of the Deccan Traps in Western India. Nor will the continents escape the tensile forces. Rifting will surely occur, and, as we have already seen, is very much in evidence.[106]

These lines may be read as foreshadowing the modern concept of sea-floor spreading—that oceanic crust is generated by the eruption of basaltic magmas—and as Joly advocating continental drift. But Joly's words left considerable room for interpretation. Was he advocating wholesale continental movement, à la Wegener? Or was he proposing a mechanism to explain dislocations along the margins of continents whose relative positions did not much change?[107] Joly's frequent references to American research and terminology—as well as to the basaltic coastal features of northern Ireland and Scotland—seemed to suggest the latter. In chapter 7 of *The Surface-History*, "The Building of the Mountains," Joly posed the question, What is the nature of the association between the oceans and the deformation of the adjacent continents? Then he answered it:

> Just this: The sea-floor has been the passive instrument of orogenesis. It has forced inwards the continental margin and deformed and rent the continental crust to its depths. The lava wells out, caught in gaping fissures beneath and forced upwards. And as the stresses increase, the sediments themselves, folded and crushed, emerge above the surface of the geosynclinal sea. The batholithic masses of continental rocks, possibly not at a very high temperature, but softened by an aqueo-igneous action, arise along with the sediments to form the heart of the youthful ranges.[108]

This answer, so clearly within the geosyncline framework, was fully compatible with permanent continents and oceans. And Joly's idea of thermal cycles had been germinating at least since 1909, well before Wegener or Taylor had proposed continental drift. However, once the idea of drift had been proposed, the connection to thermal cycles was obvious: periodic melting would facilitate the migration of continental rafts. Thus, before going to press, Joly added three addenda to his book. The first was a three-page discussion of the "Effects of Tidal Forces on Orogenesis," and the second a specific discussion of "The Question of Continental Drift." Generalized drift, Joly argued, was clearly "improbable in the last degree" when the substrate was solid or even partly fluid. However, if one focused on local zones within the ocean basins, which "under certain conditions had become greatly attenuated or rifted in favourable directions, then it seems quite possible that the stresses arising out of tidal forces . . . might have sufficed to create differential motions among the continents."[109]

The tidal forces rejected by Jeffreys as too weak to move continents through a rigid substrate might be effective in moving them through a molten one.

The third addendum was a lecture entitled "A Summing Up: The Surface Changes of the Earth." Here, Joly argued that tidal forces not only might be effective during periods of fluidity but *must* be:

> If a fluid layer intervene between the outer crust and the underlying core, a differential motion must arise. The result is at the present time negligibly small ... the tidal force not being sufficient to shear the solid substratum. But in times when this substratum is replaced by a highly mobile fluid the displacement is inevitable. Thus the continental masses would slowly replace the oceans in position respecting the deep-lying magmas.[110]

Thus did Joly in 1925 embrace continental drift. During periods of fluidity, the continents and oceans would slowly but surely change positions.

Joly did not rely solely on the argument from tidal forces: he also proposed a role for active forces arising within the earth's interior as a result of convective heat transfer:

> During the period of thermal dissipation the ocean floor will be attacked by hot and, doubtless, often superheated currents from beneath ... It is possible that magma currents ... may result in the location of fractures principally on the western coasts. The fractures will be rapidly filled in by congealing basalt. It is not improbable that the location of the Hebridean lava-flow was determined in this manner.[111]

And he concluded his *magnum opus*:

> The thesis that the continents may have shifted during geological time gains support from the foregoing history of terrestial tectonics. For we observe that in the course of this history the ocean floor has been repeatedly assailed by thermal effects tending to reduce its thickness and rigidity, and that the continents were exposed to long-continued magmatic forces acting from west to east. If the ocean floor were in times of Revolution so far reduced in rigidity as to yield to those forces, continental movements in an easterly direction would probably have taken place. We cannot now evaluate the intensity attained by the several factors concerned, and so we must leave to other sources of evidence the final verdict in this interesting theory. An affirmative answer would not be out of harmony with possibilities arising out of the present views.[112]

Periodic melting was compatible with continental drift and might even explain it.

Reaction and Retreat

The lecture that forms the conclusion of *The Surface-History* was originally presented at the Geological Society of London in 1923, and it makes Joly's loyalties clear. Yet as noted above, the body of his book could be read in terms of vertical tectonics—and many did read it that way.[113] Why did Joly leave his position equivocal, if not contradictory? The lecture at the Geological Society had been well received: for many, the connection between thermal cycles and continental drift was not merely obvious, but compelling, as it seemed to solve the mechanism problem. Several colleagues wrote notes of congratulations. W. J. Sollas, the Oxford Professor of Geology whose daughter Hertha was Seuss's English translator, enthusiastically mixed his metaphors: "I am glad you have hitched Wegener's hypothesis to your car and are driving at such a rate.

If you read your paper at the Royal [Society] may I be there to hear. I shall if no ill wind blows me out of my course. You will be admiral of this fleet if you can start the continents and island[s] sailing on the opposite tack to Wegener."[114] Writing from India, another colleague wrote, "I often see your theories quoted in writings from almost everywhere. They seem to be accepted almost universally in conjunction with Wegener's work."[115]

Joly submitted the paper to the Royal Society of London under the title "A General Theory of the Movements of the Earth's Crust," in hope of publishing it in the *Philosophical Magazine*.[116] The manuscript involved calculations of the potential of tidal forces to cause large-scale crustal movements during periods of subcrustal fluidity. To be sure, Joly admitted, these forces were inadequate if applied to a solid earth, as Jeffreys had done, but perhaps they would be more efficacious under conditions of fluidity. The point was to address the question. Alas, Joly's mathematics, like that of many scientists, was not beyond reproach, and his detractors were quick to seize upon this fact: the paper was returned by the Society secretary, J. H. Jeans—Jeffreys's collaborator—on the grounds that the reaction of three reviewers was "uniformly adverse."[117] One reviewer called the paper "of a common type, based on the view that a force no matter how small can produce as great deformations in the earth as we like, provided it acts for long enough." Here was the voice of a physicist; geologists had grounded their science on precisely that view. Another reviewer—likely Jeffreys himself—dismissed the paper in four sentences, concluding that the ideas presented were "completely fanciful . . . The author doesn't understand what a tidal force is."[118]

Joly was furious. In response to Jeans's rejection letter, he scrawled himself a note highlighting his main point with an adjacent *xxx*: "The important point is that it was the enthusiastic reception of the theory by the Geol. Soc. [London] which led to its communication to the Phil. Mag."[119] Joly's long-sought rapprochement was not to be had. Were geologists to solve the mechanism problem, they would evidently do so without the help of physicists. Joly included his lecture in *The Surface-History*, but it was never published in a refereed journal nor read at the Royal Society.

Joly's lifelong dream of reconciling physics and geology was dashed on the shoals of continental drift. The detached tone of his subsequent comments on the subject must be read in this context: they hide what was surely a profound disappointment. Thus when asked in 1928 to contribute to the expanded volume emanating from the proceedings of the 1926 AAPG symposium on continental drift (which he had not attended), he wrote a brief paper in which he tersely allowed, "continental drift is not improbable during periods of fluid substratum."[120] But although once burned, he refused to be entirely shy. After all, his theory could help resolve one of the questions often posed to Wegener: What caused Pangea (the Mesozoic supercontinent) to break up? "The necessity for thermal escape must impose certain restrictions upon continental aggregates," Joly wrote in the AAPG volume. "Thus, if all the land areas were united to form a single continent it is probable—or at least possible—that such an arrangement would not be permanent."[121]

In 1930, Joly published a second edition of *The Surface-History*. Although he now omitted the offending lecture, he included a new section entitled "Recent evidence for surviving tangential stresses in the ocean floor" and expanded the appendix on continental drift. The former described the work of the Dutch geodesist Felix A. Vening Meinesz, whose work in the Malaysian and Indonesian archipelago had recently revealed significant, uncompensated gravity anomalies, which Vening Meinesz interpreted as evidence of tangential stresses within the crust. Joly quoted the Dutch scientist: "That great anomalies should occur in this part of the earth's crust is, for the rest, in good harmony with the universally prevailing opinion that movements of the crust took place here in very recent times, and that they may even still be in progress." The reason, Joly explained, was that the anomalies were caused by "either upward or downward flexure. Upward flexure gives rise to local excess of gravitation and downward flexure gives rise to local defect of gravitation."[122] The implication, obvious albeit unstated, was that the flexures were the result of large-scale horizontal compression.

The second edition of *The Surface-History* was published when Joly was seventy-three years old; he published only a handful of further papers before his death three years later. In 1957, as continental drift was being revived by the work of of P. M. S. Blackett in London and Keith Runcorn at Cambridge, Kendal Dixon, the son of Joly's long-time friend and collaborator H. H. Dixon, revisited the theory of thermal cycles for the *Trinity College Annual*. Joly's theory, Dixon wrote,

> provides a mechanism for the displacement of continents postulated by Wegener and others, and is attractive from many points of view. It was attacked [in the 1930s] by Jeffreys on grounds whose cogency is not universally admitted; but the theory has become unfashionable today chiefly because seismic evidence provides no grounds for supposing that there is a liquid layer beneath the surface rocks. . . . It is impossible to say with confidence that Joly's theory is well-founded or that it is certainly untrue. . . . It seems quite possible that the theory may return to favour again in some modified form.[123]

Dixon's comments, while mildly prophetic, were wrong on two counts. Joly explicitly denied the necessity of a continuous liquid layer precisely to counter Jeffreys's criticism. But more important from a historical perspective, the theory had already been discussed in modified form. In the late 1920s, as Joly began to lose energy to work further on the problem, it was taken up by his younger colleague Arthur Holmes. Whereas Daly had addressed the mechanism problem by providing a powerful driving force, and Joly envisaged conditions under which small forces would be highly efficacious, Holmes developed an explanation that did both.

Sub-Crustal Convection Currents: The Work of Arthur Holmes

Arthur Holmes (1890–1965) was one of the most gifted and influential geologists of the early twentieth century (figure 4.11). A demonstrator in geology at Imperial College from 1912 to 1920, and then chief geologist for an oil exploration company in Burma (now Myanmar), Holmes was appointed Professor of Geology at the University of Durham in 1924. Like Daly and Joly, he was the recipient of numerous awards and honors. He gave the annual Lowell Lectures at Harvard in 1933 and received the Murchison (1940) and Wollaston (1956) Medals of the Geological Society of London, fellowship in the Royal Society of London (1942), and the Penrose Medal, the highest honor of the Geological Society of America (1956). He was also hugely productive. He published his first scientific article in 1911 — at the age of twenty-one — and his last book, the second edition of his widely used text, *Principles of Physical*

Figure 4.11. Portrait of Arthur Holmes. (From *Royal Society Biographical Memoirs*. 12 (1966), opposite p. 291.) Reproduced with permission of the Royal Society.

Geology, in 1964, the year before his death. In between, he wrote nearly 200 papers and books. In 1943, Holmes was appointed Regis Professor of Geology at Edinburgh, the post he held for the remainder of his career.[124]

Like Joly, Holmes worked at the boundaries between physics and geology. Born of modest means in the farming districts of Northumbria, his interest in geology dated to his high school years when a teacher introduced him to the debate over the age of the earth. Wanting to know more, he read Suess's *Face of the Earth*. An outstanding student, he was offered a scholarship to Imperial College, London, where R. J. Strutt had begun his work on radioactivity in rocks and minerals. Holmes enrolled in 1907, intending to read physics, but quickly learned the implications of Strutt's work: the earth was fantastically old, and its story was to be found in rocks. Holmes began to attend lectures at Imperial by the great geology teacher W. W. Watts and soon accepted Watts's view that "nature is the perfect expert witness."[125] He completed the examinations for the B.Sc. degree in physics in 1909; within a year he had completed the geology requirements as well and received the degree of Associate of the Royal College of Science. He also served as president of the famous De la Beche Club, whose frequent excursions educated Holmes in field geology.

Holmes then became a research student under Strutt. In 1905, Strutt had published his estimate that a certain specimen of thorianite was 2,400 million years old based on its helium content, considerably older than Ernest Rutherford had estimated using the same technique. But such estimates were not reliable, because helium was produced by various elements at different rates, and being gaseous it could escape a sample over time. In 1907, Bertram Boltwood at Yale suggested that the ratio of uranium to its solid daughter product, lead, would be more reliable, and with this he produced the oldest numbers yet for the ages of rocks. But Boltwood was more interested in the phenomenon of radioactivity than rocks, and he soon dropped the project.[126] In 1911, Strutt put Holmes to work on the problem.

Without mass spectrometers, computers, or knowledge of isotopes, Holmes pursued Boltwood's uranium/lead method and with this produced geology's first absolute time scale.[127] In 1913, he published his results in *The Age of the Earth*, a comprehensive review of the various techniques that had been used to date rocks and a presentation of the new radiometric results. Holmes's timescale provided concrete evidence for what geologists had long suspected but had never been able to prove: that the age of the earth was to be measured in billions, rather than millions, of years.[128] This "absolute" timescale established Holmes's reputation as one of the most important geologists of his generation (figure 4.12).

But geologists were not as happy with the new results as they ought perhaps to have been; having accommodated themselves to a shorter time frame, they were now to a degree committed to it. Displaying unusually acute insight for one so young, Holmes wrote in *Nature* in 1913 that "the geologist who ten years ago was embarrassed by the shortness of the time allowed to him for the evolution of the earth's crust is now still more embarassed by the superabundance with which he is confronted." But Holmes saw the moment not as a crisis but as an opportunity:

> We must not moan over the apparent difficulties with which the geologist has been faced since the advent of radium. Rather they should be welcomed. If at present some of our ideas are mutually incompatible, the discrepancies do not demand a wholesale rejection of the facts, but simply a re-interpretation of the fundamental hypotheses on which so many of our doctrines seem to hang.[129]

The Geological Systems.	Time Scale in Millions of Years.	
	Helium Ratio.	Lead Ratio.
Pleistocene	1	—
Pliocene	2·5	—
Miocene	6·3	—
Oligocene	8·4	—
Eocene	30·8	—
Cretaceous	—	—
Jurassic	—	—
Triassic	—	—
Permian	—	—
Carboniferous	146	340
Devonian	145	370
Silurian		
Ordivician	209	430
Cambrian		—
Algonkian		1000–1200
Archean	710	1400–1600

Figure 4.12. Chart of geology's first absolute time scale. (From Holmes, 1913c.)

From the Age of the Earth to Continental Drift

Holmes quickly set to work. Between measurements of uranium and lead, he somehow found time in 1911 to travel with the Memba Mineral Company to Portugese East Africa, where he collected rocks from Precambrian to Tertiary age and saw his first active volcanoes. The carbonatites of the East African rift inspired his second professional focus: petrogenesis, the origins of igneous rocks. Upon his return from Africa, he was appointed demonstrator of geology at Imperial. Besides Watts, one of his mentors there was Dr. J. W. Evans, the petrologist who would later write the forward to the first English translation of Wegener's *Origin of Continents and Oceans*. After a brief stint as chief geologist with the Yomah Oil Company in Burma, Holmes was offered a position in the newly created Geology Department at Durham University. This seemingly undistinguished appointment turned out to be a boon: with few students, Holmes had ample time for research and writing. By 1933, he had more than 100 scientific credits to his name.

At Durham, Holmes began to think about the connection between radioactivity and petrogenesis, and to read widely. He was particularly impressed by Joly. Although Holmes was personally responsible for disproving Joly's 90-million-year estimate of the earth's age, he nevertheless admired the senior scientist and was deeply influenced by his unapologetically anti-Lyellian stance.[130] In a 1926 article entitled "Contributions to the Theory of Magmatic Cycles," Holmes picked up where Joly had left off. The assumption of thermal equilibrium, he agreed, was incompatible with geological evidence. Had such a state been achieved, then "the earth would be so nearly dormant that it could have had practically no geological history at all," which obviously was not the case.[131] This left three options: it was continuously heating up, it was continuously cooling down, or it was alternately accumulating and discharging heat. "That we are here at all to investigate the problem is sufficient proof that the earth is not continuously growing hotter," Holmes noted dryly. Theories of progressive cooling, on the other hand, while boasting a distinguished pedigree, gave no real account of volcanic or orogenic activity:

> Thus there remains for consideration only the conception of the alternating accumulation and discharge of latent heat proposed by Joly, a conception based on radioactivity of the rocks and the principles of isostasy and tidal drift. Periodically repeated cycles of fusion and consolidation provide fully for a varied geological history of just the rhythmic and discontinuous type that the earth has actually experienced. No other hypothesis has accomplished so general a synthesis, and at the moment no alternative seems to be even conceivable.[132]

Holmes assisted the synthesis by tying in the most important piece of field work of the day, Emile Argand's mapping and interpretation of the Alpine overthrusts, which "demanded a mobile substratum on a regional scale."[133] In an address to the International Geological Congress in Belgium in 1922, Argand had presented a unified synthesis of Eurasian tectonics, based in large part on his own intensive field mapping and Alpine mountaineering, in which he argued that the Alpine overthrusts were merely the local manifestations of episodic crustal movements on a global scale (figure 4.13). Holmes postulated that Joly's periods of fluidity were Argand's episodes of overthrusting—that the overthrusts were *caused* by intrusion into the upper crust of the melted substra-

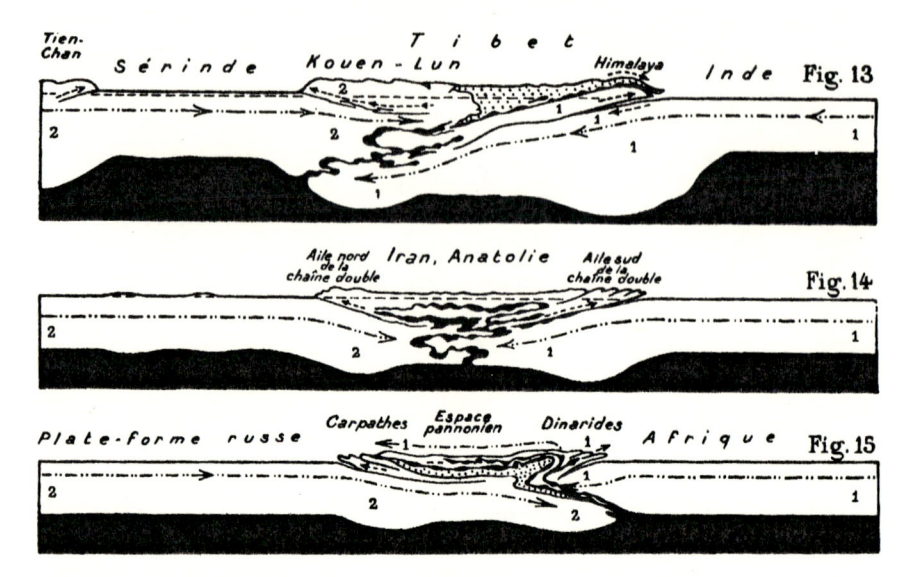

Figure 4.13. Emile Argand's tectonic synthesis. (From Argand 1922, p. 183. Copyright © 1977 Hafner Press. Courtesy of Simon and Schuster.)

tum, which provided the required push. This concept could be generalized to explain all the major stresses required to move portions of the earth's crust.

Holmes made explicit the connection between substratal melting and continental displacement. Because radiogenic heat would ultimately have to be dissipated, convection currents would develop in the molten substrate. These currents would drive the continents. In an address to the Glasgow Geological Society in 1928, published in the *Geological Magazine,* Holmes declared:

> For sound physical reasons it becomes as impossible to "sink" a continent as it would be to sink an iceberg floating in the sea. We know that the sites of considerable areas of the Atlantic and Indian oceans were formerly occupied by continental masses, and since these ancient lands are no longer there, we are driven to believe that their material has moved sideways. Evidence of lateral movement is also forthcoming from tear faults, [and] overthrust structures of the Alpine type. . . . Moving the continents back in the directions indicated by the evidence leads to a Permo-Carboniferous reconstruction similar to that made familiar by Wegener's diagrams, a reconstruction which is quite independently called for on paleo-climatic grounds. . . . It is therefore concluded that there is now convincing evidence pointing to the former occurrence of continental drift on a scale of the same order of that advocated by Wegener.
>
> Many geologists have hesistated to accept this straightforward and consistent reading of the rocks, because so far it has not been found possible to discover any gravitational force adequate to move the continental slabs. . . . Admitting that the continents have drifted, there seems no escape from the deduction that slow but overwhelmingly powerful currents must have been generated in the underworld at various times during the Earth's history . . . , namely that convection currents may be set up in the lower layer as a result of differential heating by radioactivity.[134]

Nature, then, was Holmes's witness. The geological evidence for large-scale crustal motions was overwhelming; therefore there must be an explanation for it. Convection currents arose as a deductive consequence. In a series of papers published between 1925 and 1933, Holmes developed this concept into an integrated theory of tectonics and petrogenesis, explaining such diverse phenomena as the origin of ocean basins and the formation of high-pressure rocks such as eclogite (figure 4.14). Ocean basins would form where basaltic magmas erupted to the surface in zones of crustal tension; the basalts themselves would be generated from a perioditite layer below. Eclogites would be formed under high pressure in regions of crustal compression and down-warping under geosynclines.[135]

> The sites of ascending currents would become disruptive basins. Here the accumulation of excess heat responsible for the process [of convection] would be discharged by the development of a new ocean floor. . . . Meanwhile, mountain building would have been accomplished on the continental margins or on the sites of former geosynclines. . . . The squeezing out of the more mobile parts of the substratum to form crustal magmas is an attractive side issue that opens up a new vista of possibilities, and gives a hint as to the origin of basalts.[136]

The nature of the substratum was still in doubt, but Holmes agreed with Daly that it "is probably for the most part in a glassy state; that its temperature is such that although it is rigid, it is devoid of permanent strength except possibly near the top," and with Joly that "it is and has been the main source of basic and ultrabasic magmas, including the plateau basalts."[137] But it need not, indeed should not, be wholly liquid: a completely liquid substrate would be *too* slippery—unable to exert any grip against the overlying continents. "Only currents in a highly viscous glass could get sufficient 'grip' on the continental under-surfaces to exert the required drag on the overlying material."[138] Whatever the details, it was clear that theoretical constraints could only go so far in resolving this, or any, geological problem. Drift had happened, and now there was a mechanism to explain it. The details could be left to a latter date. Quoting Argand, Holmes concluded, "Nous ne prétendons pas réduire la tectonique à la physique: c'est l'affaire à l'avenir."[139]

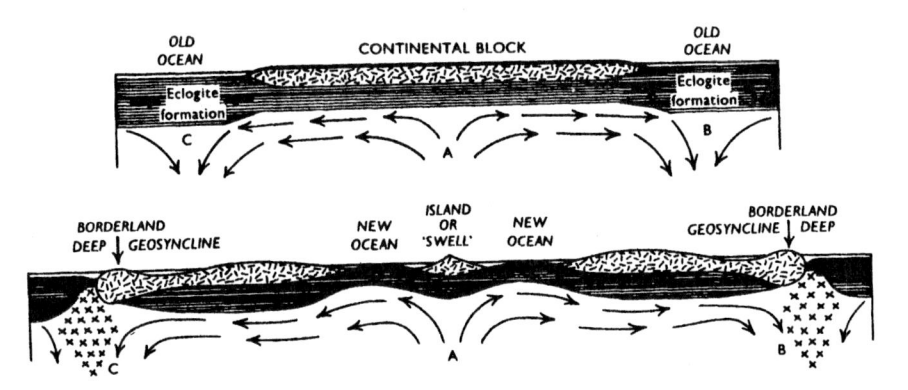

Figure 4.14. Arthur Holmes's model of continental drift by subcrustal convection. (From Holmes 1929b.) Reproduced with permission of the Geological Society of Glasgow.

A Plausible Mechanism

As both Joly's and Holmes's writings make clear, the idea of convective overturn was not new; it had long been discussed by geophysicists. There was disagreement, however, as to how significant it was in the history of the earth, disagreement grounded in whether one believed the earth were solid or liquid. In Kelvin's analysis of the age of the earth by conductive cooling, the earth was, necessarily, solid, but he assumed that there had been an early phase of convective cooling to equalize the earth's internal temperature prior to solidification. Convection as part of the earth's early history was thus widely accepted by geophysicists. But Holmes was making a different claim: that it was a continuous part of earth history up to the present. This linked him to a different tradition, to those who had argued for convective, and perhaps also advective, heat transfer in a liquid earth.[140]

Writing in 1872, for example, Joseph LeConte recalled that Alexander von Humboldt had invoked "secular currents in the earth's interior liquid" as a mechanism of homogenizing the earth's internal density distribution.[141] Although LeConte had rejected Humboldt's hypothesis as "violent," Osmond Fisher had not. In the same decade, Fisher suggested that William Hopkins's analysis of the earth's early history unnecessarily restricted the period during which convective cooling occurred: "[Hopkins,] considered that, so long as the matter of the earth retained a sufficiently high state of fluidity to admit of the circulation of convection currents, no crust could form: but that when by those means, the temperature had been so far reduced that convection became arrested, immediately a crust would be formed."[142] But, Fisher argued, if a crust formed from the outside in, then a fluid layer by definition persisted under it and convective overturn may have continued. Fisher recounted the detailed descriptions of a certain Miss Bart, traveler to Hawaii, who had described solid crusts on top of flowing liquid lava in Kilauea; such effects might be a microcosm.[143]

Fisher was sensitive to both the phenomenological arguments of geologists and the theoretical constraints posited by physicists. His own desire for a fluid layer arose primarily from his commitment to isostasy, but in 1895 he made it part of a rebuttal of Kelvin's parsimonious estimates of the age of earth. Kelvin assumed only conductive cooling: convective heat transfer would undermine his analysis and perhaps explain the discrepancy between his and geologists' estimates. It could also help explain surface crustal motions.

> If the interior be liquid, and the central parts have been constantly receiving greater accessions of heat than the parts more distant from the centre, convection-currents must have been set up, and must have played the chief part in the secular cooling. The age of the world would therefore be dependent upon the degree of their activity, which is not known. Another important consequence would follow, because the currents would necessarily have a horizontal flow in the parts of their course where the liquid had ceased to rise and had not yet begun to descend; and this would produce a horizontal stress on the superincumbent crust, acting more powerfully upon it at those places where the bottom was most uneven as beneath mountains. Thus we should be able to account for one cause, at least, of the compressing force for which geologists are in search.[144]

In the end Fisher rejected the arguments of his own geophysical colleagues. He concluded in his 1878 paper:

It is obvious that a more or less perfectly fluid substratum is required to give this capability of [crustal] slipping. Indeed, the nearest analogy . . . seems to be found in the crushing together of the adjacent edges of two sheets of floating ice. . . . As far as *authority* (that blind guide) can determine the question, we have it on both sides. . . . Physicists should [therefore] be moved to consider what elements of change in the position of the earth's crust as regards latitudes might be introduced under the influences acting upon it.[145]

In advocating convection currents in a fluid substrate as the cause of large-magnitude crustal motions, Fisher might be counted as the originator of modern geological theory, but geophysicists mostly dismissed Fisher's conclusions. Indeed, those who accepted his quantitative methodological goals were among the most intolerant of his phenomenological sympathies. By the end of the nineteenth century, most geophysicsts were working with the assumption of a solid substrate, reinforced by discoveries in the new science of seismology, and most geologists felt obliged to accept their verdict.[146] But the discovery of radioactive heat changed the context of this discussion. If Joly were right to suppose that the substrate was genuinely liquid, at least in part and from time to time, or if even, as in Holmes's view, it was merely plastic, then convection reemerged as a potential mechanism of heat transfer throughout geological history.

Nor was Holmes the only one to discuss convection currents. The idea was raised by the Alpine geologist August Rothpletz early in the century and discussed in the 1920s by A. J. Bull in England and by Otto Ampferer, R. Schwinner, G. Kirsch, and Emile Argand in continental Europe.[147] In the United States in the 1930s, University of California at Los Angeles professor David Griggs undertook a series of experiments to demonstrate the mechanism, while Andrew Lawson at Berkeley proposed that large-scale, low-angle thrusting of the sialic crust into the underlying simatic substrate—as would occur in areas of down-welling currents in Holmes's model—might in part explain uplift in coastal California.[148] Ampferer's work is particularly interesting in hindsight as he made a causal link between subcrustal currents, large-scale crustal dislocations, and the *glissement*, or detachment, of the upper from the lower crust in orogenic belts and coined the term *subduction* to refer to the process in which one section of the crust slid down below another. Griggs was retrospectively awarded the Geological Society of America's Day Medal, nearly four decades later; while Lawson, Ampferer, and Daly might rightly be credited with the idea of subduction.[149] But these comments are made in hindsight: it was Holmes whose model was most widely discussed in the late 1920s and early 1930s, and Holmes who most explicitly, unambiguously, and consistently advocated convection as the *cause* of continental drift.[150]

Not for Lack of a Mechanism

By 1929, three powerful theories—Daly's gravity sliding, Joly's periodic fusion, and Holmes's subcrustal convection currents—had been developed to explain the kinematics of drift. All were consistent with the known physical properties of the earth. Moreover, Daly and Holmes offered dynamic explanations as well. None of the theories came as an ad hoc adjustment to Wegener's proposals: Joly's theory predated Wegener's and came only gradually to be linked with it; Daly's theory incorporated evidence derived from his experience in the Appalachians and his knowledge of igneous rocks; Holmes's theory derived from geochronology and the study of radiogenic heat. All three were developed by prominent and eminent scientists; all were published in readily accessible form; and all were widely known and discussed at the time.

Together, the three theories incorporated the fundamental aspects of modern theory: a rigid moving surface riding on convection currents in a weak zone beneath, with portions recycled into the substrate. Not one of these theories required the continents to plow through the rigid ocean floor. The pieces of the puzzle, as geologists understand it today, were assembled by the end of the 1920s. Geologists had a phenomenon, they had evidence, and they had a mechanism. What was hailed as breakthrough knowledge in the 1960s—albeit on the basis of different evidence and with different justification—was proposed in the 1920s. Holmes's theory was not complete by present standards, to be sure, but neither is existing knowledge complete: there are aspects of plate tectonics that remain to be explained. The development of scientific knowledge always involves modification and development of an initial idea, which in some cases turns out to contain only the germ of the final outcome.[151] Whether Holmes's model of continental drift is precisely the same as modern knowledge is thus beside the point. The point is that *the same driving force that is generally accepted today for plate tectonics was proposed in the 1920s.* The theory of continental drift did not fail for lack of a mechanism.

Holmes corresponded extensively with colleagues across the Atlantic who in turn discussed his ideas among themselves. The attitudes of the more open-minded appear in a letter from Chester Longwell to his Yale colleague, the paleontologist Charles Schuchert, shortly after he finished reading *The Surface History of the Earth*:

> I am very glad to hear from you and to have your remarks on Joly's book. I had just finished reading it, and have prepared a review for the American Journal [of Science]. I feel quite enthusiastic about it, for I consider Joly's theory the best that has appeared to explain all the phenomena of diastrophism. As you know, there are some apparent discrepancies between Joly and the facts of geological history, but I believe these difficulties are met pretty well in an article by Holmes, published in the December issue of the Geological Magazine. Joly and Holmes together have a beautiful theory, and I believe it will be epoch-making.[152]

The epoch was not to begin for another forty years.

II

THEORY AND METHOD

> Method is metaphysics.
> —*Albert Camus*

5

From Fact to Theory

If continental drift was not rejected for lack of a mechanism, why was it rejected? Some say the time was not ripe.[1] Historical evidence suggests the reverse. The retreat of the thermal contraction theory in the face of radioactive heat generation, the conflict between isostasy and land bridges, and the controversy that Wegener's theory provoked all show that the time was ripe for a new theory. In 1921, Reginald Daly complained to Walter Lambert about the "bankruptcy in decent theories of mountain-building." Chester Longwell opined in 1926 that the "displacement hypothesis, in its general form . . . promises a solution of certain troublesome enigmas."[2] A year later, William Bowie suggested in a letter to Charles Schuchert that it was time for "a long talk on some of the major problems of the earth's structure and the processes which have caused surface change. The time is ripe for an attack on these larger phases of geology."[3]

One possibility is that the fault lay with Wegener himself, that his deficiencies as a scientist discredited his theory. Wegener was in fact abundantly criticized for his lack of objectivity. In a review of *The Origin of Continents*, British geologist Philip Lake accused him of being "quite devoid of critical faculty."[4] No doubt Wegener sometimes expressed himself incautiously. But emphatic language characterized both sides of the drift debate, as well as later discussions of plate tectonics.[5] The strength of the arguments was more an effect than a cause of what was at stake.

Some have blamed Wegener's training, disciplinary affiliations, or nationality for the rejection of his theory, but these arguments lack credibility. Wegener's contributions to meteorology and geophysics were widely recognized; his death in 1930 prompted a full-page obituary in *Nature*, which recounted his pioneering contributions to meteorology and mourned his passing as "a great loss to geophysical science."[6] Being a disciplinary outsider can be an advantage—it probably was for Arthur Holmes when he first embarked on the radiometric time scale. To be sure, there were nationalistic tensions in international science in the early 1920s—German earth scientists complained bitterly over their exclusion from international geodetic and geophysical commissions—but by the late 1920s the theory of continental drift was associated as much with Joly and Holmes as it was with Wegener.[7] And it was an American, Frank Taylor, who first proposed the idea.[8] So why was the theory rejected, the evidence seemingly dismissed?

There may not be a single answer, but historical evidence makes one conclusion clear: American methodological commitments, grounded in the context of American science, conditioned the reception of the theory of continental drift. One of these commitments had to do with the way in which theory, in general, should be approached. By the early twentieth century, the ideal of pluralism—central to so many aspects of American political and religious life—had also been embedded in the methodological commitments of American scientists. The result was a culturally rooted American philosophy of scientific method, rhetoric, and practice onto which Alfred Wegener's work was not easily grafted. American earth scientists were not opposed to theory, but they did have strongly held views as to how scientific theories should be developed and defended. From their perspective, Wegener's presentation of continental drift appeared to be flagrantly unscientific. The Americans' reaction may be put into context by first considering the rather different response of their British colleagues.

The British Response to Drift: Cautious Entertainment

Perhaps the most distinctive feature of the discussion of continental drift in the United States was its vitriolic quality, particularly as compared with its counterpart in Great Britain. There, reaction to Wegener's theory was divided but animated, with much of it cautiously favorable. A principal forum for the drift debate in the early 1920s was a meeting of the British Association in Hull, reported in *Nature* in January of 1923. The report described a "lively but inconclusive" exchange that "brought forth a great diversity of opinion regarding the validity of the hypothesis, almost the only point on which there seemed to be any general agreement being an unwillingness to admit that the birth of the north Atlantic could have occurred at so late a date as the Quaternary."[9] Rather than dismiss the idea of the birth of the Atlantic through continental drift, British geologists were arguing about when it had happened.

Much of the Hull discussion centered on the strength of the geological homologies, which struck a responsive chord among many of the participants. J. W. Evans, Arthur Holmes's mentor at Imperial College and president of the Geological Society of London, who was writing the introduction to the English translation of the third edition of Wegener's *Origin*, confirmed the existence of a "well-known similarity of the geological formations on opposites sides of the oceans."[10] So did W. B. Wright, who claimed that drift could explain why "a critical comparison of formations on the two sides of the North Atlantic shows on the whole a remarkable correspondence, both stratigraphical and paleontological, from the Archaean to the Cretaceous," and that it also solved "the hitherto unsolved problems of the Permo-Carboniferous glaciation."[11] J. W. Gregory, Professor of Geology at Edinburgh and the world's foremost expert on the East African Rift, concurred. In a 1925 review of the first English-language edition of Wegener's book, Gregory allowed that "the opposite parts of the Atlantic show a remarkable resemblance in structure and composition."[12] Many British geologists found examples from their own work to support Wegener's claims, from the "practical identity of the Old Red [sandstone] fish faunas of the Orkneys and North America" to the "similarity of lithological characters and fossil contents of the rocks [of] South America east of the Andes [with] Africa."[13] Although no one claimed that the theory had been proven, the homologies were "evidence as good as could be expected ... nearly demand[ing] some such explanation [as drift]."[14]

In a review of the second edition of *The Origin*, British geologist Philip Lake acknowledged that "a moving continent is as strange to us as a moving earth was to our ancestors, and we may be as prejudiced as they were."[15] But in an unusual rhetorical move, Lake aligned himself with the Ptolemaics and attacked the mobile earth. Like other British skeptics, Lake focused on ambiguities and incompleteness in the evidence. For example, the continental reconstructions were doubtful because the continents did not fit together like interlocking pieces of a jigsaw puzzle; they fit in an approximate way, leaving some gaps. Similarly, although Wegener had nicely reassembled the Southern Hemisphere's glacial deposits, other "boulder beds" (i.e., glacial tills) remained unexplained. Likewise, to Lake's apparent satisfaction, the Glossopteris flora were not all reunited. However, Lake conceded that "if continents have moved many former difficulties disappear." That brought him to an ambiguous end: "Wegener has performed a valuable service by drawing attention to the fact that land-masses may have moved relatively to one another. He has not proved that they have moved, and still less has he shown that they have moved in the way that he imagines. He has suggested much, he has proved nothing."[16]

The British attitude may be summed up as cautiously receptive. Many were prepared to entertain the idea of drift, and some saw in it a solution to outstanding difficulties in their specialty fields. By 1925, Wegener's ideas were being seriously considered by many British geologists, and Joly and Holmes were dedicating their formidable intellectual powers to making the theory work. In his 1925 review of the English edition of *The Origin*, J. W. Gregory wrote, albeit perhaps optimistically:

> The favorable reception of Prof. Wegener's theory is significant of marked changes of opinion as regards the structure and history of the earth. It shows that the once widely adopted view that oceans and continents have always been in their present positions no longer hampers geological interpretation, and it marks the growing belief in the effective mobility of the earth's crust.[17]

The American Response to Drift: Rejection

In the United States, reaction to Wegener's theory was almost entirely negative. Drift was the subject of articles in the *American Journal of Science*, the *Annual Report of the Smithsonian Institution*, and the publications of the Carnegie Institution of Washington—all adverse. The most important and widely known hearing Wegener received in the United States was the symposium convened by the American Association of Petroleum Geologists (AAPG) in New York City in 1926. Of all the speakers at the meeting, only one, W. A. J. M van Waterschoot van der Gracht—the convenor—spoke strongly in favor of the theory. Chester Longwell argued for reservation of judgment. All the others spoke strongly against. In part because of the almost uniformly negative response in New York, van der Gracht solicited additional, more sympathetic papers for the conference proceedings, published two years later. These came from Wegener himself and from John Joly, G. A. F. Molengraaf, and J. W. Gregory—an Irishman, a Dutchman, and a Scot.

In contrast to the Hull meeting, several Americans raised objections relating to the mechanism of drift. Van Waterschoot praised Joly's model of periodic melting as "the best available explanation," but that did not convince the Americans. Bailey Willis, Professor of Geology at Stanford, acknowledged that "Wegener's suggestion . . . has met with acceptance in Europe on the part of scientists whose views we cannot but receive with respect" but nevertheless argued that Wegener's brittle continents plowing through a fluid substrate should deform the substrate, not the continents, and so could not be a mechanism for generating continental mountain belts.[18] But this objection had been addressed by Daly, whose book, *Our Mobile Earth*, had just been published. Indeed, several speakers referred to it as providing a possible solution. And as we have seen, Holmes's modifications to Joly's model were already being discussed. If the mechanism issue was soon to be resolved, what were the remaining objections?

Many participants objected to Wegener's methodology. Indeed, the most notable feature of the New York symposium was the frequency and emotion with which Wegener's purpose and approach were attacked. The tone of these attacks is illustrated by the comments of Edward Berry, a paleontologist at Johns Hopkins University, who openly questioned Wegener's credibility as a scientist:

> My principal objection to the Wegener hypothesis rests on the author's method. This, in my opinion, is not scientific, but takes the familiar course of an initial idea, a selective search through the literature for corroborative evidence, ignoring most of the facts that are opposed to the idea, and ending in a state of auto-intoxication in which the subjective idea comes to be considered an objective fact.[19]

Bailey Willis felt that Wegener's book gave the impression of having been "written by an advocate rather than an impartial investigator." Joseph Singewald claimed that Wegener had "set out to prove the theory . . . rather than to test it" and accused him of "dogmatism," "overgeneralizing," and "special pleading." In short, Wegener was accused of being unscientific. But why? Alfred Wegener was a highly trained professional earth scientist. Why was his theory treated as if it were on the fringes of acceptability?

The hostility of American geologists to the theory of continental drift might suggest that they were hostile to theory in general. If the development of causal accounts was not a goal of American geology, then any large-scale theory would be subject to adverse criticism. This is what some historians have claimed. For example, in the introduction to *Lord Kelvin and the Age of the Earth*, historian Joe D. Burchfield suggested that Americans were content to remain apart from their British colleagues' dispute with Kelvin as they busied themselves with data collection and classification in their enormous new country. "With a vast, largely undefiled geological laboratory stretching before them," Burchfield writes, American geologists "devoted themselves to exploration and observation rather than to speculation and to theory building."[20] Burchfield was describing the late nineteenth century, but historians have often assumed that this purported pattern continued unchanged in the early twentieth. Homer Le Grand, for example, explicitly linked the rejection of continental drift to an American resistance to causal theorizing. One of the consequences of the development of plate tectonics, Le Grand has argued, was a shift in the goals of geology from being primarily descriptive to primarily explanatory.[21] The success of plate tectonic theory—

which gave causal explanations of a wide variety of geological phenomena—gave general ratification to causal-theoretical accounts, transforming American attitudes. The sociologist of science John Stewart came to a similar conclusion—that American geologists prior to the development of plate tectonics were unimaginative empiricists who eschewed large-scale causal explanations of geological phenomena.[22]

Scientists involved in the development of plate tectonics in the 1960s have helped perpetuate this view. Tuzo Wilson, the geophysicist who explained the phenomenon of transform faults, claimed that the acceptance of plate tectonics "transformed the earth sciences from a group of rather unimaginative studies based on pedestrian interpretations of natural phenomena into a unified science that is exciting and dynamic and holds out the promise of great practical advances for the future."[23] Allan Cox, a pioneer in the development of the geomagnetic time scale, likewise held that "prior to the mid-1960s, research in tectonics could scarcely be regarded as being in a state of scientific health. Experimentation and observation, not having a firm theoretical basis, were poorly focused and lacked specific objectives."[24] (One must wonder how such weak and poorly focused science produced the theory of plate tectonics!)

The glorification of present achievement through a pejorative portrayal of previous work is a common characteristic of participant history, and such sentiments have informed more recent writing as well.[25] An extreme example comes from the pen of geophysicist Robert Muir Wood. In the 1980s, Wood painted the history of geology as a Manichaean morality tale in which the progress of the interpretive earth sciences was opposed by reactionary geologists seeking to preserve the "purposelessness of summer field work."[26] With the advent of plate tectonics, Wood wrote, the old approaches were swept away by "hard" science; traditional geology, "framed according to the poetic and irrational measure of man," was gone forever. Wood's treatment suggests interesting questions about the dynamic nature of scientific standards, but any possible answers are themselves swept away by his overt (yet unexplained) hostility to the subjects of his story: his aversion to early-twentieth-century geology is so great that he equates opposition to continental drift with Nazi philosophy and Soviet communism—simultaneously! Perhaps the very fact that early-twentieth-century geologists rejected continental drift is sufficient to discredit them, but this begs the question of *why* they rejected it, as well as ignores that geophysicists, whom Wood admires, were equally if not more opposed to drift than were field geologists. Wood's emotional diatribe points to a general problem: the persistence of a historiographic cliché that American scientists in the nineteenth and early twentieth centuries were narrow utilitarians or naive empiricists focused on the collection of usable and classifiable facts.[27]

Were American Earth Scientists Opposed to Theory?

In his pioneering 1948 work *The Origins of American Science*, the historian, mathematician, and Marxist Dirk J. Struik placed these origins in the utilitarian demands of a new country. According to Struik, the potential patrons of science were less impressed by natural history and natural philosophy than with the potential uses of astronomy, surveying, and engineering in the new nation.[28] To the extent that geology captured the patrician imagination, it was for the economic value of mineralogy rather than the imaginative value of earth history. The pragmatic imperative of protecting shipping and trade in the newly united States contributed to the early establishment of the U.S.

Coast Survey—the nation's first scientific and technical agency (1807)—justified under the commerce and general welfare clauses of the constitution. But the Coast Survey was conceived of as a temporary expedient, intended to do the job of mapping the coasts and then disband: in general, American science got off to a shaky start. Thomas Jefferson, John Quincy Adams, and others argued for the authority vested in the new government to "promote the progress of science and useful arts" specifically articulated in the patent clause, but only gradually, and discontinuously, did governmental interest in science broaden. Even as it did, the focus remained on subjects of clear national concern: geographical and geological exploration, agricultural development, and a variety of areas relevant to the national defense.[29] Or so conventional wisdom goes.

The aim of the U.S. government in supporting science in the nineteenth and early twentieth centuries clearly was (and many would argue remains) to support the economic, industrial, and military needs of the nation, but this tells us little about the attitudes and motivations of the *scientists* who worked within federal scientific agencies, much less about their colleagues outside governmental walls. Unfortunately, many historians have made a leap from a governmental emphasis on utility to a more broadly descriptive and empirical—even antitheoretical—approach among scientists. Like Joe D. Burchfield, the historian Robert Bruce tied empiricism to the scale of the new nation, suggesting that the range of new phenomena demanding description left scant time for analysis. As Bruce put it dismissively, "to observe, to describe, to classify, perchance to name; then to observe again. That was the happy and simple round of most American scientists."[30] A somewhat similar view was propounded by Daniel Kevles. In his now-classic study *The Physicists*, Kevles painted a portrait of a theoretically naive community, preoccupied with experimentation—physics' epistemological equivalent to geology's observation and classification—prior to the 1920s and 1930s.[31]

Were all American scientists really arid empiricists until the mid-twentieth century? Did the governmental demand for practical results mean that American earth scientists were not interested in explanatory accounts? A great deal of historical evidence suggests otherwise. Kevles himself acknowledges that physics did not represent the best of American science in the nineteenth century. Using foreign recognition as a standard of accomplishment, Kevles points out that it "was largely in the earth and biological sciences—geology, topography, paleontology, botany, and zoology—that Americans had earned the respect of Europeans. . . . Few American physicists had won foreign acclaim." Many American *geo*physicists and physically oriented geologists had won acclaim (and not for description and classification!), including William and Henry Rogers, Alexander Dallas Bache, Robert Woodward, Clarence King, Clarence Dutton, and G. K. Gilbert. Moreover, as Kevles reminds us, the Geological Society of America won the Cuvier medal of the French Académie des Sciences for its collective contributions.[32] Yet many historians have read American physics, especially in its early highly empiricist mode, as a stand-in for American science. This is plainly a mistake.

Furthermore, as Nathan Reingold has shrewdly emphasized, one should not confuse the rhetoric of scientists when they speak to politicians with what is actually happening in their laboratories and surveys. Because so much American science in the nineteenth century was done in governmental agencies and surveys—as opposed to the academies of continental Europe or the research universities that developed in the later twentieth century—American science necessarily involved a "complex mixture of pure and applied research."[33] In retrospect, such work may look more applied than pure, particularly if the theoretical questions that motivated it have long since vanished from view.[34] If some scientists lamented inadequate support for basic research in America, this may reflect not so much a lack of basic scientific activity on their part as a desire for greater intellectual autonomy or perhaps social status—particularly as compared with the scientific "mandarins" of France and Germany.[35] Put another way, how often does anyone say "Yes, thank you, we have ample funds for our work"?

The very fact that some Americans complained about lack of support for basic research is evidence that basic research is precisely what they wanted to do. As Reingold and others have stressed, one needs to separate the issue of patronage from motivation.[36] Here, the evidence seems clear. When Americans began to develop graduate research programs in the sciences in the late nineteenth century, they did so consciously to promote theoretical research in the European style. The year 1876 is often taken as a singular one in the development of the American research university, marked by the launching of graduate education at Johns Hopkins University—the first to make research a central focus of its mission. When Ira Remson, professor of chemistry and the second president of the university, established a graduate program in chemistry, he self-consciously followed the German academic model.[37] By the turn of the century, graduate programs designed to foster basic research were well established at most major American universities.

Looking specifically at the earth sciences, the evidence of theoretical activity is even clearer. As Mott Greene has shown in his detailed history of tectonic theories in the nineteenth century, some of the most important work on this subject came from the United States. The work of the Rogers brothers on the structure and origin of the Appalachians, of James Dana on the origins of continents and oceans, and of James Hall in developing the concept of the geosyncline were major theoretical contributions, widely discussed in the United States and overseas. In the case of geosyncline theory, it was also widely accepted.[38] Toward the end of the nineteenth century, the theoretical development of geology was taken up by prominent Americans like Bailey Willis and T. C. Chamberlin. And as we have already seen, Americans were the leaders in the development of the theory of isostasy, even though the empirical data originally came from elsewhere. Clarence Dutton, who coined the term isostasy, was perhaps the most explicit of all American earth scientists in his advocacy of theory. Writing in praise of his mentor, John Wesley Powell, Dutton suggested that Powell's "industry and energy in the collection of facts . . . would alone have secured his fame," but even more important was the "analytic skill with which he grouped them together." Facts were necessary, to be sure, but "we want more than facts," he concluded; "we want their order, their relations, and their meaning."[39]

By the early twentieth century, work on isostasy and regional uplift had expanded into a lively theoretical tradition dealing with tectonics topics, as evidenced by the work of Hayford and Bowie, the competing versions of continental drift proposed by F. B. Taylor (1910, 1925), Howard Baker (1914), and W. H. Pickering (1924), and the mechanism of drift proposed by R. A. Daly.[40] Given this context, it is not plausible that Americans rejected continental drift because they were opposed to theory per se.

Theory and Practice in American Earth Science

Perhaps the most important American geological theoretician at the turn of the century was T. C. Chamberlin (1843–1928). As chief geologist of the Wisconsin Survey (1876–1882) and then of the glacial division of the U.S. Geological Survey (1881–1904), Chamberlin had demonstrated the existence of multiple, distinct glacial episodes and proposed the idea that climate change could be produced by alterations in atmospheric carbon dioxide levels associated with geological processes.[41] As the first chairman of the geology department at the University of Chicago (1892–1918), he worked on the worldwide correlation of diastrophic events, the origins of the Quaternary deposits of North America, the mechanisms of glacier motion, and the nature of oceanic circulation. In all, Chamberlin published more than 250 scientific books and articles. Much of his later work was supported by the Carnegie Institution of Washington, which bestowed upon him the title of "investigator of fundamental problems in geology."[42]

As founder of the *Journal of Geology* (in 1893), Chamberlin was well placed to promote his fundamental ideas, and the most fundamental was his planetesimal hypothesis, developed with astronomer F. R. Moulton (1872–1952).[43] In the course of his work on climate change, Chamberlin had come to doubt the nebular hypothesis for the origin of the solar system: the presence of water in the earth's atmosphere and hydrosphere contradicted the kinetic theory of gases, unless the original earth was cold rather than hot. Astronomers had recognized other problems with the nebular hypothesis: it was inconsistent with the retrograde rotations of Uranus and Triton (a satellite of Neptune) and the behavior of Phobos (the inner satellite of Mars that orbits faster than Mars rotates); it failed to explain why the angular momentum of the sun is smaller than that of the planets; and it had never resolved precisely how the nebula could throw off discrete rings to form the multiple planets and their satellites. As further insult to Laplace's idea, refined telescopic photographs demonstrated that the starry nebula, at first heralded as confirmation of his theory, had spiral rather than annular structures. How spirals condensed into planetary orbits was anybody's guess.[44]

Chamberlin and Moulton proposed that the planets formed by the accretion of small, solid particles ("planetesimals") after a close encounter between our sun and another star. Or as Bailey Willis elegantly put it: "the dust particles gathered to form the globe."[45] In a report on the 1901 meeting of the Section of Geology and Geography of the American Association for the Advancement of Science, Chamberlin's presentation was called "the most important paper [presented] from a theoretic point of view." If true, the reviewer continued, the planetesimal theory would require geologists to "revise most of our old theories."[46] The particular old theory that geologists were happiest to revise was Lord Kelvin's analysis of the age of the earth. If the primordial earth were solid, not liquid, then Kelvin's calculations of its age were fundamentally flawed. In Chamberlin's words, "most of the restrictive arguments of Lord Kelvin lose their application under [the planetesimal] hypothesis."[47] But equally important was that the theory had the effect, albeit temporary, of rescuing Dana's permanence theory from the death of secular cooling. Building on the work of Dutton and Gilbert, Chamberlin argued that the origins of the continents and oceans were to be found in the material diversity of the protoearth. A liquid earth would necessarily have convected and homogenized, but a solid earth would preserve its original heterogeneity. Gravitational contraction could pull the terrestrial materials closer together with time—accounting for the oscillations of the continents through continual gravitational readjustment—but it would not fundamentally rearrange them. One could retain continental permanence by replacing thermal with gravitational contraction.[48]

Because Chamberlin was America's most renowned "promulgator of original and fundamental hypotheses" in geology[49] and indeed one of America's most famous scientists in any discipline at the turn of the century, his attitude toward abstract knowledge is of considerable interest. How did a government field geologist become a leading theoretician? How did practical field science lead to fundamental theoretical work? At least part of the answer is that Chamberlin believed in the unity of theory and practice—for pragmatic, moral, and metaphysical reasons.

A letter from Chamberlin to George Darwin provides a window onto his beliefs. In 1906, Darwin sought Chamberlin's advice on how best to solicit governmental support for high-precision geodetic work—an area in which the Americans had been conspicuously more successful than the British. Chamberlin's response—that Darwin should neglect neither the theoretical nor the practical uses of geodesy when speaking to politicians—opposed what might be construed as the conventional wisdom (then and now):

> I have stood in a parental relation to about a score of legislative measures involving appropriations for scientific work [and s]uch experience as I have had ... quite strongly supports the view that a scientific motive, *added* to a commercial one, greatly strengthens the case, and in not a few cases is apparently decisive. A certain percentage of legislators cannot be convinced that the commercial returns from certain classes of work (i.e., such refined measurements as geodesists attempt) are commensurate with the expense involved, and it is hard to convince them that they are not right in thinking that a less close approximation, involving much less expense, would not serve practical purposes equally as well. When, however, it is made [apparent] that work of a high degree of precision, while only slightly, if at all, more serviceable for commercial purposes, subserves scientific ends in correlated fields as well as that immediately involved, then they are ready to favor the necessary appropriations.

Chamberlin's experience as a survey scientist had led him to conclude that politicians were indeed sympathetic to concerns of science, if they were expressed honestly. But it was a mistake to disguise theoretical scientific work under a false rubric of utility, because the practical needs of the country could generally be addressed more directly and cheaply by other means. Furthermore—and again contrary to many people's assumptions—it was the scientific rather than the commercial benefit that could be construed as democratically distributed. "Some legislators are even ready to vote money to advance science when they are hesitant to take similar action for commercial reasons," Chamberlin pointed out, "on the ground, not unjustly taken, that science is universally beneficent, while commercial work is usually helpful merely to individuals or classes." If one subscribed to a universalistic ideal of scientific knowledge, as Chamberlin did, then theoretical science was not at odds with American democracy. To the contrary, it served the benefit of all. In any event, he concluded, there is "nothing more genuinely flattering to a legislator than the delicate assumption that he and his constituents are interested in the advancement of science."[50]

As an example of a confluence of theoretical and utilitarian needs, Chamberlin cited the recent San Francisco earthquake, which had dislocated the boundaries of highly valued California real estate. Scientists were proposing to undertake precise geodetic work to ascertain the exact nature and extent of the dislocations, and this would also be of "no small economic importance."[51] But most important to Chamberlin personally was his belief that a properly grounded scientific worldview could be spiritually liberating. Whereas Kelvin's flawed thermodynamic view of the earth implied degeneration and eventual death by cooling, Chamberlin's planetesimal theory was compatible with regeneration and rebirth and provided an alternative to apocalyptic thinking.[52]

> The public is now very generally depressed by needless apprehension of great impending disasters, if not a universal and final catastrophe, apprehensions derived from the narrow and pessimistic views of the past. From my point of view, which is doubtless a partial one, a contribution of supreme value to the happiness and well being of mankind is likely to grow out of rectified views . . . derived from the prosecution of the earth sciences.[53]

The Carnegie Institution of Washington (CIW), the most important patron of theoretical earth science in early-twentieth-century America, agreed with him. Nearly identical views on the interrelation of theory and practice were promoted by Charles Doolittle Walcott (1850–1927), one of the CIW's founders.[54] As the third director of the U.S. Geological Survey (1894–1907), Walcott was unusually frank about the mundane nature of much scientific work, particularly work with practical motivations, as government work generally was, and he suggested that theoretical advance was both the motivation and the recompense for daily scientific drudgery. Walcott wrote in an address to the Geological Society of America in 1901:

> For the sake of indicating the distribution of formations having economic value, the geologist performs a great deal of routine labor—observing phenomena of familiar kinds, grouping them in well-known categories, and making a record for future use, chiefly by others. . . . In performing these various duties he uses well established methods and merely applies to new materials the known principles of science. Such work is necessarily monotonous, and the active mind would soon lose interest and its operation become perfunctory were it not for the possibility of discovering new principles. . . . Geologic science . . . makes applied geology . . . possible, and reciprocally, the practice of geologic art opens the way to progress in geologic science.[55]

The CIW represented an alternative to survey science: conceived as a "sheltered haven for basic research," its scientists were in principle unencumbered by practical demands.[56] Nevertheless, CIW geologists also argued for the ethos of science in the public service. In 1937, CIW Director John Merriam asked Frederick Wright, the petrologist who had traveled with Daly to Africa, to develop a position paper on the issue of public relations. The secretary of the U.S. National Academy of Sciences from 1931 to 1951, Wright (1877–1953) was a distinguished scientist and close confidant of Merriam.[57]

Wright argued that theory versus practice was a false dichotomy, and he spoke from direct experience. During the first World War he had supervised the production of optical-quality glass for the U.S. Army, work which resulted in a number of widely cited scientific papers.[58] He was also an inventor who held several patents: among his accomplishments he counted a device for distinguishing between cultured and natural pearls and a method to decrease the "scratching proclivities" of scouring powder.[59] (Nor did his colleagues dismiss these practical applications; Wright's memorial in the *American Mineralogist* emphasized the importance of his theoretical work on the ternary system $CaO\text{-}Al_2O_3\text{-}SiO_2$ in the development of Portland cement.) Reflecting on his own experience integrating theory and application, Wright concluded that "these two objects are intimately linked and actually represent different phases of the same problem." If any American scientific institution might have pursued abstract knowledge in disregard of practical uses, it should have been the Carnegie, but Wright argued that the institution would suffer if it did, because the life of this "semi-public" institution was inescapably linked to public support of it.[60]

Merriam accepted Wright's position and circulated the paper widely. Without public support, Merriam agreed, great men would choose to work elsewhere, and cooperation with university departments and government scientific agencies would be impossible. The purpose of the CIW was no different from Andrew Carnegie's other endowments: all were dedicated to "progress of the human race." The task of the CIW scientist, therefore, was two-faceted: "a] to contribute to progress in science through research of the highest quality; and b] to interpret the results of research in terms of their value to mankind."[61] Interpretation, in this context, meant use. Research and practice were synergistic, not competing.

Patrons and Practice

The strength of American theoretical geology and the commitment of American geologists to the unity of theory and practice are linked to the strong institutional support that geology received from American governmental agencies.[62] A network of interrelated yet independent institutions supported American geology in the nineteenth century: the U.S. Coast and Geodetic Survey, the U.S. Geological Survey (USGS) and its precursors, the Smithsonian Institution, and the various state surveys, which supported such luminaries as James Hall (New York) and T. C. Chamberlin (Wisconsin). Steady employment made it possible for men like Hall and Chamberlin to dedicate their lives to uninterrupted scientific work without the enabling personal fortunes held by many British colleagues or the prestigious academic chairs held by French and German colleagues.

At the same time, these sources of American geological employment—governmental agencies with practical mandates—necessarily had an impact on the work that Americans did and the way they did it. This distinctive pattern of patronage fostered a distinctive blend of theory and practice.[63] But, as Wright's comments demonstrate, the intermingling of practical and conceptual work continued in the twentieth century even with the founding of the Carnegie Institution and research programs at universities. One reason is obvious: the founders of the Carnegie had nearly all served in some capacity in governmental scientific service.[64] The same was true of professors in the leading geology departments: at Stanford, Professor J. C. Branner had worked first with the Arkansas Survey; at the University of Wisconsin, Charles van Hise had worked for that state's survey and then for the USGS. Charles Schuchert had been an apprentice to James Hall in New York and worked at the U.S. National Museum in Washington before becoming a professor at Yale. Reginald Daly, at Harvard, had been a field geologist for the Canadian Department of Mines on the survey of the 49th parallel, while his future Stanford colleague, Bailey Willis, worked the other side of the border.[65]

Even as research was increasingly done in private universities in the 1920s and 1930s, much of the available financial support still came from federal agencies. For a number of years after leaving the USGS, Chamberlin received funds from it; other academic geologists, such as Chester Longwell and Eleanora Bliss Knopf at Yale, pursued summer field studies under the Survey's "while actually employed" (WAE) program. Others, particularly geophysicists, undertook summer work on behalf of petroleum companies. But if the leaders of American geology believed that theory and practice were two sides of the same coin—that it did no good to argue for one without the other—then why have so many historians and later geologists developed the impression that they were somehow opposed to theory? What was it about American geologists that seemed to suggest that they were against theorizing even while they were doing it?

An Antiauthoritarian Logic of Discovery

The private letters of American earth scientists at the start of the twentieth century are filled with discussions of interpretive issues. However, they are also filled with copious details about particular geological problems: particular stratigraphic units, particular fossils—in short, particularities of every kind. The diligent collection of observational data and the development of theoretical ideas comprised their daily scientific lives. But what was the relationship between them? Did theory guide observation or vice versa? On this point, many Americans expressed themselves explicitly: they rejected theory-driven science and subscribed to an antiauthoritarian logic of discovery.

American geologists were suspicious of privileged theoretical systems that claimed universal applicability, and of the individuals who expounded them. Such behavior was considered inimical to scientific discovery. The most famous—nay, infamous—example of unhealthy theoretical monism was the "neptunist" program of the eighteenth century, which was frequently cited as a cautionary parable. Neptunism, associated with the German mineralogist Abraham Werner and his followers, was a theoretical framework in which rock strata were interpreted as the sedimentary deposits from a primeval, universal sea. The theory had many conceptual strengths: it highlighted the role of geological succession as evidence of earth history, it broadened the recognition that the age of rocks could be gleaned from their relative positions, and it laid the groundwork for regional correlation. It was internally consistent, it was simple, and it had tremendous explanatory power.[66] But it was also incomplete, and at the start of the nineteenth century, a rancorous debate developed between the followers of Werner and those of James Hutton and John Playfair. Hutton and Playfair rejected the idea of a universal ocean and emphasized instead the role of the earth's internal heat in the formation of crustal rocks. The resulting dispute was immortalized by geologists as the "neptunist–plutonist" debate.

The conflict prevailed primarily in Great Britain, where there were strong advocates on both sides, and particularly in Hutton's own Edinburgh, famous for its "greenstones"—rocks that appeared not to have formed by sedimentary processes but by intruding as melts into existing sedimentary layers. Americans, however, were soon drawn in, as New England geologists like Edward Hitchcock found themselves mapping greenstones within a few miles of the site of Benjamin Silliman's new geology program at Yale.[67] Silliman himself had been present in Edinburgh in 1805 at the peak of the debate, having been sent overseas by Yale University President Timothy Dwight to learn natural science in preparation for becoming Yale's first professor of geology. This set the stage for Silliman's training of the first generation of American geologists. As historian David Spanagel has put it, "What [Silliman] brought home from this experience served as the starting point for all subsequent American theoretical discourse in geology."[68]

Silliman found relief from the heat of controversy "in the cold bath of the Wernerian ocean," but, in time, Americans tilted toward the plutonist school, which ultimately "won" the debate.[69] Writing in the *American Journal of Science* in 1823, Hitchcock asked rhetorically:

> Does the greenstone of Connecticut afford evidence in favour of the Wernerian or of the Huttonian theory of its origin? Averse as I feel to taking a side in this controversy, I cannot but say, that the man who maintains . . . the original hypothesis of Werner . . . will find it in his interest not to . . . examine the greenstone of this region, lest . . . he should be compelled, not only to abandon his theory, but to write a book against it.[70]

As the neptunist school broke down, Werner's disciples were faulted in retrospect for an excessively general and excessively theory-driven interpretation. Geologists in both the United States and Great Britain reacted against deductive methodology, which by valorizing theory seemed to have contributed to the neptunist–plutonist impasse. But Americans in particular seemed to wish to distance themselves from the methods and mistakes of European geology. Writing retrospectively in the centenary volume of the *American Journal of Science*, Joseph Barrell, professor of geology at Yale and a widely admired scholar, wrote:

> The Plutonists, although time has shown them to have been correct in all essential particulars, were for a generation submerged under the propaganda carried forward by the disciples of Werner. . . . The authority of the Wernerian autocracy caused its nomenclature to be adopted in the new world [even while] strong evidence against its interpretations was to be found in the actual structural relations displayed by the igneous rocks.[71]

In Barrell's view, monistic theoretical systems were no better than propaganda; Werner's system had led to the persistence of inappropriate terminology that hindered the interpretation of American geology. Even more offensively, the neptunists had regarded "the order of superposition and the consequent sequence of age . . . as settled by Werner in Germany and not requiring investigation in America."[72] To Barrell, this was both insulting and false: even the most cursory examination of American successions showed them to be different from European ones, while the structural issues were radically different. American geologists were determined to develop their own science on their own ground—literally and metaphorically—and a starting point was the rejection of European theoretical "autocracy."[73]

Even when Americans credited the lasting accomplishments of European scientists, they still cautioned against excessive respect for European authority. Robert Woodward, reflecting on the history of mathematical theories of the earth at the end of the nineteenth century, concluded that the greatest impediment to scientific progress was the "contentment with . . . the completeness of knowledge of the earth" that followed the work of individual geniuses like Newton and Laplace. To be sure, these men had made enormous contributions, but their work had been followed, according to Woodward, by periods of stagnation caused by excessive acceptance of their scientific authority.[74] A similar view was expressed some years later by Bailey Willis, writing home to his wife after visiting the great Alpine geologist Pierre Termier, Director of Geological Mapping, in Paris. "Termier is an *authority*," Willis underscored, "[and his] young geologists . . . cannot decline to accept the theory of 'grandes nappes.'"[75]

If Europeans subscribed to unhealthy scientific authoritarianism and autocracy, then the American alternative would be scientific egalitarianism and democracy. But how did one achieve scientific egalitarianism? A starting point, at least rhetorically, was to deny that one was proposing a final solution to the issues under discussion—or perhaps to deny that one was even taking a stand at all. Thus Hitchcock concluded his report on the New England greenstones in the subjunctive: "And were I to adopt any hypothesis in regard to the origin of our greenstone," he wrote counterfactually, "it would not be much different from" that proposed by the Huttonian school.[76]

By the late nineteenth century, a rhetorical posture of theoretical nonpartisanship was entrenched in American geological writing. In the *Geology of the High Plateaus of Utah*, Clarence Dutton—who had earlier praised John Wesley Powell for his skill in organizing facts into meaning—claimed that he personally "doubted the propriety of embodying in a work devoted to the statement of observed facts any views of a speculative nature." Dutton stated that his ideas on the origins of vulcanism were merely a "trial hypothesis" (albeit one that had received support and encouragement from colleagues), and he assured his readers that he was far from taking a final stand.[77] "It may be," he wrote, "that the ultimate cause of vulcanism will eventually be traced to the shifting of vast loads of matter from place to place on the earth's surface, but at present this subject has not been investigated . . . with sufficient method and system to admit of safe generalization."[78]

T. C. Chamberlin summarized American views in the title of his 1919 article, "Investigation vs. Propagandism." Good scientists were investigators, not propagandists, and their rhetoric reflected this belief. In his eulogy to Chamberlin, Bailey Willis took pains to point out that Chamberlin's solution to the problem of glaciation was "not final, [but] does take its place, as he would wish it to, among the working hypotheses that may be reasonably considered."[79]

Theoretical Pluralism: The Ideal of Multiple Working Hypotheses

In calling Chamberlin's glacial theory a "working hypothesis," Willis was referring to Chamberlin's own philosophical program—the most famous American articulation of an antiauthoritarian logic of discovery—the method of multiple working hypotheses. This method, as first advanced by G. K. Gilbert but named and widely promoted from the turn of the century onward by T. C. Chamberlin, consisted of the attempt to weigh observational facts in light of competing explanatory options. The method was an explicit prescription: the geologist going out into the field was to approach the problem of discovery not through the lens of an established theoretical viewpoint but through a prism of conceptual receptivity that refracted multiple explanatory options.

The appropriate stance was not one of testing one hypothesis but of weighing various hypotheses.[80]

Chamberlin likened the scientist's struggle for impartiality to the parental striving to divide love fairly among offspring; the goal of the method was to find a way to evaluate competing hypotheses on equal terms.

> In developing the multiple hypotheses, the effort is to bring up into view every rational explanation of the phenomenon in hand and to develop every tenable hypothesis relative to its nature, cause, or origin, and to give to all of these as impartially as possible a working form and a due place in the investigation. The investigator thus becomes the parent of a family of hypotheses; and by his parental relations to all is morally forbidden to fasten his affection unduly upon any one.[81]

A good scientist was like a good parent: fair and equitable. For the parent, fairness was a goal unto itself; for the scientist, it was the means to the end of intellectual receptivity. Of course, scientists do feel protective toward their own favored hypotheses—just as parents feel protective toward their offspring—but this impulse is diminished when the scientific "children" are raised by the whole village. Most important, one is freed to entertain the possibility that both sides in a debate may be partly right. Chamberlin elaborated:

> Where some of the hypotheses have been already proposed and used, while others are the investigator's own creation, a natural difficulty arises, but the right use of the method requires the impartial adoption of all alike into the working family. The investigator thus at the outset puts himself in cordial sympathy and in parental relations (of adoption, if not of authorship) with every hypothesis that is at all applicable to the case under investigation. Having thus neutralized so far as may be the partialities of his emotional nature, he proceeds with a certain natural and enforced erectness of mental attitude to the inquiry, knowing well that some of his intellectual children (by birth or adoption) must needs perish before maturity, but yet with the hope that several of them may survive the ordeal of crucial research, since it often proves in the end that several agencies were conjoined in the production of the phenomena. Honors must often be divided among hypotheses.[82]

The possibility of multiple causes was especially pertinent for geology, because there had been so many debates in which both sides were right in what they affirmed and wrong in what they denied. To Chamberlin's mind this was no accident: geological processes were rarely "unitary phenomenon explicable by a simple single cause."[83] The neptunist–plutonist debate was a case in point: that impasse arose at least in part because the neptunist program was construed as a unique explanation of petrogenesis rather than one cause among many. The method of multiple working hypotheses thus made a virtue out of reality: geologists had always had a proliferation of explanations for phenomena, but this could now be viewed as an accomplishment rather than a failure. Chamberlin himself had shown that the Quaternary glacial drift deposits of North America had multiple sources, and multiple factors had conjoined to make the Great Lakes. It was a mistake to assume uniqueness in explanation; the method of multiple hypotheses helped keep this insight in focus.[84]

Chamberlin realized that no scientist could achieve pure theoretical egalitarianism any more than a parent could achieve pure affective equality; his method was an ideal, a "principle" toward which one should strive. In essence, it was a recognition of the philosophical problem of underdetermination—to which the great French scientist and philosopher Pierre Duhem was turning his attention at about this time—but Chamberlin's concern was less with metaphysical issues than with the practical problem of fostering scientific discovery.

Training the Next Generation

As Barrell's comments suggest, a central and persistent question for American geologists was how to develop an independent science, not derivative from European sources, but uniquely grounded in their own experience. For Chamberlin, the issue was reified by the task of developing a graduate geology program at the University of Chicago. The question, as posed by Chamberlin, was how to foster mental independence among his charges, so that the work of the next generation would "be individual and independent, not the mere following of previous lines of thought ending in predetermined result." The choice of science as a topic of study was a start, because individual thought was most readily fostered in subjects where "much . . . remains to be learned." (Geology satisfied this criterion particularly well; it was replete with such topics and indeed presented "few of any other class.")[85] But an appropriate methodology was needed as well.

Chamberlin approached the problem of discovery in the same manner as he approached the study of the earth—historically—and argued that intellectual development could be viewed as a social and cultural progression. In the distant past, when knowledge was very limited, it was possible for a single man to bring within his compass all that was known about a discipline. Under such circumstances, there "grew up an expectancy on the part of the multitude that the wise and the learned would explain whatever new thing presented itself," and up to a point this expectation was reasonable. But as more knowledge came to the fore, it became increasingly unreasonable, yet the wise man felt compelled to offer explanations nonetheless: "Pride and ambition on the one side and expectancy on the other joined hands in developing the putative all-wise man whose knowledge boxed the compass and whose acumen found an explanation for every new puzzle which presented itself." Although no one would be so foolish as to claim in 1897 that he knew everything there was to know, certain bad habits—certain unhealthy "intellectual predilections"—remained as vestiges of this earlier evolutionary condition. The worst of these was the habit of "precipitate explanation," which devolves into ruling theory. Chamberlin explained:

> As in the earlier days, so still, it is a too frequent habit to hastily conjure up an explanation for every new phenomenon that presents itself. . . . The habit of precipitate explanation leads rapidly on to the birth of general theories. When once an explanation or special theory has been offered for a given phenomenon, self-consistency prompts to the offering of the same explanation or theory for like phenomena when they present themselves and there is soon developed a general theory explanatory of a large class of phenomena similar to the original one. In support of the general theory there may not be any further evidence or investigation than was involved in the first hasty conclusion. But the repetition of its application to new phenomena, though of the same kind, leads the mind insidiously into the delusion that the theory has been strengthened by additional facts.[86]

Ruling theory was a self-deception which, as Barrell argued in the case of neptunism, tended to become self-perpetuating. Each application of the theory was taken as ratification of it, irrespective of whether other explanations might better account for the facts. Some scientists had recognized this problem and would claim to hold their theories only provisionally, but this was flimsy camouflage. Once any theory is held in a preferred position, Chamberlin argued,

> there is the imminent danger of an unconscious selection and a magnifying of phenomena that fall into harmony with the theory and support it, and an unconscious neglect of phenomena that fail of coincidence. The mind lingers with pleasure upon the facts that fall happily into the embrace of the theory, and feels a natural coldness toward those that assume a refractory attitude. Instinctively, there is a special searching-out of phenomena that support it. . . . The mind rapidly degenerates into the partiality of paternalism. The search for facts, the observation of phenomena and their interpretation, are all dominated by affection for the favored theory until it appears to its author or its advocate to have been overwhelmingly established. . . . A premature explanation passes first into a tentative theory, then into an adopted theory, and lastly into a ruling theory.[87]

Chamberlin did not name names, but those of Werner, Elie de Beaumont, von Buch, and Suess would have come readily to his American readers' minds. All-encompassing theoretical systems were considered a characteristic of continental European geology; Germans in particular were associated with Goethean theories of the *Weltall*.[88] But if continental theorists had gotten stuck in phase one of intellectual evolution—ruling theory—the British had become stuck in phase two: colorless observation. In response to the excesses of ruling theory, some had responded with an excessive empiricism characterized by repression of all explanatory attempts. As far as Chamberlin was concerned, this cure was worse than the disease. "The advocates of reform," he wrote,

> insisted that theorizing should be restrained and the simple determination of facts should take its place. . . . Because theorizing in narrow lines had led to manifest evils, theorizing was to be condemned. The reformation urged was not the proper control and utilization of theoretical effort but its suppression. We do not need to go backward more than a very few decades to find ourselves in the midst of this attempted reformation.[89]

Chamberlin was of course referring to Charles Lyell, who in the opening chapters of his *Principles of Geology* represented observation and theorizing as contradictory activities and castigated the latter as one of the "prejudices" that had "retarded the progress of geology."[90] Lyell's history was more rhetorical than historical: as several historians have pointed out, Lyell vilified specific theories he disapproved of while praising as "fair induction" others he wished to advance, not least of all his own.[91] Nevertheless, Lyell's acerbic contentions about high theory were shared by many of his British colleagues, who, partly in response to the neptunist–plutonist debate, had made enumerative induction—simple generalizations of data—the official methodology of the Geological Society of London.[92] In 1897, the year that Chamberlin's article on multiple working hypotheses was published in the *Journal of Geology*, the British geologist Archibald Geikie published his *Founders of Geology* in which he continued to promote the rhetoric of a strictly empirical geology.[93] But the point was not to excise theory but to harness it. "The vitality of study quickly disappears when the object sought is a mere collocation of unmeaning facts," Chamberlin concluded.[94] The exaltation of fact was no more desirable than its suppression.

The third phase of human intellectual history was, predictably, Chamberlin's own time. From the thesis of ruling theory and the antithesis of pure observation there emerged a new synthesis: the method of multiple working hypotheses. Its goal was to navigate between the risks of dogmatic deductivism and the infertility of naive inductivism. Like Lyell, Chamberlin was a historical geologist, not a historian, and his characterization of each side was didactic and exaggerated for effect. Actually, continental geologists had made enormous empirical contributions and British geologists had made many theoretical ones. Like many rhetoricians, Chamberlin constructed his opponents out of straw, but they served to make his point nonetheless: the ways of the old world had run their course. Out of the new world, a new methodology had arisen that pointed the way out of the unproductive dichotomy of blinkered theorists on the one hand and dry catalogers on the other.

A Methodological Ideal and a Rhetorical Reality

To what extent did American geologists subscribe to Chamberlin's ideal? The available evidence suggests they did so to a remarkable degree.[95] Chamberlin's ideas were widely cited and republished for many years to come. His original formulation was published in *Science* in 1890 and republished in the *Journal of Geology* in 1897, and closely related articles appeared in *Popular Science Monthly* (1905) and the *Journal of Geology* (1919). His lasting impact is evidenced by the republication of his 1890 piece in the *Journal of Geology* in 1931, in *Scientific Monthly* in 1944, and again in *Science* in 1965.[96] And although Chamberlin was most widely known in his lifetime as a glaciologist and originator of the planetesimal hypothesis, his enduring legacy was his methodological precept. After his death, the method of multiple working hypotheses appeared in many introductory textbooks and was taught in many American classrooms as *the* method of geology; as recently as 1990, geologists were still wrangling over its rightful role in investigation.[97]

Furthermore, as already mentioned, Chamberlin did not invent the method; he articulated and named what was already being advocated and practiced by others, most notably by G. K. Gilbert.[98] As historian Stephen Pyne has pointed out, Gilbert was unlike Chamberlin in many ways: he was inclined more toward mathematics and mechanics than toward fossils and formations, yet like Chamberlin he had experience as a survey geologist and was concerned with how to organize the massive amounts of uncontrolled information one encountered doing fieldwork in uncharted territory.

Gilbert had pondered deeply what specific methods, if any, scientists rely on. He concluded that observation without inference is impossible, but scientists strive to separate the two as far as humanly possible:

> Scientific observation is not sharply distinguished from other observation. It may even be doubted whether there is such a thing as unscientific observation. If there is a valid distinction, it probably rests on the following. . . . [In] scientific observation . . . the investigator endeavors to discriminate the phenomena observed from the observer's inference in regard to them, and to record the phenomena pure and simple. I say "endeavors," for in my judgment he does not ordinarily succeed. His

failure is primarily due to subjective conditions; perception and inference are so intimately associated that a body of inferences has become incorporated in the constitution of the mind. And the record of an untainted fact is obstructed not only directly through the constitution of the mind, but indirectly through the constitution of language, the creature and imitator of the mind. But while the investigator does not succeed in his effort to obtain pure facts, his effort creates a tendency, and that tendency gives scientific observation and its record a distinctive character.[99]

If the first characteristic of science was the attempt to discriminate between observation and inference, the second was the grouping of observations toward the goal of "understanding [the] deeper relations which constitute the order of things." Gilbert explained:

It is the province of research to discover the antecedent of phenomena. This is done by the aid of hypothesis. A phenomenon having been observed, or a group of phenomena having been established by empiric[al] classification, the investigator invents an hypothesis in explanation. He then devises and applies a test of the validity of the hypothesis. If it does not stand the test he discards it and invents a new one. If it survives the test, he proceeds at once to devise a second test. And thus he continues.[100]

Gilbert thus articulated the hypothetico-deductive method of science. But he also pointed out that the investigator "is not restricted to the employment of one hypothesis at a time." Like Chamberlin, Gilbert saw benefit in entertaining several hypotheses at once, "for then it is possible to discover their mutual antagonisms and inconsistencies, and to devise crucial tests—tests which will necessarily debar some of the hypotheses from further consideration."[101] The best geologist was not the one who had seen the most rocks, but the one who was "rich in hypotheses, [for i]n the plenitude of his wealth, he can spare the weakling without regret."[102]

Gilbert believed that hypotheses were developed primarily by analogical reasoning, and he found an analogy for his own methodological view in surveying. As a survey scientist, Gilbert often had to locate himself in unknown terrain. This was normally done by triangulation—taking three (or more) bearings on known points to circumscribe the unknown location. The scientist, Gilbert argued, was like the surveyor who makes measurement to inaccessible points by means of bearings from different sides: "Every independent bearing on the Earth's [problems] is a check on other bearings, and it is through the study of discrepancies that we are to discover the refraction by which our lines of sight are warped and twisted."[103] Gilbert's survey colleague, Charles van Hise, made a similar point in 1904: "If only the human mind were sufficiently powerful, and means of expression adequate, the ideal method of treatment would be simultaneous consideration and exposition of all possible points of view."[104]

Fourteen years later, as continental drift was about to be discussed in America, Joseph Barrell complained that Americans had not achieved this multiplicity of perspective but had focused instead on the ever "larger accumulation of observed facts." Barrell urged his colleagues to look beyond the details of their local field areas. "Important though these details are from the standpoint of the working geologist, their study is comparable to the close study of the technique of a painting whose motive and artistic effects are visible only from a distance. It is the larger conception which determines the expression of the details."[105] The problem, of course, was that not all geologists were as full of ideas as Gilbert, Chamberlin, and Barrell wished them to be.

But armed with multiple hypotheses or not, Americans took the first step in Chamberlin and Gilbert's method: the presentation of facts.

An Inductive Logic of Justification . . .

If deductive theoretical systems reduced to a species of prejudice, the alternative was an inductive logic of justification. Equitable consideration of explanatory frameworks necessarily presupposed foreknowledge of the facts waiting to be explained; this was what "ruling theorists" forgot. In theory-driven science, Chamberlin argued,

> interpretation leaves its proper place at the end of the intellectual procession and rushes to the forefront. . . . It is an almost certain source of confusion and error when it runs before serious inquiry into the phenomenon itself. A strenuous effort to find out precisely what the phenomenon really is should take the lead and crowd back the question, commendable at a later stage, "How came this to be so?" First, the full facts, then the interpretation thereof.[106]

One cannot gauge the extent to which American geologists entertained competing interpretations in their daily labors, but one can see that they followed Chamberlin's facts-first prescription in their published work. In American geological literature of the late nineteenth and early twentieth centuries, observations were presented first and were clearly delineated from the interpretive frameworks that followed. And this format was followed even by the most highly theoretically inclined among them. A striking example is Clarence Dutton's *Geology of the High Plateaus of Utah*. The first third of this 300-page treatise consists of detailed descriptions of structural and volcanic features. In each of the first four chapters, a discussion of the origin of the features described is saved for the end of the chapter—or not offered at all. Only after this copious documentation did Dutton pause for a chapter entitled "Speculations on the Causes of Volcanic Action," before embarking on eight more chapters of geological accounting. Another example is Gilbert's *Lake Bonneville*, whose table of contents outlines 141 pages of descriptive text before the word "hypothesis" appears. There, Gilbert paused for a presentation of evidence that might rule out certain interpretations, but his preferred account—isostatic rebound after desiccation—was saved for the penultimate chapter, 200 pages later.

A notable feature of both these works is their descriptive titles, which give little suggestion of the major theoretical questions discussed within. Compare these with De la Beche's *Researches in Theoretical Geology* (1834), Heim's *Untersuchungen über den Mechanismus der Gebirgsbildung* (1878), or Suess's *Die Entstehung der Alpen* (1875). Indeed, the word *Entstehung*—origin—appears like a mantra in the titles of German geological monographs of the late nineteenth and early twentieth centuries. Even Lyell's famous *Principles of Geology* suggested that geology *had* principles, unlike American monographs whose titles might seem to suggest that geology was a catalog of sites.

Perhaps the most striking demonstration of titular circumlocution emerged from the pen of Bailey Willis. Willis was known to his colleagues as a "lover of theory" for his extensive discourse on theoretical issues; the introduction to his most famous monograph cited the influences of "the two great leaders in the modern advance of geological philosophy, Chamberlin and Suess," from whom he had "derived the highest inspiration."[107] Willis was known for his booming voice and strong opinions, yet this lively man's theoretically inspired monograph was blandly entitled *Research in China, Volume 2 (1907)*. The title was given to Willis by his sponsors at the Carnegie Institution—there were other volumes by other authors—but Willis's own sub-title, *Systematic Geology*, was hardly more revealing, except insofar as it reflected its systematically inductive structure. Beginning with the Archean and working up through the geological time scale, Willis recounted the distribution and sequences of the rocks of China. His theoretical discussion was reserved for the final chapter. There, he claimed that the facts were "inadequate to a satisfactory discussion"—that is, inadequate to compel a unique causal interpretation—so he offered the "alternative . . . of drawing inferences which are sometimes little more than suggestions." Protestations notwithstanding, Willis dedicated the final eighteen pages of his monograph to developing these inferences.

Willis's principal inference—or hypothesis as he then called it—was that the continent of Asia was composed of independent structural pieces, or elements, that moved vertically relative to one another. Certain elements were areas of sustained or recurrent uplift, others of recurrent subsidence, and could be referred to as positive and negative elements (figure 5.1). Between them lay the zones of continental flexure. There was also evidence of horizontal dislocation at the edges of the continents. Following Hayford and Chamberlin, Willis attributed this to a suboceanic "spread" of material flowing from the denser suboceanic areas to the less dense subcontinental areas. "The general result," Willis wrote, "is plastic flow in the rigid and solid rock masses. [If] such flow is a persistent condition . . . we may ascribe [to it] those accumulated stresses that have sufficed to produce the recurrent pronounced effects of diastrophism."[108] Willis concluded that all these effects were cyclical: geological history was characterized by short intervals of orogeny interrupting more extended periods of stability. Stable periods would be characterized by peneplanation of the landscape and transgression of the oceans to form extended epicontinental seas; the result would be a rather uniform earth which, among other things, could account for the uniformity in fossil fauna so well described by Suess.

Willis's "hypothesis" was a well-developed concept that explained many of the observations Wegener was soon to address and drew on similar assumptions about the rheology of the subcrust. But Willis distanced himself from his beliefs and closed his monograph with a rhetorical disavowal of commitment:

> In closing, it is well not to lose sight of the fact that this chapter is one of hypothesis. The analysis of Asia into continental elements, the interpretation of successive geographic conditions, the attempt to explain the arrangement of the [fold axes], and the statement of a source of tangential thrust are contributions to theory. They are, so far as I can now judge, consistent with the known facts, but they must stand or fall quantitatively and qualitatively by the test of the great host of facts which are yet to be gathered . . .[109]

Willis's inductive format and provisional rhetoric were variations on the themes discussed by Gilbert and Chamberlin. The rhetorical alternative to dogmatic theory was a stance of provisionalism, receptivity, and open-mindedness. The facts-first structure of argumentation implied that the theoretical conclusions had emerged freely as inductive generalizations from the mass of available facts; the more facts there were, the more evident it would be that the generalizations offered were not precipitate.

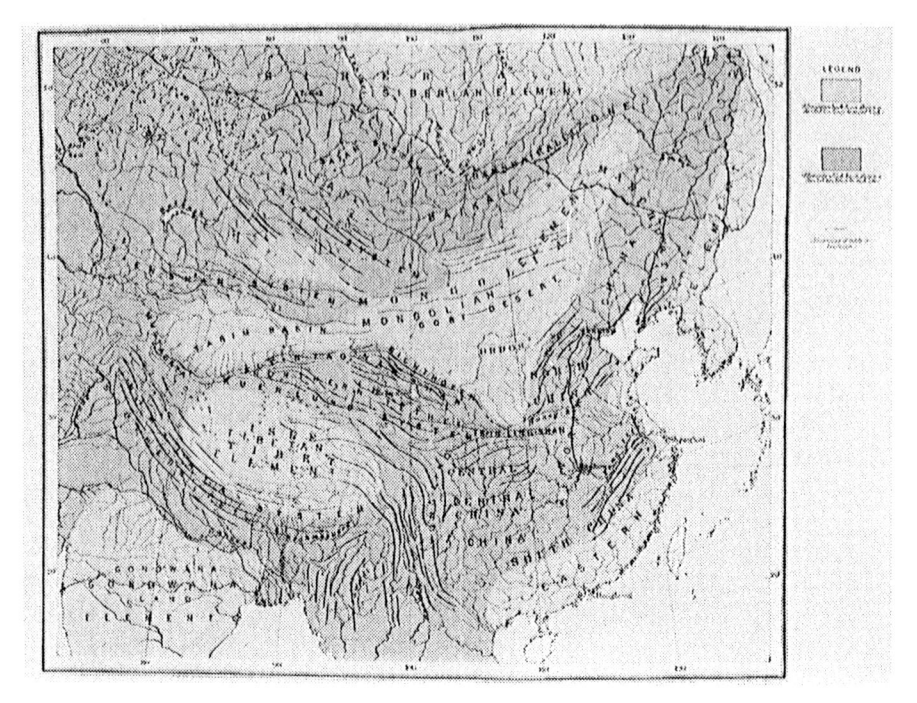

Figure 5.1. Willis's elements of continental structure. In the early twentieth century, Bailey Willis proposed a model of vertical tectonics involving independent motion of structural pieces, or elements, based on the tectonics of Asia. "Positive" elements (light grey) experienced sustained or recurrent uplift, "negative" ones recurrent subsidence (dark grey). Between them lay the zones of continental flexure. In addition, suboceanic "spread" of material flowing from the denser suboceanic areas to the less dense subcontinental area caused horizontal dislocation at the edges of the continents, and all these effects were cyclical: geological history was characterized by short intervals of orogeny interrupting more extended periods of stability. (From Willis 1907, p. 122; originally published by the Carnegie Institution of Washington.)

. . . and a Form of Scientific Practice

American earth scientists were empiricists: they believed that reliable knowledge is grounded in observation. Facts necessarily came first; theory came later (and admittedly for some geologists, sometimes not at all). If historians have concluded that American geologists were against theory, this may be why. But American geologists were not reflexively anti-theory, nor were they naive. Gilbert understood that "perception and inference are . . . intimately associated"; Chamberlin's program was an attempt to obviate the "unconscious selection . . . of phenomena that fall into harmony" with theoretical commitment and "unconscious neglect of phenomena that fail of coincidence"; Barrell argued that the "larger conception . . . determines the expression of the details." For each of these men — all leaders in North American geology — the point was to maintain a useful self-consciousness of the difficulties involved, to err neither one way nor the other. Their methodological stance was an attempt to grapple with the fundamental question of the relationship between fact and theory.

Contemporary philosophy of science takes the goal of separation of perception from conception to be illusory. In recent years, it has become a philosophical truism that observations are "theory-laden," leading many people to dismiss inductive science as either sophomoric or disingenuous. If all observations are theory-laden, then science *cannot* be truly inductive. From this viewpoint, the inductive format of American theoretical geological writing would be a rhetorical charade at best, a form of stupidity at worst. But there is another way to view the situation: that the segregation of observation from inference was a form of scientific practice originating in field experience.

The laboratory scientist necessarily creates an experiment with something in mind, but most earth scientists at the turn of the century were field-workers.[110] These men (and rare women) went out into the field with sets of classification schemes and explanatory resources that embodied theoretical presuppositions, to be sure, but once there they frequently encountered phenomena for which they did not have established conceptual frameworks.[111] Theory existed on a variety of levels; field observations necessarily invoked it on local and analytical but not necessarily synthetic levels. One *could* describe and classify a sequence of rocks or a pattern of deformation while deferring an account of the events that produced it. As historian Rachel Laudan has argued—and as already discussed in chapter 3—causal theory could be set apart and often was set apart, for both pragmatic and epistemic reasons.[112]

Gilbert's experience with the dry beds of the former Lake Bonneville is a good example. His initial contribution was the recognition and documentation of a previously unknown pattern of uplift; this did not require a belief in the existence of isostasy, nor even knowledge of the concept. If scientists like Gilbert and Chamberlin offered explanatory accounts, it was because they wanted to, not because they had to. Furthermore, as survey scientists, which Gilbert and Chamberlin and so many of their North American colleagues were for some or all of their careers, the inductive format of data-gathering and description was almost inescapable. The point of these surveys was to *survey*—to examine, appraise, and generally take a comprehensive view of the geological phenomena encountered. Only after such appraisal did one know what was in need of explication. Thus when Chamberlin argued, "first, the full facts, then the interpretation thereof," he was offering both a prescription and, to some extent, a description.

The segregation of observation from interpretation in practice is clearly evident in the field notebooks of Reginald Daly. On his 1922 South Africa trip, Daly kept detailed records of the rocks, minerals, fossils, and formations he observed, but only on the even-numbered pages of his books. The otherwise blank odd-numbered pages were reserved for intermittent theoretical "break-outs," where Daly discussed interpretive issues (figure 5.2). Typically, Daly's break-out notes take the form of a discussion with himself. For example, after a visit to the Drakensburg lava flows, Daly pondered the origin of the pipe amygdules—long, thin tubes at the bases of the flows. How do these form, he wondered, and why are they restricted to the bottoms of the flows? What do they tell us about the origin and emplacement of magmas? Daly wrote on the odd-numbered page facing his field observations: "Try hypothesis that pipe vesicles merely rep[resent] primary magmatic gas, extra rich, trapped at lower contact. Similar vesicles escaped to air in overlying portion of flow." Lower on the same page he added, "[but then] why no pipes in Hawaiian pahoehoe flows? See if any observed . . ."[113] Daly's

My hypo: that pipe vesicles merely rep. primary magmatic gas, extra rich, trapped at lower contact. Similar vesicles escaped to air in overlying portions of flow.

Horizontal layers of vesicles due to rise (later) in lowest 1"± chilled phase as latter stilled.

Magma must have been gas-rich ∴ so fluid.

Why no pipes in Hawaiian pahoehoe flows? See if any observed.

be completely surrounded by non-vesicular or little-vesic. rock, as if rising en masse & so trapped by viscosity.

Make expts with Canada balsam & pipette (Du Toit)

Note – du T. describes enstatite andesites in central eruptions connected with Drakensberg series.

Horse to Witzieshoek Hotel, Bar. 5865' = true elevation.

See carious weathering of Cave ss = 2" to 4' basins with raised rims – Silica dissolved by pool water within – re-deposited in ss of rim (Du Toit) – Wright photos these.

See a 200'-deep valley in Cave ss filled with several β flows.

Gen. Concl'n – Time-separated flows quite clear in lowest 500' of Drak'berg series & prob. at 11761 beacon, down 100' or 200'.

But between apparently mono-lithic flows 200 to poss. 1000'+ thick – Wright & Du Toit doubt flows ever over 200' or so thick.

Clearly rapid upbuilding of Drak. series except at very bottom, where ss interbeds.

No aa lava in Drakensberg!

– See vert. dikes cutting Cave ss.

Heavy cloud-mist on Drak. all day. Lucky to have had 3 perfect days. High W wind on "berg" yesterday!

interpretive reactions were not recorded as part of the flow of observational evidence and correlations deduced from them; they were placed on the opposing page, as if theory and observation were engaged in a dialogue, perhaps even a debate.

Daly's phrase, "try hypothesis," which recurs throughout his notebooks, suggests an iterative conceptual process along the lines outlined by Gilbert and Chamberlin. And Daly did indeed follow Chamberlin's method. After visiting the volcanics, the group moved on to the Drakensburg scarp, a huge physiographic discordance at the edge of the Karroo Basin. Daly wrote:

Cause?
1) initial axis of Karroo basin? Du Toit thinks this . . .
2) displacement of Drak[ensburg] magmas a la Bushveld lacc[olith].
3) elastic rise of surface in eroded area
4) isostatic rise of ditto
5) sinking of original volc. plateau[114]

Here was Chamberlin's method of multiple working hypotheses in practice.[115]

But Daly's comments on the scarp appear on an *even*-numbered page—seemingly contradicting the pattern of segregating causal explanation from phenomenological account. However, there is a discernible logic at work here: the origin of the Drakensburg scarp was a site-specific problem. Its origins may or may not have had any bearing on the origins of other scarps in other places. In contrast, the origin of pipe amygdules was a general question, related to the composition of magmas and their mechanisms of emplacement. Daly seems to have been distinguishing between the interpretation of specific geological features and the resolution of general theoretical questions. The origin of the Drakensburg scarp was a matter of direct inference, akin to the origin of the thick accumulations of sediments in the Appalachian geosyncline. In contrast, the mechanism of magma emplacement was beyond direct observation, akin to the cause of geosynclinal uplift.

The Drakensburg discussion was nevertheless demarcated from the general flow of observation with the word "Cause?" Elsewhere such discourse was labeled "Discussion" or "Conclusion." Daly's habits of recording show that both small and large-scale inference were segregated from direct observation, but the latter warranted more separation than the former. In short, the bigger the inference, the bigger the physical segregation. One can think of this as separating levels of theory, or as separating the accessible from the inaccessible, and it was surely both of these things, but it was something more as well: it was a separation of *work* from *thought*, of practice from theory. The structure Daly followed is most consistently interpreted as an attempt to separate

---

Figure 5.2. *(facing page)* Reginald Daly physically segregated observation from theoretical inference in his field notebooks. On the right hand side, in a page from his South Africa trip in 1922, he describes the Drakensburg lava flows; on the left he proposes a hypothesis about the origin of pipe vesicles within them: "Try [the] hypo[thesi]s that pipe vesicles merely rep[resent] primary magmatic gas, extra rich, trapped at lower contact. Similar vesicles escaped to air in overlying portions of [the] flow.... Magma must have been gas rich [therefore] so fluid." He adds, as an afterthought, "[But] why no pipes in Hawaiian pahoehoe flows? See if any observed...." (From RADP 1922, Field notebooks, Box 2:Folder II, pp. 46–47.)

the continuous daily work of observation and correlative inference from the intermittent thought processes this work provoked.

In publication, the segregation of fact from inference was buttressed by a complex if not always consistent pattern of linguistic demarcation. At the Carnegie Institution, Director John Merriam told his scientists that "the Institution can properly give its support to the study and testing of [great] hypotheses" and instructed them to do this "by clear separation of data into fact, theories and suggestions."[116] But how did his scientists decide which was which? American geologists had a consensus as to what constituted a fact; they were less unified on the subject of theory and suggestions. Facts consisted of direct observational statements of geological phenomena, usually in the field, but also in the laboratory. One observed the order of rock sequences, the lithological characteristics of the rocks, and the identity of the contained fossils. In the lab, one examined rocks and fossils in greater detail and classified them accordingly. The resulting observation statements were facts. However, Daly's notebooks reveal a second, broader sense of fact, namely the correlations and interpretations made by experts of direct observations. The existence of the Drakensburg Scarp was a physiographic fact; although its origin was uncertain, its existence was not. It was an example of what Americans in the early twentieth century referred to as a "definitely observable fact."

Another sense of fact occupied the edges of what was "definitely observable"—namely, an inference based on generally accepted geological diagnostics. The example that Dutton had stressed was the conclusion that the Paleozoic strata of the Appalachians had been deposited in shallow water. In Dutton's discussion of geosyncline theory and its implications for isostasy, he recounted the "surprising fact," evident to all who worked in the area, that the thick Paleozoic strata "were all deposited in comparatively shallow water."[117] Field geologists had mapped the strata, measured their thickness, and seen characteristics diagnostic of shallow water throughout. Because these diagnostics were generally agreed upon, the conclusion of shallow-water deposition was considered a fact. No one disputed this fact, and James Hall built geosyncline theory upon it.

An example closer to the problem of continental drift is the "fact" of Gondwana. In the mid 1920s, Charles Schuchert wrote to Arthur Holmes that "Gondwana is a fact."[118] Schuchert was an advocate of neither contraction nor continental drift, although he struggled mightily with the question for over a dozen years. What he meant was *paleontological* Gondwana, the well-documented and generally accepted correlations of homologous fossils across the globe in late Paleozoic to middle Mesozoic times, which necessitated some form of biological contiguity to explain them. For Schuchert, whether the fossils were finally explained by contraction, continental drift, or some other mechanism, the faunal homologies were facts. Observations, classifications, and correlations were facts, because experts agreed upon their existence, if not on their interpretation.

In contrast, the disputed ground of causal explanation gave rise to a complex, diverse, and inconsistent terminology. Chamberlin chose the word *hypothesis* rather than *theory* to describe provisionally entertained explanations, because the former was "associated with a better controlled and more circumspect habit of the mind."[119] Others thought of hypotheses as stepping stones to general theory. For example, when John Merriam entertained the idea of continental drift in the mid-1920s, he described

Wegener's idea as a "hypothesis initiated with a view to constructing an adequate theory of . . . the Earth."[120] Here Merriam used the word theory in a positive sense, as a form of knowledge toward which one was working. William Bowie made a similar distinction that he pressed to his own advantage, referring to continental drift as a "very interesting hypothesis" but calling isostasy a "theory or principle, since it seems beyond question."[121] Bowie called his own causal account of earth oscillation a "theory, in harmony with isostasy" because it was based on the "facts" established by geodetic evidence.[122]

Bowie's claim that isostasy was based on established facts was an implicit response to Chamberlin's warning against precipitate explanation. In Chamberlin's prescription, theory was only justified after "a full accounting of the facts." Thus both Bowie and Hayford emphasized that such an accounting had been accomplished. In his vice-presidential address to the American Association for the Advancement of Science in 1910, Hayford had declared that "the existence of isostasy is now established by evidence which is available in print. The time has come to speculate upon the manner in which the isostatic adjustments are produced, to look for the relations between the known condition, isostasy, and other known facts."[123] But was isostasy a fact or a theory? George Becker had praised Hayford for establishing the "actuality" of isostasy; Gilbert had acknowledged the success of the "hypothesis of isostasy" in reducing anomalies of the vertical. In England, Harold Jeffreys had praised Hayford and Bowie for placing the "theory" on its "present footing" while John Joly had credited them with elevating "theory" into "fact." Was isostasy a principle, a theory, a hypothesis, or a fact? Hayford begged the question: it was a "known condition."

There was of course no sharp line where facts ended and theory began. Tensions existed over how vigorously one should pursue theory, and no two individuals necessarily agreed as to how many data were needed to consider something "established by evidence."[124] Perhaps for this reason, American earth scientists sometimes spoke as if they considered speculative discourse less scientific than data collection. When American geologists wished to indulge in an admittedly speculative discussion beyond the bounds of "definitely established" facts, they earmarked these discussions as contributions to "philosophical geology."[125] Reginald Daly often claimed that it did "no harm to speculate," but as Chester Longwell's review of Daly's *Our Mobile Earth* suggests, "conservative geologists" did not necessarily agree.[126]

But whether American geologists called their explanations hypotheses, theories, speculations, or philosophical geology, and whether or not they were apologetic about it, they did pursue theoretical accounts. If their rhetorical stances seem strained in retrospect, it may reflect the historical moment. Gilbert, Chamberlin, and their colleagues were men of firsts: they founded journals and built university departments; they worked for the U.S. Geological Survey in its earliest years; they mapped portions of the United States that had never been mapped before. As they passed the torch to the next generation—or resisted passing the torch—codification of method may have seemed to them increasingly imperative. Yet even as they attempted to explain what what they had done, or tried to do, they were also attempting to answer the unanswerable: What is the relationship between observation and inference? How should one account for human creativity and unexplained insight? Where does one draw the line between justified inference and unbridled speculation? Often it seemed as if geologists wanted to have it both ways. Writing in the spring of 1922, Charles Schuchert accused Wegener of getting carried away with his ideas: "Continents are not like little corks on oceanic currents flowing around to suit the fertility of the German imagination!" Yet only four weeks later, while organizing the 1922 Geological Society of America symposium on mountain-building, he concluded that scientists did not live on facts alone: "Of course, we can drop all theories and end up too scientific, but not for me. I must have some imagination."[127]

Good Science Is Hard Work

The American emphasis on documentation of observational detail was above all an attempt to do the best possible inductive science. It may well have reflected in part the "frontier effect" described by Burchfield and Bruce—there *was* much to be described in America and hasty generalizations were likely to prove wrong. But there was much to be described in Europe as well, and Americans were by no means unique in writing long treatises; Suess's most famous work comprised four volumes. But his was a compendium of work of other investigators; American monographs were normally compilations of one's own field mapping. The documentation of copious quantities of geological detail reflected a belief that good science was built on hard work—the diligent finding, recording, and documenting of the features of nature, upon which any reliable theory would necessarily be founded.

Writing in 1893 on the question of the origin of mountain ranges, Joseph LeConte presented a gradualist view of the scientific process to complement his gradualist view of earth history:

> The citadel of truth is not usually taken at once by storm, but only by very gradual approaches. First comes the collection of carefully observed facts. . . . Next comes the grouping of these facts by laws more or less general. . . . At first the grouping includes only a few facts. The explanation or theory lies so close to the facts as to be scarcely distinguishable from them. It is a mere corollary or necessary inference. It is modest, narrow, but also in the same proportion certain. Then the group of explained facts becomes wider and wider, the laws more and more general, and the theory more daring (but in the same proportion also perhaps more doubtful).

The result was a "gradual approach toward perfect knowledge" founded on the hard work of compiling carefully observed facts.[128]

Using an analogy close to geologists' hearts—the "winning" of gold by miners—Charles Schuchert advised his students that geology was "an intricate subject, and one of the most inspiring, but its gold is won only through the hardest kind of work." Schuchert's own paleontological work, he boasted, was "not the result of a sudden impulse, but the accumulation of at least twenty years of hard thinking and plotting."[129] Some might argue that paleontology consisted of more plodding than plotting, but for American geologists, to be called a hard worker was a compliment, not an insult. T. C. Chamberlin was widely recognized by his colleagues as a brilliant man, perhaps a genius, but when he died Bailey Willis denied that he had come to his theories through the flash of insight. Rather, like all good scientists, he came to them through careful and gradual analysis of facts. Discussing the planetesimal hypothesis, Willis wrote that "this fundamental discovery," like all of Chamberlin's important contributions, was "the result of painstaking analysis of the realities. It was in no sense a speculative concept. It was not a flash of genius. Rather it was the child of that genius which is characterized by intensive attention to details and facts."[130]

The prominence that Chamberlin gave to theoretical insight led one reviewer of his textbook to worry that it might give students the wrong idea about science. While acknowledging that Chamberlin's up-to-the-minute discussion of theoretical points made it an excellent resource for seasoned professionals, the reviewer suggested that it might mislead a novice:

> To the young student with an insufficient basis of facts and limited experience in their interpretation, the prominence given to hypothesis, though otherwise an excellent feature, may be dangerous, possibly leading him to neglect the detail of the science for this more attractive field. Geology is a science which has suffered much in the past from ungrounded speculation, since speculation is easier than investigation.[131]

The notion that lasting rewards come from hard work suggests an obvious parallel with broader ideological patterns in American culture. Historian Richard Hofstadter's classic claim about American political belief—that "a preference for hard work [was] better and more practical than commitments to broad and divisive abstractions"—could well have been written of American geologists.[132] Indeed, Hofstader used the pragmatic quality of American science as an example of the broad rejection of ideological generalities in American cultural life. The point he missed—as later commentators have noted—was that in science, no less than in the rest of American culture, the ideology of Americans was to have no ideology.[133]

Throughout the nineteenth century in America, tensions existed over whether science, with its cultivation of a small cadre of knowledgeable authorities, was somehow undemocratic. Joseph Henry, the director of the Smithsonian Institution and first president of the National Academy of Sciences, initially opposed the formation of the Academy because it might be considered "at variance with our democratic institutions."[134] What was needed, as Nathan Reingold has argued, was to find an ideology that would permit American science to flourish in an antielitist culture, and this is precisely what Chamberlin did by incorporating egalitarian ideals into the very methods of science.

Multiple working hypotheses reflected an ideal of conceptual democracy, in which each hypothesis would find its place in the sun, every idea given a chance to compete on equal terms. Not that all theories were equal (any more than most Americans believed that about all men!)—but rather that they all deserved an equal chance. If European science was autocratic and dogmatic, American science would be pluralistic and receptive. If Americans achieved commercial success through hard work, they would achieve scientific success the same way. And if Europeans achieved scientific rank through privilege or personal fortune, Americans would achieve it through diligence and commitment. And this was not purely a myth or an abstract ideal: although many American scientists hailed from elite backgrounds, many others did not. Chamberlin was the son of a middle-class midwestern preacher; Charles Schuchert, a self-educated son of impoverished immigrants, never attended college yet became a professor at one of America's foremost universities.

While Chamberlin promoted a vision of science behaving democratically, he also believed that science could advance democratic behavior. The antidogmatic cast of mind could be translated into general terms, because "greater care in ascertaining the facts, and greater discrimination and caution in drawing conclusions," would be a virtue in civic as well as scientific life. Suggesting an alternative to his own dogmatic religious upbringing, Chamberlin concluded his 1890 article in *Science*:

> The general application of this method to the affairs of social and civic life would go far to remove those misunderstandings, misjudgments, and misrepresentations which constitute so pervasive an evil in our social and political atmospheres, the source of immeasurable suffering.... I believe that one of the greatest moral reforms that lies immediately before us consists in the general introduction into social and civic life of that habit of mental procedure which is known as the method of multiple working hypotheses.[135]

Chamberlin was not alone in promoting the scientific mind-set as a means to improve American society. When Charles William Eliot became president of Harvard University in 1869, he explicitly promoted science in the general curriculum as a means to "nurture an antidogmatic cast of mind." As Daniel Kevles explains, quoting Eliot:

> The replacement of the "dogmatic and overbearing" with the "receptive" mind would be "a most beneficent result of the study of natural history and physics."... Young men, tempered by science to a high-minded devotion to truth, would carry high-mindedness beyond the daily practices of their lives to their outlook upon the world. Give science more weight in the curriculum, in short, and Harvard's graduates were bound to become virtuous citizens.[136]

Ira Remson likewise argued for chemistry in the Johns Hopkins University curriculum as an antidote to religious sectarianism. When one learned to be open-minded through the practice of science, one would become open-minded about theological matters as well. In fact, Remson argued, science could improve religion because it "may modify, and has modified, dogmatic theology."[137] But to teach the virtues of open-mindedness in religious and other matters, science had to demonstrate such virtues itself—and this is what Chamberlin's method entailed. Without theoretical receptivity, the very justification for science in general education collapsed.

By the end of the nineteenth century, a pattern of theoretical nonsectarianism was well established, not only in geology but also in other branches of American science.[138] As American physics began to develop into a coherent community in the early twentieth century, its workers were self-consciously skeptical of large-scale theoretical programs. Indeed, physicists went further than earth scientists and adopted explicitly agnostic theoretical programs, like Percy Bridgman's operationalism.[139] (Bridgman was a friend and colleague of Daly at Harvard, but Daly never accepted his friend's extreme antiinterpretive position.) In biology there is similar evidence of a distinctive American pattern of research. In a study of late-nineteenth-century embryology, Jane Maienschein found a sharp contrast between working styles in American and German laboratories. In contrast to German embryologists, for whom theories had pride of place, American embryologists tended to pursue a gradual and cumulative approach to gaining knowledge, accepting the possibility of multiple causation and generally offering theoretical explanations "only at the ends of their papers."[140] Maienschein suggests that American biologists may have been influenced by T. C. Chamberlin, a plausible suggestion given the wide dissemination of his writings in the American scientific community.

A Violation of Standards

Given this historical context—of theoretical pluralism coupled with an inductive methodology, a suspicion of highly ambitious theories and an emphasis on hard work—one can see why Americans so reacted to Alfred Wegener's argument for continental drift. In content, in manner, and in purpose, Wegener's work contradicted the edifice and rhetoric of practice that Americans had laboriously constructed and articulated over the course of nearly a century. Wegener violated the norms of American geological practice. The theory of continental drift was universalist and comprehensive, it was presented as the result of sudden inspiration rather than long labor, and the format in which he presented it violated the American pattern of putting facts first.

Wegener's treatise was "grand theory," and he said so explicitly. His goal was to effect a broad synthesis in the style of Suess, whose work he admired for its "bold simplicity of concept and wide diversity of application."[141] Wegener freely admitted prior allegiance to Suess's contraction theory, which was "for a considerable period a sufficient epitome of our geological knowledge." Indeed, he suggested, the "simplicity of its basic principle and the multiplicity of its applications still give it a strong hold."[142] Americans who had been raised on Dana's version of contraction theory and later accepted Chamberlin's modification of it would have been uncomfortable with this account of the history of their discipline, and they would have become more uncomfortable as they read on. Contraction theory, Wegener continued, had failed in global application and needed to be replaced by an alternative but equally global theory, and this he now proposed. Drift was to provide a "new conception of the nature of the earth's crust and the movements occurring therein," uniting "all the facts of . . . geological history."[143]

Americans considered causal theory to be sensitive ground upon which one was expected to tread carefully. General theories were not necessarily bad, but they were often bad because they tended be too general and therefore false. The broader their scope, the more likely this was to be the case. From the American standpoint, Wegener was approaching the territory of excess universalism, and the resultant ground was likely to be similar to that trod by his famous compatriot, A. Werner, whose name was eerily similar. And Wegener was anything but cautious in his assertions. In the very first pages of his book, he freely acknowledged that he was "convinced of the fundamental correctness" of his idea. Unlike Americans who rhetorically distanced themselves even from their own theoretical offspring, Wegener stood proudly behind his first and only son. He also freely admitted that he had come to his conviction not by dint of long labor, but *by accident*.[144] In the opening chapter of the 1924 edition of *The Origin*, the one read by most North American geologists, he explained:

> The first notion of the displacement of continents came to me in 1910 when, on studying the map of the world, I was impressed by the congruency of both sides of the Atlantic coasts . . . In the autumn of 1911, I became acquainted (through a collection of references, which came into my hands by accident) with the paleontological evidence of the former land connection between Brazil and Africa, of which I had not previously known. This induced me to undertake a hasty analysis of the results of research in this direction in the spheres of geology and paleontology, whereby such important confirmations were yielded that I was convinced of the fundamental correctness of my idea.[145]

Wegener had not only discovered the evidence of land connections "by accident" but also convinced himself of the correctness of his central idea by "hasty analysis" of those particular lines of research that provided support for his idea. He could hardly have said anything more likely to inspire American indignation. For Americans, coming to an important idea "by accident" looked like a summary dismissal of the role of hard work in the formation of reliable scientific knowledge. Wegener's analysis of research results in "this direction" looked like Joseph Singewald's "selective search through the literature for corroborative evidence." Chamberlin too had explicitly warned against the unconscious selection of facts that fit preconceived theories. And Wegener freely admitted to *conscious* selection of such facts!

If Wegener lacked circumspection in presenting the origins of his theory, he was even more reckless with regard to its current epistemic status. Throughout his book, he referred to the supporting evidence of drift not as data or observations, nor even as geological facts, but as "proofs." For example, in discussing the geological congruences on either side of the South Atlantic, Wegener described the jigsaw puzzle fit of the continents as providing "confirmation of the correctness of our views, [supported by] numerous other proofs."[146] The word *proof* occurs repeatedly throughout his text, in phrases such as the "numerous proofs" of drift, the "proofs given in this chapter," and, particularly, the "paleontological and biological proofs."[147] And rather than suggest that the evidence was best explained by the idea of drift, he claimed that the said proofs demonstrated the necessity of it. He was equally direct in his arguments against prior theory. The contraction theory had experienced "absolute contradiction" and must be "completely rejected." The same was true of permanence theory, whose proponents "arrive[d] at false conclusions" based on their erroneous premises.[148]

To be sure, Wegener suffered the slings and arrows of translation: a variety of different words and phrases were reduced by his translator, J. G. A. Skerl, to the single word *proof*. For example, the phrase *die paleontologischen und biologischen Belege* might have been better rendered in English as "the paleontological and biological evidence."[149] Similarly, the German *besagen*, translated by Skerl as "prove," might have been more fairly expressed as "say," "show," or, perhaps, "demonstrate." The German *Nachweis* and *Beweis* were both given in English as "proof," although in context they might have been conveyed as "strong evidence." Clearly, some of Wegener's linguistic distinctions were lost by his translator. But Wegener did supervise and review Skerl's work, as did J. W. Evans, author of the preface to the English edition. In his own introduction to the translation, Wegener stated that the modifications made since the previous edition were intended to cast the book into a "new, and ... more convincing form"—not to tone it down or correct previous errors.

There seems little doubt that Wegener was convinced of his theory and believed his task was to persuade others. But the art of persuasion requires understanding of the presuppositions of the potentially persuaded; Americans tried to convince by feigning lack of conviction. Thus, even if Wegener's words had been faithfully translated, they would not have been received by his American colleagues as he intended. *Proof* may well have seemed fine to Wegener as he read Skerl's translation, but it was not fine for his American readers.

Nuances of wording aside, Wegener's treatise was explicitly deductive. His was a causal-theoretical account, in which a basic principle was used to illuminate a wide range of geological evidence, not the other way around. Wegener began with an idea.

"He who examines the opposite coasts of the South Atlantic Ocean," he wrote in his opening line, "must be somewhat struck by the similarity of the shapes of the coastlines of Brazil and Africa." This one basic, even trivial, observation, summarized in a few sentences, was the "starting-point" for a new "conception of nature." By the second paragraph of his text, Wegener was developing his theory according to an explicitly deductive logic. Geological evidence necessitated some mechanism of prior land connection; geophysical evidence proved the impossibility of sunken continents. Ergo, both contraction theory and permanence theory were inadequate conceptions of the globe.[150]

If there were plausible alternative interpretations, Wegener did not pursue them. For him, the evidence was sufficiently compelling as to make discussion of alternative hypotheses moot. The clearest example of this occurs in his analogy of the reunited pieces of a torn newspaper. The structural correspondences between re-united continents yielded definitive "proof . . . of the validity of our supposition," he argued, because the exactness of the match "obviously [leaves] no course but to conclude that the pieces were once actually attached in this way."[151] This led to his final point: that the theory of drift placed firm constraints on models of the structure and rheology of the subcrust. Some of Wegener's opponents would attempt to argue that mechanical considerations refuted the hypothesis, but Wegener turned this argument upside down. Since the arguments in support of drift were compelling, one could "henceforth assume them to be correct, and on this supposition . . . deal with" the questions of viscosity and rheology.[152]

Defending American Pluralism

By declining to consider alternative interpretive frameworks, Wegener violated the American ideal of applying democratic principles to the development of scientific theory. Thus American geologists—far more than their British or other European counterparts—unleashed a host of methodological critiques at Wegener. The accusation that he advocated a cause may be understood as a reaction to his violation of the American standard of putting facts first, the accusation of dogmatism as arising from the prescribed stance of Chamberlin's methodology. One might suppose that Wegener would have understood the nature of the American criticism against him and responded to it. But in his final edition, Wegener became more rather than less adamant. In particular, his discussion of the geological evidence became more rather than less firm. Perhaps Wegener thought geologists would not argue with their own evidence, but the result was a treatise that to American ears was more dogmatic than ever.

In attempting to build a global theory, Wegener, according to Schuchert, had "generaliz[ed] too easily from other generalizations" and forgotten proper scientific method.[153] In language reminiscent of Francis Bacon, Schuchert reminded Wegener that "facts are facts, and it is from facts that we make our generalizations, from the little to the great."[154] Rollin T. Chamberlin, son of T. C.—who died in 1928—argued that Wegener's own dogmatism made dogmatic rejection of his theory more acceptable than it otherwise would be.[155] Even Chester Longwell, who was privately sympathetic to the theory, argued that the scope and ramifications of Wegener's theory—and perhaps the romantic impulses of Germans in general—required it to be held to even stricter than usual methodological standards:

> Certain demands are made of this new and romantic speculation before it is admitted into the respectable circle of geological theories.... Very naturally, we insist on testing this hypothesis with exceptional severity; for its acceptance would necessitate the discarding of theories held so long that they have become almost an integral part of our science.[156]

At best, Wegener could have been accused of being incautious; in fact, he was accused of violating the scientific method. When one put theory first, one violated the principle that demarcated science from other forms of human explanatory effort. G. K. Gilbert had questioned whether there even was such a thing as *unscientific* observation; what distinguished science was the attempt to discriminate phenomena from the observer's stance in regard to them. Of course, from Wegener's perspective, this was a moot point, because the observations he invoked had been made by other people. In retrospect, this might have been taken as evidence in support of his objectivity: Wegener did not go out to find facts in harmony with his theory, since other people had found them for him. But Americans did not view it this way, because one was supposed to find one's facts for oneself. As Charles Schuchert concluded, finishing his Baconian axiom with a particularly American corollary, "it is wrong for a stranger to the facts he handles to generalize from them to other generalizations."[157] If success came from hard work, then each man was responsible for making his own success from his own hard work. If Wegener's hasty deductivism were not grounds enough for criticism, there was finally the issue of where Wegener found time to develop his theory properly: while lying in a hospital bed in 1915, recovering from wounds inflicted while fighting against democracy![158]

As the Baconian overtones of Schuchert's comments imply, American methodological preferences were derived from the British empiricist tradition. The pattern of putting facts first, building up theories gradually, and linguistically demarcating areas of acknowledged interpretation are all of a piece with the empiricist inheritance that Americans shared with their British counterparts. That Americans followed an inductive logic of justification is thus not particularly surprising; the crucial point is that Americans had taken the problem of induction seriously as a methodological issue and laid out a course that they came to see as uniquely theirs. Their national identity as a distinct scientific community was embedded in their commitment to theoretical pluralism (a commitment that their British colleagues did not necessarily share). It does not matter whether their program can be epistemologically defended in retrospect or whether they satisfied their ideal in practice. What matters is that Americans had this ideal, it was at the core of their understanding of what it meant to be scientific, and Alfred Wegener had violated it.

6

The "Short Step Backward"

In 1922, Harry Fielding Reid, a founder of American seismology, encapsulated the American perspective on scientific method in a review of Wegener's *Origin of Continents and Oceans*.[1] "There have been many attempts to deduce the characteristics of the Earth from a hypothesis," he wrote, "but they have all failed. There is the pentagonal system of Elie de Beaumont, the tetrahedral system of Green . . . [continental drift] is another of the same type." The history of hypothetico-deductive reasoning in geology was a history of failure; progress was to be made another way. "Science has developed," Reid concluded, "by the painstaking comparison of observations and, through close induction, by taking one short step backward to their cause; not by first guessing at the cause and then deducing the phenomena."[2]

An obvious reading of Reid's comment is that Wegener's faulty methodology led him to faulty conclusions. No doubt Reid thought so. But another reading is the implicit suggestion that a different approach—a different presentation—might have elicited a more favorable response. Charles Schuchert's comment that it was "wrong for a stranger to the facts to generalize from them to other generalizations" suggested that if evidential support for drift could come from someone who was *not* a "stranger to the facts," then Americans might be more disposed to entertain the theory. This thought had occurred a few years earlier to Reginald Daly and Frederick Wright.

A Test of Continental Drift

When Daly and Wright returned from South Africa in the autumn of 1922, each pondered the question of continental drift. Daly, whose proclivities ran to theory, developed a mechanical account of drift; Wright, an experimentalist, began to think about a possible test. The key empirical evidence was the alleged similarities between the Karroo formations in South Africa and age-equivalent rocks elsewhere in the world—evidence that had earlier motivated Suess's idea of Gondwanaland and now supported Wegener's theory of drift. But how similar were these rocks, really? Suess and Wegener had based their ideas on compilations of published literature; neither man had studied any of these rocks in person. In fact, there had never been a direct comparative study of the so-called Gondwana beds. Stratigraphic and paleontological homologies could thus be considered a *prediction* of the theory—one that had yet to be confirmed or denied. Daly and Wright discussed the issue and, on the first of November, Wright sent a letter to John Merriam, director of the Carnegie Institution, suggesting a comparative study of the Gondwana beds of the Southern Hemisphere. On behalf of Daly and himself, Wright began:

> The hypothesis of continental shift as first proposed by F. B. Taylor and later in greater detail by A. Wegener is arousing the keen interest of geologists the world over. In a review of Wegener's book on "Die Entstehung der Kontinente und Ozeane" in *Nature* (vol. 109, p. 203, 1922) it is asserted that "The revolution in thought, if the theory is substantiated, may be expected to resemble the change in astronomical ideas at the time of Copernicus."[3]

Wright explained that at the recent British Association meeting at Hull, where drift had been debated in detail, "each speaker emphasized the importance of ascertaining by actual field work the validity of the correlations made, especially by Wegener, of specific formations now widely separated, but, according to his hypothesis, formerly contiguous." A revolutionary idea was on the table, and the next step was to test it. The best way to do this was by direct examination of the critical evidence by an appropriately qualified scientist, and the best man for the job was their former traveling companion, Alexander du Toit. Wright explained:

> The validity of the Taylor–Wegener hypothesis rests primarily on field evidence ... from the southern hemisphere and is based on correlations between the Karroo rocks (especially Permo-Carboniferous) of South Africa and their equivalents in South America. It appears that an expert field geologist can by a careful study of these formations in South Africa and South America determine the actual signifcance of these correlations. ... The man best qualified to undertake this work is Dr. A. L. du Toit, chief geologist of the Irrigation Department of the Union of South Africa. Dr. du Toit has worked for twenty years as a field geologist on the Geological Survey in South Africa, ... he knows better than anyone living the details of the geology of South Africa, is admittedly the keenest field geologist in the Union, and has thought for many years on problems of intercontinental correlation. ... An expedition by him to South America, with the publication later of a monograph on the Karroo fomations of the southern hemisphere ... would aid materially in settling the question of the validity of the Taylor–Wegener hypothesis. ... Here is an opportunity which should not be missed.[4]

Wright thus proposed that du Toit, who already knew the geology of South Africa, should go to South America to test the theory of continental drift at the place where the evidence was best.

Wright's proposal hardly suggests membership in a scientific community that eschewed the development of explanatory theories. To the contrary, John Merriam had encouraged Wright to develop the proposal, not least because as a paleontologist Merriam was intrigued by the fossil correlations. As Wright explained in a letter to Daly, "Dr. Merriam is particularly interested in this matter, especially because many years ago he spent a considerable amount of time in correlating the fossil fauna of the southern hemisphere."[5] But, to Wright's dismay, Merriam rejected the proposal. The problem was not the idea, but how Wright had framed it. Wright recounted events in a letter to du Toit a few weeks later:

> I am enclosing copies of two letters which I sent to Dr. Merriam. In the first, which I sent up to Daly for his signature, the Wegener hypothesis was emphasized. This form of approach did not please Dr. Merriam, and after conference with him I submitted the second letter, which is more in line with Dr. Merriam's conception of the situation, and also, I believe, with your own.[6]

Wright's closing was a polite fiction — du Toit was motivated by the Wegener hypothesis and would soon become the world's foremost advocate of it. Wright was conveying to du Toit what Merriam had conveyed to him: that the project must be approached from a pluralistic perspective. Merriam had not rejected the idea of du Toit's working on the Gondwana correlations, but he *had* rejected Wright's framing of the problem as a *test of continental drift*. This is evident from Wright's revised proposal, which begins as follows:

> In accord with your request for more information I take pleasure in presenting, herewith, an outline of some of the results that may be expected from a comparative study by A. L. du Toit of the Karroo, Gondwana, and other equivalent formations of the southern hemisphere. From a geological viewpoint, the most prominent feature of the southern hemisphere is the Karroo system of South Africa and the homotaxic systems of India (Gondwana), of Australia, of South America, and of the Falkland Islands. These widely separated deposits form a closely connected, conformable series of beds of upper Paleozoic and Lower Mesozoic age and show a remarkable community of lithological characters and of conditions of sedimentation. The flora (and to a lesser extent the fauna) of the successive stages is also identical, many of the species ranging over the whole of this immense area of hundreds of thousands of square miles in the southern hemisphere.[7]

In the first version, Wright had announced a grand hypothesis and proposed a test of it. The tone was confident, important. In the revised proposal, Wright began by discussing the observational data and possible interpretive frameworks, and the tone was far more circumspect:

> This brief recital of features peculiar to the Gondwana–Karroo epicontinental deposits may suffice to indicate the great variety of geological problems of fundamental importance which they offer to the geologist. Among these problems are the following: Conditions that produced the Permo-Carboniferous glaciation in the southern hemisphere. Extraordinary development of the Glossopteris flora over such a vast area. Does this require the existence of land connections between the land masses of the southern hemisphere (Gondwanaland, Lemuria) which have since sunk; or were the lands then directly connected, having later drifted apart to their present positions (Taylor–Wegener hypothesis)? What were the conditions that led to the wholesale intrusion and extrusion of molten basalt? What are the characteristics of plateau basalts and what were the physical conditions obtaining during the extrusion? These and many other questions are probably best answered by intensive studies on these systems.[8]

Here was the American methodology of science in practice: Merriam had directed Wright to follow an inductive format and the method of multiple working hypotheses.

Wright—who had earned his Ph.D. in Heidelberg—had failed to follow the American pattern of pluralistic interpretation, which Merriam now enforced.[9] Du Toit was not to privilege the Wegener hypothesis but to place equal emphasis on alternative explanations. In the phraseology of Wright's revised proposal, furthermore, the resolution of theoretical questions might occur, but this task was posed as secondary—indeed, incidental—to the primary task of geological documentation:

> It occurred to us, that if [du Toit] were to spend possibly six months in South America studying the Karroo equivalent beds, . . . making collections of fossil flora and fauna and of the different rock types, and investigating the field relations with reference to mode of formation and stratigraphic succession, he could then prepare a monograph on the Karroo, Gondwana, etc. of the southern hemisphere which would be more authoritative than anything that has yet appeared and would be a notable contribution to the science of geology. Mooted questions such as Gondwana land, climatic conditions, the Taylor–Wegener hypothesis would naturally be discussed and evidence presented which might prove conclusive one way or the other.

In addition to being structurally inverted, the revised proposal was twice as long as the first, extended by more information on the geology of the Southern Hemisphere. Wright elaborated on the necessity of direct field observation by a local expert, albeit with the right international credentials:

> Up to the present time comparative studies of these homotaxic beds have been made chiefly by geologists of the northern hemisphere who have had little first-hand experience with the actual formations. We believe that results of considerable value could can be obtained by a comparative field study of these rocks and of their floral and faunal content by a geologist who has grown up, as it were, in these regions. Such a geologist is Dr. A. L. du Toit of the Geological Survey of the Union of South Africa. After a thorough European training he has spent the past twenty years in the field on the Karroo system of South Africa. He has also studied the beds equivalent to the Karroo in Australia and Tasmania. He has published many reports on the Karroo system of South Africa; he has made large collections of fossils from the Karroo, and is at present writing a monograph on the Karroo flora. During our trip to South Africa we spent some time with him in the field and were greatly impressed with his ability as a keen field geologist. We considered him easily the best field geologist in the Union. . . . He knows the Karoo as no one else does.[10]

If reliable scientific knowledge was created by experts working diligently in the field, du Toit was the relevant expert. He would go to South America and spend six months investigating and documenting the facts of the case. Merriam approved the revised proposal and submitted it to the Carnegie Board of Trustees.

In early 1923, the Trustees approved and funded Wright's revised proposal for a "comparative study by [A. L. du Toit] of conditions recorded in certain geological formations of similar type occurring in South America, South Africa, Australia and India . . . for the purpose of securing an interpretation of the meaning of the similarity of these formations and of their contained records."[11] Du Toit was *not* being sent to South America merely to make a catalog of rocks; he *was* being sent to secure "an intepretation of the[ir] meaning." But this interpretation was not to be prejudiced by prior conceptual commitment. Having received Wright's letter enclosing the two forms of the proposal, du Toit already understood his patron's terms, but Merriam made sure. "Without any preconceived notion as to the outcome of the investigation," Merriam wrote to du Toit, "we consider it exceedingly important to have such a careful study made.... We shall look forward with the greatest interest to your results."[12] Merriam's letter arrived in Pretoria on June 6, 1923, the day of du Toit's departure for Cape Town en route to Brazil. The contrast between Merriam's prescription for proper scientific research and du Toit's own intellectual commitments made the latter openly apprehensive. He confessed in reply:

> On the eve of my departure I am just a bit oppressed with the magnitude of the task, the immensity of the area, and the complexity of the various problems at issue.... While at present favoring the Hypothesis of Continental Disruption [drift], I am endeavouring to keep an open mind on the subject and to examine everything from an "orthodox" viewpoint as well.[13]

It took over two years for du Toit to complete the monograph of the results of his five-month excursion in Brazil, Uruguay, and Argentina; writing was squeezed in between official duties at the Irrigation Department. (He did, however, return to the Carnegie Institution £45 of unspent funds.) In February 1926, du Toit wrote to Merriam to discuss the issue of publication, outlining the signficance of his results. The monograph would be the first comprehensive account of the geology of South America in English, it would be the first full account of the geology of Argentina in any language, and it would give a critical evaluation of stratigraphic and paleontological homologies with respect to the problem of continental drift. Recalling Merriam's earlier advice, du Toit carefully explained his philosophical position: "If the arguments veer rather too much in the direction of the 'Displacement hypothesis' it is because the facts are so strikingly in favour ... At the same time, I am attempting to keep an open mind."[14] Merriam retained his neutral stance while encouraging du Toit to submit the manuscript:

> Although I am not committed to any hypothesis, it is clear to me that the situation which you have examined is one of exceptional interest. Such is, I am sure, the opinion of all American geologists. In these days of extraordinary developments in science one must be open-minded toward any hypothesis initiated with a view to constructing an adequate theory of formation and history of the earth.[15]

Du Toit submitted his manuscript in December, and Merriam passed it on to Wright for review. Wright was pleased with it. There was an impressive "mass of detailed observations" compiled with clarity into a comprehensive account; the morphological details of fossils, stratigraphy, and sedimentological features were scrupulously reported and illustrated (figure 6.1). No doubt recalling Merriam's earlier reactions, Wright also scrutinized du Toit's stance. Du Toit had, in fact, taken a stand: the agreement of lithological and stratigraphic characteristics was so extreme—even in the fine details—that they could only be made sense of by the displacement hypothesis. South Africa and South America must have been, by du Toit's reckoning, at one time no more than 400–800 kilometers apart. But Wright reassured Merriam that du Toit had not gone overboard; his tone was sufficiently measured: "In this respect, it is suggested [by du Toit] that the observed relations seem to be most easily explained on the displacement hypothesis, but this hypothesis is not insisted upon. In other words, the report is in no sense wild."[16]

Merriam was satisfied and replied to Wright with a summation of his views on the problem of hypothesis-testing. "It is extremely important to have published con-

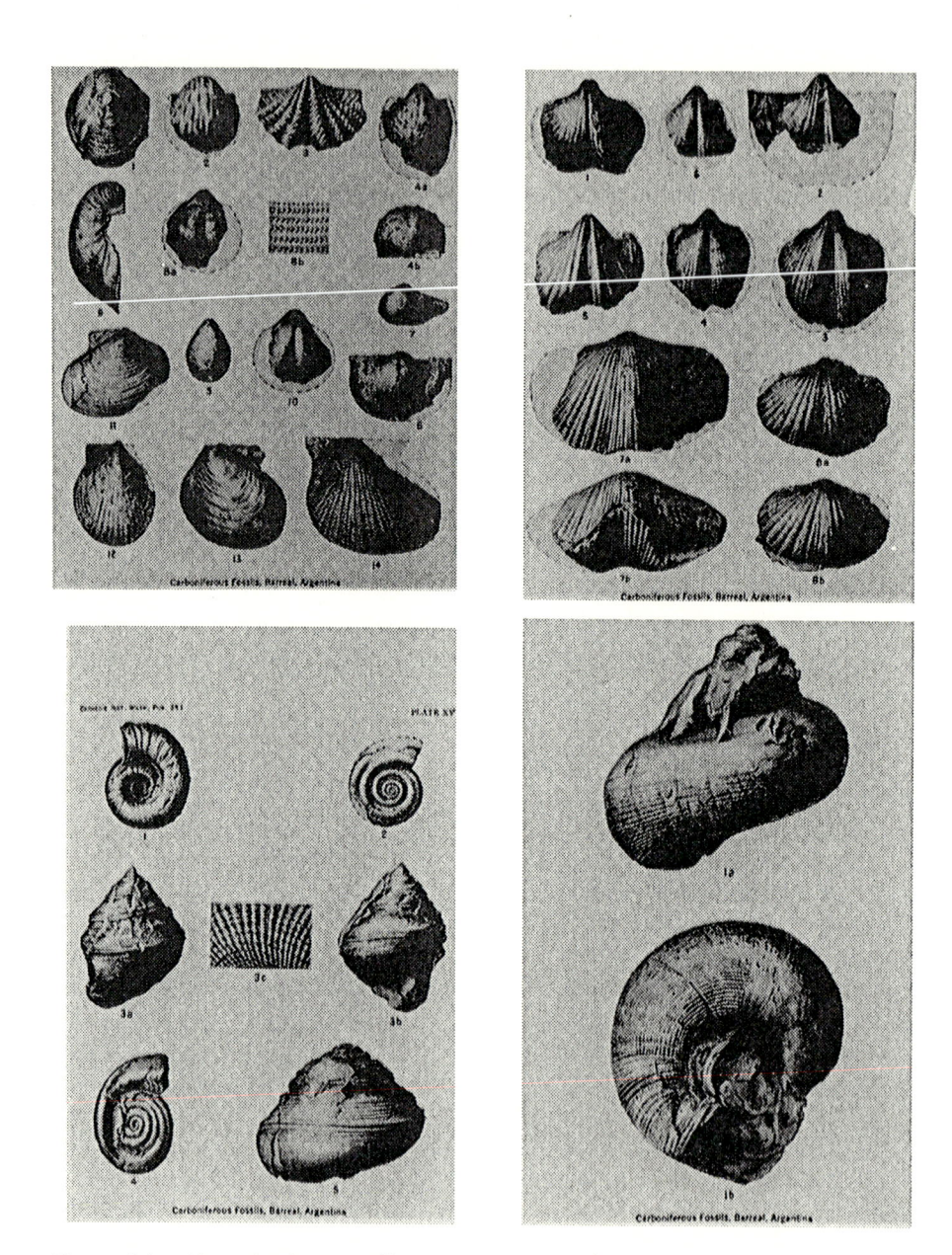

Figure 6.1. Alexander du Toit's illustrations of key fossils from South America. These fossils revealed the depositional ages and environments of the units being mapped and thus provided support for du Toit's stratigraphic correlations with South Africa. (From du Toit 1927; originally published by the Carnegie Institution of Washington.)

cise statements of data relating to a great hypothesis" he declared. "I believe that the Institution can properly give its support to the study and testing of such hypotheses and that this can be accomplished by clear separation of data into fact, theories, and suggestions."[17] Content that du Toit had done this, Merriam agreed to publish his work. The monograph, *A Geological Comparison of South America with South Africa*, was published in Washington, D.C., in 1927.

A Geological Comparison of South America and South Africa

Du Toit organized his monograph inductively, as Merriam demanded, but the tension between his own beliefs and his patron's methodological commitments is evident. Also evident is a disparity with respect to what each considered orthodoxy. Like Wegener, du Toit assumed that geological "orthodoxy" meant Suess's theory of sunken continents, an idea that had to be rejected on geophysical grounds. But this was precisely what Americans had *never* believed; Americans held to the concept of permanent continents and were therefore stymied not on geophysical but on paleontological grounds. Thus, from the opening arguments, du Toit's position would have sat awkwardly with most American readers. Chapters 2–4, however—which consituted the body of du Toit's work—were explicitly descriptive, as Merriam wanted. Du Toit carefully documented the details of the geological phenomena he had studied in South America and the parallels to South African structures and successions (figure 6.2). Following the American model, theoretical discussions were for the most part set aside. However, much as he tried, du Toit was hard pressed to deny his own enthusiasm. Concluding his second chapter, on the "Geological Framework of the South Atlantic," he could not help but allow that the overall stratigraphic parallels strongly suggested that the "two continental masses were in the past geographically closer to each other, which is actually the keynote of the Displacement or Disruption Hypothesis. Evidence for this momentous assertion will be found in the following pages."[18]

Indeed it was. The next eighty pages of his text documented the major stratigraphic systems and their structures and fossils according to geographic distribution and without reference to continental drift. But when du Toit compiled these stratigraphies into a geological history—the standard goal of any monograph in regional geology—the impossibility of separating his observations from their theoretical implications became manifest. The geological history of the Afro-American land mass, as he called it, could hardly be written without reference to the problem that had motivated the study. He wrote, then, in uncharacteristically awkward prose:

> That these two continents were intimately connected during several geological epochs will, I venture, be acknowledged after the perusal of the preceding pages, though the manner of such union would admittedly be speculative. In preparing this review an attempt was made first of all to write the historical account, irrespective of any hypothesis as to the manner of such union or of the ultimate mode of separation of the land-masses though it became evident, as the data were assembled, that they pointed very definitely in the direction of the displacement hypothesis, and that they could most satisfactorily be interpreted in the light of that brilliant conception.[19]

Du Toit settled on a compromise: to treat the geological history from the viewpoint of drift, "but only in such a way to bring out particular features the significance of which

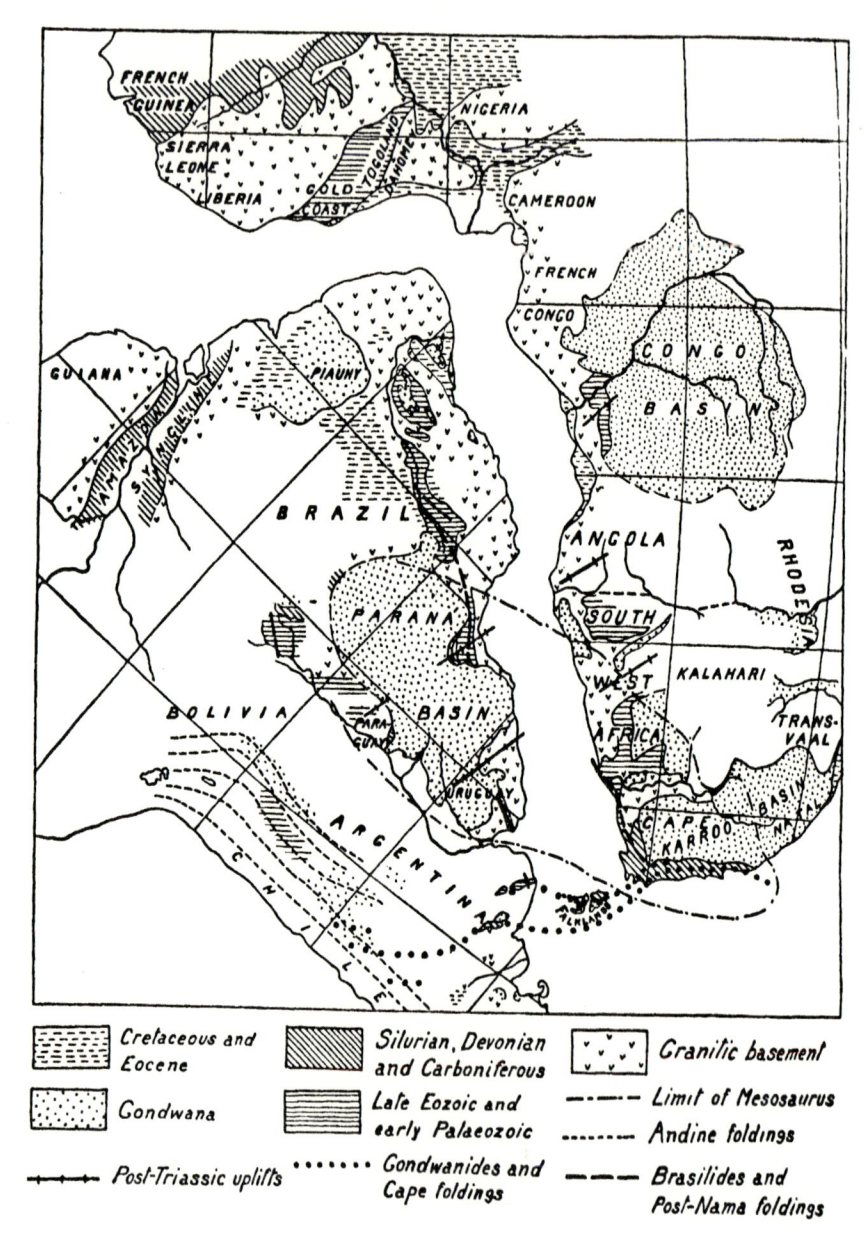

Figure 6.2. Du Toit's suggested reconstruction of Africa and South America. On the basis of facies analysis, he concluded that the continental coastlines as manifested today were not precisely contiguous in a reunited Gondwana but were within a few hundred kilometers. This conclusion accords closely with contemporary knowledge of the continental shelves. (From du Toit 1927; originally published by the Carnegie Institution of Washington.)

would otherwise be overlooked." Drift itself could then be discussed independently, "without prejudicing" the case.[20] This compromise permitted him to summarize the geological history epoch by epoch, for the most part in a theoretically neutral manner, but also to point out various places where key data supported the thesis of drift. Most telling was the evidence of the glaciation of the Carboniferous, where thick tillites suggested a glacial center "situated out in the present Atlantic."[21] Without drift, the shape and size of the required ice cap were peculiar, and far greater than anything ever posited for the Northern Hemisphere; with drift, the tills could be brought together "in the most simple manner, within an oval perhaps no larger than the African continent." Also important was the Carboniferous geology of the Falkland Islands, which "display[ed] not the slightest affinities with Patagonia" but was strikingly similar to the rocks of the Cape Province (figure 6.2). The Falklands, du Toit concluded, were an "outpost of Gondwana."[22]

No discussion of South American geology would be complete without considering the Andes, and here du Toit tipped his hand.

> Under the displacement hypothesis, the South African continent is viewed as having proceeded to drift westward during the latter part of the Mesozoic, crumpling up in its path the marine sediments bordering the Western side of the Brazilian "shield" . . . The Tertiary foldings and overthrustings, directed usually towards the east, that characterize the Cordillera and extend from Venezuela to Tierra del Fuego, find their explanation in logical fashion under this hypothesis.

Venturing still farther into speculation, du Toit suggested that marine vulcanism could also be tied to drift: "In the broadening gap between Africa and South America the presumably basaltic ocean floor . . . would normally have been in tension, and foundering of sections thereof is a possibility that can be invoked to explain the volcanicity of the South Atlantic basin."[23]

In his final chapter, du Toit revisited the point he had raised early on: that there were simply too many similarities to be coincidental. Any one homological relationship might be dismissed as a fluke, but why, without drift, would there be so many? The displacement hypothesis explained them all: it matched congruent coastlines and brought the similar formations together in a continent of familiar size; it brought the pattern of Permo-Carboniferous glacial deposits into coherence; it made sense of the spatial distribution of arid regions at the end of the Triassic; and—perhaps most important—it explained why the homological relations came to an end in the late Cretaceous.[24] No singular similarity would compel the conclusion of drift, but, collectively, the homologies were "so numerous as . . . almost to exceed the bounds of coincidence," involving as they did "vast territories . . . and embrac[ing vast] times."[25]

Although du Toit argued throughout his book that the evidence of drift was most compelling in its collectivity, he focused his closing arguments on the one line of evidence he considered most important. This was the pattern of facies changes. The word *facies* indicates the diagnostic lithological features of a group of rocks and a depositional environment: the evidence of shallow water so central to the theory of geosynclines, for example, or the evidence of alluvial deposition indicated by cross-bedded sandstones and conglomerates. If one compared the facies patterns of the homologous formations on either side of the Atlantic, du Toit argued, one found a stunning thing:

these formation often exhibited less change across the Atlantic Ocean than they did over much shorter distances on land. He explained:

> Consider the case of two equivalent formations, the one in South America beginning on or near the Atlantic coast at A and extending westward to A' and the other in Africa and starting similarly near the coast and extending eastward to B'. . . . More than one such instance can be designated, where the change of facies in the distance AA' or BB' is greater than that found in AB, although the full width of the Atlantic intervenes between A and B. In other words, these particular formations along the two opposed shores tend to resemble one another more closely than either one or both of their actual and visible extensions within the respective continents.[26]

Point B on the African coast was more like point A on the South American coast—now thousands of kilometers away—than it was like Point B' only a few kilometers inland. The inescapable conclusion was that the present configuration of these formations was not that which prevailed when the rocks were laid down. Points A and B were once much more nearly contiguous.

Du Toit considered this evidence to be his most important for two reasons. First, it could be quantified, albeit in a qualitative way. To a first approximation, one would expect the facies changes within a formation to be a function of distance: the farther apart two points in space, the more different the rocks at that locale ought to be. And yet, one found the opposite result. The implication of a South Atlantic rift was further supported by

> the central Atlantic rise beneath the ocean, with its surprisingly symmetrical position nearly midway between the Old World and the New. Interpreted *mathematically*, the great regularity of these . . . features, extending through the entire length of the South Atlantic, would betoken an *enormously high probability* that such features had owed their origin to one and the same set of tectonic forces at a relatively late geological period.[27]

Second, the lithological correlations could not be explained by any competing hypothesis. Here du Toit was taking seriously the most important aspect of Chamberlin's methodology—the point emphasized by Gilbert—that consideration of multiple hypotheses helped to focus one's sights on data that could discriminate between them. He argued that the facies data did just this. Homological fossils could be equally well explained by the hypothesis of sunken continents—indeed, this is how they had been explained for many decades—or by a configuration of lesser land bridges, or by effective biological migration and dispersal. But the pattern of facies distribution necessitated direct physical proximity.[28] The lithological evidence demanded drift.

But how was one to *explain* drift? No doubt Daly had raised this point with du Toit in South Africa, and du Toit's answer, like Daly's, was that the cause of drift was a separate issue, to be decoupled from the phenomenon:

> It is not proposed to discuss here the physical basis of [the displacement] hypothesis. . . . The intention is merely to set forth some of the data regarding Africa and South America and to state the conclusions to be drawn therefrom, that are distinctly awkward of explanation of the current and orthodox view of "land bridges," but which, on the contrary, appreciably favor the "hypothesis of continental disruption."[29]

Other scientists would have to deal with the North Atlantic, the Indo-Australian correlations, or other issues raised by the theory. But for du Toit's own data, the answer was clear: southern Africa and South America had once been linked and then been torn asunder.

The year was 1927. A full statement of the facts had now been made—by a man who was fully acquainted with them—and presented in accordance with the normative ideology of American geology. Du Toit had done what Reid prescribed: from a painstaking comparison of observations, he had stepped back to the cause of continental drift. Other scientists, however, were focusing on other observations, and when they stepped back from their observations, they ended up in a different place.

Another Step Backward: William Bowie Adheres to Isostasy

Wegener had proposed continental drift as a reconciliation between historical geology and isostasy, and it was scientists who worked in these areas who therefore most engaged with the idea. But ironically both isostasy and historical geology were used in arguments against it. The man who made the case on behalf of isostasy was the leader of American geodesy in the 1920s, William Bowie (1872–1940). In June of 1927, Bowie wrote to Charles Schuchert that many were "now thinking well of the Wegener hypothesis in the absence of any theory or hypothesis, other than this, which seems to give a logical account of surface changes."[30] But before mailing this typed letter to Schuchert, Bowie paused and penciled in—"I am not one of them." As a practicing geodesist, trained as a civil engineer, Bowie had little vested interest in traditional geological knowledge and might have been well placed to argue the case for continental drift. But by the time of his letter to Schuchert—the year of publication of du Toit's monograph—Bowie had already staked his claim on another theory, based on observations of his own.

Known today as the namesake and first recipient of the American Geophysical Union's annual William Bowie Medal for "unselfish cooperation in research," Bowie was active in scores of different national and international geophysical initiatives. The son of an elite Maryland family whose ancestry was traced to fourteenth-century English knights and eighteenth-century American colonials, Bowie received a general education at St. John's college in Annapolis before studying engineering at Trinity College, Connecticut, and Lehigh University, Pennsylvania. In 1895, he joined the U.S. Coast and Geodetic Survey, where he rose rapidly as the protegé of John Hayford, succeeding him as chief of the Geodesy Division in 1909. A founding member of the International Union of Geodesy and Geophysics, Bowie served as its president from 1933 to 1936; as a member of the U.S. National Academy of Sciences' Committee on Oceanography, he helped to secure the Rockefeller Foundation funds that endowed the Wood's Hole Oceanographic Institution. Through these various activities, Bowie became a close colleague of the leading figures in American geology, including Schuchert, Merriam, Willis, Daly, and Wright. Nor were his activities restricted to academic circles: he strongly believed that if science were to be supported by the public purse, then the public must understand science; he thus wrote many popular articles and gave frequent radio addresses.[31]

As chief of the Division of Geodesy for nearly thirty years, Bowie kept the public purse in mind, and throughout his tenure his overriding concern was the practice of geodesy at his agency. In the mid 1920s, the Survey was deluged with requests for triangulation and leveling data—from mining and petroleum companies laying claims, from state and local governments building roads and bridges, and from its sister agency, the U.S. Geological Survey, which was embarked on a major program of transcontinental geological and topographic mapping. Requests for geodetic data also came from the Carnegie Institution, with which the Coast Survey was cooperating in the study of ground displacements in regions of seismic activity; from the International Astronomical Union, which was attempting to track variations in the earth's magnetic axis; and from the Survey's own Division of Charts for its continual updating of bathymetric data in coastal waters. Bowie strove to accommodate these requests in a period of high inflation and low government appropriations by constantly looking for means to improve the productivity of his division. Toward this end, he introduced the portable steel leveling tower, obviating the need for an erected wooden edifice at every geodetic station, and he was the first director to purchase trucks rather than horses for his field crews.[32] Motivating these pragmatic innovations was another overriding professional goal: to complete the primary triangulation network for the United States—the geodetic equivalent of the golden spike. Bowie accomplished this in the mid 1930s when additional funds became available through the Works Progress Administration. Bowie put men to work holding rods and chains across the country, and the United States was for the first time connected by a complete network of continuous triangulation.

By vigorously pursuing instrumental and technical improvements, Bowie was following closely in Hayford's footsteps. But Bowie's interests were broader. From 1912 onward, he published and lectured widely on the demonstration of isostasy and its geological implications. (His lifetime publication list numbers nearly 400 books and articles.) As the geodetic evidence of isostasy grew, Bowie became increasingly interested in its geological significance. As Hayford had put it in 1911, the existence and comprehensive nature of isostasy were facts "established by abundant geodetic evidence, ... [and] may not be removed or altered by showing that difficulties are encountered when one attempts to make them fit existing theories geological or otherwise."[33] For Hayford, the task of finding a geological theory consistent with these facts was someone else's problem, but Bowie made it his own, and in the mid 1920s he found an answer in the fissiparturition theory of George Darwin and Osmond Fisher. Initially Bowie thought that ripping the moon from the earth was a "pretty wild" idea, but he subsequently came to consider the Darwin–Fisher hypothesis the most logical one he had seen, because it satisfied his preeminent concern: to account for the facts of isostasy.[34]

Of course Fisher's theory was consistent with the facts of isostasy: it had been invented to be. But as Fisher had seen it, this was only part of the problem: having postulated a low-density crust floating in a higher density substrate, one needed an account of how this configuration had come about. Darwin's theory of fissiparturition provided an answer: the low-density continents were the earth's original crust, which split apart when the moon and earth were sundered. The high-density oceanic crust was secondary, formed as magmas welled into the fractures. The Darwin–Fisher hypothesis thus gave a geological account of how and why a bipartite crust had come into existence. It answered the question that G. K. Gilbert had identified in 1892 as the critical problem of tectonics: the differential elevation of continents and oceans (see figure 2.4).

In his various writings on the subject, Bowie repeatedly stressed that the critical fact— the critical observation—was that isostatic equilibrium was complete and local: it was everywhere, and it was not merely approximate. Bowie reasoned that this meant that the earth's surface features must have reached their present configuration early in earth history, so that there had been adequate time for equilibration. The Darwin–Fisher hypothesis satisfied this constraint.[35] The major features had formed early in earth history and—from the point of view of isostasy—the rest of geological history was commentary.

From 1922 to 1926, Bowie published a series of articles in which he developed these ideas, and in 1927 he published *Isostasy*, a book in which he sought to present in one place "the results of the investigations leading to the proof of the isostatic condition of the earth's crust, and [to] apply the isostatic principle to some of the major geologic problems." The bulk of the book was a recapitulation of his published work with Hayford, but in the final chapter, he offered his own "theory, in harmony with isostasy, to account for major changes in the elevation of the earth's surface."[36] The crux of his theory was the assumption that equilibrium was the norm, perturbances the exception. This seemed to him necessitated by the results of his and Hayford's own observations. The critical question for him thus became: What are the forces that *temporarily* disturb the isostatic balance? For Bowie, one part of the answer was clear. The rigidity and strength of the earth's crust decreased with depth—the existence of isostatic equilibrium required that this be so. This suggested that the crust was weak; it *responded* to stresses exerted on it—like a bridge responding to the shifting weights of vehicles moving across it—rather than being the *source* of any important force. Stresses arose in the earth's exterior, and the interior responded. Reviving the idea first developed by Herschel and more recently advanced by Dutton, Bowie suggested that the external stresses arose from the shifting of material over the earth's surface by erosion and sedimentation. He explained by analogy:

> If two rafts of logs were floating in a lake and if some [logs] were taken from one raft and added to another, the hydrostatic equilibrium would be temporarily disturbed, but would soon be restored. Some water would be pushed aside from under the augmented raft and a like amount would move up under the depleted one. This phenomenon familiar to all is just about what occurs . . . in the earth; but the movement of deep material does not adjust itself so rapidly . . . as does water.[37]

The earth's crust was passive, like a raft floating on a stream. The substrate was yielding, like water, but it responded over geological rather than human time frames. The erosional transport of materials is equivalent to the shifting of logs on metaphorical rafts: streams and rivers move sediments from mountains to continental margins, and the earth responds isostatically to the changed load (figure 6.3).

Like Herschel and Dutton—as well as Gilbert, Fisher, Hayford, and Wegener—Bowie understood that the fact of isostasy implied the necessary correlative of a plastic substrate, able to respond to the shifting load above: "The crust sinks down under the sediments and it rises up under areas of erosion. In order that this may be, the subcrustal material must be plastic to forces acting for hundreds of years. *This must be so since isostatic equilibrium exists.*"[38]

Like the earlier advocates of isostasy and the more recent advocates of continental drift, Bowie accepted the idea of a mobile substrate. Indeed, he insisted upon it. But Bowie's earth was passive, with no internal motive power of its own. It was a structure responding to external stresses imposed upon it, and like any well-built structure, it responded primarily in an elastic manner. This engineer's perspective was made

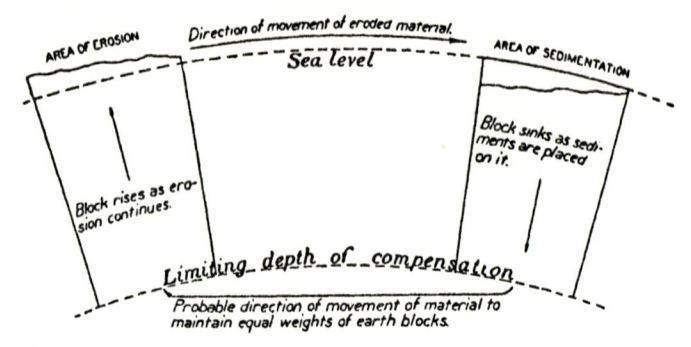

Figure 6.3. William Bowie's theoretical model of isostatic adjustment in response to erosion, transport, and deposition of sediments. In Bowie's model, following earlier suggestions of Herschel, Dutton, and Hayford (see figures 2.5 and 3.4), areas of sedimentary deposition are compressed in response to loading, while areas of erosion expand in response to unroofing. Subcrustal flow from regions of sedimentation and compression to regions of uplift and erosion helps to maintain the isostatic balance. (From Bowie 1927c, p. 128. Copyright 1927 by E. P. Dutton. Used by permission of Dutton Signet, a division of Penguin Books USA, Inc.)

explicit two years later, in an article entitled "The Earth as Engineering Structure." "We know the earth is not a stable structure," he declared, "it is yielding continuously to forces which are acting on it."[39] Some parts are overloaded and some are underloaded, and occasionally the structure ruptures—hence earthquakes. Bowie went on to specify four sets of forces acting upon the earth, all arising from the erosional transfer of crustal materials: (1) loading of the crust in regions of sedimentation; (2) elevation of the crust in regions of erosion; (3) thermal expansion of depressed crust penetrating into deeper, hotter, portions of the Earth; and (4) thermal contraction of uplifted crust cooling off. These four forces, he believed, could account for all major geological instabilities, including earthquakes, volcanoes, and marine transgressions and regressions. Trained as an engineer, Bowie had treated and solved the earth's instability as an engineering problem.

Bowie knew that others had different views, and in a 1929 paper, "Possible Origin of Oceans and Continents"—an obvious rejoinder to Wegener—he explicitly advocated the Darwin–Fisher hypothesis as an alternative to continental drift. Indeed, he suggested that it was the alternative that one would naturally arrive upon if one considered the facts of isostasy. "With isostasy proved," he wrote, "we must go back one step further in the earth's geological history and try to unfold a rational hypothesis regarding the configuration of the earth's surface."[40] That, he suggested, was fissiparturition.

Unlike many of his geological colleagues, Bowie was clearly sympathetic to theory-driven science, but he recognized that overt advocacy of a particular theory might prove counterproductive. He thus retreated to a rhetorically diffident stance, allowing that "all of this is speculation." But, as geologists all acknowledged, some speculations are better than others; the point was to figure out which ones were better supported by evidence. Here, Bowie was unequivocal: the geodetic evidence supported the Fisher hypothesis.[41] But what of du Toit's evidence? What of *all* of the geological evidence of continental drift? Here and elsewhere, Bowie discussed only one line of evidence—the one he called Wegener's "strong point"—namely, the jigsaw-puzzle fit of the continents.[42]

> On a globe, we notice that the two shores of the Atlantic Ocean are so nearly parallel that they look like the banks of a great river. It was this similarity of conformation of the coast lines of the Atlantic Ocean that led Alfred Wegener to advance his hypothesis that North and South America are drifting away from Europe and Africa. It will be noticed that Australia and the East Indies could be pushed westward into the Indian Ocean [as well]. All of this seems to be corroborative evidence in favor of the hypothesis of Osmond Fisher.[43]

To this engineer, concerned above all with questions of equilibrium and stability, the evidence of rupture was the most compelling part of Wegener's story. But rather than read it as confirmation of drift, Bowie read it as evidence of an entirely different theory: one that placed continental rupture early in earth history and made it a single, unique event. In doing so, Bowie was ignoring the paleontological and lithological evidence—and the paleoclimatic evidence—evidence of which we know he was fully aware. How could he do this? How could the president of the American Geophysical Union dismiss whole areas of earth science research? From his private correspondence, it is evident that Bowie did not discuss the other evidence because he had already decided that drift was impossible, not for lack of a mechanism, but because of what it would mean for isostasy.

From Calculational Strategy to a Theory of the Earth

By 1935, the debate over continental drift in America was just about over, and Bowie revealed this implicitly in a paper usurping Wegener's title entirely. In "The Origins of Continents and Oceans," published in the *Scientific Monthly*, Bowie began, "There have been many hypotheses regarding the origin of oceans and continents, but most of them seem to conflict with the laws of physics and mechanics and therefore seem to be untenable."[44] In earlier correspondence, Bowie had expressed the same idea. Writing to Schuchert in the autumn of 1928, after reading Arthur Holmes's review of papers read at the American Association of Petroleum Geologists' recent symposium, Bowie rejected drift without discussing the geological evidence:

> I really cannot figure out how the continents can drift about in an aimless sort of way . . . [and] Holmes brings out a new thought which is even more impossible than Wegener's hypothesis. That is that the submerged ridge through the Atlantic Ocean is the place at which North and South America separated from Europe and Africa, the latter two continents drifting eastward and the Americas drifting westward. I do not see how the same force, operating to send one mass westward, could make another go eastward. I believe that we need to apply elementary physics and mechanics to the continental drift problem in order to show how impossible that drifting would be.[45]

Holmes's force was of course convection, which is accepted today as the driving force of crustal motions. And by modern sights Holmes was right about the mid-Atlantic Ridge being the site of continental breakup. So which laws of physics and mechanics was Bowie thinking of? The answer is revealed in another letter, written to Schuchert the previous year: *"If we have the Airy isostasy, then there must have been horizontal movements forming roots to the topographic features."*[46]

The mechanics to which Bowie referred were those that controlled the *structure* of the continents. Bowie's theory—with its emphasis on vertical motions arising from erosional transfer of materials—was a direct consequence of the Pratt–Hayford model of isostasy. And so was his rejection of continental drift. Pratt had proposed that isostatic equilibrium was achieved by density differences, and Hayford had used this assumption as the basis of his figure of the earth. Bowie extrapolated this to suggest that the density differences in the crust were a response to erosional loading and unloading. Loading in areas of sedimentation, geosynclines, compressed the crust below it; unloading in areas of erosion, mountain belts, permitted crustal expansion. The Pratt model could be explained causally by vertical compression and expansion of the crust in response to varying superincumbent loads. Bowie elaborated in a published article based on his presidential address to the Philosophical Society of Washington that same year:

> Since the earth's crust is so weak as not to withstand the loading and unloading caused by sedimentation and erosion, we must conclude that some of the geological theories that are based on the idea of a very rigid crust, carrying horizontal thrusts for hundreds and even thousands of miles, must be modified or abandoned. In this particular [aspect] it would seem that isostasy has its most important bearing. Isostasy in itself is not an active agency; it is a condition of rest and its proof leads to the logical conclusion of a very weak crust and thus restricts the field within which hypotheses and theories may be formulated to account for surface changes.[47]

Bowie's adherence to the Pratt model made him assume that isostatic compensation had to be accommodated within the crustal prisms—by expansion and compression of the crust—and therefore that the crust must be very weak. And a weak crust could not transmit lateral stresses over any appreciable distance. Bowie's opposition to drift was a direct result of his commitment to Pratt isostasy.

As we have seen, there was an alternative view of isostasy: the Airy model, in which equilibrium was achieved by the existence of deep crustal roots below topographic highs. And the Airy model had its proponents. Among them were University of California at Berkeley Professor Andrew Lawson, who was beginning to advocate the Airy model in order to account for field evidence of horizontal compression in the Coast Ranges of northern California. Writing to California Institute of Technology seismologist Harry Wood, Bowie acknowledged Lawson's work. Bowie wrote in the spring of 1927: "Geologists are inclined to advocate the 'roots' hypothesis, while the geodesists are inclined towards the uniform thickness hypothesis."[48] Among geologists, Bailey Willis was perhaps the most vocal advocate of the Airy model; Lawson most clearly connected it to field evidence of horizontal compression in mountain belts. Wood followed Lawson, but Bowie remained unconvinced.[49]

Bowie understood that crustal roots would require the existence of horizontal stresses originating within the earth to maintain them. That is, *the existence of crustal roots implied some form of lateral compression to create and sustain them.* Not surprisingly, this was the model to which Wegener and many other European geologists subscribed.[50] The Airy model seemed to lead directly to continental drift. Roots of mountains implied horizontal compression to create and maintain the roots; horizontal compression implied the deformation of the continents to form isostatic roots. One seemed to follow inevitably from another.

For Bowie to have accepted continental drift, he would have had to abandon the Pratt model of isostasy. But Bowie's seminal accomplishment as a scientist—the basis of his status in the American scientific community—was his demonstration, with Hayford, of isostasy based on the Pratt model. The premise of a uniform depth of compensation was the simplifying assumption that made their calculations possible, calculations which led to what Sir Harold Jeffreys called "one of the outstanding scientific achievements of our time."[51] As a result of Hayford and Bowie's accomplishment, Hayford's eulogizer could write in 1931, "there are few today who will deny the existence of isostasy for any considerable portion of the Earth's crust."[52]

But why had Hayford used the Pratt model? Mathematically, Pratt and Airy were equivalent, and geologically there was little basis for preference. The answer is not a mystery. Hayford had been explicit about the basis for his choice: it was an instrumental strategy. In *The Figure of the Earth*, he wrote, "The assumption [of Pratt isostasy] was adopted as a working hypothesis, because it happens to be that one of the reasonable assumptions which lends itself most readily to computation."[53]

Hayford did what all scientists are taught to do: he applied Ockham's razor. Given two possible interpretations, he used the simplest one—simplicity in this case being defined as ease of calculation. But his choice was more than just a routine application of a widely used but little analyzed principle. Hayford was a highly self-conscious scientist, and he believed that the choice of methods—the use of particular techniques and strategies—was fundamental to what he had done.

> Very early in the investigation it was realized that it would be necessary to compute the topographic deflection for each station, and that the computation to serve its full purpose must extend to a great distance from the station. It was also realized that to make such computations by any method known to have been used hitherto would be impossible on account of the great expenditure of time and money involved. It was necessary, therefore, to devise some new method of computation, or to modify old methods, so as to make these computations feasible.[54]

Hayford's task was Herculean—to analyze the topographic deflections for hundreds of stations by accounting for topography for hundreds of kilometers around each one. His ingenious solution was to divide the land surface into compartments and use specially designed templates to calculate the mean elevation for each compartment around each station. The assumption of a uniform depth of isostatic compensation was similarly motivated: the computations of isostatic corrections would have been otherwise impossible by "any method known to have been used hitherto."

Hayford's point, here and elsewhere, was that the use of timesaving devices was not merely a utilitarian response to constricted governmental budgets—although it certainly was that as well—but served the interests of science. Data in geophysics were never sufficient, so the path to new knowledge necessarily involved creative shortcuts.

> The complete solving of geodetic problems awaiting solution would require millions in dollars and scores of years. The real problem before the geodesist, therefore, is to make as much progress in his attack on a sensibly infinite problem as can be made with finite available money and time. Every device which saves money and time without serious loss of accuracy is a means to a greater advance into the un-

> known, and therein rests its real value. Conversely, every unnecessary refinement or complication which is allowed to remain in the methods of computation produces not simply a comparatively useless expenditure of money and time, but is also a serious drag, preventing real advance. Such unnecessary refinements or complications may give satisfaction to the computer and the investigator and may be frankly sanctioned if the real purpose of the investigation is to furnish such satisfaction. They should not be sanctioned if the real purpose is the much broader, better one of securing as much as is feasible of real increase in the sum of human knowledge.[55]

This latter comment was an obvious dig at European colleagues—often better trained in mathematics than their North American counterparts, but not necessarily better scientists for that—and perhaps at those Americans who emulated them.[56] Hayford continued:

> Continuous and close attention to the economics of the problem has led to real advance in the research ... The spirit of the investigation and the motives which control in fixing methods will be missed by the reader who does not realize this. The methods used represent an effort to utilize available finite resources in the ways which are most effective in increasing the sum of human knowledge in regard to the figure and size of the Earth. Such are the reasons for the unusual prominence given to time-saving devices and methods in this publication.[57]

In other disciplines, a crucial experiment—a "golden event"—may change scientific thinking, but in geodesy knowledge rests in the totality of data.[58] A single geodetic station is of virtually no significance. Hayford's insight was to understand this profoundly, and therefore to defend rather than apologize for the creative shortcuts he had taken.

Was Hayford sincere or was he disingenuous? Perhaps he was making a virtue out of necessity. The context of his work—the enormous pressure for geodetic data in the United States of the late Gilded Age—certainly made his philosophy pragmatic. But if so, this pressure worked to the benefit of American science, for his results were internationally applauded. Perhaps the important point is this: Hayford's approach could be considered distinctively American—with its obviously pragmatic flavor, its emphasis on results rather than method—and it led to what was widely discussed as an *American* contribution to international science. American geodesists felt proud of his accomplishment. If continental drift threatened it, then it would be American geodesists who would be most sensitive to this threat.

But did Hayford think that the Pratt model was true? Did he think that the depth of isostatic compensation in the earth was actually uniform? Not as far as we can tell. In his writings, Hayford never suggested that the Pratt model was a realistic representation of the material constitution of the earth. Indeed, he never discussed the issue at all. As far as the evidence goes, the Pratt model, like Hayford's templates, was an operational strategy, a calculational device. Accordingly—and unlike most geological treatises—*The Figure of the Earth* contains no picture of the earth: it is a mathematical figure, not a visual one. There is no illustration of the earth according to the Pratt model. There is no cross section showing the columns of the earth with varying density but uniform depth. The text is accompanied only by graphs, charts, and tables. Hayford's exclusive reliance on these graphic visual aids is consistent with isostasy as a *mathematical* theory; it was no more necessary for Hayford to illustrate the Pratt model than for him to illustrate his pen. But in the hands of William Bowie, things changed.

Representation as Reified Belief

When Hayford left the Coast Survey in 1909 to become the first Dean of the College of Engineering at Northwestern University, Bowie began to disseminate the results of their joint work. Initially, like Hayford, he relied primarily on mathematical arguments: Bowie's early papers contain many tables of isostatic reductions but few illustrations. But as Bowie began to write increasingly for a geological audience—and subsequently for a popular one—he began to include illustrations. And these illustrations reveal a fundamental shift in his thinking relative to Hayford's.

In *Isostasy*, Bowie illustrated the two theories of isostasy by reference to Archimedes' principle of flotation. Pratt isostasy could be visualized by imagining a series of blocks, composed of different materials, floating in a vat of liquid mercury (figure 6.4). All the blocks have the same mass and cross section, but their heights vary inversely as their densities. Airy isostasy, in contrast, could be visualized by imaging a series of blocks all composed of the same material, in which the height of the block above the surface of the mercury is proportional to the depth to which the base of the block extends (figure 6.5).

But here Bowie departed from Hayford: his next picture was a realistic representation of a cross section of the earth according to the theory of isostasy. But which isostasy? The answer, of course, is Pratt. Bowie did not present a realistic representation of the Airy hypothesis, because for him it was not realistic: it was "just" a hypothesis, of which he made no use. The truth, for Bowie, lay in the Pratt conception, and this is what he therefore drew (figure 6.6).[59] Indeed, the caption for this drawing does not state that this is how the earth would appear if the Pratt model were correct; rather, it states that this is how the earth would appear *if we could cut a cross section through it.* If we could cut open the earth, this is what we would see. In the representation of the theory, Bowie's beliefs were reified. Bowie believed in the Pratt model of isostasy. And this belief led to the deductive consequence that continental drift was impossible. What began as a calculational strategy was now a belief about reality.

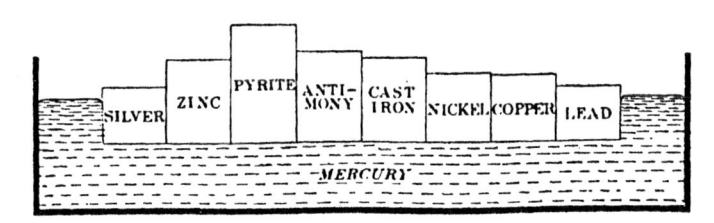

Figure 6.4. Bowie's illustration of Pratt isostasy. In the Pratt model of isostasy, equilibrium is achieved by variations in density proportional to variations in elevation: the highest mountains are composed of the lightest crust (pyrite), the lowest valleys of the densest crust (lead). The depth at which isostatic compensation is achieved is uniform. A better analogy is loaves of bread: Imagine a series of loaves, each with the same weight of dough but with varying amounts of yeast, set on a table to rise. The loaf with the most yeast rises the highest and therefore ends up the least dense, but the base of all the loaves is the same. (From Bowie 1927c, p. 22. Copyright 1927 by E. P. Dutton. Used by permission of Dutton Signet, a division of Penguin Books USA, Inc.)

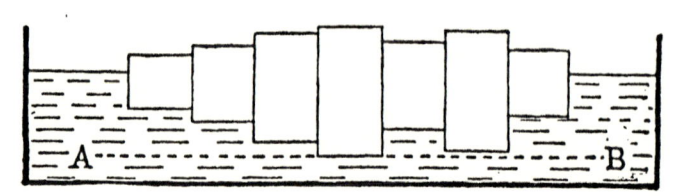

Figure 6.5. Bowie's illustration of Airy isostasy. In the Airy model of isostasy, the crust has the same density throughout, and the continental blocks sink down into their substrate proportionately to their overall height. This model, called the "roots of mountains" hypothesis, is based on Archimedes' principle of flotation and describes the phenomenon by which light objects float in denser media—hence the well-known fact that the visible portions of an iceberg is only the "tip." Bowie's illustration of Airy isostasy was borrowed from the Yale geologist and close colleague of Charles Schuchert, Chester Longwell. (From Bowie 1927c, p. 24. Copyright 1927 by E. P. Dutton. Used by permission of Dutton Signet, a division of Penguin Books USA, Inc.)

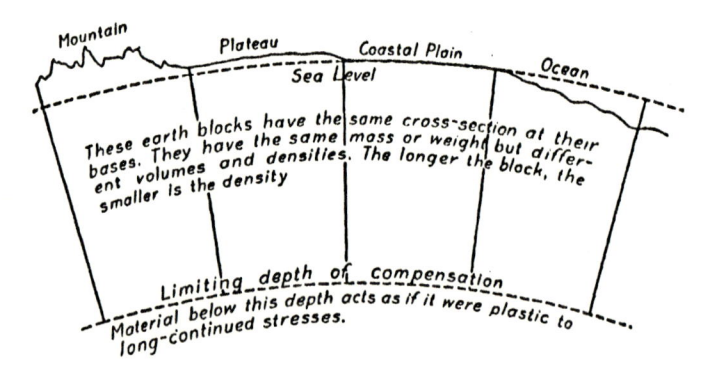

Figure 6.6. Bowie's representation of the structure of the earth's crust according to the Pratt model of isostasy. In his caption to this figure, Bowie wrote: "If the earth's crust should be cut into blocks of equal horizontal cross section by vertical planes, each block would have very nearly the same mass as each of the other blocks. The blocks would exert the same pressure on the subcrustal material." Thus he implied that this is what the earth would really look like if we were to cut through it. Pratt isostasy is no longer just a mathematical convenience; it is a belief about how the earth really is. Note also that Bowie explicitly states that "material below [the limiting depth of isostatic compensation] acts as if it were plastic to long-continued stress." The idea of a plastic substrate was not, therefore, in dispute. What was in dispute was the implication of crustal structure for tectonic processes, and here the assumption of the reality of the Pratt model had important consequences. (From Bowie 1927c, p. 25. Copyright 1927 by E. P. Dutton. Used by permission of Dutton Signet, a division of Penguin Books USA, Inc.)

Why did Bowie believe in the physical reality of a uniform depth of isostatic compensation when his mentor had adopted it merely as a operational assumption? For Bowie had not forgotten that he had.

> To make the application of the theory of isostasy at all feasible, Hayford had to make certain assumptions of general application. These [included] that the compensation extends to a certain depth, and that this depth is the same for all parts of the crust. . . . While they were necessary to simplify the computations, neither Hayford nor the writer believed them to be strictly true.[60]

Bowie knew that a uniform depth of compensation was an assumption, and he did not even initially assume that it was true. On this occasion he continued:

> By the use of the assumptions, however, it was possible to prove that the theory of isostasy is true to a remarkable degree. . . . The corrections for isostatic compensation as thus computed were applied to the deflections of the vertical and to the values of gravity with the remarkable result that for the area of the United States deflection anomalies were on an average reduced to 10 per cent of what they would be if there were no isostatic compensation and the average gravity anomaly to from 10 to 15 per cent of what it would be on the rigid-earth hypothesis. These results indicate very clearly that the United States, both as a whole and locally, is in practically complete isostatic adjustment and that the computation of the isostatic effect is substantially correct.[61]

Bowie thus happily announced: "The theory of isostasy has been proven."[62] The success of the model seemed evidence of its truth. And he concluded, "In this book, [I] have assumed that the Pratt theory of isostasy is the only one so far advanced that is sound, and [my] conclusions, therefore, are based on it."[63]

Because the model worked, Bowie came to believe it was true. The success of the Pratt–Hayford model seemed to validate it. In retrospect we can see the obvious error of his logic, but why else do scientists accept the veracity of their models and theories? Scientists accept theories because they work: because they explain things that were not explained before, because they make correct predictions, or because when one accepts them as true, one can solve problems that could not be solved before. This is what Bowie did. But the difficulty is obvious: it is the logical fallacy of affirming the consequent. False theories sometimes work. The Ptolemaic system of astronomy enabled scientists to predict the motions of the planets, and the Pratt model enabled John Hayford and William Bowie to calculate the isostatic adjustment and gravitational anomalies of the United States. An idea initially accepted for pragmatic reasons became a form of scientific practice, which became a belief about the physical constitution of reality. And that belief had consequences: it was incompatible with continental drift.

7

Uniformitarianism and Unity

William Bowie settled on a theoretical position that accounted for isostasy and the jigsaw-puzzle fit of the continents but ignored the facts of historical geology. And yet, as we have seen, he interacted and corresponded with historical geologists, particularly with Charles Schuchert (1858–1942). Of all the American geologists who ultimately rejected the theory of drift, Schuchert was perhaps the one who engaged the problem the most seriously. As America's foremost historical geologist, Schuchert was well placed to argue the case for or against drift, and he grappled with the question of continental connections for at least fifteen years. In the end, however, Schuchert, like Bowie, rejected continental drift. Just as Bowie argued against drift because of beliefs grounded in the exigencies of geodetic practice, Schuchert ultimately argued against drift because of beliefs grounded in the exigencies of geological practice. For Bowie, the practice was Pratt isostasy, for Schuchert, it was uniformitarianism.

Charles Schuchert rejected continental drift because he interpreted it to be incompatible with uniformitarianism. However, he did not reject it because he could not see drift taking place, as might be supposed. Uniformitarianism has meant many things to many people, and to Charles Schuchert in the late 1920s, it meant—rightly or wrongly—an essentially steady-state earth, whose details were forever changing but whose large-scale patterns and relationships remained the same. And this seemed to him to deny the possibility of major changes in the configuration of the continents. Moreover—and perhaps more importantly—for Schuchert, as for most historical geologists, uniformitarianism was a form of scientific practice, a means of *doing* historical geology. It was, in fact, *the* primary means of doing historical geology. For Schuchert, abandoning uniformitarianism was nearly tantamount to abandoning historical geology altogether. Not surprisingly, he declined to do this. Like William Bowie, Charles Schuchert settled on a theoretical position that preserved his scientific practice. But whereas Bowie's theoretical ideas had little staying power, the alternative that Schuchert embraced influenced a generation of geologists to believe that drift was not so much impossible as unnecessary.

Grappling with Gondwana

Schuchert was to historical geology what Bowie was to geodesy: the acknowledged American leader. Socially, however, the two men were a study in opposites. Schuchert's long scientific career traversed as far a distance as was possible to go—socially, economically, and intellectually. In contrast to Bowie's upper class Episcopal inheritance, Schuchert was the son of impoverished Catholic immigrants from Bavaria. The eldest of six children, he was forced to leave school at the age of 13 to help his father establish a tiny furniture business. In 1877, fire struck, and Philip Schuchert suffered a breakdown; Charles rebuilt the business and became the family provider. But his heart was elsewhere. Since the age of six, he had pursued a hobby of fossil collecting and, through self-guided reading, had taught himself to identify the abundant brachiopods he found near his home. Later, he took night classes in lithography and became interested in its application to scientific illustration and analysis. In 1878, Schuchert began to attend meetings of the Cincinnati Society of Natural History, where he became a friend of E. O. Ulrich, curator of the Society's collections. The young amateur introduced his mentor to lithography and convinced him to seek remunerative work preparing lithographic illustrations for various state geological surveys. In 1884, fire struck the furniture factory again and Ulrich convinced Schuchert to become his paid assistant. Schuchert worked with Ulrich for three years, preparing specimens and producing scientific lithographs.[1]

Schuchert's work and collection (which eventually numbered over 10,000 specimens) came to the attention of James Hall, State Geologist of New York, who hired him to assist in the production of *Paleontology of New York, Vol. VIII* (1889–1891). In Hall's sphere, Schuchert was exposed to the larger geological problems of stratigraphy, mountain-building, and the relation of fossil evolution to earth history and began to develop contacts in the professional geological community. In 1893, he obtained his first professional position, assistant paleontologist at the U.S. Geological Survey, and in 1894 he was appointed assistant curator of stratigraphic paleontology at the U.S. National Museum in Washington, D.C. On behalf of the museum, Schuchert participated in numerous international collecting expeditions and became an expert in the global distribution of fossil species. He also became widely recognized as a leading brachiopod expert; his 1895 pamphlet, *Directions for Collecting and Preserving Fossils*, became a curatorial standard.

In 1904, Schuchert was offered the tripartite position from which he achieved his greatest level of scientific accomplishment and professional leadership: professor of paleontology, Yale University; professor of historical geology, Sheffield Scientific School; and curator of geological collections, Peabody Museum. This appointment— teaching in America's oldest geology program and curating the country's most important invertebrate collection—solidified Schuchert's stature as the leading figure in American historical geology and paleontology. During this time, he published prolifically on brachiopods, paleogeography, and stratigraphic correlation, and with his Yale colleague L. V. Pirsson he wrote *A Textbook of Geology* (1915), which became a classroom standard for decades. He never married, but, like many scientific men of his era, he had a longtime woman assistant, Clara M. LeVene, who attended to both personal and scientific matters and was the coauthor with him of several books and articles, including the great bibliographic catalogue *Brachiopoda* (1929).

Schuchert served as the first president of the Paleontological Society (1910), as president of the Geological Society of America (1922), and as a vice-president of the American Association of Science (1927). He was a member of the National Academy of Sciences and an editor of the *American Journal of Science* and received numerous scientific awards and medals, including the Penrose Medal of the GSA. Although he retired from teaching in 1926, he remained extremely active: when he died in New Haven at the age of 84, he was at work on the third and final volume of his landmark *Historical Geology of North America* (1935–1943, the last volume published posthumously). A man whose career was forged entirely on self-education and apprenticeship, Schuchert was awarded honorary doctorates by Harvard and Yale Universities; his legacy is recalled annually by the Geological Society of America's Schuchert Medal for contributions to paleontology.[2]

Ironically, the organizations he presided over helped to consolidate the professionalization of American science that eventually excluded the self-taught scientist. Schuchert's career thus highlights the tremendous changes that occurred in American science near the turn of the century, beginning as it did in an era when few scientists had formal scientific training and ending in a period when few did not. Schuchert was one of the last professional geologists in America with no formal education in any scientific field.

At the 1926 American Association of Petroleum Geologists (AAPG) symposium on continental drift, Schuchert spoke harshly and at length against the theory, and his words were frequently quoted in later years.[3] But behind the scenes, something more subtle and complex was occurring for Schuchert—and had been for some time. Twenty years before, the demands of teaching had led Schuchert into the area for which he ultimately became most famous: paleogeography. Having never taken a college course himself, he was vexed by the difficulty of conveying to students the massive amounts of geological detail needed to understand the distribution of fossil species in relation to earth history. To solve the problem, he began to compile maps illustrating the distribution of fossils with respect to global structure and stratigraphy. Initially descriptive, the maps became increasingly synthetic; ultimately, the data of geology and paleontology were used to reconstruct the limits of land and sea in each geological period.

Through the work of paleogeographic reconstruction, Schuchert became acutely aware of the problem of fossil homologies—the evidence for Gondwana. As an autodidact and literate in German, he had read Suess and was impressed by his arguments. By the time he came to teach at Yale, there was no question for Schuchert but that continental connections had existed in past geological periods. By the time Wegener proposed continental drift, Schuchert understood that there was considerable petrological evidence to support the paleontological point. "It can be truthfully said," Schuchert acknowledged at the AAPG meeting, "that Wegener's hypothesis has its greatest support in the well known geologic similarities on the two sides of the Atlantic, as shown in strikes and times of mountain-making, in formational and faunal sequences, and in petrography."[4] Schuchert thus concluded—despite his harsh criticism of Wegener's methodology—that "we must be open-minded toward at least some unknown amount of continental displacements."[5] But as a professor of geology at Yale, he was also heir to Dana's tradition of continental permanence, which was clearly at odds with both sunken and drifting continents. Schuchert thus personally experienced the logical conflict so clearly laid out by Wegener. Indeed, no one, perhaps, experienced the conflict more acutely than he. His dual loyalties are revealed by the dedication of his *Paleogeography of North America*: to Suess *and* Dana.[6] Suess's theory was consistent with the facts of paleontology and Dana's was compatible with isostasy, but neither could claim both. For a time, it seemed to Schuchert that the dilemma would be solved by his friend and Yale colleague Joseph Barrell. To understand the development of Schuchert's theoretical views, we need to look closely at the work of this important colleague.

Isostasy Reexamined: The Work of Joseph Barrell

Born in New Jersey, Joseph Barrell (1869–1919) was the son of amateur naturalists and farmers Henry and Elisabeth Barrell. As a child he was said to have read the *Encyclopedia Britannica* for hours at a time and loved to collect rocks and plants on the family farm. After studying mining engineering at Stevens Preparatory School and Lehigh University, he earned a master's degree in geology and practical astronomy (i.e., surveying) from Lehigh in 1897 and a Ph.D. in geology from Yale in 1900. Barrell taught geology and biology at Lehigh for two years before returning to Yale as an instructor of geology in 1903. In 1908, he was promoted to professor of structural geology, the position he held until his death.

Although his entire professional career was spent in academic positions, Barrell gained practical experience and financial compensation through summer work in mining geology. The most important of these experiences, financed by the U.S. Geological Survey, was his field study of mineralized granite plutons in Montana. The monograph that resulted, *Geology of the Marysville Mining District* (1907), documented the structural relations of the Marysville granite batholiths and their country rocks and used them to develop a general theory of igneous intrusion. The rocks around the batholiths were extensively metamorphosed and locally melted; Barrell suggested that this had been caused by the heat of the intrusion, which had emplaced itself by melting and assimilation of the adjacent rock. Barrell's theory of "magmatic stoping"—still accepted today—made him internationally famous.

When Schuchert joined the Yale faculty the year after Barrell, the two became confidants. Although Schuchert was eleven years the elder, Barrell's was the more prodigious intellect: his formal education and profoundly penetrating mind made him both an intellectual resource and a personal inspiration to Schuchert, and to all who knew him. But Barrell was struck down in his prime, dying from spinal meningitis complicated by pneumonia just a few days after his election to the National Academy of Sciences. His death devastated his colleagues. Widely viewed as one of America's leading geologists—perhaps the leading theorist of his generation—Barrell was described by one eulogizer as a man whose "intellectual power was so obvious and so continuously displayed that twenty years of intimacy have left me with the impression of a mind rather than a man."[7] T. C. Chamberlin thought Barrell would have become the "chief [American] exponent in the subject of geologic sedimentation, metamorphism, and the geologic bearings of isostasy and the genesis of the earth" and was already "one of the most promising leaders in the darker problems of earth science." William Morris Davis, the Harvard professor who helped found the discipline of geomorphology, described his premature death as an "overwhelming disaster for American geology."[8] Davis may have been right, because at the time of his death Barrell was working on the same problem as Alfred Wegener.

After completing the Marysville work, Barrell had become interested in the relationship of isostasy to geological problems. The problem of magmatic stoping was in part an isostatic problem: Did magmas rise because of density differences? Barrell thus extended his inquiries into general tectonic questions. In a series of eight papers entitled "The Strength of the Earth's Crust," published in the *American Journal of Science* in 1914 and 1915, he discussed the geophysical implications of the theory of isostasy. Augmenting Hayford's and Bowie's mathematical deductions with geological evidence, Barrell furthered the argument, earlier made by G. K. Gilbert, that isostasy necessitated a flexible, plastic underlayer beneath the earth's rigid outer crust. Barrell coined the term *asthenosphere* (zone without strength) for the plastic underlayer and *lithosphere* (zone of rock) for the rigid zone above it. The asthenosphere, he reasoned, must be involved in large-scale tectonic processes.

Bowie's and Hayford's demonstration of isostasy in the framework of the Pratt model—which placed isostatic response within the crustal columns—had led them to the conclusion that the earth's crust was exceedingly weak. Indeed, they concluded that it had virtually no strength at all. (This became part of Bowie's argument against the possibility of continental drift: that the crust was too weak to sustain tangential stress.) But, although Barrell was sympathetic to the isostasists' argument, the situation, he argued, was more complicated than they allowed. Tangential stressses were a *reality*— the geological evidence for their existence was overwhelming—so the crust *must* be able to transmit them. River deltas suggested a similar conclusion: if the crust were perfectly weak, then a large sedimentary load would rapidly depress it and there could be no accumulation of sediments to form a delta.

This contradiction had led to the development of two schools of thought: a "weak crust" school, represented among geologists by Clarence Dutton, and a "strong crust" school, led by T. C. Chamberlin. Barrell suggested that here, as in most cases, the truth lay somewhere between the two extremes. G. K. Gilbert had suggested as much in his famous demonstration of isostatic rebound after the dessication of Lake Bonneville: the lake had not yet fully recovered from its former position, suggesting that the crust was still accommodating a portion of the stress.[9] Gilbert's conclusion was consistent with the most famous example of isostatic adjustment: Pleistocene glacial rebound. The delayed response of the Fennoscandian shield to the removal of glacial ice demonstrated that the crust had some rigidity, because in a perfectly isostatic world, rebound would be immediate, and geologists would never know it had happened.

Hayford's mistake, Barrell suggested, was in the assumption that compensation was entirely or even mainly accommodated in the crustal columns. An alternative explanation was that the weakness lay not in crust, but below it. The asthenosphere was thus critical to reconciling the contradictory evidence of geology and geodesy. As Louis Pirsson summarized the matter some years later:

> The maintenance of such isostatic conditions, in spite of opposing geological activities, is held to imply the existence of a zone of undertow below the zone of compensation, which is both thick and weak to shearing stresses. Geologically, such a zone,

called the asthenosphere—the shell of weakness—must have important bearings, which are treated in the later portions of [Barrell's] work.[10]

Barrell drew on stratigraphy, sedimentology, petrology, and paleontology, as well as on isostasy and mechanics, to develop his theoretical ideas, and he urged his colleagues to do the same. The details of observable features had to be considered in terms of their hidden connections to underlying earth structure: "Important though these [surface] details are from the standpoint of the working geologist," he wrote, "their study is comparable to the close study of the technique of a painting whose motive and artistic effects are visible only from a distance. It is the larger conception which determines the expression of the details."[11] For Schuchert, the larger problem was to reconcile the paleontological details of Gondwana with the geophysical problem of getting rid of Gondwana when its demise was demanded. Conversations with Barrell had convinced him that Suess's account could no longer be maintained: large areas of continental crust could not be sunk away at paleontological convenience. If isostasy were a fact, then sunken continents were not. Paleontologists needed another explanation.

In 1926, as he prepared his notes for the AAPG symposium, Schuchert wrote to Arthur Holmes recollecting the history of his own thinking on the matter:

> With the death of Barrell I lost the worker on whom I could try my paleogeographic data to get at their probable geophysical explanation. [The Gondwana] land bridge must have existed all through Paleozoic time into the Mesozoic, and seemingly it began to founder in early Cretaceous time. It used to be one of my difficulties to explain what made it founder. Then Barrell came to my rescue.[12]

The rescue consisted of this. Barrell had closely analyzed Hayford's and Bowie's data—indeed, he was one of a handful of geologists who *had* looked closely at these data—and become convinced that the geodesists' conclusions were overstated.[13] Although agreeing that the crust was broadly in isostatic equilibrium, Barrell argued that there were local areas of significant deviation, and these deviations were the key to crustal tectonics.

Hayford and Bowie had proved that isostasy existed—that is, that geodetic and gravitational anomalies were smaller with the assumption of isostasy than without it. For geodetic purposes, the remaining discrepancies—the "residuals"—were not particularly important, and they were scarcely the point of the study. But for geological purposes, the residuals were critical, because they revealed the locations of stress in the crust and the degree to which the crust was able to sustain and transmit such stress. The areas where isostasy was not perfect were the areas where geological processes were most likely to be taking place; what mattered least for geodesists was what mattered most for geologists.

Geodetic data, however, were difficult for geologists to assess, because, as Barrell put it mildly, "they are cultivated with much labor."[14] Furthermore, the very nature of mathematical work tended to lead one to ignore small discrepanices in favor of larger patterns—in short, to oversimplify:

> The mathematical mode of investigation [in geodesy has] both advantages and disadvantages. The advantages lie in giving quantitative results and in [allowing for] the test of the accuracy of the trial hypothesis by means of the method of least squares. A disadvantage lies in the necessity of erecting simple hypotheses in place of the

complex realities of nature, in order to bring the data within the range of mathematical treatment. The precision of mathematical analysis is furthermore likely to obscure the lack of precision in the basal assumptions and through the apparent finality of the results tends to hide from sight other possibilities of solution.[15]

It was these other possibilities that most concerned Barrell—possibilities that might help to resolve Schuchert's conundrum. Thus, while crediting the importance of Hayford and Bowie's work, Barrell declined to accede to the "finality" of their results.[16] To start, he questioned the conclusion that isostatic compensation was local (figure 7.1). On what basis had Bowie and Hayford come to this conclusion? Evidently not much, Barrell explained.

Hayford and Bowie had calculated the mean residual gravity anomalies for four cases: compensation distributed over 1, 18.8, 58.8., and 166.7 km around each station. The first assumption (1 km) represented local compensation; the others they labeled K, M, and O, respectively. If isostatic compensation were local, then the mean residuals would be smallest for the local case. If, on the other hand, compensation were regionally distributed, then the mean residuals would be smaller in one of the other three cases, depending upon how broadly the compensation was distributed. Barrell recounted Hayford and Bowie's results: for twenty-two stations in mountainous regions with below-average elevations, the resulting mean residuals were 0.017, 0.017, 0.018, and 0.019 dynes. For eighteen stations in mountainous regions with above- average elevations, the means were 0.018, 0.018, 0.017, and 0.020. What was the significance of these numbers? The mean residuals were scarcely different in the different cases. Barrell dryly concluded: "The present writer does not see in these computations any support for the hypothesis of local compensation of topography."[17]

For the eighteen stations with above-average elevations, the smallest residuals occurred for case M—58.8 km—but the calculated differences were vanishingly small. "These figures merely show," Barrell explained, "that, to the outer limit of zone M[,] and probably to the outer limits of zone N[,] one method is as good as another for purposes of computation." More important, the calculated differences were trifling compared with what might be expected from errors in observation, computation or,

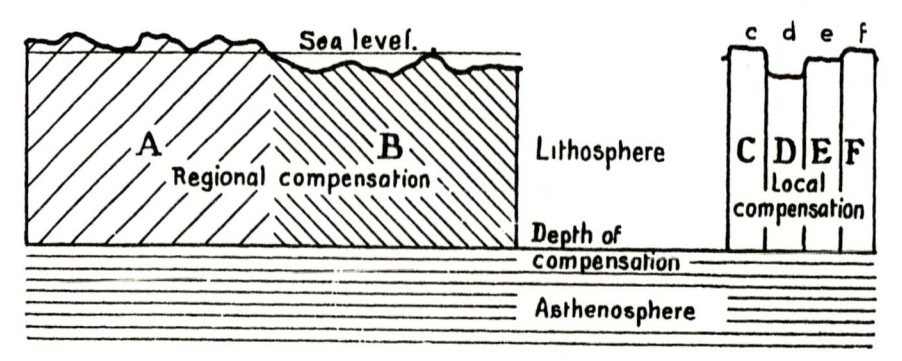

Figure 7.1. Local versus regional isostatic compensation. Local compensation means that crustal loads are accommodated by adjustments in the column of rock directly below the load—like putty sinking beneath a stone. Regional compensation means that the weight is spread over a larger area, which requires a stronger crust. (Barrell 1919b, p. 296.)

most fundamentally, "the preliminary hypothesis regarding . . . the depth and distribution of compensation."[18]

Barrell had performed what in modern parlance is called a sensitivity analysis and found that the residual anomalies were insensitive to the range over which compensation was distributed. Given these data, isostastic compensation might be local, but then again it might not. And if it were not, then the crust need not be so weak as Bowie supposed. The clincher came from Bowie's own data. Bowie had plotted the residual gravity anomalies over the forty-eight contiguous states and shown that departures from equilibrium were not randomly distributed; they could be *contoured* (figure 7.2). There were regional departures from equilibrium, which contradicted the assumption of local compensation. Recalling the earlier arguments of G. K. Gilbert, Barrell explained, "If the strength of the crust was so small that it was able to support notable departures from compensation over areas of only one square degree or less, then these large unit areas could not exist."[19] The crust would immediately warp, upward or downward as required, and only minor, local deviations would persist. "It would seem, therefore, that the geodetic results[,] instead of indicating local compensation . . . show on the contrary a ready capacity of the crust" to maintain regional departures from isostatic equilibrium.[20] The crust was *not* too weak to sustain stresses; on the contrary, the evidence of regional stress was right there in the geodetic data. As Barrell put it: "Isostasy, then, is nearly perfect, or is very imperfect, or even non-existent, according to the size and relief of the area considered."[21]

If isostasy was not local, was it complete? Barrell closely examined this conclusion as well, and he again concluded that Bowie and Hayford had overstated their case. Returning to Bowie's contour map, Barrell pointed out that the anomalies were not systematically related to topography or geology, as they ought to be if their only source was recent, as yet uncompensated, erosion and deposition. In fact, the anomalies showed "a common disregard of physiographic provinces, structural provinces, and geologic formations."[22] This suggested that, against Bowie's later claims, isostatic anomalies could not be explained exclusively, or even primarily, by the shifting of loads on the earth's surface. In fact, it could not be explained by anything on the surface at all. The sources of isostatic anomalies lay within the earth: "Over the greater part of the United States the distribution of anomalies appears to depend more upon the internal than the external departures from regional uniformity and complete isostasy. The internal heterogeneities of mass are therefore presumably greater than the shiftings of mass due to external activities."[23]

Barrell's reference to internal heterogeneity may be interpreted in various ways, but the evidence suggests he was thinking primarily of vertical mass transfer causing vertical uplift. Like Dutton and Gilbert and many other American geologists, Barrell lived East but theorized West. Western U.S. geology was so much more dramatic—so much more *conspicuous*—than Eastern U.S. geology, and perhaps its most conspicuous feature was the evidence of vertical tectonics. The Colorado Plateau, the block faulting of the Basin and Range, the uplift of Lake Bonneville: these features demonstrated the importance of vertical motions rather than "surficial compression as the cause of these [recent crustal] movements," and this suggested upward migration of magmas from depth: "The rising of great bodies of magmas to high level in the zone of isostatic compensation, their irregular distribution, the great quantities of heat and gasses which would invade the roofs are suggested by the observed evidences of re-

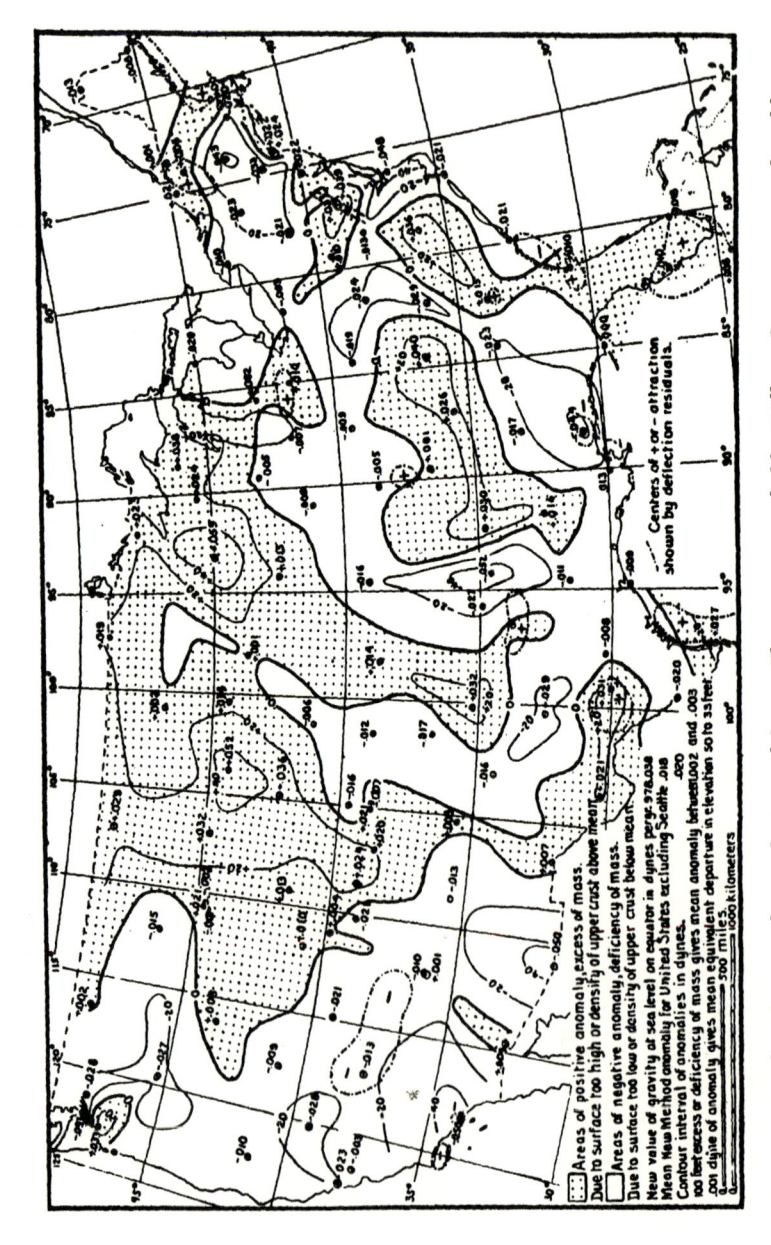

Figure 7.2. Gravity anomalies in the United States. This map, compiled by William Bowie, was reproduced by Joseph Barrell to argue that the residual anomalies showed no coherent relationship to surface physiography and therefore were not caused by surface features but by heterogeneities or forces internal to the earth. The stippled areas represent regions of positive gravity anomaly. (Barrell 1914b, p. 153.)

gional igneous activity at the surface as the probable causes of the changes in density and regional vertical movements."[24] Vertical mass transfer within the earth was a principal cause of disturbances of both the earth's surface and its isostatic equilibrium.

The evidence of vertical dislocations led finally to the critical question of the depth of compensation: Was it really uniform? Barrell once more raised a red flag and argued that the answers were present in Hayford's own data. Although Hayford's uniform depth of compensation was an assumption, not a conclusion drawn from data, it could be tested by data, and Hayford had done so. Dividing his geodetic stations into four geographic groups, Hayford had separately calculated the mean values of the squares of the residuals for five different cases ranging from a zero depth of compensation to an infinite depth. The results were very different for the different areas, seemingly suggesting that the depth of compensation was *not* uniform (figure 7.3). But the results were different in a disordered way: areas near each other were no more similar than areas that were far apart. Hayford took this variation to weaken "the evidence that there is any real variation in the depth of compensation over the whole area investigated."[25] Barrell begged to differ. The residuals were much greater in some areas than in others, and some were much larger than the average values for the United States as a whole. This suggested that the averages were masking local variations, and the same might be true of the depth of compensation. One certainly could calculate an average depth of compensation for the United States as a whole, but like all averages, the resulting number masked whatever variability existed.

PROBABLE DEPTHS OF COMPENSATION

GROUP	No. OF RESIDU- ALS	Solution B Infinite Depth	Solution E 162.2 Km.	Solution H 120.9 Km.	Solution G 113.7 Km.	Solution A 0.0 Km.	MOST PROB- ABLE DEPTH Km.
United States (all observations)...............	733	146.50	14.05	**13.73**	13.75	25.77	**122.2**
Group 12 (parts of Minn., N.Dak., S.Dak., Neb., Kan.)...............	36	196.57	**7.00**	7.47	7.59	11.46	**305**
Group 5 (Mich., Minn., Wis.)	52	34.97	**23.60**	23.64	23.67	27.53	**152**
Group 8 (parts of Utah, Nev., Cal.)..........	42	128.97	22.27	18.79	**18.25**	35.78	**66**
United States (residuals multiplied by 1.327 to compare with Group 8)	....	194.40	18.65	**18.23**	**18.25**	34.21	....

Figure 7.3. The depth of compensation over four areas of the United States. John Hayford's raw data suggested that the depth of compensation over the United States might not be uniform, but he had dismissed these data as too scant to warrant modification of his highly functional assumption. (Barrell 1914d, p. 293.)

Having refuted the assumptions of "high isostasy"—of complete and local isostasy, of compensation accommodated within a weak crust, and of a uniform depth of compensation—Barrell presented his own alternative.[26] The crust was strong: it had to be to explain the regional maintenance of isostatic anomalies and to account for the horizontal tectonics evident in the Appalachians and Alps. But below it, the substrate was weak: it had to be to explain the existence of regional isostatic equilibrium and account for the vertical tectonics of the American West. It was thus that Barrell came to his idea of the asthenosphere: beneath the rigid crust, there was a zone of "low rigidity," not sharply defined, but gradually developing as a function of increased temperature at depth, which acts as the "hydraulic agent" balancing the forces of sedimentation and tangential pressure against uplift and erosion.[27] The stage was set for an isostatic theory of continental subsidence that would solve the problem of Gondwana.

Fragmenting the Continents

In a manuscript written in 1917, which he discussed at length with Schuchert, Barrell suggested that the solution to the problem of continental fragmentation lay in the isostatic effects of basalt intrusion.[28] He argued that massive intrusions of basalt into the upper and middle crust could sufficiently increase the density of local areas to cause them to subside. If these basalts were introduced into the crust from the underlying, high-density substratum, then isostatic readjustment would have to occur (figure 7.4). Continents would be fragmented, and areas that were previously connected would be separated. Ocean basins would form where continents had once been.

Figure 7.4. Joseph Barrell's model of continental fragmentation by basalt loading. Barrell suggested that continental breakup could be explained by large-scale basaltic intrusion, which would weigh down the crust sufficiently to cause it to subside. The subsided portion—loaded with basalt—formed an ocean basin. Barrell's theory was thus, like Wegener's, an attempt to account for the faunal evidence of Gondwana in a manner consistent with isostasy. In this diagram, Barrell uses his idea to explain the formation of the North Atlantic. B = Baffin Island, G = Greenland, I = Iceland, F = the Faroe Islands, and H = the Hebrides. The dark area at the base of the profile (zone 5) represents the basic substrate beneath the crust from which basaltic magmas are derived. Zone 4 is the asthenosphere, a zone of intermediate chemical composition—probably diorite—penetrated by basaltic magmas from below, where the bulk of isostatic adjustment takes place. The dashed line (3) represents the mean depth of isostatic compensation, which, Barrell notes, "has in reality no sharp boundary." Zones 1 and 2 are the surficial layers of the crust; where they are penetrated by high-density basaltic magmas, they subside to form ocean basins. (Barrell 1927, p. 286.) Reproduced with permission of the American Journal of Science.

Barrell made use of his concepts of asthenosphere and lithosphere to explain isostatic adjustment: the plastic asthenosphere was the source of the basaltic magmas, whereas the lithosphere was the zone within which isostatic compensation occurred. To reestablish isostatic equilibrium after intrusion of the high-density basalt, the lithosphere would have to subside. Readjustment would occur gradually, as the intrusions solidified in the lithosphere and "slow lateral displacement of material" occurred in the asthenosphere.[29] Subsidence would be evidenced by normal faulting at the surface, eventually leading to the formation of a new ocean basin. Upstanding remnants would form the mountain belts so conspicuous on rims of ocean basins, or the oceanic islands within them. Thus continental crust *could* be sunk, but only by first transforming it into oceanic crust.

In proposing such a view, Barrell was implicitly refuting Dana's permanent continents and oceans. As a man whose professional credentials and Yale loyalties were indisputable, he was perhaps better placed than anyone to say that the time had come to reexamine this particular American fidelity. "Most Americans hold strictly to Dana's theory of the permanency of continents and ocean basins," he acknowledged, "whereas European workers in general stand by the older view that ocean basins are broken-down portions of the granitic shell. The writer accepts the European view."

Barrell supported this pronouncement with a detailed discussion of the geological evidence of fragmentation, past and present. First, he pointed out, geologists already accepted the idea that "great intercontinental troughs, such as the Red Sea and the Caspian, [were made] by fracture of their margins and subsidence of their floors," and it was well known that the floor of the Red Sea was basaltic.[30] Was it not plausible that the Red Sea was an incipient ocean basin? Second, within the continents, there were rift valleys associated with magmatism: the East African Rift, Death Valley, the Gulf of California. Each of these was formed primarily by block-faulting, and each was now below sea level. Third, when one examined ancient rocks, there was overwhelming evidence of land masses now vanished. The most conspicuous example was the North Atlantic: on the British side, Paleozoic sediments pointed to a source from the northwest; on the American side, same-aged sediments pointed to a source from the northeast. Barrell's own calculations of the volume of sediment required to supply the Upper Devonian alone proved that the source was no mere oceanic island but a land mass of continental scale.[31]

Then there was the biological evidence. Here, Barrell relied on what he had learned from Schuchert: the striking feature of faunal assemblages, particularly in the Paleozoic, was that their present geographic proximities were no key to past affinities. In some geological intervals, fauna of northern Europe were more like those of America than like those of southern Europe; in other intervals Canadian fauna were more similar to those in northwestern Europe than to those in the rest of America. "At long intervals," Barrell surmised, "the geosynclinal seas widen, so that their faunas comingle in the great interior waters and become cosmopolitan. At other times these seas do not coalesce and the faunas remain provincial." And likewise on land. The Pennsylvanian period was famous for coal swamps whose flora were virtually identical in the United States and western Europe; even more famous were the Glossoptera flora of the Permian of the Southern Hemisphere. The Permian sequences were particularly telling, as they contained not only identical flora over India, South Africa, South America, Australia, and Antarctica (and strikingly missing from the Northern Hemisphere) but also distinctive glacial deposits that bespoke a continental ice sheet similar to that which covered the Northern Hemisphere in recent times. And the nature of the flora—hardy ferns—was consistent with a chilly climate. "The spores of these plants could not have crossed thousands of miles of ocean water, and it is equally incredible that a nearly identical *assemblage* of species could have arisen independently in separated continents." A single common species might be dismissed as convergent evolution, but an identical assemblage was not a coincidence. "These and similar facts," Barrell concluded, "have led many geologists of Europe and America to a belief in the great Paleozoic continent of Gondwana."[32]

But how was Gondwana to be accounted for? Land bridges popping up and down would not do. "It is often too easy to explain any real or fancied organic interrelationships by such a hypothesis, but sound science demands that there should be a careful examination to see if the facts to be explained can not be harmonized without recourse to bridging and rebridging deep oceans."[33] Thus Barrell turned to his isostatic argument. "To suppose, then, that an ancient continent has partially foundered, requires that the crust must here have become on the average from 3 to 5 per cent denser"—precisely what Barrell's basaltic loading would do. Add enough basalt to a column of rock, or tranform the granitic crust through magmatic stoping and assimilation, and one could increase the density of the crust enough to make it subside. No one disputed that the opposite had occurred: the uplift of the Colorado Plateau was widely held to be the result, at least in part, of regional density decrease due to the intrusion of hot magmas and gases and accompanying thermal expansion of the cover rocks: "Vast volumes of magmas are needed to account for a change of density in the whole crust sufficient to cause an isostatic uplift of from 1 to 2 miles, and these volumes must therefore be at such a depth that large local surface outpourings would not be possible as proof of their presence."[34] If one could accept the idea of hidden magmas uplifting the Colorado Plateau, one could accept the idea of hidden magmas depressing the Gulf of California or the Red Sea.

There was ample evidence of past large-scale basalt intrusions associated with down-warped or down-faulted portions of the crust. Barrell cited the Sudbury intrusion of Ontario, where "it appears that the floor subsided under it during or after the intrusion"; the Duluth gabbro, underlying the southwestern portion of Lake Superior; and the huge Keweenawan basalts of this same region, which filled a crustal depression with more than 20,000 feet of basaltic lavas in Precambrian times. More recently, there were the basalts of the Deccan Traps and the Columbia River Plateau and, most recently of all, Iceland, with its geysers and active volcanoes. Last but not least, there were the lunar maria. Although the geology of the moon was scarcely known, recent photographs from the Yerkes Observatory strongly supported the idea that the maria were lava plains—seas of basalt. Perhaps Galileo's appellation was not as much of a misnomer as was commonly thought: if the lunar lowlands were basaltic, like the earth's ocean basins, then they *would* have been seas, if only the moon had had water. And their lack of cratering indicated that the maria were younger than the surrounding uplands. Barrell suggested that the maria, like the ocean basins, had formed comparatively recently and were perhaps still forming. In any event, the point was clear: the moon's major features were not permanent, and neither were the earth's.

Returning to the earth, the critical point for Barrell was that geological evidence of fragmentation coincided with areas of known geodetic anomalies. For Schuchert, the critical point was that the geological evidence of large-scale basalt intrusion coincided with the paleontological evidence of foundered continents. For example, in Scotland, Ireland, Iceland, and Greenland there were copious volumes of Oligocene basalt coincident with fossil evidence requiring an early Tertiary separation of North America and Europe. Barrell proposed that these basalts and the fossils associated with them were the remnants of the ancient continent that once connected northeastern Canada and northwestern Europe. "Between these [now] widely separated regions, it seems probable that still larger areas of basalt must occur, which are now at the sea bottom."[35] Incipient examples of the fragmentation process could be found today in Death Valley, the Red Sea, and the East African Rift—all three locations where recent basalt intrusion and concomitant subsidence had occurred. For Barrell, the discovery of present-day examples was satisfying on uniformitarian grounds, but for Schuchert the critical evidence was that which spoke to the breakup of Gondwana. Here, the theory worked beautifully: the enormous basalt flows of the Deccan traps of India were evidently the remnants of "the eastern portion of Gondwana." As he explained some time later to Arthur Holmes, Schuchert was thrilled when similarly huge volumes of basalt were later found in Brazil: they were the other half of the Deccan Traps, the remnants of western Gondwana.[36]

"Gondwana Is a Fact, but I Still Have Finally to Get Rid of It"

Barrell died before he had a chance to publish his manuscript on continental fragmentation, but Schuchert believed that he had found the resolution to the problem of Gondwana. In 1922, Schuchert became president of the Geological Society of America, and he decided to present Barrell's theory at a special symposium at the society's annual meeting to take place in Ann Arbor, Michigan. Arguably this would be the high point of Schuchert's career: as president of the nation's largest geological society, he would present the theory of his cherished colleague that would solve one of geology's largest problems. Schuchert corresponded extensively with colleagues in advance of the meeting to be sure of their attendance. It was to be—he ardently hoped—a "grand show"; the program gave Schuchert's symposium top billing.[37] But to his chagrin and embarrassment, Schuchert was publicly criticized, the theory dismissed. The isostasists vociferously objected. As Schuchert recounted to Holmes four years later, both Bowie and Andrew Lawson, the well-known Berkeley professor, "got on [their] feet and told me at open meeting (Geol. Soc. America) that Barrell's loading wouldn't work."[38]

The crux of their objection, as recounted by Chester Longwell, was that "vertical intrusion can only transfer matter from one part of the lithosphere to another, without altering the average density and so disturbing the isostatic balance."[39] If basalts were displaced vertically within the lithosphere, there would be no net change in mass and therefore no change in density of the continental block. If, on the other hand, the basalts came from outside the lithosphere, then they would disturb the isostatic balance, and this ought to have been evident in geodetic data. This, of course, was exactly Barrell's point: that the basalts came from the asthenosphere, they did disturb the isostatic balance, and this disturbance was evident in geodetic data. Barrell's long series of articles on the strength of the earth had been written to prove this point. Had he been alive, Barrell might have argued that the coastal areas where he detected the most notable discrepancies were precisely where they should be according to his fragmentation hypothesis. But he was unable to press the point.

With Barrell unavailable to rescue him a second time, Schuchert put the problem on hold. For several years he continued his work on Permian paleogeography—his own special section of Gondwana—and from time to time discussed the problem with Longwell, Barrell's successor at Yale. Longwell liked the theory, and he thought that publication of the full manuscript, with the details of Barrell's argument in his own words, might produce a better result than Schuchert had obtained at Ann Arbor. In early 1926, when the American Association of Petroleum Geologists was organizing a highly publicized symposium on continental drift and William Bowie was propounding a theory of continental fragmentation that completely ignored paleontology, Schuchert and Longwell decided to revive Barrell's proposal and to publish his manuscript posthumously. As an editor of the *American Journal of Science*, Schuchert had a ready outlet, and so Barrell's paper appeared there the following year.[40] Longwell wrote an introduction in which he addressed the criticisms made at the 1922 meeting. Schuchert, meanwhile, continued to search out alternatives. He had recently begun a correspondence with Arthur Holmes, and he turned now to Holmes as he had turned to Barrell before.

Like Alfred Wegener, Joseph Barrell had developed a theory to reconcile historical geology with isostasy. Unlike Wegener, however, Barrell had analyzed the geodetic data in scrupulous detail. Was it possible, as Bowie and Lawson claimed, that Barrell had got isostasy wrong? Schuchert wrote privately to Holmes to ask his opinion. "If his work stands the test of geologists," he wrote hopefully in the summer of 1926, when he and Longwell first decided to revive Barrell's work, "we can get rid of Gondwana and other ancient lands."[41] Now, in the spring of 1927, Schuchert sent Holmes the edited manuscript. As before, the crux of Schuchert's argument was the "fact" of Gondwana—the indisputable evidence of faunal homologies—and the need for a theoretical account of it. He wrote in his cover letter to Holmes:

> To me the South Atlantic land bridge between Brazil and Africa = Gondwana (the western part) is a fact for Paleozoic and much of Mesozoic time. We must have a means of getting rid of it in early Cretaceous time, and I asked Barrell to get rid of this land for me. The outcome is the enclosed paper—the last of Barrell's manuscripts left us. Would you please read this paper and tell me if the theory will work? If it will work, maybe you will care to say so in a short paper to be published in the Amer. Jour. Sci.

Lest there be any doubt about his own leanings, Schuchert added a postscript: "I might add that vast dolerite-basalt mass rose into the Karroo and the Triassic of Africa and Brazil, either in late Triassic or early in the Jurassic. This rising of basalt might be used to explain the loading of Gondwana, its final breaking-up at the end of the Jurassic and during early Cretaceous time."[42]

Holmes, an igneous petrologist who had worked in Africa, was well aware of the details of the Karroo sequences. That Gondwana had broken up, that it had done so in Cretaceous times, and that basalt was somehow implicated in the mechanism of its breakup: these were things that Holmes took for granted, things he had already said in print. The issue was how and why they had happened. Here, the isostasists' objections were sustained by Holmes. He wrote back to Schuchert promptly:

> I am sorry to say that I cannot agree with Longwell in his Introduction [to Barrell's paper] that "the results outlined by Barrell are in full accord with isostasy." If you load a continent with basalt (or sediment or ice) the *original* surface subsides, but the *new* surface is higher than the original one, provided that the density of the added material is lower than that of the material displaced at depth. It does not matter at all where the added material comes from. . . . To transform a continent into an ocean . . . the light sial[ic crust] must be got rid of somehow from the original vertical column. That is to say, it must be displaced sideways—by stretching, by fissuring and moving apart of the sides, or by magmatic corrosion over the base of the sial producing light magma which flows away.[43]

Holmes concluded his letter with an offer to elaborate the point in print, to explain why Wegener's theory—or something like it—seemed to him the best solution.

> I shall be glad to write a paper for the Am. J. Sci. on these topics if you think it not too speculative. I have been hoping to find a clear road through all our tectonic difficulties, but every hypothesis raises fresh difficulties of its own. However, on the whole I find a combination of Wegener's ideas with magmatic convection currents inside the earth on a gigantic scale to provide the energy seems best to fit our needs.[44]

Holmes's position could not have been clearer: he was explicitly advocating continental drift to his American colleague. But Schuchert was not yet prepared to abandon Barrell, and he tartly declined Holmes's offer: "At present I will not ask you to write the paper suggested in my former letter."[45] Instead, he handed the correspondence over to Longwell, who tried once more to save Barrell. Longwell replied to Holmes a few weeks later:

> Your argument against Barrell's theory would hold under certain assumptions; but not under the assumptions made by Barrell, as I believe you will agree after careful re-examination of his paper, particularly of his Figure I [figure 7.4]. [Barrell's] sial is thicker than his lithosphere; or in other words his asthenosphere includes a considerable thickness of the dioritic shell that overlies the basic substratum. [Therefore] the upper part of this weak zone consists of rock less dense than does the basic substratum. [Barrell] assumes throughout that the added material has *higher* density than the material displaced [at] depth. As the mechanism he has proposed continues to function the difference in density between added and displaced material would logically increase, because under the growing load of basalt the most acidic part of the sial, near the original surface, would eventually sink into the asthenosphere.[46]

These gymnastics (figure 7.5) would not satisfy Holmes, and he responded firmly in a letter to Schuchert a few weeks later. Barrell's mechanism would not work, but of course *no* mechanism of vertical tectonics would work. The problem had to be solved by horizontal movements. "It is impossible (within the conditions and limits of present day knowledge) to get rid of lands that formerly occupied the sites of present oceans except by moving them sideways," Holmes declared. "I can see no alternative at all to continental drift, and I have come to that conclusion from a position of strong prejudice against such processes."[47]

Holmes's argument was not merely negative—that drift was the best of a bad bunch of theories. He believed that the theory worked. It had explanatory power. It even had predictive power, for Holmes had discovered an effect that Daly had predicted in *Our Mobile Earth*. Holmes wrote:

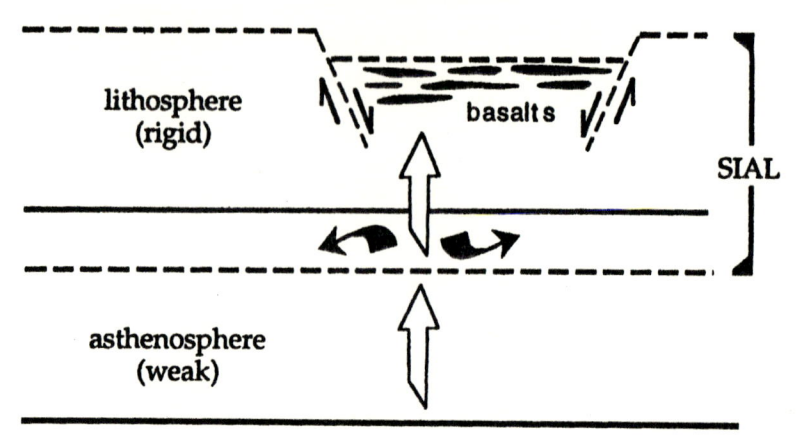

Figure 7.5. The mechanics of continental fragmentation. William Bowie and Andrew Lawson criticized Barrell's model as inconsistent with isostasy, because igneous intrusion could only cause sinking if the intruded material were more dense than the substrate. If the substrate were basalt, as most geologists assumed, then loading basalt on basalt would have no net effect. Longwell pointed out, however, that Barrell had argued for an asthenosphere that was partly dioritic—less dense than basalt—so that the mechanism could work. (Illustration by Melody Brown.)

> One of the prettiest proofs that bodily movements of the [continental] shields can occur or rather have occurred, is provided by ... recent work on the geology of Europe [that] shows that at the same time as the Caledonian orogeny was going on in Scandinavia and Britain (at the Silurian–Devonian time boundary), the Ural strip was fissured and stretched and became the site of deep-sea sediments and intrusions of peridotites. That is to say the Russo-Fennoscandian platform moved away from the Urals towards Scandinavia and produced just the sort of effects that Daly advocates in his "Mobile Earth."[48]

In the margins, Holmes penned a little sketch (figure 7.6). Daly had argued that, if drift were true, one should see evidence of fissuring at one end of a continental block at the same time as one saw compression at the other end; Holmes had found just such a situation in central and northern Europe. In 1927, Arthur Holmes was using continental drift to understand the geology of the earth.

Unless he could think up a new explanation, Schuchert was left with drift, or back where he had started a decade before. Unhappily, he put on a brave face. "Geologic theories are now in the state that biologic ones are regarding the causation—or how—of evolution ...," he finally replied to Holmes. "Out of all this boiling of ideas we should soon arrive at a working hypothesis."[49]

Like William Bowie, Charles Schuchert was neither opposed to theory in general nor ignorant of the specifics of the theory of continental drift. Nor was he irrationally fixated on older ideas. On the contrary, he had long since abandoned both Dana and Suess and badly wanted a new theory to believe in.[50] He corresponded with scientists around the United States and the globe to discuss theoretical issues, he organized a major symposium on causal explanations of mountain-building, and he arranged

shields can occur, or rather have occurred, is provided by Buonoff's recent work on the geology of Europe. He shows that at the same time as the Caledonian orogenesis was going on in Scandinavia and Britain (at the Silurian -- Devonian time boundary), the Ural strip was fissured and stretched and became the site of deep-sea sediments and intrusions of peridotites. That is to say the Russo-Fennoscandian platform moved away from the Urals towards Scandinavia, and produced just the sort of effects that Daly advocates in his Mobile Earth". *Probably the Siberian sheef also moved towards the Yenisei geosyncline - but the stratigraphy is not detailed enough for completeny.* Very many thanks for your further information about the

With cordial greetings and all good wishes to yourself,
Yours very sincerely,
Arthur Holmes

Figure 7.6. The creation of the Ural Basin by continental drift. In a letter to Charles Schuchert in November 1927, Arthur Holmes explained how continental drift could explain the creation of the Ural Basin and drew a marginal sketch to illustrate it, thus demonstrating the theory's explanatory power. (CSP, Holmes to Schuchert, Nov. 12, 1927, Box 20: Folder 171, p. 2.) Reproduced with permission of Yale University Library.

for the posthumous publication of Barrell's theory, all in an attempt to account for Gondwana and its evident disappearance in late Mesozoic times. Permanence theory failed to do this. Fissiparturition failed to do this. And now, it appeared, Barrell's theory failed, too.

If Bowie's theoretical position had accounted for the faunal homologies, Schuchert might have accepted it. But Bowie's concern had been to explain his geodetic data; in doing so he virtually ignored the facts of historical geology, facts, that is, in the sense that Schuchert used the term: as seemingly unavoidable inferences from empirical evidence. When Schuchert called Gondwana a "fact," he was referring to *paleontological* Gondwana—the observation of homological assemblages, homologies that seemed to compel biological contiguity. Schuchert could not separate—linguistically or epistemically—the inference of contiguity from the observation of identity. Although he called Gondwana a fact, as a self-conscious autodidact, he was also sensitive to the demands of his colleagues (far more so than Bowie) and he understood that an acceptable theoretical account had to be consistent with geophysical principles, as then understood. Most likely it would come *from* geophysicists, or at least physical geologists. As he put it to Holmes in a letter in the autumn of 1927, "I am calling on you geophysicists and geologists to get rid of certain lands when the faunal evidence calls for their exit."[51]

By the late 1920s, continental drift came to be understood as just such a theory: a theory, like Barrell's, that would reconcile the fact of Gondwana with the constraints of isostasy. And now Schuchert had Holmes, widely acknowledged to be one of the brightest of his generation, certainly equal to Barrell, arguing the case for continental drift. "Gondwana is a fact," Schuchert concluded his lament to Holmes, "but I still have finally to get rid of it, or else go into the Wegener following, and heaven knows I will not trail into that trap!"[52] But why not, now that Holmes had provided a mechanism? Why was continental drift a "trap"? Why, in spite of Holmes's arguments and his own compelling paleontological evidence, did Schuchert remain unconvinced?

"I *Want* to Give Up the Theory"

When van Waterschoot van der Gracht, vice-president of Marland Oil Company and head of the program committee of the American Association of Petroleum Geologists, wrote to invite Schuchert to participate in the symposium on continental drift that he was organizing for September of 1926, the former penned in a postscript to his invitation: "I know you don't like the Wegener theory: that is why I all the more want you! *Please come.*"[53] But Schuchert was not nearly so firm in his resistance as van der Gracht thought, nor as his published comments from the symposium later suggested.[54] In his correspondence with Arthur Holmes that summer, Schuchert's opposition had weakened considerably. Unless he could salvage Barrell or come up with a new solution, he would be heading rapidly into the Wegener following. "Can you keep me from sliding away from Barrell?" he punned in another letter to Holmes. For Schuchert was sliding indeed: he continued his letter with a long discussion of Joly's theory of periodic fusion, with its implication that substratal melting could facilitate horizontal mobility of the continents. "I gather," he pondered, "that there may, under [Joly's] theory, be a very great deal of sliding, and that North America may have gone around the world many times without knowing it! We appear to be drifting into the Wegener camp from the east, very much like the Lutherans invading the Holy Roman Church. Surely there is going to be a quarrel!"[55]

Schuchert's letter focused on the central question: Can we get away from a differential movement of the continents? The answer appeared to be no.

> A few days ago, I read Steinman's stimulating paper in the Geologische Rundschau [Schuchert wrote], "Gibt es fossile Tiefseeablagerungen?" and I am ready to accept his view that we have such [differential movements], and that they are very widely distributed in the Apennines and the Alps. If, then, Tethys has been pushed together 550 km, as Steinman says, does it not follow that Africa has flowed to the northwest something like 400 miles? Even so, I can not see why Brazil cracked away from Africa, forming a vast fissure, and then why this vast continent flowed several thousand miles away to the west. [On the other hand,] If it did so flow, then seemingly we can explain all the more easily the Permian glaciation and the Andine [South American] faunas of Tethyan affinities.[56]

G. Steinman, Professor of Geology at Bonn, had argued that the Alpine movements were caused by the closure of a proto-Mediterranean ocean—the Tethys, which previously separated Eurasia from Africa—and that certain unusual Alpine rock sequences (later called ophiolites) were shards of ocean floor trapped by inter-continental collision. Schuchert saw the logical consequences of these views and abruptly interrupted himself: "But I am not working along these lines!"[57] If not working, however, then definitely thinking, for he concluded his letter in characteristic style: "Help me along on Gondwana! Give me either land to stand on, at least for a time, or cause me to swim among the ammonites with eyes to see how the Tethyian forms got to Peru and Chile without going south into the Atlantic."[58] A month later Schuchert wrote to Holmes again, this time discussing Frank Taylor's just-published paper, "Greater Asia and Isostasy."[59] "What is interesting is his continental drifting," Schuchert told Holmes. "See it."[60]

If Schuchert was on the verge of sliding into the Wegener camp, Bowie pulled him back from the brink. Bowie was absolutely insistent that large areas of the ocean basins could not move far one way or the other—up, down, or sideways. But he did suggest an alternative. Although large areas of ocean could not move much, small areas might. "I see no reason," he opined to Schuchert in June of 1927, "why shallow portions of the oceans should not have been above sea level at one time or another. I think that the ridge that runs through the northern Atlantic may have been, at one time, higher in elevation than it is now." The mechanism lay in the thermal effects of isostatic readjustments:

> There can be some upward and downward movement of the earth's surface as a result of heating and cooling from sedimentation and erosion. If the temperature gradient is the same throughout the crust as it is for the first mile, sinking of the crust beneath 3000 feet of sediments would result in a thermal expansion of about 300 feet, after sedimentation had ceased. On the other hand, if 3000 feet of material were eroded from an area, there would later be a downward movement of 300 feet, merely from loss of heat.

This is where Bowie made the link to Pratt isostasy: "Of course, this is on the idea that we have the Pratt isostasy. If we have the Airy isostasy, then there must have been horizontal movements forming roots to the topographic features that have been given great elevation."[61] Bowie's allowance—that if Airy isostasy was correct, then horizontal movements would be not only possible, but necessary—seems to have slipped past both of them. As far as their letters suggest, neither man pursued this idea. Bowie declined to worry about it, while Schuchert began to focus on the idea that if large continental areas could not sink, perhaps small ones could.

This seed planted by Bowie took some time to germinate. Meanwhile, Schuchert was as busy as ever, despite his advancing age. His comments from van der Gracht's symposium were published in a proceedings volume in 1928, for which he received many laudatory comments from colleagues. He completed a critical review of du Toit's monograph for the *Bulletin of the Geological Society of America*—arguing mainly about the details of his faunal correlations—and gave an address on the subject at the University of California.

Schuchert's public pronouncements on drift were all negative, but they in no way resolved his internal dilemma. The problem of Gondwana remained. Mostly, he remained optimistic, writing to Bowie in the autumn of 1928 that Holmes was "searching deeply" and that he, Schuchert, was confident that "in a few years more we should arrive at something definite about continental drifting."[62] To Holmes, he wrote, "I want you to find the mechanism of why borderlands and continents sink out of sight, and I have faith that you will find it some day, and then not by the Wegener method."[63]

Several of his colleagues agreed. Howard Baker, a former student and author of his own version of drift theory, wrote presciently, "I suspect that developments to come will be more interesting than those that have appeared so far and a little judicious poise now may come in handy a few years hence. There has been an amazing change in geological thought in the last twenty years and it can hardly stop all of a sudden."[64]

But at times it did seem to stop, and Schuchert felt deeply discouraged. In the summer of 1929, he penned a long, rather sad letter to J. W. Gregory, the Scottish geologist famous for explicating the structure of the East African Rift:

> I have long been thinking about this subject, and it may surprise you when I say that I have for nearly ten years been trying to break away from holding to a greater Gondwanaland ... What scares me are the conclusions of our isostastists like Chamberlin, Lawson, [and] Bowie ... that there is no known means by which a continent can be sunk into the oceanic depths. For several years, I worked on the late Professor Barrell about this matter, and he finally wrote a paper about how Gondwana could be weighed by material rising out of the asthenosphere and so sink out of sight. But all of the workers in isostasy say that mere rising of heavier materials into the lithosphere will not do it, but if I could show flow from lateral regions the thing could be done. Of course, no one can show such lateral flowage. In spite of all the ... American criticism about Gondwana, and the fact that I *want* to give up the theory, so far I have not been able to explain away what I know of the marine faunas beginning with the Devonian and closing with the Lower Cretaceous. ... Many years ago I intended some day to write a book about Gondwana and have gathered many notes about it. The problem grows ever greater and now I shall probably never attempt the job.[65]

But Schuchert refused to give up. At the end of 1930, while sick in bed over the Christmas holidays, he read Holmes's most recent contribution, "Radioactivity and Earth Movements," published in the *Transactions of the Geological Society of Glasgow*. The article was the summation of the ideas on the role of convection currents that Holmes had been developing since 1926, ideas that he had discussed with Schuchert in earlier letters. Holmes explicitly proposed that convection currents could explain the origin of ocean basins and provide the driving force for continental drift. Schuchert was impressed. "It is along the lines laid down by Holmes that geologists will eventually solve the causes of deformative mountains and how they are made," he wrote contentedly to Longwell early in the new year.[66] Longwell, meanwhile, had written to Schuchert on the very same day, and their letters crossed in the mail. He agreed that Holmes was a "brilliant fellow" whose hypothesis was "far superior to Wegener's," chiefly because it provided "some motive power for separating and deforming the continents," but he backed off his earlier unrestrained enthusiasm, perhaps because of the negative publicity Wegener had received in the interim. "Of course, his suggestions are purely speculation and at present at least they are not subject to proof." Nevertheless, he intended to write to Holmes to congratulate him on the paper.[67]

On the same day that Schuchert wrote to Longwell praising Holmes, he also wrote to the German geologist Erich Haarman (who had proposed his own tectonic theory somewhat similar to Daly's) criticizing Wegener:

> Probably I am a conservative or better, in regard to the ultimate causes of mountains, I wish to remain noncommitted. ... but I completely reject the wholesale sliding theory of Wegener. No paleontologist can accept continental movements of

the extent of thousands of miles. The one theory that I lean more to than any other is that of Holmes as described in his "Radioactivity and Earth Movements."[68]

This is odd. How could Schuchert accept Holmes's convection theory yet *on the very same day* reject continental drift, when the whole point of convection was to provide a mechanism for drift? And why should he reject drift as a *paleontologist*, when paleontology formed the backbone of the argument for formerly contiguous continents? Schuchert's letter to Holmes begins to provide an answer.

> I am delighted to have your explanation of how borderlands have foundered [by sinking in regions of down-welling convection currents] and as borderlands are of various sizes, it is only another step (even if a big one) to the sinking of the land where now is the Scandic Sea, or the downbreaking of Lemuria to make the Indian Ocean, and finally even the breaking up of the Ethiopian-Brazil land-bridges.[69]

This was a misreading of Holmes's theory. Holmes's mechanism for the foundering of continental borderlands by sinking above descending currents could not be used to sink entire continents, because the continents were the sites of ascending or laterally migrating currents. Schuchert was clinging to sunken continents, and he explained why in his next paragraph.

> Excuse me if I say so, as a marine faunal paleontologist and stratigrapher, that the weakest link in your convection hypothesis is that of continental drift. To move the British Isles in late Paleozoic time from the tropics to their present position is to fly in the face of all the marine faunas. These biota are not tropical, while those of the Permian of northern Europe remind one far more of cooler waters than warm ones. Where in these lands are the coral faunas and reef limestones, the bryozoan reefs and the great thicknesses of limestones such as we look for in tropical regions? There are such now in the Mediterranean countries, and according to your hypothesis they must have originally been formed in the southern temperate regions. And yet, in the temperate regions the waters would have been too cool to have permitted the making of great limestones and bryozoan reefs, and for the existence of an abundance of ammonites![70]

Geologists will recognize in this a uniformitarian demand. But it is uniformitarianism of a particular—even idiosyncratic—kind. Schuchert was demanding that ancient faunas and their associated climatic zones should show the same relationship to latitude as do modern ones. In Wegener's reconstruction, the British Isles lay in the tropics in late Paleozoic time, but Schuchert pointed out that the fossil assemblages did not display the types of biota that one expected in association with tropical climes on the basis of present associations. "There are such *now* in the Mediterranean countries," he argued, referring to corals and reef limestones, so where were the equivalent fauna in the Mediterranean of the Permian? Despite all the evidence of climatic change in geological history, Schuchert's view of climatic zones was essentially static. Waters in low latitudes were always warm, in high latitudes always cool. He thus concluded:

> As a paleontologist, I have no objection to your moving about the continents about a few to several hundred miles, *but when you demand thousands of miles [of drift], you drifters say to us workers on marine faunas that all that is known about faunal distribution is worthless. This means that the present distribution of marine faunas cannot be trusted, and cannot be applied to the life of the past.*[71]

The crux of Charles Schuchert's objection to continental drift was not isostatic or mechanical; *it was not a geophysical objection of any kind*. It was a *uniformitarian* objection. Schuchert believed that the relation of climate to latitude must have remained more or less stable throughout geological history. If continents had moved through thousands of kilometers of latitude, this should be evident in the fossil record in terms of climatic indicators. One should see reefs at low latitudes, ferns and mosses at high ones. If not, then the continents cannot have been at those latitudes. Continental drift was not possible, not because of any mechanical objection, but because it would mean that the present was not the key to the past after all.

This objection makes no sense by contemporary lights: today we accept that the earth's continents have migrated across the globe, and that climatic zones have changed as well, sometimes in concert with these migrations, sometimes not. The equator has not always been stiflingly hot; the poles have not always been under glacial ice. Geologists today would not assume that the relation of climate to latitude in the Paleozoic need have been the same as it is now. Yet this is what Schuchert believed. He subscribed to a substantive uniformitarianism: the belief, expounded most famously by Charles Lyell, that the earth today is more or less as it has always been.

Preserving the Practice of Uniformitarianism

The present is the key to the past—this is the slogan that appears in so many introductory geological textbooks as the encapsulation of uniformitarianism, the fundamental principle of geology. But what does it actually mean? Geologists have long argued the point. Like most successful slogans, this one has meant various things to various people. In the late twentieth century, as Steven Jay Gould cogently pointed out some years ago, it has mostly been interpreted as a methodological principle: that the laws of nature are invariant with respect to time and space, and that geologists should not invoke the miraculous. But such a claim is scarcely distinctive to geology: science by definition depends on the assumption of the invariance of natural laws and on the methodological commitment to the exclusion of supernatural explanation. If geologists have thought uniformitarianism to be the fundamental principle of their science, it surely must go beyond this point—which in any case was accepted and implemented by most working geologists well before Lyell.[72]

Gould suggested that the other part of uniformitarianism—the substantive half—was the postulate of uniformity of rates or material conditions. Lyell believed that earth history was not progressive: that geological change, although ubiquitous and constant, had no *net* effect. Most famously, he argued against any major change in the rate of geological change or the intensity of geological forces: Observable processes, acting in their presently observable manner, were sufficient to account for the earth's history. There was no need for episodes of unusual intensity to account for the rock record.[73]

Gould's analysis is trenchant, but most geologists have not been so clear in their thinking. Nor is substantive-methodological the only pair of axes along which a historical analysis may be drawn, nor along which historical debates *were* drawn.[74] As historian Martin Rudwick has pointed out, geologists argued at length about Lyell's proposals precisely because they could be interpreted in so many ways: Was earth history progressive or not? Were presently observable processes sufficient to account for earth history or did other factors become increasingly relevant as one went further back in time? Was change always gradual or could it be saltatory? And what justification was there for assuming the present to be characteristic, considering that it was just a moment in the long course of geological time? Lyell took an extreme position: that all change was gradual, that the history of the earth showed no progression, and that presently observable processes could account for the entirety of earth history. (He did, however, allow for limited saltatory change, as long as it had no cumulative directional effect.)

Rudwick argues that geologists for the most part applauded Lyell's attempt to "extend the heuristic value of actualism"—that is, to focus attention on presently observable processes—but rejected its use in the service of an extreme steady-state interpretation.[75] While this statement is no doubt true, it masks an important point: even while Lyell's steady-state view was widely criticized, and in some respects clearly false, it nevertheless greatly influenced generations of geologists.[76] The scope of Lyell's exegesis, the power of his textual rhetoric, and the widely acknowledged importance of his gradualist worldview in shaping the ideas of Charles Darwin—all these resulted in Lyell having a profound and lasting effect on geologists' thinking. Sudden change, large change, dramatic effects—all these came to be seen as suspect, things to be avoided in geological explanation as far as possible.[77]

To use modern vocabulary, Lyell changed the default assumptions of geology. In the twentieth century, the burden of proof would be on those who argued for rather than against large-scale or sudden change.[78] Many geologists—of whom Schuchert was one—came to feel that good science required one to aim in the direction of Lyellian explication, as far as possible. One ought to look to the present to understand the past. One ought *not* to suppose that things were substantively different in other times. Uniformitarianism shaded into a belief in uniformity.

Here, then, was the essence of Schuchert's complaint that if drift were true, then "the present distribution of marine faunas cannot be trusted, and cannot be applied to the life of the past." Drift appeared to Schuchert to be antiuniformitarian, not because it could not be *seen* taking place at present (although this may have been relevant for other geologists) but because it implied a wholesale reconstitution of floral and faunal patterns. Continental drift seemed to imply that the interrelationship between latitude, climate, and faunal assemblages was being constantly reconfigured in unpredictable ways, ways that made the present the key to nothing at all.[79]

This accounts for what seems otherwise to be a bizarre element of Schuchert's logic: that he was willing to accept crustal dislocations of a few hundred kilometers but not of a few thousand. In the proceedings of the AAPG conference, Schuchert claimed that he was "wholly open-minded toward the idea that the continents may have moved slowly, very slowly indeed, laterally, and differently at different times." Indeed, he argued that "every student of tectonics . . . must have said to himself again and again that there has been actual differential continental displacement."[80] Yet, repeatedly in both his private correspondence and his published articles, he rejected what he called "wholesale sliding."

As early as 1923, Schuchert had corresponded with G. A. F. Molengraaf, Daly's and Wright's traveling companion in South Africa, on this subject. Schuchert knew of Molengraaf's support for drift and wondered if perhaps Daly had given him a false impression of the situation in the United States. "Taylor's continental drift theory no one paid much attention to until Wegener made so much of it, and then it was the Europeans who took it up," Schuchert wrote. Among Americans, he continued, Daly "stands nearly alone." Schuchert then explained why:

> That the supercrust does move, all are willing to admit, and that parts of it may have moved a few hundred miles, some of us also admit. But . . . to admit that North America and South America have drifted to the west up to 5000 miles can only be acceptable to forgetful or deranged minds.[81]

Schuchert made the same claim in a letter to Frank Taylor two years later.

> That the continents, i.e. parts of them, do move is certain, and that in places the horizontal movement attains to some hundreds of miles is also fairly certain. But that the whole of Asia is moving southward we do not see. If so, it must be doing so at least since the Devonian! . . . [And] if the earth has captured the moon [as Taylor proposed, to explain the formation of the Pacific basin] why as late as the Cretaceous? I should look rather for it early in the history of the Earth.[82]

Schuchert's substantive uniformitarianism and the logical constrictions it entailed are in full evidence here. On the one hand, he denied that large-scale drift had occurred. On the other hand, he argued that if it *had* occurred, it must have done so *throughout* geological history.

Eight years later, he was still arguing the point. "I know that parts of the continents do move hundreds of miles," he wrote in 1931, "but I completely reject the wholesale sliding of Wegener."[83] Reginald Daly failed to comprehend the distinction. "You do not give a lot of rope when you say 'hundreds of miles,'" he complained. "Where is the logical stopping point, once you go that far? I mean, in the physics of the thing?"[84] But, as Schuchert's correspondence with Holmes makes clear, the stopping point lay not in physics but in the practice of historical geology. A few hundred miles of drift could be tolerated, because that would leave faunal zones intact. But a few thousand miles would mean a radical rethinking of his uniformitarianism. Thus Schuchert concluded an earlier letter to Daly, "I have no faith in the sliding theory of Wegener, but more in the one of Joly, and because the latter does not disturb my paleogeography!"[85]

It might be objected that crustal dislocations could be *seen*. Schuchert acknowledged as much in several of his letters; Andrew Lawson had recently demonstrated that the western side of the San Andreas Fault was moving northward. But Lawson's results could be accommodated by the allowance of a few hundred kilometers of movement and, in any event, as Schuchert had put it to Molengraaf, "what the rate is per thousand years no one has as yet the faintest idea."[86] Schuchert's position in 1933 was thus the same as it had been in 1923: moderate-scale movements, yes; wholescale drift, no. His position was a compromise between the compelling geological evidence of lateral movements and a uniformitarian commitment to the integrity of faunal zones.

Laws and Rules

In the proceedings of the AAPG meeting, published in 1928, Schuchert declared that "we are on safe grounds only so long as we follow the teachings of the law of uniformity in the operation of nature's laws."[87] But he was not, in fact, talking about the laws of nature at all. He was talking about a different kind of uniformity: a *literal* uniformity in the history of the earth. Uniformitarianism shaded into uniformity, both conceptually and linguistically. This was Lyell's legacy: the default assumption that the world on average remained more or less as it had always been; that change was cyclic or rhythmic but not secular or cumulative.[88] Mountains were raised up and erosion wore them back down again; seas transgressed the land and then regressed again; new species evolved and old ones disappeared, but the earth remained a biological plenum. Such were the uniformitarian assumptions under which late-nineteeth-century geologists operated, assumptions that by the mid-twentieth century had become reflexes. But why? Why did geologists accept what might be construed as a rather startling assumption?

Geologists and historians are familiar with the opposition of uniformitarianism to catastrophism, the idea that the earth history had been punctuated by intermittent cataclysmic events, different in both degree and kind from presently observable events. It was the mid-nineteenth century polymath William Whewell who gave uniformitarianism and catastrophism their names, thereby focusing attention on them, and perhaps unwittingly implying that these were the only two interpretive options available.[89] But Whewell's historiography was incomplete, for there had long been a third option: progressionism.[90] Progressionism—the idea that the history of the earth was directional and controlled primarily by progressive cooling from an initally molten state—had been advocated in the first half of the century by geologists such as Elie de Beaumont and De la Beche, following the nebular hypothesis of Laplace and the work of Fourier on heat diffusion.

Progressionism and catastrophism were closely linked ideas: progressive cooling implied that earlier geological events were necessarily more cataclysmic than later ones—that geological processes were generally winding down. In the early twentieth century, therefore, many geologists (including De la Beche and Elie de Beaumont) were both catastrophists and progressionists. But whereas catastrophism per se lost credibility in the late nineteenth century, progressionism received additional impetus from the development of thermodynamics, whose second law seemed to suggest that thermal decay was inevitable.

An apparent corollary of the second law of thermodynamics was that the past had witnessed forces of greater intensity than the present, as the earth's internal heat had decreased and entropy increased. But this conclusion was in direct conflict with Lyell's perspective.[91] For a time, this contradiction was used as an argument against Lyell within the geological community. But when Lord Kelvin drew on the progressionist viewpoint to declare that the earth could not be more than a few tens of millions of years old, the situation changed. Kelvin's demand for a very young earth and his arrogant dismissal of geological evidence that spoke to the contrary discredited all forms of progressionism in the eyes of historical geologists, who closed ranks against it. Geologists defended uniformitarianism as the alternative to a viewpoint which seemed, at least in its extreme manifestations, to deny the very legitimacy of historical geology as a science.

If that were not enough, a second reason is this: If one did *not* make the uniformitarian assumption that the present was representative of the past, then how *was* one to interpret the relics of the past? The world was full of complicated and diverse phenomena whose meanings were little understood: How did one make sense of fossil data? How did one make sense of stratigraphy? How did one "do" the science of historical geology? The answer was by analogy: by comparing fossil organisms to living forms, ancient rocks to modern environments. *Historical geology was an analogical, not a nomological, science.* It relied on postulated parallels between the modern and the ancient, the living and the dead, the near and the far, the visible and the inferred. Uniformitarianism was an articulation of this reality, a declaration that in the absence of laws one could still practice geology as a science.[92]

But what did it mean to "practice" geology as a science? Above all, it meant that geologists should agree on how interpretations were forged from diverse and incomplete field data. This is, of course, the problem of induction: How does one generalize from data? How does one know if one's generalizations are valid? Philosophers since Hume at least have argued this point, but for geologists this was more than an abstract problem: it was the very stuff of their discipline. It is not without reason that historians of science have focused on controversy in geology: the history of geology has been a history of controversy—the neptunist–plutonist controversy, the Devonian controversy, the Highlands controversy, the Taconic controversy—each of these named not in retrospect by historians but by the participants themselves.[93] Each was a dispute about the interpretation of field data—about the meaning of physical evidence—and each occurred precisely because there were no hard and fast rules by which such interpretations could be made. Uniformitarianism is the closest geologists have come to such a rule. It was geology's answer to the problem of induction. Rather than being a trivial restatement of the invariance of natural laws, it was a fundamental assertion of geology's right to be a science.[94]

A third reason geologists have held uniformitarianism to be the fundamental principle of their discipline is the role of catastrophism in the history of geology. Historians have argued about the proper place of the catastrophist school of thought in the history of geology—or even if there was such a school—but there seems little doubt that by the end of the nineteenth century, the uniformitarian program was credited by Anglophone geologists as crucial to the advancement of their science, as the victory of a guiding principle instantiated through the elimination of theoretical dogma and the removal of scriptural reference as a source of geological explanation and evidence. The eighteenth century association, particularly in England, of geology with theology in general, and of catastrophes with scriptural exegesis in particular, led geologists in the nineteenth century to try finally and firmly to eliminate all mechanisms that might smack of religious apologia. Sudden change, dramatic change, change unaccounted for by the normal processes of daily geological life—these were all too close to miracles for most geologists' comfort.

In the early twentieth century, the professionalization of geology apart from theology was complete, yet competition over the intellectual territory of earth history remained, at least in the United States, and tensions were often not far from the surface. Even as Schuchert wrote to Holmes that scientists were debating the "how of evolution," other portions of American society were again debating the "whether" of it. For as several historians have noted, the continental drift debate coincided with the trial of John Thomas Scopes for the teaching of evolution in a Tennessee public school. On one level, these debates were clearly independent; in all of Schuchert's voluminous scientific correspondence, there is scarcely a mention of the Scopes trial. Indeed, as far as historical evidence shows, few American geologists seemed to consider these events to be scientifically serious. Certainly no one claimed that continental drift was in any way pro- or antiscriptural. Nor is there evidence that defenders of drift had any particular religious or antireligious bent.

And yet, from a geologist's point of view, the trials in Tennessee could hardly have been considered good news, and it raised the specter of theologically motivated catastrophism all over again. Schuchert made this point explicitly in the journal *Science*, in a review of *The New Geology* by George McCready Price, a leader of the American antievolution movement. Price's subtitle, *A Text-book for College, Normal Schools and Training Schools; and for the General Public*, revealed both his ambitions and the threat that he posed. By Price's own admission, the book was a refutation of professional geology, which he called a "travesty on the real science of geology." The alternative he proposed was founded on the premise that the biblical account of Noah's flood was more or less correct. This, of course, was precisely what geologists thought they had laid to rest with the doctrine of uniformitarianism, and with the rise of professional societies, university departments, and published lists of who was who in American science. As Schuchert took pains to emphasize, Price was "unknown to the membership list of the Geological Society of America and to the pages of [James] Cattell's 'American Men of Science.'"[95]

The notion of an alternative geology, reconstructed from whole (ecclesiastical!) cloth, undercut all that professional geologists thought themselves to have achieved. Indeed, it undercut the very premise of a science founded on experience and observation of the natural world, and it did not go unnoticed that Price entirely lacked both formal scientific training and field experience. In the words of historian Ronald Numbers, Price was "an armchair scientist" who consciously minimized the importance of field experience.[96] His reasons for doing so were obviously self-serving, and the conclusions he drew from the safety of his living room were sufficient to discredit him in the eyes of most geologists. But in denying the legitimacy of field evidence, Price challenged not only the conclusions of geology but also its methods. Not surprisingly, Schuchert—himself self-educated—rose to the challenge:

> Geology [is] a science that has been built up through the labors of the geologists of the world during more than a century of observation, study, and criticism. The things that have long been accepted as the fundamentals of this science—immensely long geologic time and an orderly development of organisms from single-celled individuals into the most complex community of cells as exemplified in mammals and man—are rejected by Price and in their place is substituted his "new catastrophism." This is our old friend, the "universal deluge" . . . So much for the "new" geology.[97]

Although Price's ideas were never taken seriously within the scientific community, he was highly influential among American fundamentalists. As Numbers emphasizes, his books sold widely, he wrote extensively for American religious periodicals, and when the Scopes trial convened in Dayton in July 1925, William Jennings Bryan invited him to serve as an expert witness. Price was unable to attend, but at the trial Bryan cited him as one of the two scientists he respected.[98]

Schuchert was likely reminded by these events of the intellectual territory he had traversed from his own Catholic roots. Perhaps this is why he contributed to the American Civil Liberties Union's Tennessee Evolution Case Defense Fund (with strict instructions not to bother him with any other matters) and served on the board of the Science League of America, an ad hoc group organized to respond to the antievolution movement. Perhaps also this is why, when he chose metaphors to describe the continental drift debate, they were the metaphors of religious conversion.

The drifters were like the "Lutherans invading the Holy Roman Church," Schuchert wrote in his private correspondence. He called himself "orthodox," called Holmes "heterodox," and reassured his British friend that he would "soon come back into the fold."[99] Schuchert's own life was one of leaving the fold, one that could be viewed as a long progressive conversion: from impoverished immigrant origins to a professorship at Yale; from a childhood steeped in Catholicism to leadership of a science that officially refuted the miraculous. Perhaps the Scopes trial reminded Schuchert of the distance he had come, of the beliefs and faiths, as well as the friends, he had left behind.[100] Perhaps it raised the question of how secure his scientific "truths" really were. Most likely it reinforced his commitment to the practices that underlay his hard-won scientific beliefs.

To what degree Schuchert was ever a religious believer is not clear, but when he allowed the possibility of changing his scientific perspective, he couched it in the language of religious reformation: "It does not necessarily follow that our 'orthodox' views of today will be those of tomorrow," he wrote to E. W. Berry in 1928. "The question [is, do we have] a 'new religion' to knock out the orthodox view?"[101] But what was geological orthodoxy for Charles Schuchert? Not permanence theory; he had abandoned that a decade before. Geosynclines? Perhaps, but they did not resolve his paleontological problems. By the early 1930s, no theory adequately solved Schuchert's concerns or accounted for all the theoretical and evidentiary constraints he faced. The result, not surprisingly, was theoretical agnosticism—catholicism with a little *c*. As he put it in the spring of 1931: "Regarding the permanency of the ocean basins, I hardly know what stand I will finally take."[102] And yet, he still called himself "orthodox." Given this, and the diverse theoretical views held by other American geologists, it seems most plausible to argue that orthodoxy was not so much a particular theory as a way of thinking *about* theory. Orthodoxy was *practice*. One aspect of that practice, as we have already seen, was the ideal of multiple working hypotheses (an orthodoxy of heterodoxy!). Another aspect was uniformitarianism.

American geologists in the early twentieth century not only subscribed to uniformitarianism, they credited themselves with its development as a reasonable rather than an extremist program. "The doctrine called, rather inaccurately, uniformitarianism," T. C. Chamberlin wrote in 1924, "was brought across the Atlantic and naturalized in part, but given a saner interpretation."[103] The "saner" interpretation was a loosening of Lyell's insistence that the rates of geological processes were invariant, and the addition of the idea of rhythms or cycles, of which T. C. Chamberlin, G. K. Gilbert, and Joseph Barrell were all advocates.[104] This modified uniformitarianism took hold and became an essential element in American geology. Under the uniformitarian principle, the geologist Charles Berkey wrote in the *Proceedings of the American Philosophical Society* in 1933, "the whole geologic field has been opened to interpretation on rather definite and simple rules." It was thus that Schuchert declared that "safe ground" was the ground of uniformity: safe, meaning that geology as a science was safe so long as there was agreement about the basis of interpretation. From this perspective, we can understand how it was possible for Americans to face theoretical pluralism with equanimity and even embrace it: American geologists were comfortable with a wide range of theoretical options so long as they appeared to be consistent with uniformitarian practice.

The theory of continental drift did not violate the laws of nature as they were understood in the late 1920s, but in Schuchert's view it did challenge, and seriously threaten, the uniformitarian principle that was at the root of his science, cordoning it off from both scriptural interpretation and physical reductionism. The defense of uniformitarianism was a defense of geology qua geology, rather than as applied physics on the one hand or applied theology on the other. By the twentieth century, uniformitarianism had become a form of standardized practice, a practice that facilitated consistent interpretations of field data. The laws of nature had been of little help to historical geologists; what was of help was the assumption that what one saw at present could be used to explain the past, the *assertion* that making inductive inferences from the present was a reasonable thing to do. Uniformitarianism was a heuristic principle that ensured that when historical geologists took that "short step backward," they all ended up in the same place.

A Uniformitarian Solution

Lord Kelvin had been spectacularly wrong in his application of the laws of physics to the age of the earth, and Schuchert now suggested that Holmes was perhaps similarly wrong about radiogenic heat. Holmes no doubt understood this implication, since he had personally contributed to overturning Kelvin's chronology, and in response to Schuchert's objections about the Paleozoic marine faunas he wrote back politely suggesting that there was another way to look at the situation: "Surely the real discrepancy here is not between marine faunas and continental drift as between marine faunas and the climate evidence on which [we] mainly rely. How *can* you account for the Permo-Carboniferous glaciation without assuming that the glaciated parts of the fragments of Gondwanaland originally lay much nearer the South Pole than they do now?" Holmes saw the obvious flaw in Schuchert's logic and tried to work around it. Deductions could be wrong if derived from faulty premises, and inductions could be wrong if derived from faulty data, so how was one to settle the debate? "Let us say that there is evidence for drift and evidence unfavourable to it," Holmes concluded, trying to be diplomatic. "The balance in favour is then weighed down [i.e., tipped] by the belief that there is an excess of heat to get rid of, and drift appears to be the only mechanism that solves this difficulty."[105]

Schuchert was unconvinced. To admit that the climate zones might be other than they were today was to suggest that his entire practice of historical geology, founded on the analogical interpretation of past geography, climates, and fauna by reference to the present day—to presently observable patterns, processes, and relationships— was amiss. His most basic methodological commitment—the basic methodological principle of historical geology—would have to be reconsidered. This was not an acceptable outcome. "This all means that you must find some other way of getting rid of the radioactive heat where it is not wanted," Schuchert concluded, tongue only partly in cheek. "*Have the chemistry or any old theory absorb it!* . . . *You will find a way.*"[106]

If there is truth in jest—if Schuchert's overriding commitment was to uniformitarianism—then an acceptable resolution would have to satisfy this commitment. And so it did. The solution to the problem of Gondwana that Schuchert finally adopted was based on an application of uniformitarianism, developed through a collaboration with Bailey Willis.

Willis (1857–1949) was one of the great personalities of American geology. A structural geologist, experimentalist, explorer, illustrator, and writer—the son of a poet—he had been educated in Germany and at Columbia and had worked on nearly every continent, traveling with Bowen to the East African Rift, organizing expeditions to remote corners of China, studying earthquakes on behalf of the Carnegie Institution in Chile. After several years as a geologist with the U.S. Geological Survey, he became a professor at Stanford University, where he had tried to institute northern California's first earthquake building codes (later he would quip that although he had not changed the building codes, he did manage to raise the cost of earthquake insurance).[107] In 1931, at the age of seventy-four, Willis had been an emeritus professor for nine years and would continue to write and work for another eighteen. A prolific writer, he began a voluminous correspondence with Schuchert (now 73), culminating in a resolution to the problem of Gondwana.[108]

They began writing after Schuchert had reviewed Willis's latest book, based on his trip to Africa for the 1929 International Geological Congress. Willis replied in characteristic style: "Your generous review of *Living Africa* almost persuades me to read the book. I have a notion, however, that I could not learn much from the author. It would be more profitable to discuss the problems with you."[109] Scientifically, the two men could hardly have been further apart, at least for two North Americans: Willis was a vociferous adherent to permanence theory. In a widely cited paper, published in *Science* in 1910, Willis had emphatically declared, *"the great ocean basins are permanent features of the earth's surface and they have existed, where they are now, with moderate changes of outlines, since the waters first gathered."*[110]

Early in his career, Willis had gained fame from a series of innovative physical experiments illustrating the efficacy of horizontal compression in folding rocks, designed to demonstrate the role of tangential pressure in creating the mappable structures of the Appalachians (figure 7.7). But with time he came to agree with American colleagues that horizontal compression was an incidental effect of vertical motion. In his well-received *Researches in China*, published in 1907, Willis proposed that the Asian continent—and by implication every continent—was composed of discrete crustal blocks or "elements" that had moved consistently upward or downward, with crustal deformation accommodated by flexure in between the blocks, subcrustal flowage below.[111] With the work of Hayford and Bowie, however, Willis abandoned the idea that crustal elements could sink and embraced permanent continents and oceans. Isostasy "constitute[s] an insuperable obstacle to the former existence of continental lands where there are ocean basins," he explained in a letter to Schuchert in January 1932, and "I cannot but regard this evidence as quite as cogent as that of the migration of floras and faunas."[112] Gondwana horrified him. Nevertheless, he had claimed to be "open to conviction if adequate reasons for its one time existence can be cited to outweigh the difficulties of its disappearance." He floated an idea to Schuchert that he had briefly suggested in the AAPG proceedings: isthmian connections caused by basaltic intrusions in the ocean basins. "If the idea that the margins of oceanic deeps are liable to be built up and to be uptruded takes in your mental processes I would ask you to consider the possibility of an isthmian connection between India and Africa

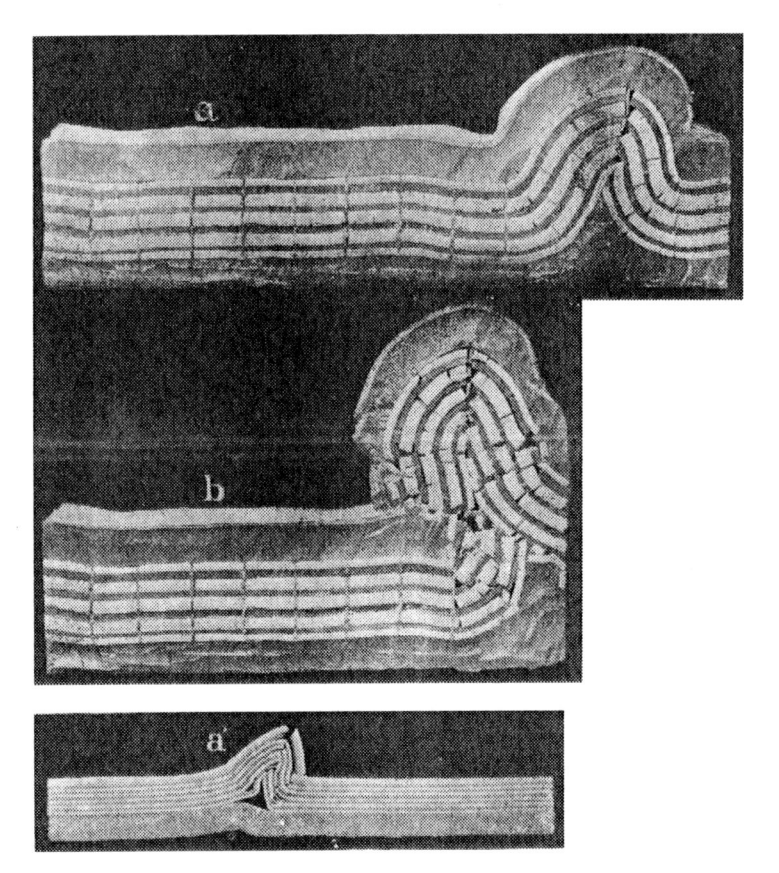

Figure 7.7. Willis's mechanics of Appalachian structure. Early in his career, Bailey Willis had performed a series of innovative scale-model experiments to demonstrate the role of horizontal compression in generating the structure of the Appalachians. In time, however, Willis, like most Americans, came to see the effects as a by-product of dominantly vertical crustal motions. (Willis 1893, p. 263).

via the Maldives, . . . the Seychelles, and Madagascar. Would not such a connection serve the requirements of migrations and be more like what we see in various more or less complete connections today?"[113]

Schuchert was receptive, for here was the uniformitarian perspective he was looking for: an explanation "more like what we see . . . today." Rather than sink down pieces of the *continents*, perhaps one could lift up sections of the *oceans*—just as one saw the results of uplift in oceanic islands. This was akin to the idea that Bowie had already planted in Schuchert's mind—that small portions of the oceans could rise and fall— and it did take in Schuchert's mental processes. But first he wanted to make sure that Willis understood and truly appreciated the paleontological evidence. He replied to Willis in a detailed, three-paged letter:

> My difficulties you do not know and can not appreciate because you are not a paleontologist. . . . Let me put [them] to you. . . . off present east Africa, there was, at least in Mesozoic time, a great geosyncline that is proved by the marine faunas of

east Africa and western Madagascar.... This syncline and its complement, the Indo-Madagascar geanticline [i.e, a large upwarping], have gone into the depths of the Indian Ocean, and what remains above sea level is the islands you mention in your letter to me. To me these are well established facts on the basis of the distribution of the ancient and living fauna and cannot be gainsaid.... This, however, is only the beginning of our difficulties... To me, it is established that in Paleozoic time the eastern present shore of North America extended out into the Atlantic at least 300 miles, since all of the 13,000 feet of Devonian clastics of Maryland came from the east.... If the oceans have swallowed great marginal parts of the continents, Lemuria, and tremendous areas between Indo-china and Australia and New Zealand, why not an Atlantic land bridge? The magnitude of the job is no greater than in the cases admitted and just mentioned.[114]

Schuchert's strategy was to emphasize the necessity of land bridges, as inferred from faunal evidence, and to lay the burden of interpretation on his colleague. "Just as I am talking to you now, I used to talk to Barrell... on the basis of multiple hypotheses," he recalled to Willis. And he pointedly refused to accept the argument that land bridges were physically impossible:

Remember, I am not raising these lands—they are primordial—I am only pointing out their existence and seeking among you geomorphologists a way to get rid of them. It is up to me to prove that the faunal connections as I see them do actually exist. We paleontologists... see no clearer than you geomorphologists, but the latter act nowadays as the physicists used to do in the days of Kelvin, when they said that the age of the earth could only be 25 million years. And now they tell us we can have two billions![115]

Willis ignored the criticism and focused on the flattering comparison to Barrell. "You are in a position to answer many questions of fact authoritatively," he replied to Schuchert, "and I shall be delighted if I can from my point of view throw upon our common problems something of the light Barrell used to furnish."[116] If fact and theory were segregated in rhetoric and writing, they would also be segregated in the persons of Schuchert and Willis: Schuchert would provide the facts, Willis the theory. The latter began by emphasizing areas of interpretive agreement. These included (1) that the continents are dominantly granitic; (2) that granite is an igneous differentiate "uptruded" from within the earth, and the continents must have formed from repeated granitic eruptions during Archean and Proterozoic time; (3) that isostasy requires granitic masses to remain elevated above the sea floor; (4) that basalts comprise the ocean floor and require a force other than isostasy to explain their presence on the continents; and (5) that geosynclines are compressed areas of continental margins, suggesting the emanation of forces from the oceanic deeps. These points led to an inescapable conclusion: Bowie's view—that tectonic processes were the result of internal adjustments to external mass transfer—was wrong. There must be internal forces to explain compression of geosynclines and basalts on the continents.

Willis suggested that these forces were the key to solving the problem of land bridges: if the earth's internal forces were sufficient to "crush up a continental margin, [they] should be equally competent... to squeeze out the contents of a magma basin and force the melt to any height, regardless of relative specific gravities and isostasy. That is the origin of the sub-oceanic ridges, as I see it," he concluded, "and also of oceanic islands, like the Hawaiian chain." Despite isostasy, basalts were erupted at crustal levels; oceanic ridges and island chains were made this way. But there was one critical difference between granitic and basaltic eruptions: granites, once intruded into the crust, are isostatically balanced, and so there remain. Basalts, on the other hand, are isostatically imbalanced in the upper crust and should tend to sink back into the depths. "A basaltic elevation is by reason of the uptrusive [sic] action raised above its isostatic position. It should seek to return to it. It is upheld by the rigidity of its foundations only."[117] Daly had made this argument in *Our Mobile Earth*: that basaltic crust was isostatically unstable and should tend to founder back into the depths. Daly had made this instability a mechanism for pulling continental slabs; Willis would make it into a mechanism of sinking land bridges.

Following the uniformitarian program, Willis looked to modern examples. The most obvious was the isthmus of Panama, connecting North and South America. Panama proved that a narrow ridge, scarcely above sea level, could connect continents, so all that remained was to show that such a ridge might later sink. Were there similar features now submerged? Willis suggested that the submarine ridge extending from Cape San Roque in South America to the mid-Atlantic and thence to Africa was one — the South Atlantic land bridge that Schuchert needed: "It occupies the position in which Schuchert says he must have a trans-Atlantic bridge from South America to Africa to separate the faunas. It seems to be tectonically logical and faunally demonstrated." By shutting off oceanic circulation between the equator and the poles, such a bridge might also help to explain the Permian antipodean glaciations, while other bridges or chains of islands could explain Lemuria and the rest of Gondwana. "Are we not now getting closer together?" he asked happily, concluding: "I am sorry that we did not do it sooner."[118]

Willis's bold and stylized approach took Schuchert somewhat off guard. Atypically, he delayed replying while considering the argument. Finally, he responded: "I have had your very kind letter . . . now for over a month, and have read it several times, and the more I peruse it, the more I like it, and now I want to exclaim, 'Welcome, Brother-in Arms — we either live or go down together!'" With the letter, Willis had sent a preliminary paleogeographic map illustrating the configuration of Permian land bridges; Schuchert affirmed the collaboration by enumerating a few details that needed adjusting. A few bridges could be narrowed, others he wanted to widen, but "not enough to alarm you."[119] Because the Permian was a period of special interest to both of them, Schuchert suggested that they focus their collaboration on production of a Permian paleogeographic map with a discussion of the principles behind its construction. Schuchert would defend the paleogeography and Willis would explain the physical principles. As ever, Schuchert checked things out with Longwell, who considered the model "not so very improbable." Longwell was suspicious of Willis's reputation for speculation and disliked his neologisms — particularly "uptrusion" — but he allowed that isthmian links "could have existed just as well [in the Permian] as the Isthmus of Panama at the present time."[120] At least on uniformitarian grounds, the model worked.

Willis had no problem with Schuchert's adjustments, provided he agreed to the heuristic principle: "Once a deep, always a deep. Once a continent, always a continent."[121] Narrow basaltic ridges might be upthrust *between* the deeps — along the sites of present oceanic ridges or island chains — or along continental margins, but that was all. Continents remained continents, and the abyssal depths remained abyssal. Willis thus appeared to be drawing Schuchert back into the fold of American theoretical orthodoxy, except that he did not see it that way. Borrowing on Schuchert's religious metaphors, Willis jokingly called his position "the depths of . . . heterodoxy," presumably for its eclectic mix of uniformitarianism and progressionism. But mostly he emphasized the uniformitarian nature of his solution. In a letter to Arthur Holmes on December 20, he summarized the essence of his theory.

> My article . . . deals with the Permian connection between Africa and South America, using the Tertiary dynamics of the Caribbean deep . . . as a modern example. . . . The [South Atlantic] deeps were active during late Paleozoic and Permian time and the argument is that they produced the same effect upon their framework as has been produced by the out-thrusting from the Caribbean during this later episode of orogenic activity. You see I stick pretty close to Lyellian principles.[122]

Willis had started a correspondence with Holmes at Schuchert's urging. Schuchert wanted to know whether Willis's model could be linked to Joly's and Holmes's magmatic cycles, to account for the source of basaltic intrusions and the periodic transgression and regression of the oceans, thus generating a truly causal theory. "Can you make [their] cyclic heating and consequent submergence and emergence of the continents agree with your ideas?" he asked Willis. "And can you fit in as well Holmes's subcrustal flowage minus the Wegener sliding circus?"[123] Willis's short answer was no, because he rejected the global cyclicity implied by Holmes.

Willis believed that the basic outlines of the earth were constant — a uniformitarian position — but the deeps, he thought, were still deepening — a progressionist position. "The heat engine is still at work," he declared to Schuchert. As for periodicity, he rejected it in the global form that radioactive heat models seemed to require, but he believed that there was periodicity within each individual ocean basin and its continental margins. "The history of the North Atlantic was recorded in Europe and the eastern half of North America. That of the South Atlantic may be read in Africa and South America. The parallelism from east to west across each basin is remarkable and demonstrates the unity of the basin and its opposite shores."[124] Tectonic events seemed to be the same on opposite sides of ocean basins, suggesting basin-wide forces, but Joly and Holmes implied a global unity for which he saw little evidence. In any case, Willis was suspicious of Holmes because of his background in physics, and he was perhaps smarting from Schuchert's implication that Willis, as a "geomorphologist," was somehow aligned with Kelvin, or even worse, Jeffreys. Willis hardly saw the situation that way, given his overwhelming geological credentials and his extensive criticisms of Harold Jeffreys, both in private and in print. It was Holmes, not Willis, who took the side of the "mathematical physicist," and postulated "an ideal, unnatural earth, of perfectly uniform organization," he insisted.[125] Jeffreys had declared drift impossible, Holmes declared it unavoidable, and Willis dismissed them both.

Schuchert was troubled by this, for he still needed Joly and Holmes, as a letter from William Bowie made clear. In early July, Bowie wrote to congratulate Schuchert on the second edition of *Outlines of Historical Geology*. While generally praising the book, he expressed surprise that Schuchert believed in "great shifts of continental blocks." Bowie was referring to the 200–400 miles of lateral displacement that Schuchert accepted for the formation of the Alps and the Applachians — not to continental drift — but he disputed whether even this much lateral movement was possible. "I cannot see what would be behind these great movements to push the material forward and I cannot understand comparatively thin strata being moved so many miles forward against the shearing and frictional resistance and the resistance to deformation that the forward moving crust would have to undergo."[126] Joly and Holmes provided an answer—the answer on which Schuchert was evidently now relying—that crustal movements occurred during periods of fluidity when frictional resistance was greatly reduced. But if magmatic cycles were not to be incorporated into Willis's model, then how would geologists answer Bowie's complaint?

Willis was by now "impatient of delay in throwing our combined weight in the balance of truth" and forwarded to Schuchert his own recent correspondence with Holmes by way of explaining why he rejected the latter's views.[127] Schuchert approved of Willis's attempt to resolve their theoretical differences in private, prior to publication, and wrote back to Willis that he hoped they would all three work "together in a friendly correspondence . . . rather than by thrashing out our differences in the public prints." Meanwhile, he continued to press the point with Willis. Radioactivity had to be considered, he continued in his reply, not "as a heat engine to slip the continents all the more easily around the surface of the earth, according to fancy, but as the very causation of the mountains, the rising magmas pushing inward the continental margins." But Schuchert also clearly sought resolution, for he closed his letter by demurring and making reference to his own lack of education in physics and chemistry, "But I am a paleontologist, and due to my training can not enter these geophysical and very difficult problems."[128]

On July 21, Schuchert forwarded to Willis his completed paleogeographic map of "all Permian times," content that "these small land bridges . . . are easily admitted, and do not compare with the necessary geophysics of oceanic land bridges like those of Gondwana." Whether it was wishful thinking or whether the elderly Schuchert was simply getting confused, he referred to Willis's model as "your convection-subcrustal flowage theory."[129] In *Researches in China*, Willis had invoked subcrustal flowage in response to the vertical motions of crustal elements; perhaps Schuchert recalled this and assumed that Willis still held this view. Whatever the reason, Willis exploded in reply: "My *subcrustal flowage theory*! You even link it with *convection*!" Calming down, he continued:

> Now my dear Schuchert, convection and flowage are effects of gravity and require certain special conditions of mobility to carry on. I long since abandoned gravity as an effective cause of movements. Convection I regard as an impossibility in the solid earth. And rock flowage, where it does occur, is an effect of recrystallization due to high temperature and differential stress. Pray don't get Wegener and Willis mixed in your mind![130]

With that, Schuchert dropped the idea of convection for the last time. Bowie's complaint would be left unanwered, and the various discussions of alternative mechanisms would be left hidden in private correspondence.

Gondwana Dwindles Away

The result of Schuchert's and Willis's correspondence was a pair of papers published back-to-back in the *Bulletin of the Geological Society of America* at the end of 1932: "Gondwana Land Bridges" by Charles Schuchert and "Isthmian Links" by Bailey

Willis, with reprints issued under one cover.[131] For Schuchert the work was a conceptual breakthrough, finally resolving the problem of Gondwana. Gondwana *did* exist until the end of the Cretaceous, but as a group of continents connected by borderland bridges and upthrust suboceanic ridges. The solution rested on the principle of extending the present configuration of the continents and oceans into the past, with modification to provide connections where required. "The Caribbean-Panama offers a clue to the riddle," Willis summarized in early 1932. "Its framework is an upthrust ridge, which has united two continents in the past and does so now. The frameworks of other caribbeans lie beneath the oceans and have similarly united other continents in their period of activity."[132]

Schuchert recapitulated the history of the concept of Gondwana and the faunal evidence for land bridges, and he presented his map of Permian paleography with inferred connections forged across regions of contemporary ridges and island chains (figure 7.8). He assured his readers that future sonic sounding of the ocean basins would reveal the details of these features, inasmuch as the Navy had been making rapid progress of late with the recently invented sonic depth finder. Willis presented detailed maps of three specific land bridges—the "Brazil-Guinea" ridge, the "Africa-India" ridge, and the "East Indian" ridge—connected on the basis of contemporary physiography and bathymetry but presumed to have been emergent at appropriate intervals in the past, and he explained how these bridges were raised and how they subsided. The isthmian links were the result of upthrusting within oceanic basins, just as mountain chains along continental margins are (figure 7.9). He coined the phrase "cordillera of the sea" to suggest the genetic link between submarine and continental mountain chains. Finally, he noted how isthmian links could help to explain the Permian glaciation by altering the circulation of oceanic currents: the Antarctic ocean was entirely hemmed in throughout Permian times (figure 7.10). This was in direct response to a challenge from Holmes to "give some reasonable explanation of 'Gondwanaland' glaciation without continental or crustal drift."[133]

Willis was extremely happy with their papers, looking back on their "exchanges of views [as] a very happy episode of my scientific relations."[134] But Schuchert had remaining doubts, albeit not expressed in print. In May of 1932, just before their papers were submitted, the two had exchanged final drafts to ensure interpretive consistency. Both men noted a sticking point concerning the Atlantic land bridge: Willis wanted to subsume it into his general model of upthrust basaltic ridges, but Schuchert insisted it had to be a granitic mass of continental dimensions to account for both the quantity and the type of Paleozoic sediments derived from it. "May I ask you once more to add to your paper—the pages on Gondwana—an explanation of how a granitic Gondwana—a continental Gondwana—can be gotten rid of?" Schuchert wrote upon reading Willis's final draft. There was also the problem of how long the Atlantic bridge had lasted: Willis's upthrust basaltic ridges were supposedly isostatically unstable—hence they subsided and disappeared—but the Atlantic land bridge apparently lasted for over 200 million years. "You may want to say that the Atlantic land bridge never was a mass of granitic rocks, but that it always was of basaltic ones," Schuchert suggested, but then "I want you to comment on why the bridge stood all the vast time from the Pre-Cambrian into the late Cretaceous."[135]

Willis reminded him in response that there was really no evidence for a "*continuously maintained* connection," but rather only for a long period in which con-

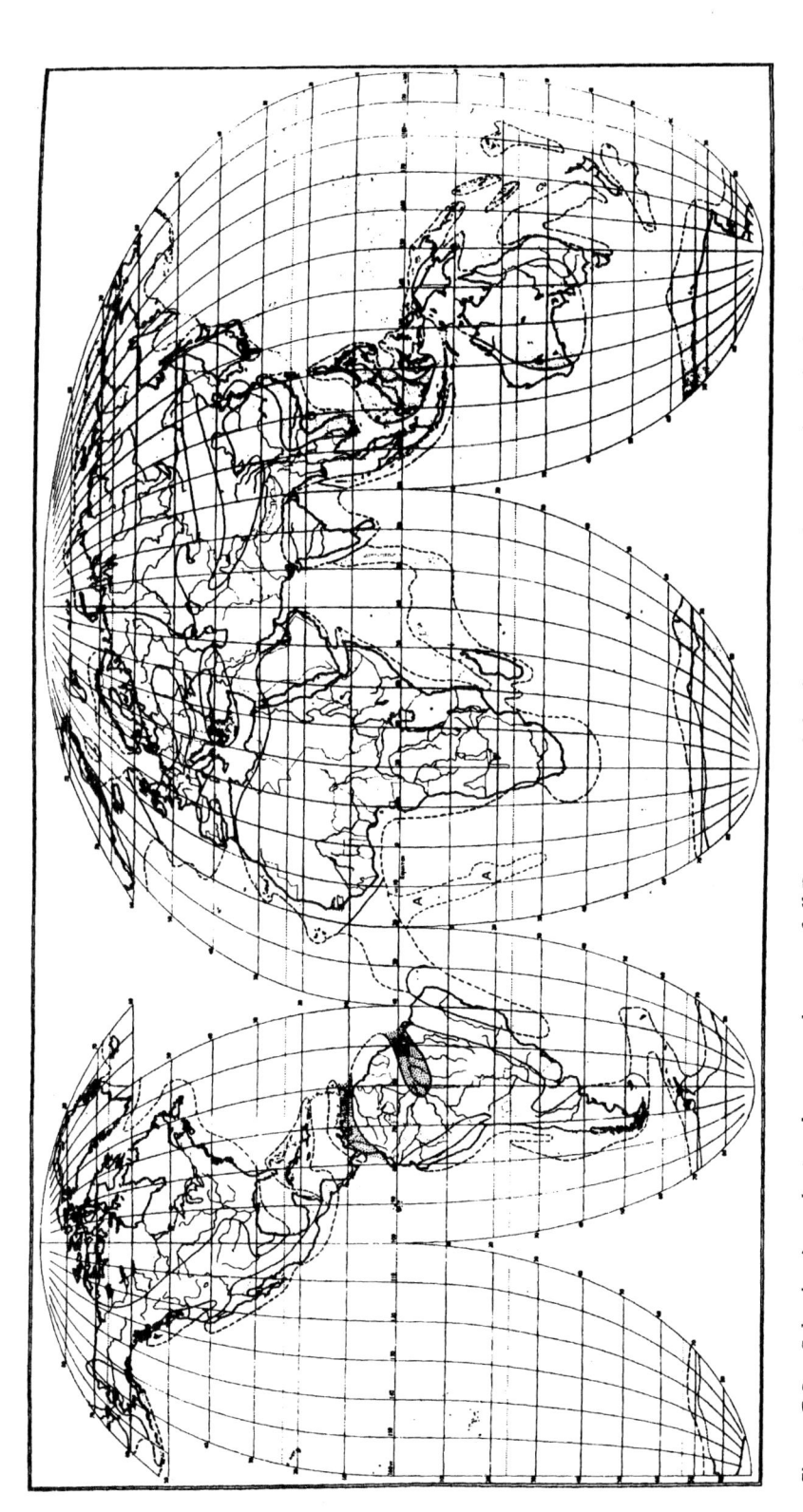

Figure 7.8. Schuchert's synthetic paleogeography map of all Permian time. Schuchert summarized in visual form the faunal evidence for land connections in late Paleozoic time and indicated where sunken land bridges might still be evident, as indicated by contemporary bathymetry. The point was to show that small changes in sea level, or small degrees of uplift, could cause presently submerged oceanic ridges and chains of islands to become emergent, thus explaining Permian faunal correlations. (From Schuchert 1932a. Reproduced with permission of the publisher, the Geological Society of America, Boulder, Colorado, U.S.A. Copyright © 1932, Geological Society of America.)

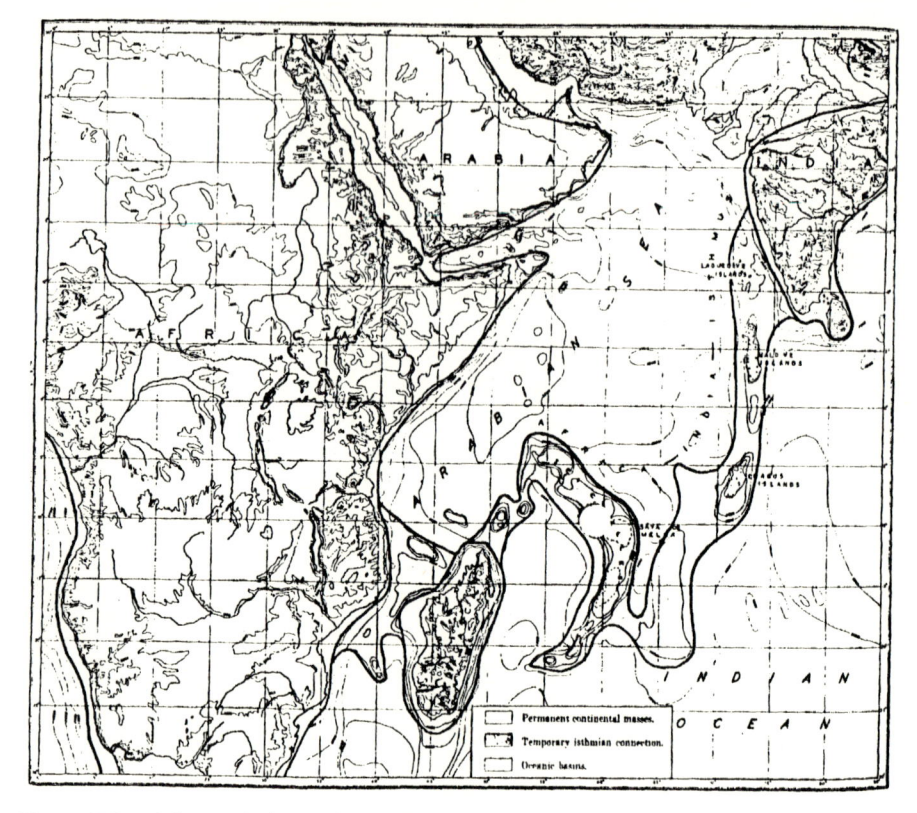

Figure 7.9. Isthmian links. Bailey Willis extended Schuchert's account to provide detailed reconstructions of former land bridges as required by faunal evidence. Here he illustrates the "Africa–India Isthmus"—a bridge between East Africa and India via Madagascar, the Seychelles, and the Maldives, connected across zones of bathymetric highs. Willis argued that these highs were the remnants of a submarine mountain chain that was once emergent but later sank in accordance with isostasy. (From Willis 1932. Reproduced with permission of the publisher, the Geological Society of America, Boulder, Colorado, U.S.A. Copyright © 1932, Geological Society of America.)

nections were at least intermittently maintained. And the latter was easy to explain: the same orogenic forces that uplifted the land bridges in the Paleozoic could have done so intermittently throughout much or all of the Atlantic's history. Willis asked Schuchert to add to his discussion the phrase "since isthmian links are possible at any dynamic time."[136] Schuchert was not entirely convinced—after all, where did the felsic Paleozoic sediments come from if the land bridges were basaltic?—but he agreed to Willis's editorial changes and dropped the question of the Atlantic bridge composition.

Why Schuchert gave way to Willis on this and other important theoretical points remains an unanswered, perhaps unanswerable, question. Willis's solution satisfied Schuchert's uniformitarian demands, but it left unresolved a number of issues. The Atlantic land bridge was one. The cause of upthrusting was another. Nor was there

Figure 7.10. Bailey Willis's reconstruction of Permian ocean circulation. A major strength of the theory of isthmian links, Willis argued, was that it also accounted for the Permian glaciation of the southern hemisphere. In his and Schuchert's reconstruction, the Antarctic Ocean was entirely hemmed in, thus cutting off circulation with warmer waters and explaining why it became subject to glaciation. (From Willis, 1932. Reproduced with permission of the publisher, the Geological Society of America, Boulder, Colorado, U.S.A. Copyright © 1932, Geological Society of America.)

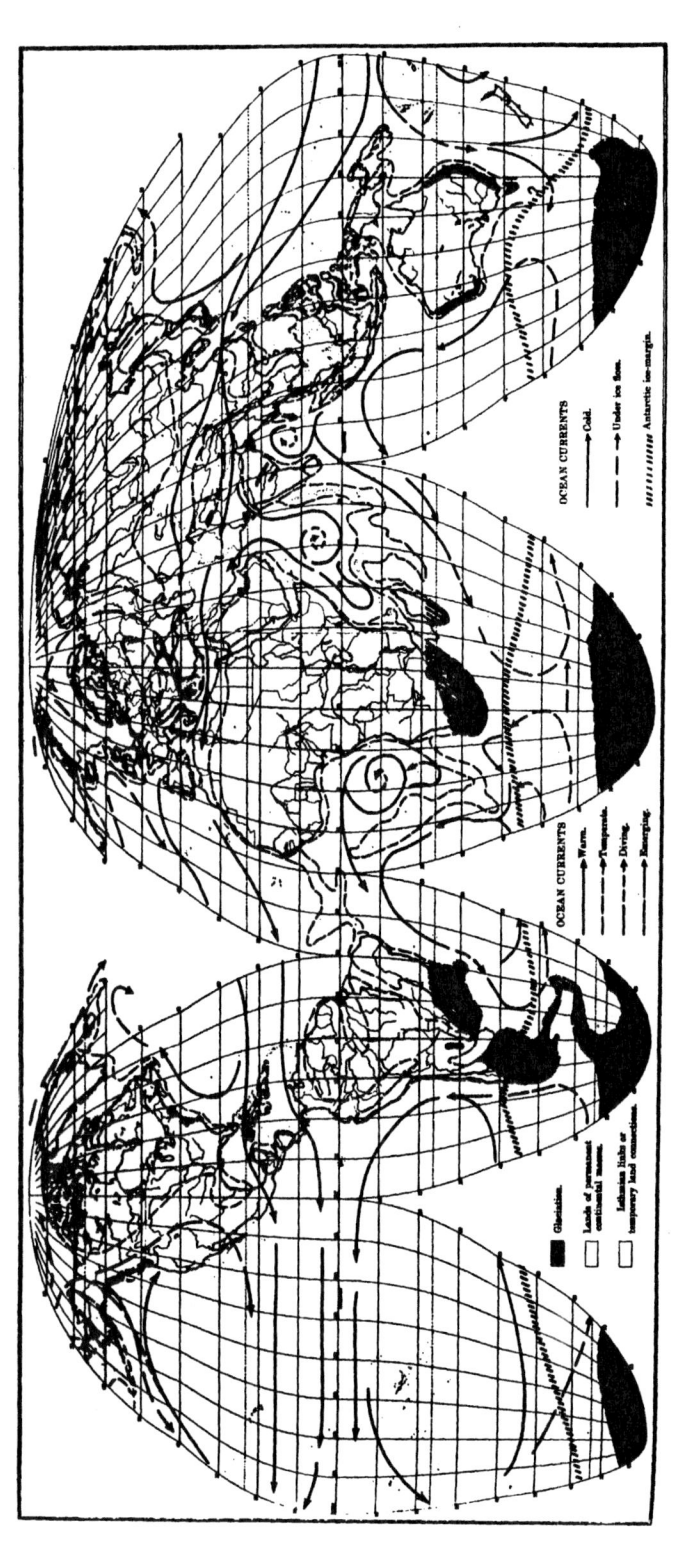

any independent evidence that the postulated land bridges ever existed. In that regard, they could well have been considered equally as speculative as Wegener's continental perambulations. But Schuchert did give way, and he swept the remaining problems under the rug. "Of course we have our differences," he concluded, attributing these differences to their divergent starting points, "you being a strict interpreter of permanence while I am a liberal, a follower of our mutual friend Suess." But both men had come a very long way from their starting points: Schuchert was seventy-three years old and had been working on this problem for a long time. Perhaps he had simply run out of steam. Perhaps he felt this was a reasonable compromise. Or perhaps he was simply tired of disagreement. His final letter to Willis before submitting his paper suggests so: "I hope the geologists of the world will agree to a man to accept our views as to how the transoceanic bridges may be made and unmade; not the *unreasonable* ones asked for by biogeographers crossing all parts of the oceans, but the *reasonable* ones asked for by paleontologists! Ha! Ha! Ha! We live or die together, nicht wahr?"[137]

If the geologists did not agree to a man to accept isthmian links, many of them did, particularly in the United States. Schuchert's and Willis's resolution of the problem of Gondwana effectively marked the end of active debate over continental drift in the United States. The faunal homologies had been removed from the list of arguments in favor of drift, and without them Wegener's argument was drastically undermined. Isthmian links explained the homologies while violating neither isostasy nor uniformitarianism. A compromise had been found and no recourse to drift was required. Thus it was that when Arthur Holmes came to Harvard in 1932 to deliver the Lowell lectures, his support for continental drift seems to have had little impact on American thinking.[138]

The hypothesis of isthmian links was patently ad hoc: it had been proposed by Willis for the purpose of explaining—or explaining *away*—the faunal evidence that Wegener had used to support drift, Suess to support sunken continents. But it evidently satisfied the concerns of American geologists, because isthmian links were widely accepted. The feeling among American geologists—that Schuchert and Willis had indeed found a reasonable solution to their problems—was perhaps best summarized by E. W. Berry, the paleontologist who had earlier accused Wegener of "autointoxication." He wrote to Schuchert to congratulate him early in 1933: "I cannot and never could divorce the continents without some land connections of this kind," Berry explained, "and I could not explain the present distribution by means of the ideas of continental drift.... The new isthmian links will stand much closer inspection than the larger 'land bridges' such as Gondwana Land."[139]

From this point on, then, for the better part of three decades, American geology students were taught that flora and fauna had migrated among ancient continents via narrow, intermittently emergent land bridges.[140] Not surprisingly, between the mid-1930s and the mid-1950s very few articles were published in America on the subject of moving continents.[141] In some places, the idea was banished. One geologist, Joseph Poland, recalled that at Harvard in the late 1930s continental drift was discussed only "on the back stairs."[142] Steven Jay Gould has recounted how at Columbia in the 1950s advocates of drift were publicly ridiculed.[143] Certainly, few Americans in the 1930s expressed much regret that the debate seemed to have achieved closure. An exception was Howard Baker, Schuchert's former student and author of his own version of drift theory, who in January of 1933 lamented Gondwana's passing with a rare reference to the economic disaster around them. "Times are hard and rigid economy is the thing, but it does seem tough to see a fine big Gondwana Land dwindling away to a paltry isthmian link . . . Gondwana," he concluded, "is on the wane alright."[144]

III

A REVOLUTION IN ACCEPTANCE

Such is the infelicity and unhappy disposition of the human mind in the course of invention that it . . . first will not believe that any such thing can be found out; and when it is found out, cannot understand how the world should have missed it so long.

—*Francis Bacon*

8

Direct and Indirect Evidence

At the California Institute of Technology in the mid-1940s, a young Henry William Menard—later an expert on submarine physiography and director of the U.S. Geological Survey—learned about continental drift from Beno Gutenberg. For although most American earth scientists considered the question of drift settled, many Europeans did not. Among them was "Dr. G," famous for his pioneering work on microseisms (the continual seismic disturbances that form the background "noise" of seismographs) and deep-focus earthquakes, who had come to Caltech from Germany in 1930. In 1939, he edited *Internal Constitution of the Earth*, part of a series entitled Physics of the Earth sponsored by the National Research Council.[1] Gutenberg's chapter, "Hypotheses on the Development of the Earth's Crust and their Implications," focused on the evidence for a plastic crustal substrate and "currents" within it. More than just an idea, he argued, subcrustal currents were necessary—in the past and at present—to account for both isostasy and horizontal crustal dislocations: "Many writers have expressed the belief that the strength of the interior of the earth prevents any currents today. The results of geophysical research, however, leave no doubt that such currents still exist. . . . [either] as a consequence of changes produced by disturbances at the surface [or as] the primary cause of movement at the surface."[2]

Gutenberg's course at Caltech reflected these views.[3] The strength of the crust was "enough to support [the] highest mountains," he explained in class, but isostasy demonstrated that this strength "decreases downwards, and below 40 km or so plastic flow may occur." This flow was implicated in both geological and seismological processes. Among the forces causing earthquakes, for example, Gutenberg suggested "elastic rebound as a release of shear due to sub-crustal flow and contraction of the crust & possibly differential movements in the crust from continental drift." He noted that the energy release associated with earthquakes was "of the same order of magnitude as that due to temperature gradient," which suggested that the most likely cause of plastic flow was internal temperature differentials.[4] One preexamination review sheet asked students for the meaning of isostasy and of "Wegener's hypothesis."[5] Menard wrote in his notebook: "The lighter sial layers of the crust may be considered to be floating on the denser sima layer which over long periods flows and behaves like a liquid in that respect.... Wegener's hypothesis is that, since Carboniferous times, when they were all closely joined, the continents have drifted apart. The configuration of the land was roughly as illustrated."[6] His illustration is reproduced here (figure 8.1).

Reading assignments in Gutenberg's class included David Griggs's 1939 paper, "A theory of mountain building," in which Griggs suggested that the earth's temperature gradient "is enough to set up convection currents," as recorded in Menard's index cards. Following Griggs's argument, Menard noted: "Theories on forces available for m[oun]t[ain] building: 1. tidal force 2. polflucht [pole flight] force 3. coriolis force 4. compression due to thermal contract[ion]. 5. *viscous drag.*" Menard summarized Gutenberg's views on the efficacy of these forces:

Tidal force too small by 10^{-12}

Polflucht force   "   " 10^{-4}

Coriolis   "   "   " 10^{-8}

Thermal contraction would require cooling $\geq 1500°$ in Tert[iary] times[7]

Viscous drag was the preferred answer for both Griggs and Gutenberg.[8]

Gutenberg also assigned a pair of papers by M. King Hubbert (later famous for his work on the depletion of oil reserves), one of which had formed the mathematical basis for the scaling of Griggs's experiments.[9] Hubbert cut to the quick. Geologists and geophysicists had been at cross purposes for nearly a century, trying to resolve "how an earth whose exterior is composed of hard rocks can have undergone repeated deformations *as if* composed of very weak and plastic materials."[10] Seismic and astronomical data suggested that earth materials were rigid; geological and geodetic phenomena suggested that they were weak. It was time, Hubbert declared, to put this conflict to rest, for it was "more apparent than real." The solution lay in thinking seriously about scale, both in time and space.

Just as the rigidity of earth materials was time-dependent—as Wegener had pointed out with his sealing wax analogy—Hubbert argued that it was scale-dependent: the larger the area over which stresses were applied, the more likely a material was to behave in a viscous manner. Scale models and laboratory experiments exaggerated the earth's rigidity. Bailey Willis's experiments on the structure of the Appalachians were a case in point: although they had made him famous and had stimulated his colleagues to think about the mechanisms of crustal deformation, the results bore little resemblance to the real Appalachians. Instead of displaying the smooth folding seen in rocks, the experimental slabs had broken up into brittle pieces with gaping holes between them (see figure 7.7). While its heuristic value was evident, it was difficult to see what an experiment like that proved. To model the behavior of earth materials realistically, one had to correct for time and scale; Hubbert laid out a mathematical theory of how to correct for them. Menard summarized: "Principle of physical similarity i.e. if length (and $\therefore$ area (l^2) and volume (l^3)) is reduced, then corresponding reductions should be made in viscosity."[11]

Griggs had followed this principle in the construction and design of his experiment, and the results, according to Hubbert, were "in remarkable accord with the major tectonic features of many existing mountain ranges."[12] Hubbert concluded:

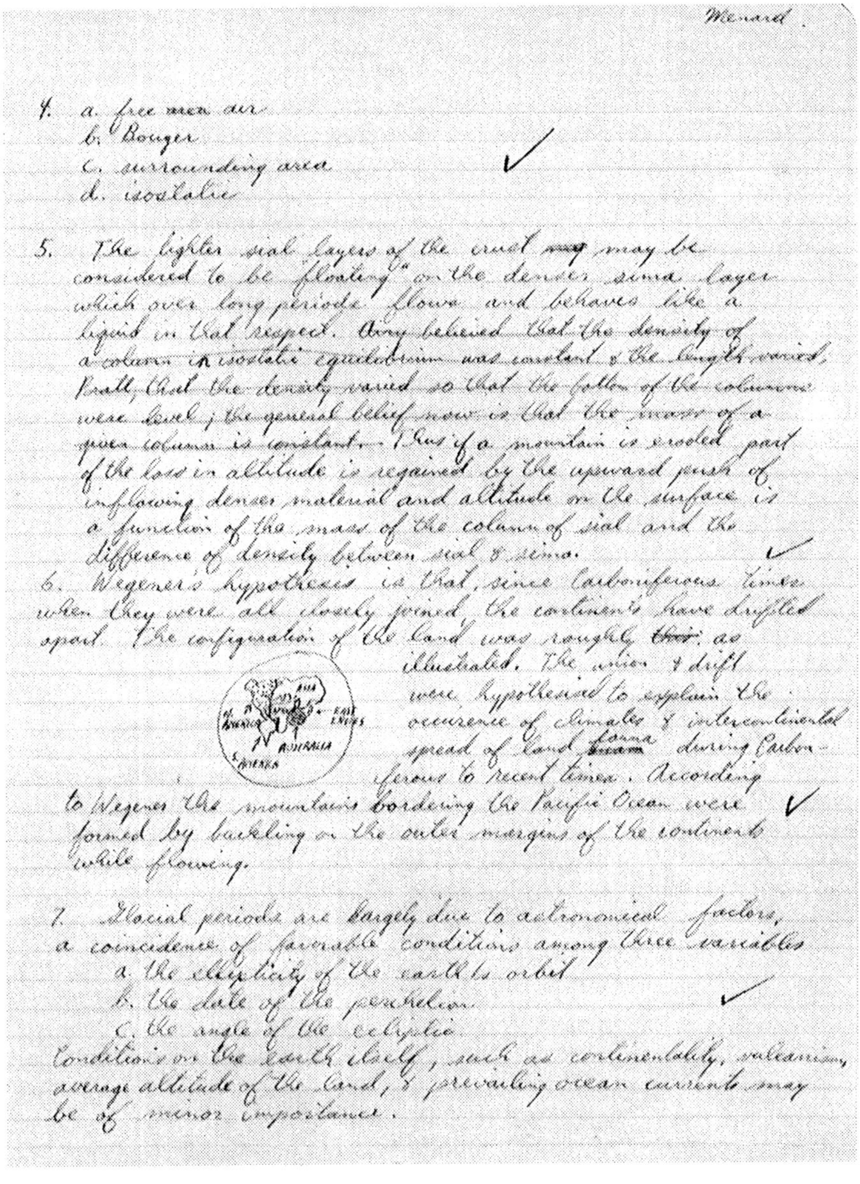

Figure 8.1. William Menard's notes on Beno Gutenberg's study sheet at the California Institute of Technology, mid-1940s. Question 5 asked, "What is the meaning of *isostasy*?" Question 6 asked, "What is Wegener's hypothesis?" In response to the latter, Menard wrote: "The union and drift [of the continents] were hypothesized to explain the occurrence of climates and the intercontinental spread of land fauna during Carboniferous to recent times. According to Wegener the mountains bordering the Pacific Ocean were formed by buckling on the outer margins of the continents while flowing." (Archives of the Scripps Institution of Oceanography, UCSD, Henry William Menard Papers, 1939–1986, Box 33, Folder 5.)

> It now becomes manifest that much of the conflict... between the evidences of rigidity and those of fluidity was not a conflict of data but of thinking. Our direct experiences with features of the surface of the earth are unavoidably limited to quantities commensurate with those of the human body—lengths of a few meters, times of a few years, forces of a few pounds, etc. In such terms, the crystalline rocks of the earth appear infinitely rigid and enormously strong.... It is only when we analyze the expected behavior of such materials on a scale of geological time and space and translate these results into terms within the domain of human experience that we realize how enormously wrong we have been. We learn that the resemblance of the behavior of rocks on a length scale of thousands of miles and a time scale of millions of years is not to that of the rocks with which we are familiar but rather to that of the viscous liquids and weaker plastics of our personal experience.... In geological space and time, the behavior of the earth built to the specifications of the staunchest adherents of the hard-rock school would leave little to be desired by the most incorrigible apostles of fluidity.[13]

Geologists might reasonably have argued that it was not they who had been wrong, but their geophysical opponents, who insisted on rigidity in the face of overwhelming evidence to the contrary. But, until Hubbert, all the mathematics had been on the geophysical side. Anyway, the message to Gutenberg's students was clear: contra Jeffreys, large-scale continental displacements were possible.

Gutenberg was unusual in the breadth of his scholarship and in his willingness to discuss continental drift so seriously in print. Among American academics in the late 1930s and 1940s, drift was only rarely discussed in published papers. To some extent it was suppressed. When T. Wayland Vaughan, director of the Scripps Institution of Oceanography, wrote to Charles Schuchert on behalf of the National Research Council to invite him to contribute a chapter to the *Physics of the Earth* volume on oceanography, published in 1932, he specifically asked him to include "a brief statement of the principles or hypotheses that have been advanced to account for the changes in the relations of land and sea, [including] the effects of loading and unloading due to transportation of material from the land to the sea, radial shrinkage and tangential thrust, and the continental drift hypothesis."[14] Schuchert ignored Vaughan's request and wrote entirely in the framework of modified permanence to which he himself had only so recently come.[15]

For the most part, drift was seen as unsupported, and, in light of isthmian links, unnecessary. Without the faunal data, drift became merely one theory among many to account for mountain-building. Thus when Menard completed his military service in 1945 and returned to pursue his Ph.D. in geology at Harvard, drift was discussed not in the context of paleontology but in that of structural geology. Marlin Billings, Harvard's renowned structural geology professor, outlined four possible causes of crustal compression: contraction, convection currents, igneous intrusions, and continental drift. (Like Schuchert, Billings decoupled convection from drift.) Billings's discussion of drift was based on the 1924 translation of Wegener's *Origin of Continents and Oceans*, and he recycled the old objections to it in disregard of later work that had partly answered them. Billings's final examination in January of 1948 thus demanded, "What principal facts must a theory of orogeny explain? What are some of the objections to the theories most commonly advocated? What additional data do we need for a satisfactory theory?" Menard responded that "drift has been questioned," because

1. The drift has not been proved.
2. The margins of the continents are thought to be folded by pushing into a

viscous basic layer although the more "fluid" basic layer shows no evidence of rippling.
3. What formed the early Paleozoic orogenies before the drifting occurred?

As for additional data, "geophysicists must obtain more information on the nature and distribution of the earth shells, the strength and physical characteristics of the earth, and the effect of radioactivity in earth processes. Geologists must do more detailed field mapping to unravel the exact sequence of events in the orogenic cycle."[16] (For this answer, Menard received eight out of ten points.) Yet, even while Billings recited old objections, he also taught his students about Holmes's convection model and Griggs's experiments (figure 8.2). Harvard students in the 1940s were not ignorant of the argument for convection currents as the driving force of continental drift.

Billings adhered to the method of multiple hypotheses—indeed he followed it to perhaps an illogical extreme—and presented drift as one of several possible tectonics theories. In a final lecture, he outlined no less than nineteen (figure 8.3). Meanwhile Daly covered continental drift in his tectonics seminars and was perhaps the only teacher in the United States to have students actually read Wegener[17] (figure 8.4). Attitudes were different in Daly's classroom; Daly accepted periodicity and entertained the idea of convection currents. While Menard waffled through Billings's exam, he wrote decisively for Daly:

> An adequate and acceptable theory of mountain-building must be capable of explaining . . .
> (1) the periodicity of mountain-building
> (2) the genesis of tangential forces in the earth's crust suitable for shortening the crust & forming folded mountains
> (3) the reversal in the movement of geosynclines from subsidence to upheaval
> (4) the formation of arcuate ranges and island festoons
> I believe these criteria are met by the theory of mountain making by sub-crustal flow resulting from thermal convection currents. . . . These currents rise to the bottom of the crust, drag along the bottom of the crust for an indefinite distance and descend to depth for reheating.[18]

When Daly retired, he was succeeded by geophysics professor Francis Birch, a follower of Harold Jeffreys and an adamant opponent of drift.[19] With Daly's retirement, the North American geological community lost its strongest and most consistent defender of moving continents. And yet, the case was never entirely closed. For, in the 1920s, two research initiatives had been launched that bore directly on the matter, neither of which convinced scientists at the time that continents were moving, but both of which laid the groundwork for the work that did.

Looking for Proof

Marlin Billings taught his students that drift had "not been proved," but what *would* have constituted proof? This question vexed Wegener himself and his answer is well known: he thought that actualistic evidence would be decisive. The faunal evidence, the climatic evidence, the stratigraphic evidence—these were all, ultimately, indirect

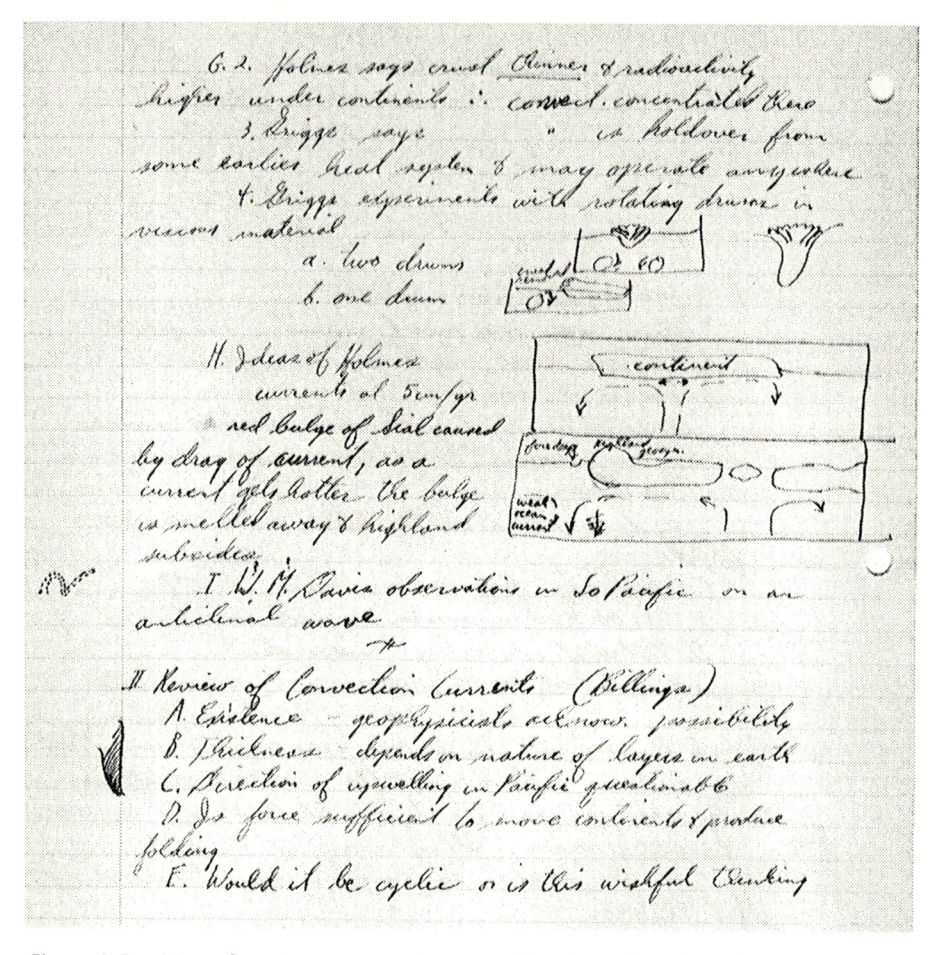

[Handwritten notes image]

Figure 8.2. Menard's notes on convection as a driving force of continental drift. At Harvard, where Menard did his doctoral work after completing his B.S. and M.S. at Caltech, at least two professors, Marlin Billings and Reginald Daly, discussed convection as a driving force of continental drift. In these notes, from the graduate seminar in tectonics in the winter of 1947, Arthur Holmes's theoretical model and David Griggs's experimental work are discussed. Note the rifting of continents under the up-welling current, as proposed by Holmes, and, in the uppermost sketches, the notation "crust removed" at the edge of the oceanic crust at the site of the down-welling current. (Archives of the Scripps Institution of Oceanography, UCSD, Henry William Menard Papers, 1939–1986, Box 33, Folder 11.)

and circumstantial. And they could be explained by other means. But if one could show that the continents were actually moving today, this would satisfy the uniformitarian impulse to look for explanation in presently observable processes. It would be *direct* evidence. Indeed, Wegener had argued that this was the distinguishing feature of his theory, because "it can be proved by accurate astronomical determinations. If continental displacements were active throughout such long periods of geological time, it must be assumed that they are continuing even at the present day."[20] And if they were continuing, then they could be measured.

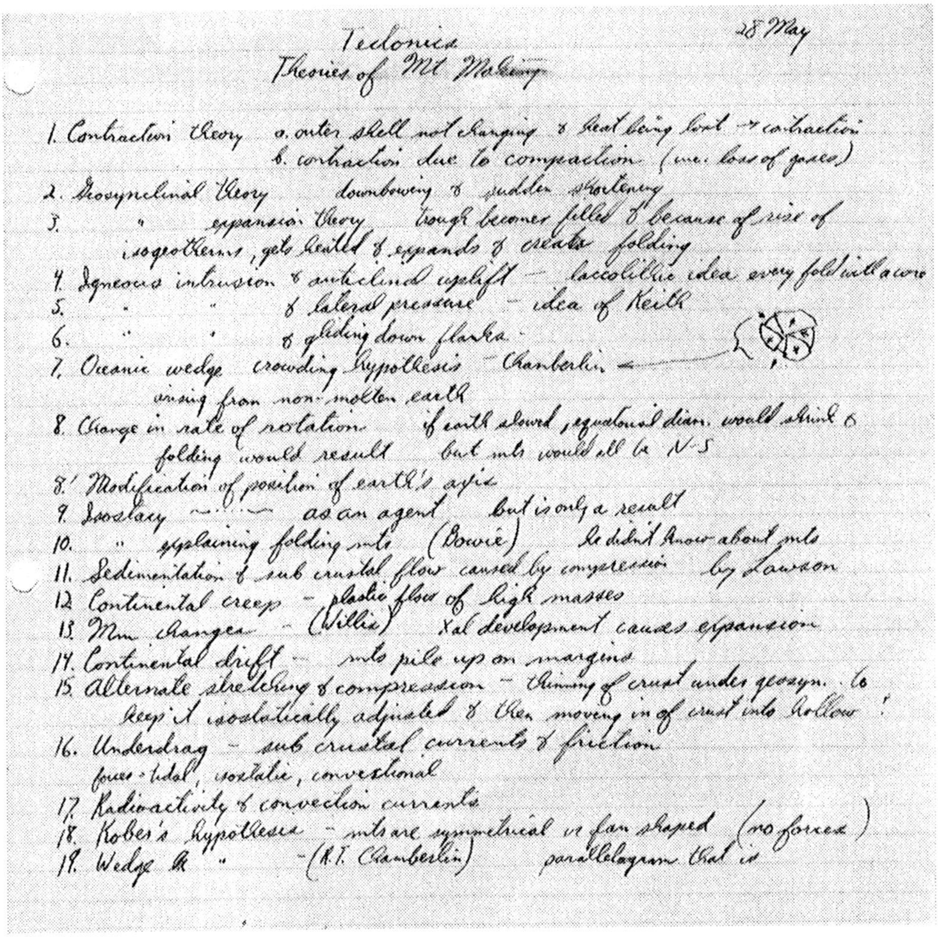
Tectonics — 28 May
Theories of Mt Making

1. Contraction theory a. outer shell not changing & heat being lost → contraction
b. contraction due to compaction (i.e. loss of gases)
2. Geosynclinal theory downbowing & sudden shortening
3. " expansion theory trough becomes filled & because of rise of isogeotherms, gets heated & expands & creates folding
4. Igneous intrusion & anticlinal uplift — laccolithic idea every fold will a core
5. " " & lateral pressure — idea of Keith
6. " " & gliding down flanks
7. Oceanic wedge crowding hypothesis Chamberlin
arising from non-molten earth
8. Change in rate of rotation if earth slowed, equatorial diam. would shrink & folding would result but mts would all be N-S
8' Modification of position of earth's axis
9. Isostasy ~ as an agent but is only a result
10. " explaining folding mts (Bowie) he didn't know about mts
11. Sedimentation & sub crustal flow caused by compression by Lawson
12 Continental creep — plastic flow of high masses
13 Min changes — (Willis) Xal development causes expansion
14. Continental drift — mts pile up on margins
15. Alternate stretching & compression — thinning of crust under geosyn. to keep it isostatically adjusted & then moving in of crust into hollow ?
16. Underdrag — sub crustal currents & friction
forces - tidal, rotatic, convectional
17 Radioactivity & convection currents
18. Kober's hypothesis — mts are symmetrical or fan shaped (no forces)
19. Wedge th " — (R.T. Chamberlin) parallelogram that is

Figure 8.3. Multiple working hypotheses of mountain-making, Harvard 1947. Marlin Billings took the method of multiple working hypotheses to a perhaps illogical extreme. In these notes, from his graduate seminar on tectonics in May of 1947, Billings differentiated no less than nineteen different orogenic theories. Continental drift (no. 14) is differentiated from "underdrag" (no. 16) and "radioactivity and convection currents" (no. 17).

On one level, these *were* separate hypotheses: Wegener had originally proposed drift without convection currents, and some geologists accepted the convection currents without necessarily linking them to continental drift. But on another level, Billings's separation of these "hypotheses" was artificial and anachronistic, because both Holmes and Griggs had argued unambiguously that these were but the elements of a single coherent theory in which drift of the continents was caused by "underdrag" generated from convection currents. (Archives of the Scripps Institution of Oceanography, UCSD, Henry William Menard Papers, 1939–1986, Box 33, Folder 11.)

```
Tuesday, April 30

    Isostasy, and its Relation to Mountain Making:  Textbook of Geology, by
                                 Longwell, Knopf, and Flint, Second Edition, 1939
                                 pp. 389-403
                                 Structural Geology, by Charles M. Nevin, Third
                                 Edition, 1942, pp. 237-256
                                 Geologic Structures, by Willis, Third Edition
                                 Revised, 1934, pp. 442-445.
                                 The Deformation of the Earth's Crust, (1933) by
                                 Bucher, pp. 25-61

Tuesday, May 7

    Continental Drift:  Structural Geology, Third Edition, 1942, by Nevin,
                           pp. 277-280
                        The Deformation of the Earth's Crust (1933) by
                           Bucher, pp. 69-76
                        Theory of Continental Drift; a Symposium (1928). In-
                           troduction, (The Problem of Continental Drift) by
                           W.A.J.M. Van Waterschoot van der Gracht, pp. 1-75.
                           Also read abstracts of the succeeding chapters.
                        The Origin of Continents and Oceans, by Alfred Wegener,
                           Third Edition (1922). Translated by J. G. A.
                           Skerl (1924). Peruse Table of Contents and
                           glance through the volume sufficiently to gain an
                           idea of Wegener's mode of attack on the problem.
```

Figure 8.4. Reginald Daly's tectonics reading assignments, spring 1946. In contrast to Billings's pluralistic presentation, Reginald Daly focused on a few specific theories. Daly had his students skim the Skerl translation of *The Origin of Continents and Oceans* and read van Waterschoot van der Gracht's summary of the problem in the American Association of Petroleum Geologists' symposium. (Archives of the Scripps Institution of Oceanography, Henry William Menard Papers, 1939–1986, Box 33, Folder 11).

John Joly had argued otherwise—that drift might have occurred sporadically during periods of substratal melting—but Wegener grounded his optimism in the work of J. P. Koch, the Danish explorer who had analyzed the longitude measurements of three successive expeditions to Greenland and concluded that it was moving westward relative to Europe.[21] As discussed in chapter 3, however, these calculations were based on lunar measurements of longitude, which were known to be inaccurate. A better alternative, Wegener argued, was wireless telegraphy. From the time required to send a radio signal from one continent to another, one could determine the distance between them. If the continents were moving, then the transmission time would change with time. Already there was some telegraphic evidence of longitude change, but it was scanty and inconclusive. J. W. Evans, author of the introduction to Wegener's third edition, agreed; the verdict of geologists on the lunar data was "non-proven," but telegraphy, "carried out say ten years hence, should enable a definite verdict to be obtained."[22]

In the fourth and final edition of *The Origin of Continents and Oceans*, geodetic evidence became Wegener's first line of argument. These data, now expanded and moved forward in the book, were putting "the discussion on a whole new footing."[23] At least some geophysicists agreed, for even while Wegener was preparing to return to Greenland to obtain better longitude determinations there, William Bowie had been promoting an international collaboration to test the theory by longitude measurements worldwide.

The Worldwide Longitude Operation: Direct Evidence of Drift

"SCIENTISTS TO TEST 'DRIFT' OF CONTINENTS" ran a banner headline in the Sunday *New York Times* on September 6, 1925 (figure 8.5). The feature article, written by Bowie, explained:

> Radio telegraphy will be used in the winter of 1926–27 to learn whether the continents are drifting around like icebergs in the sea or are firmly rooted to the interior of the Earth. It may seem strange to many that there is any doubt that the continents are rigidly fixed in place, but certain scientific men, especially in Europe, believe that the continents are wandering.[24]

Bowie summarized the biological, geological, and geographical data behind the Wegener hypothesis which, he explained, has been "the subject of much animated discussion in recent years." He continued: "If the hypothesis is proved to be true, then some of the most puzzling questions as to the distribution of land animals over the Earth's surface can be accounted for." For Bowie the strong point of the hypothesis was not the geological or paleontological evidence, but the jigsaw puzzle outlines of the continental coastlines, which "fit together on a map almost as well as two parts of a sheet of paper torn in two." Such an important and provocative idea deserves scrutiny, he suggested, and therefore "geodesists and astronomers of the world have decided to make a very severe test of the Wegener hypothesis."[25] An editorial writer of the day agreed: "[Drift] has seemingly enough support to make it a live scientific question and to suggest that it is worth while to find out whether we are drifting continentally, and if so, whither." Rejecting the prevailing American isolationism, he continued:

> The drifting apart of continents is not ominous, physically. It is only a drifting apart in their consciousness of a common responsibility for what goes on the whole way round the planet, though they are no longer tied together in one great continental mass, that should give us concern. Fortunately, the very means by which the ... geologist proposes to test the physical drifting apart, the radio, is one of the instruments that will help to hold the nations of the world together spiritually as their lands were once held physically.[26]

Those who had pursued radio telegraphy for positioning, however, were concerned less with keeping the world together than with preparing for the times when it came apart. The U.S. Navy had been experimenting with telegraphy as a means to determine longitude since the 1870s. The U.S. Coast Survey had pioneered the use of radio-telegraphy to determine longitude, and with the acquisition of the Philippine Islands in 1898 the perceived need for accurate measures became all the more pressing. By the early 1900s, the U.S. and other military forces were using telegraphy to determine longitudes, and there may have been some discussion between German and American geodesists about joining forces to corroborate each other's results.[27] If so, World War I put a halt to such plans, but after the war the idea was promoted again by Gustave-Auguste Ferrié, the brigadier general in charge of communications for the French Army, a member of the Paris Académie des Sciences, and the man responsible for converting the Eiffel Tower into a radio antenna to keep distant French forces in contact with Paris.[28] By the 1920s, radio was keeping the public in contact as well.

An international collaborative effort to measure longitude by radio telegraphy was discussed at the General Assembly of the International Astronomical Union (IAU) in

Figure 8.5. "SCIENTISTS TO TEST 'DRIFT' OF CONTINENTS." This article by William Bowie, from the Sunday *New York Times* of September 6, 1925, describes plans for the Worldwide Longitude Operation, an attempt to detect continental drift by looking for changes in radio wave transmission times. Bowie emphasized the need for direct evidence to support the "suggestive" faunal homologies. (Copyright © 1925 by the New York Times Co. Reprinted by permission.)

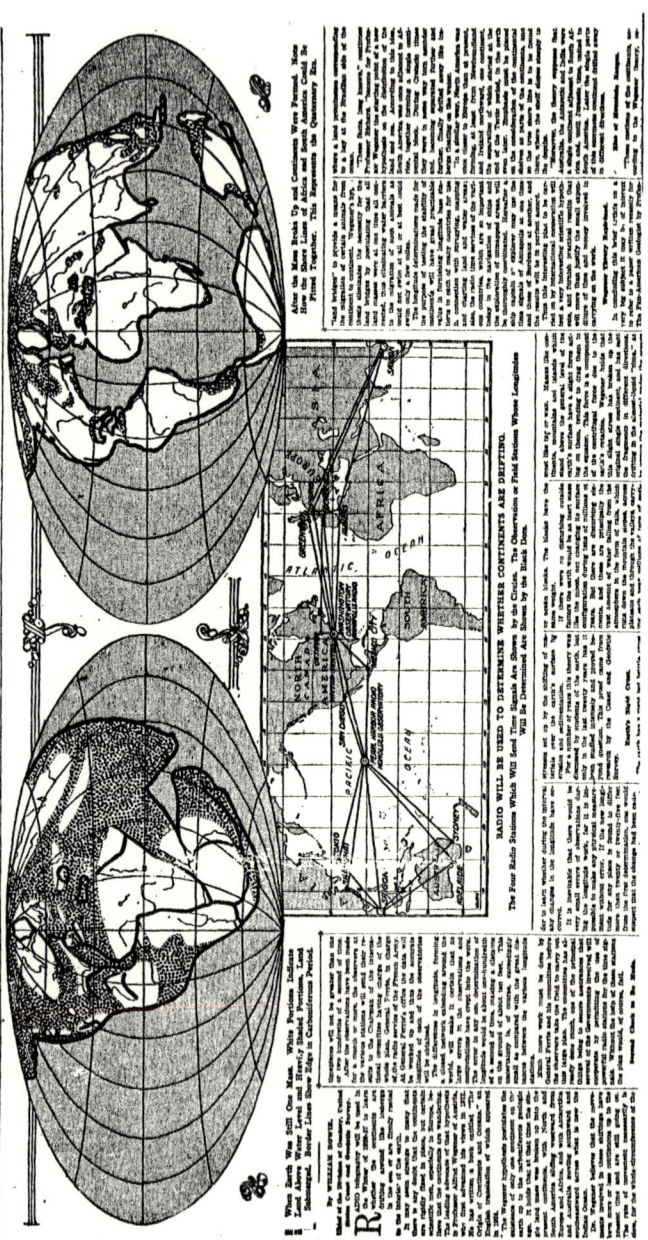

Rome in 1922 and again at the International Union of Geodesy and Geophysics (IUGG) meeting in Madrid in 1924; plans were apparently solidified at the second IAU general assembly, held at Cambridge, England in July 1925.[29] Following its history, the idea seems to have centered originally on developing a precise means of determining longitude. At the 1922 meeting the project was discussed under the title "Wireless Longitudes," and proceedings of that meeting refer only to the formation of "a longitude network around the globe, and to study in advance the problems that preliminary work would indicate."[30] But at the same time, the IAU was also involved in studying changes in *latitude*, including those in California along the San Andreas Fault, and somewhere between these various meetings the connection between longitude and Wegener's theory was made. By 1925 Bowie was describing the longitude operation in the popular press as a test of continental drift.

Three stations were chosen as the basis of a primary global polygon: Algiers, Shanghai, and San Diego. Forty-one additional stations in twenty-five countries would include Greenwich, Paris, and Washington, D.C. Measurements at San Diego and Washington would be performed cooperatively by the U.S. Naval Observatory and the Geodesy Division of the Coast Survey; additional readings would be taken by the Survey at Honolulu and Manila. For the Survey, these points would provide fixed references to become part of the primary geodetic control survey; for the Navy, they offered the prospect of accurate navigation at sea by means of radio transmissions. The entire project would be coordinated by a central office in Paris, associated with the French military: "The observers at the various stations will send their results to the Chairman of the International Committee having charge of the whole plan, General Ferrié, in charge of the radio service of the French Army. At General Ferrié's office, the data will be worked up and the accurate longitude of each of the observations will be obtained."[31]

Geodesists grounded the worldwide longitude project on an analogy with a triangulation net. In surveying, one detects errors by closing the loop: returning to one's starting place and correcting for the closure error. So it would be with the longitude operation. By closing the network—from Algiers to Shanghai to San Diego and back again—errors would be detectable and correctable. Small errors would be eliminated and any real motion of the continents discerned. "The difference in longitude," Bowie explained, "forming a closed network extending around the world, will make it certain that no large errors in the observations and computations have crept into the work." Longitude measurements in October and December 1926 would create a baseline to be compared against measurements in future years. If, after 5 or 10 years, "the new longitude for any place is found to differ more than twenty or twenty-five feet from the first determination, one would suspect that the change had been caused by earth movements."[32]

Bowie also summarized the project for scientific readers in a report on the IUGG's 1924 meeting in Madrid published in the *American Journal of Science*. Recasting the project as if it had always been conceived as a test of continental drift, he wrote: "Much has been said in recent years about the separation of continental and island land masses, but little or nothing is available to test the hypothesis. . . . The Section of Geodesy, realizing the importance of the stability of land masses . . . decided to plan and have executed in the very near future a world longitude net."[33]

It might seem strange that Bowie, who argued in his private letters to Charles Schuchert against continental drift, should espouse the idea in public. Indeed, the two men took opposite stances: Schuchert arguing dogmatically against the theory in public, but weighing it seriously in private; Bowie arguing against it in private but promoting it in public. Bowie's position may be accounted for in part by the opportunity the longitude project gave him to publicize his own accomplishments in isostasy, insofar as the two theories were intimately linked. "The Wegener hypothesis is really an outcome of the proof of isostasy," he reminded *Times* readers, because drift was premised on the conclusion, derived from geodesy, that the crust of the earth was composed of "brittle rocks underlain by material which, though solid and rigid to short continued stresses, will yield like putty in the course of time." Here Bowie saw the situation as clearly as he ever did: if isostasy were true, then drift might be as well (albeit not necesarily in the form to which he was loyal). Therefore, he concluded philosophically, "it would seem to be unscientific to reject untested [such] a hypothesis."[34]

But this hardly seems sufficient explanation for a man who only a short while later insisted that drift was impossible in light of "elementary physics and mechanics." In his ample published work, Bowie consistently used the plasticity of the substrate as an argument for *vertical* motions, not for horizontal ones. Why spend time organizing a major international effort to search for something he did not believe existed?[35] One obvious answer is that Bowie wanted to *refute* the theory: a man who conclusively disproved continental drift would surely be acclaimed. But in the case of the worldwide longitude operation, negative results would be of limited value. A positive result—a definite, measurable change in the position of the continents—would be proof of drift, but a negative result would leave open the question whether the rates of motion were smaller than Wegener imagined, or whether Joly was right that drift occurred intermittently throughout geological history rather than continuously. A negative result would not disprove the theory.

Perhaps Bowie's own claim should be taken on face value: that, no matter how unlikely, drift was important enough to entertain and perhaps pursue.[36] Or perhaps he was irrational and acting in ways that were contradictory. Although either of these suggestions might be correct, neither can be sustained by historical evidence. In none of his copious scientific papers and correspondence from the 1920s did William Bowie ever suggest that he believed continental drift were true or even that he was seriously considering the prospect. Nor does the evidence suggest that he sought glory through refutation. What historical evidence does demonstrate, however, is that Bowie was a man committed to the advancement of his discipline. Having made a seminal scientific contribution in the early stages of his career, he dedicated the later stages to the advancement of geodesy. Bowie had global ambitions, not so much for himself personally, but for his science.

Bowie's mentor, John Hayford, had consistently stressed the importance of *quantity* in geodetic data. His demonstration of isostasy and determination of the Hayford spheroid had been accomplished above all through sheer bulk of information. Some other fields achieved theoretical advancement through instrumental or computational refinements, but in geodesy, Hayford had argued, "unnecessary refinement" was a "drag, preventing real advance." His own seminal contribution was based not on refinement but on effective approximation. And his case was illustrative. In geodesy, large quantities of moderate-quality data were more informative than small quantities of even the most exacting work. In geodesy, more data meant better science.

Bowie pursued this idea energetically, first at the Coast Survey, where he emulated Hayford in technical and procedural innovation aimed at streamlining data gathering and analysis, and later in his efforts to advance international collaboration. Having demonstrated isostasy on the basis of data gathered in the United States, the obvious next step was to improve geodetic data collection worldwide. In Africa, South America, Australia, and large parts of Asia there was scant geodetic information. Hugely important geological features, including the East African Rift, the Red Sea, and even most of the Himalayas, where isostatic compensation had first been detected, had scarcely been studied. Bowie considered it urgent to promote international collaboration in geodesy to fill in the gaps. Toward this end, he helped to promote the work of the International Astronomic Union, which convened its first general assembly in Rome in 1922. He also helped to found the International Union of Geodesy and Geophysics in 1919 and served as its president from 1933 to 1936. Believing that advancement would come from the interaction of geodesy with geophysics, he argued strongly for the inclusion of both under one rubric.[37]

Bowie also worried about funding for geodesy within the United States. As the chief of a Division with an explicit operational mandate, Bowie had to argue for his appropriations each year and to make the case on the basis of practical imperatives. His annual budgets never included line items for scientific projects: research was always squeezed in between the cracks. Hayford summarized the situation in the opening paragraphs of *The Figure of the Earth*: "The computing division (thirteen persons upon an average) is engaged principally in obtaining from the field observations the best available lengths, azimuths, and positions of lines and points fixed by triangulation, and in furnishing these values to persons who are to use them as controls for surveys. Research work, such as that here reported on, is but a small part of the activity of the division."[38]

In the 1920s, the situation was worse rather than better. The demand for geodetic data was increasing, but budgets were not. Bowie did not share these concerns with academic colleagues—he was an intrinsically optimistic man, and rarely expressed negative feelings—but his annual office reports from the 1920s betray a growing frustration, and he began to look for other ways to support his science.[39] One way was international collaboration. Another was through closer ties with the U.S. military. The U.S. Coast Survey had always had close, if at times tense, ties with the U.S. Navy; the worldwide longitude operation provided an opportunity for Bowie to make use of these ties and improve upon them.[40]

The proof of isostasy had come out of practical work; perhaps the proof of continental drift would as well. The connection was obvious: the Navy needed accurate positions for its stations, and if those stations were moving, the Navy needed to know it. Radio waves might also be used to locate ships at sea, and any operation that promised an improvement in techniques of precise positioning was of potentially huge strategic value. These "practical results," Bowie thus wrote, "will more than justify the small expenditure of time and money."[41] The first task of the IUGG committee, therefore, was "to secure assurances that several of the Governments involved will cooperate in the use of powerful radio stations for the time signals. Without the help of these stations the plan would, of course, fail."

In the end, the worldwide longitude operation did fail, although not for lack of governmental cooperation. The expected accuracy of radio wave positioning was 1/100 of a second of a degree of longitude, equivalent to ten feet—approximately three meters—on the ground. Today, scientists have measured the rates of continental movements at a few centimeters per year. With measurement errors of three meters, it would take more than one hundred years to detect continental motions. Bowie and his colleagues were looking for a rate of drift that was too high. Not surprisingly, the results of the worldwide longitude operation were inconclusive. As Charles Whitten, a geodesist who joined the U.S. Coast and Geodetic Survey in 1930, put it in retrospect, "The results really didn't show anything."[42] Measurements made in 1933 revealed no discernible differences from those made in 1926, except those attributable to operator error.[43] But although generating no data useful to resolving the question of moving continents, the worldwide longitude operation may have generated useful institutional contacts, because—before the first set of measurements was even under way—Bowie was making plans for another collaborative venture. And this one *would* make a difference to debates about the earth.

USS *S-21* Expedition: Gravity as a Surrogate for Crustal Motions

Two weeks after his article on the worldwide longitude operation, Bowie had another piece in the *New York Times*, this one headlined "SCIENTIST TO WEIGH THE 'FLOATING' EARTH CRUST"[44] (figure 8.6). In this article, he recounted the work of the Dutch geodesist Felix Vening Meinesz, who had recently developed an improved gravimeter for measuring gravity at sea.

Bowie had immediately recognized the significance of Vening Meinesz's work. The confirmation of isostasy, while hailed as one of the great scientific accomplishments of the period, was in fact incomplete. There was a major lacuna in the data base: a total absence of gravity data from the ocean basins.[45] As Bowie explained, isostasy had been "proved," but "only by measurements made on the continents and islands." The state of the crust beneath the oceans had never been examined. Even the available land-based data were almost entirely from North America and Europe. From the point of view of gravity, most of the world was unexamined. The interpretation of isostasy as a universal condition of the earth's crust could hardly be sustained if that universe omitted more than three-fifths of the globe.

Did isostatic equilibrium obtain over the oceans? This was a fundamental question, one that Joseph Barrell had raised in 1919. In his reexamination of Hayford's data, Barrell concluded that the evidence of isostatic equilibrium was weakest in coastal regions, suggesting that isostasy might *not* apply at sea. Hayford had understood this; in 1928, Fred Wright recalled that "the late Dr. Hayford told me years ago that if a gravity apparatus for use at sea were available, a single ship could, in the course of a year, contribute more important data bearing on the figure of the earth and the theory of isostasy than have been collected in the last generation on land."[46]

The obstacle to Hayford's ambition was technical. Gravity measurements were based on the well-known relation between the local acceleration of gravity and the period of a pendulum, $g = 4\pi^2 L/T^2$, where T is the period of the pendulum and L is its length. If the bulk earth were a regular spheroid with topography superimposed on

Figure 8.6. "SCIENTIST TO WEIGH THE 'FLOATING' EARTH CRUST." In this article, also from the *New York Times* (Sunday, September 20, 1925) William Bowie recounted the work of the Dutch geodesist Felix Vening Meinesz, who had recently developed an improved gravimeter for measuring gravity at sea as a test of the theory of isostasy. Bowie, along with scientists from the Carnegie Institution of Washington, hoped to bring Vening Meinesz to the United States to begin a program of marine gravity surveys in conjunction with the U.S. Navy. The plan came to fruition in the USS S-21 submarine expedition to the Caribbean in 1928. Plans to extend the program using the civilian ship *Carnegie* were quashed when the ship burned; subsequent work in the 1930s was continued in collaboration with the Navy. (Copyright © 1925 by the New York Times Co. Reprinted by permission.)

it, then one could predict the value of gravity at any location based on its elevation. Differences between measured and predicted values would reflect either divergence from the calculated figure of the earth or uncompensated isostatic anomalies.[47] However, this conclusion presupposed no external disturbances. The only forces acting upon the pendulum should be the acceleration of gravity and the centrifugal acceleration caused by the earth's rotation. Random accelerations encountered on board a ship would render measurement impossible. Attempts to measure gravity at sea had failed; the isostatic condition of the oceans remained unknown.[48] However, in the early 1920s, the situation changed.

Felix Vening Meinesz (1887–1966) was a civil engineer by training, employed by the Geodetic Commission of Holland. While American geodesists labored to overcome black flies, rattlesnakes, and the Ozark Mountains, Vening Meinesz and his Dutch colleagues struggled with ground vibrations induced by storms and waves in coastal Holland. In the early 1920s, Vening Meinesz designed a gravimeter that would work in unstable conditions. Two pendula of nearly the same vibrational period swung in the same vertical plane, and the difference in their angles of elongation was photographically recorded. If the two pendula are equally affected by extraneous horizontal acceleration, then the difference between them is due to the acceleration of gravity at that location. That is, the difference between the two pendula is equivalent to a single "virtual" pendulum free of horizontal disturbance. (In application, the device used three pendula in the vertical plane, with the outer two set in motion to create two pairs of pendula swinging in opposite phase. In addition, Vening Meinesz's gravimeter contained three dummy pendula to record temperature and humidity inside the apparatus, and to record the motion of the entire device.) The Meinesz apparatus overcame the effect of horizontal acceleration that otherwise disturbed the motion of a single pendulum at sea.[49]

In 1923, Vening Meinesz tested his device on the Netherlands *K-II* submarine expedition to Indonesia. A second cruise to Indonesia in the opposite direction, the *K-XIII*, which traveled westward from Holland via the Panama Canal, resulted in a complete circumnavigation of the globe.[50] On the latter expedition, Vening Meinesz completed a detailed survey of the enigmatic Java Trench, the area that Molengraaf attributed to the convergence of two giant crustal slabs. In all, over 200 measurements of gravity at sea were obtained, a staggering level of productivity.[51] These expeditions caught Bowie's attention, and he discussed them with Arthur Day, director of the Geophysical Lab at the Carnegie Institution of Washington.

This was Hayford's ambition realized, with the potential to solidify beyond doubt— or radically challenge—the theory of isostasy. In the spring of 1928, Bowie and Day approached C. S. Freeman, superintendent of the U.S. Naval Observatory, with a plan to invite Vening Meinesz to measure gravity aboard a U.S. Naval submarine. The Navy would supply the submarine, the Carnegie Institution would invite Vening Meinesz and pay for his trip, and the Coast Survey would provide the base station for calibrating the gravimeter and the personnel and computational expertise to perform the data reductions.[52] The Carnegie Institution would also supply Fred Wright to assist with the investigations. Wright was a man of tremendous breadth: alongside his research in petrography and geochemistry and his facilitating du Toit's research in South America, he was also an inventor with a gravimeter of his own. While investigating volcanic processes, Wright had designed a torsion gravimeter to determine density differences in lavas from a healthy distance. In 1926 he had applied for a patent on his instrument, and he was anxious to learn the details of Vening Meinesz's.[53] In June 1928, Navy Secretary Curtis Wilbur agreed to the proposal.[54]

From Hayford's and Bowie's perspective if isostatic disturbances were to be found, they should be in areas of recent loading, such as river deltas. This was a direct consequence of the conclusion that isostasy was complete and local: disturbances should only occur in areas where surficial material was being actively transferred. On land, this meant regions of erosion; at sea, regions of sedimentation. Positive gravity anomalies were therefore expected in coastal regions where large quantities of sediment were accumulating, but not over the deeper portions of the ocean, which were thought to be ancient, fully equilibrated features. But Vening Meinesz's cruises had suggested otherwise. Measurements over the Nile Delta revealed no positive anomaly despite the huge volume of sediment accumulating there. Furthermore, unexpected negative anomalies had been found over the Java Trench.[55]

Vening Meinesz's preliminary interpretation of the Java Trench was that it was a recent or active down-warping in the crust that was not yet adequately compensated. This suggested that, contrary to Bowie's views, there *were* major stresses present in the crust.[56] Bowie's initial impulse was to explain away the missing anomaly over the Nile Delta by suggesting that isostatic adjustment was even more rapid than had hitherto been supposed. Quoting from the geologist and Arctic explorer Fritjof Nansen, in reference to the isostatic rebound of Fennoscandia, Bowie suggested that perhaps "the earth's crust . . . approaches its level of equilibrium much more closely than even the most extreme advocates (like Hayford) of perfect isostasy have considered to be possible."[57] But Bowie was clutching at straws: Nansen's own work on Fennoscandia had shown it was a delayed reaction to the removal of ice five to ten thousand years before. And even if isostatic responses were nearly immediate, the negative anomalies over the Java Trench remained to be explained.

And there was an explanation available: continental drift.[58] Wright explained in his preliminary report of the cruise, sent to John Merriam in February 1929:

> Gravity measurements over the land areas of the Earth prove that the earth's crust is everywhere in a state approaching equilibrium. Wherever large departures do occur, they indicate excess or deficiency of load and these in turn produce stresses in the Earth's crust. It is an axiom in geology, as in other matters, that extremes are temporary in character; high mountain masses are not eternal but are soon worn down and effaced. If they are actually extra loads on the Earth's crust, they give rise to abnormally high gravity values. . . . [Thus g]ravity anomalies above a certain value serve to locate unstable portions of the Earth's crust where movements are taking place so rapidly that compensation has not kept pace with the disturbance . . . A knowledge of these factors, especially of the order of magnitude of the stresses active in mountain building, earthquakes, and other crustal movements, is fundamental geologic theory.[59]

If the crust were moving, isostatic anomalies would result. Conversely, isostatic anomalies not attributable to sedimentation were prima facie evidence of crustal motion.

The itinerary of the *S-21* expedition was designed to address these theoretical issues. As Eleanor Lamson, head of the computing section that analyzed the results, stated, the goal was "to include as many stations as possible which would assist in solving the geophysical and geological problems in and near the West Indies."[60] The plan was to cross the Mississippi Delta to check for an expected positive anomaly, to cross three zones of unusually deep water—the Nares, Bartlett, and Sigsbee Deeps—to check for negative ones, and to determine the normal value of gravity over the bulk of the Caribbean Sea and the Atlantic Ocean. The Caribbean Deeps were especially close to the hearts of Coast Survey scientists, as the discovery of the Sigsbee Deep by their colleague Charles Sigsbee was considered a major accomplishment of nineteenth-century hydrography. As historian Thomas Manning has said, the discovery of great depths was a passion of nineteenth-century hydrographers, and Sigsbee's discovery was one of the deepest.[61] Sigsbee's work was followed by that of hydrographer John Bartlett. The Sigsbee and Bartlett Deeps perplexed both geologists and geodesists. Existing tectonic theories gave no account of heterogeneities in the ocean floor; until the late nineteenth century, they had never needed to. Nor did isostasy explain it. But there was a possible answer. Perhaps Molengraaf's suggestion for the East Indies also applied to the West Indies. Perhaps these zones of unusually deep water were down-warpings caused by converging crustal slabs.

Vening Meinesz and Wright had these issues in mind as they set sail on October 2, 1928, from Hampton Roads, Virginia. The cruise was a huge success: they covered a distance of 7,000 miles in just under two months and completed forty-five gravity stations at sea and five more in port at Hampton Roads, Key West, Galveston, Guantanamo, and St. Thomas. During the inevitable downtime as well, while on board Vening Meinesz gratefully received a copy of Du Toit's newly published *Geological Comparison of South Africa with South America*, sent to the ship by Carnegie President John Merriam.[62] No doubt Wright and Vening Meinesz discussed the problems posed by large-scale crustal migration, and in the final report of the cruise they discussed the relationship between gravity anomalies and crustal dislocations. Based on Vening Meinesz's early work, there appeared to be real isostatic anomalies—i.e., not attributable to surface topography—and this implied that large stresses were present in the crust, most likely caused by recent or ongoing crustal motion. Indeed, one could calculate the magnitude and direction of those stresses:

> It is probable that real isostatic anomalies occur, caused by departures of isostatic equilibrium of the earth's crust. These departures in equilibrium can only be caused by stresses in the earth's crust or in the subcrustal layers, which prevent the crust from adjusting itself in floating equilibrium. If, for instance, the crust is subject to tangential stress (horizontal compression), its position may be on a higher level than that which corresponds to free floating equilibrium because of the curvature of the earth. This gives rise to an excess of mass that reveals itself in positive isostatic anomalies. A formula may be deduced giving a relation between this positive anomaly and the tangential stress which is the cause of it; thus it is possible to ascertain the magnitude of the stress if the anomaly is known.
>
> If, therefore, we succeed in determining the true isostatic anomaly for a certain part of the earth's crust, we may obtain data on the trend and the magnitude of the tectonic stresses existing in that region.[63]

The scientists' language was obtuse but their meaning was not: their work was a test of crustal mobility. While it was perhaps not as direct a test as changes in longitude were, it was a test nevertheless, for gravity measurements provided direct evidence of stresses in the crust, not in the distant geological past, but at present. Gravity measurements were a surrogate for direct detection of crustal motions.

Results of the S-21 Expedition: Airy Isostasy After All?

The S-21 expedition corroborated Vening Meinesz's earlier results and demonstrated that the condition of the crust was considerably more complicated than Hayford and Bowie thought. First, there was no anomaly over the Mississippi Delta. According to the model of complete and local compensation, anomalies should only occur in areas of active sedimentation where the loading of sediments would produce a transient positive anomaly until the crust had time to respond. In the Mississippi region, it was estimated that more than 12 billion tons of sediment were accumulating every year; therefore a significant positive anomaly was expected. None was found.

Second, there were large negative anomalies over the Nares and Bartlett Deeps. According to Hayford's and Bowie's views, ancient geological features should be fully compensated. Yet the Bartlett Deep appeared to be virtually uncompensated, the Nares only partly compensated. This suggested that they were youthful features. Furthermore, the negative anomaly associated with the Nares Deep extended beyond its length. This implied that the anomalies were not caused by the mass deficit of the Deeps but rather that both anomaly and Deep were caused by something else, presumably movement of the earth's crust. If so, then the ocean basins were not as permanent as the permanentists supposed.

Third, there was a large systematic positive anomaly over almost the entire flat-bottomed portion of the Gulf of Mexico — an area they labeled the Gulf of Mexico plate — and this anomaly increased abruptly by an order of magnitude at the edge of the continental shelf. These findings utterly contradicted the expectation of general isostatic equilibrium. Here was an anomaly with no topographic expression, a result completely at odds with theoretical expectations and for which there was no apparent explanation.[64]

Besides being contrary to theoretical expectation, these results also appeared to be mutually exclusive. On the one hand, the complete compensation over the Mississippi Delta suggested an elastic crust, responding almost instantaneously to sedimentary load. On the other hand, the positive anomaly over the Gulf of Mexico plate implied a rigid crust, able to transmit and sustain regionally extensive stress. The challenge that they posed to the theoretical framework of isostasy was encapsulated by Wright, in the final report of the S-21 expedition: "The theory of isostasy has been so well established that it is not easy to understand actual excess loads on the Earth's crust over areas of such vast extent. If stresses active in the crust are responsible for them, then the engineering difficulty arises of explaining their maintenance over such great areas."[65] To maintain regionally extensive stress, one needed a strong crust. To produce rapid isostatic response, one needed a weak crust. It appeared that the crust was strong and weak at the same time.

Or maybe isostatic compensation was not located in the crust after all. Perhaps the crust *was* rigid, and compensation was restricted to the plastic substrate below, as the Airy model implied, as Gilbert and Fisher had advocated decades before, and as Wegener argued to account for continental drift. Bowie had acknowledged this possibility in a 1924 paper, "A Gravimetric Test of the 'Roots of Mountains' Theory," one of the few places where he ever questioned the assumptions of the Pratt model. He explained:

All great mountain systems occupy areas which were once at or below sea level; therefore, if the isostatic equilibrium obtained in the past as it does today, [then either] there has been an expansion of the materials of the crust . . . without any disturbance of the subcrustal matter, or there has been a thickening of the crust without any change in density of the crustal matter, but with some of this matter forced downward into the space occupied by the subcrustal matter which has been forced horizontally from its position. Only in one of these two ways—a change in density or the formation of roots by the thickening of the crust—could the isostatic balance be maintained.[66]

There were thus two requirements for Airy isostasy to hold. The first, as Bowie emphasized in his private correspondence, was the existence of horizontal compression to generate the crustal roots. The second, which he noted here, was a sufficiently fluid substrate to permit the formation of these roots. "Can the subcrustal material be so plastic as to permit such a sinking of crustal matter and can the crushed crustal matter retain so much residual rigidity as would be necessary to resist the hydrostatic forces tending to arrange materials in layers according to densities? The writer does not think so."[67]

Bowie's position had always been consistent: the crust was weak, the substrate was somewhat plastic but not very much so, and isostatic compensation was accommodated by changes in density in crustal materials, with a limited amount of shifting of substratal material. The *S-21* results challenged this view. Positive anomalies without topographic expression required a strong crust, which implied a weak substrate, which implied that the Airy model might be correct. Perhaps there *were* roots to the mountains. Perhaps there *was* large-scale horizontal compression. But were the gravity results consistent with continental drift? Lateral drift should induce positive anomalies by compression on the leading edges of migrating continents. But the *S-21* data appeared to be at odds with this interpretation, too, because Wegener's theory of *westward* drift implied that positive anomalies should occur on the western sides of continents. Anomalies on the eastern side should be negative. Vening Meinesz's earlier Pacific cruise had detected a weak positive anomaly off the Pacific coast of North America, as Wegener's theory predicted, but the *S-21* expedition found a positive anomaly on the eastern side of the North American continent, suggesting an eastward drift. North America appeared to be moving east and west at the same time.[68] "The evidence, as far as it goes," Wright dryly concluded, "is not in favor of [Wegener's] hypothesis."[69] But it was not in favor of a static ocean basin, either.

Rethinking Isostasy

The results of the *S-21* expedition led Vening Meinesz and Wright to two conclusions at odds with the mainstream of American thinking. First, they concluded that major regional stresses *were* present in the oceanic crust. The ocean basins were not a passive substrate for floating continental rafts, nor were they a fully compensated region of higher-than-average crustal density. They were geologically active provinces sustaining regionally extensive stresses. Second, they concluded that, if the crust did sustain significant stress, then it could not be as weak as the Hayford–Bowie school supposed. It must contain "some residual strength." These conclusions implied a rethinking of isostatic processes and their relation to geological change.

A change in Wright's perspective is evident in his final report of the cruise, written with Vening Meinesz and published a year later, particularly in a passage where he discusses the question of geological extremes. In his unpublished preliminary report to John Merriam, quoted above, Wright cited the general opinion that "in geology, as in other matters, extremes are temporary in character." His tone in this preliminary report suggested that he shared this opinion and doubted that the S-21 cruise would find major uncompensated gravity anomalies. But if he did hold this opinion before the cruise, after the cruise he longer did. The assumption of the impersistence of extremes was a dubious truism of the geological literature, he now argued, and rather than considering it axiomatic, geologists should be treating it as a testable hypothesis, because it was testable, albeit not in the laboratory. The magnitude and persistence of orogenic forces could be determined by looking in the field for their direct effects:

> In view of the fact that the magnitude of orogenic forces is quite beyond direct study in the laboratory, it is necessary, if we would evaluate them, to study their effects in the field where they are now active. We know from a study of the rocks themselves what changes [these stresses] produce and how large are the masses they can move. But the mechanical relations are so complex and the quantities involved so prodigious that we can not, by any direct method, measure the order of magnitude of the forces themselves. Gravity measurements afford the only available approach to this problem which is fundamental to geological theory.[70]

Wright's comments were a direct indictment of colleagues who presumed to understand the mechanics of the earth's crust based on grossly scaled-down experiments or theoretical models rife with unverified assumptions. Wright was making Marshall Kay's argument: rocks *had* moved, therefore they *could*. And gravity measurements could tell something about how. Vening Meinesz and Wright thus concluded their study by prescribing an explicit research program in gravity detection, in conjunction with the U.S. Navy or other navies.[71]

The proposed program began immediately, as Vening Meinesz embarked in 1929 on another Dutch cruise to the East Indies, which confirmed and expanded his earlier results.[72] Meanwhile, the U.S. Coast and Geodetic Survey organized an immediate land-based expedition to Haiti, Cuba, and Puerto Rico, while Bowie made arrangements for future submarine expeditions. However he felt personally about evidence that the Pratt–Hayford model might be crumbling, he fully supported this research effort, writing that the S-21 was "only the beginning of the use of American submarines on gravity surveys."[73] He was right. Within a few months, the Hydrographic Office announced plans for additional submarine-based investigations.[74] But Bowie was getting on in years; the Navy–Princeton Gravity Expedition to the West Indies and the Navy–American Geophysical Union Expedition would be organized instead by Princeton University professor Richard Field.

From Gravity Anomalies to Tectogenes

Richard M. Field (1885–1961) was a man who lived up to his name: he offered the first field course ever to earn Princeton University credit. The International Summer School of Geology and Natural Resources operated out of a specially designed Pullman railroad car in which Field, his students, and colleagues went across the United

States and Canada each summer from 1926 to 1937. His own research was in sedimentology and stratigraphy; since 1927 he had been studying reef formation in the Bahamas, where he performed preliminary gravity measurements to decipher the structure of the islands. Field was also chairman of the American Geophysical Union (AGU) Committee on the Geophysical and Geological Study of the Ocean Basins, where he came to know Bowie. (He later served as the first official AGU president, from 1938 to 1941).[75] After the success of the *S-21*, Bowie suggested that Field and Princeton take over the gravity work.[76]

Two expeditions were completed between 1932 and 1937. The first, aboard the USS *S-48*, in 1932, included Vening Meinesz, who trained Field's young colleague Harry H. Hess, recently hired at Princeton as a Procter Fellow in geology.[77] The second expedition, in November 1936 to January 1937, officially cosponsored by the AGU on the submarine *Barracuda*, included a young geophysicist from Lehigh University, Maurice Ewing—later the first director of the Lamont–Doherty Geological Observatory. Ewing was already becoming known among his colleagues for his pioneering work in seismic refraction studies and would later become famous among both scientists and Naval officers for his discovery, with J. Lamar Worzel, of underwater sound channeling.[78] On the *Barracuda*, Hess did the soundings, Ewing the gravity measurements.

The results of these expeditions corroborated both the *S-21* work and Vening Meinesz's ongoing studies in the East Indies, and the coherence of the data emboldened the geologists to make strong claims. The Caribbean, Field wrote in his notes for a National Academy of Sciences meeting in early 1933, was a tectonically active region experiencing both compressional and tensional stress caused by "the eastward movement of the Caribbean block and the shearing of that block from the block north of it." The basin was not a static depression but had been characterized by a "long and complicated tectonic history, analogous to continental examples."[79] In the official report of the expedition, Harry Hess drew a diagram with a large, heavy arrow indicating the direction of movement of the Caribbean "plate." He also reinterpreted the earlier gravity work and suggested that the Bartlett Deep was a pull-apart structure arising from transcurrent faulting along the edge of the moving block (figure 8.7).

Hess emphasized that the negative anomalies were not randomly distributed but formed distinct linear belts: in the Caribbean, they were associated with the Deeps (the Bartlett Deep had now been shown to be in excess of 4,000 fathoms, nearly five miles).[80] In the East Indies, they were associated with the Java Trench and the latter was strongly correlated with earthquake epicenters, which, Hess concluded, was "strong corroborative evidence that it represents the line of major tectonic disturbance." He also noted a correlation with volcanoes, which occurred in "arcs parallel to the strip at about 100 miles distance, and are always on the concave side."[81] Negative gravity anomalies, earthquakes, and volcanoes all added up to one thing: these were belts of compression, where stresses were accumulating, the crust was buckling, and rocks were melting.

In notes for a meeting of the AGU Committee on Geophysics and Tectonics at the end of 1931, Field had compiled a list of questions and a chart outlining possible interpretations of the crustal structure of the Bahamas (figure 8.8). But by the time they returned from the Caribbean, Hess—younger and perhaps more impatient, perhaps influenced by Vening Meinesz—had settled on a single hypothesis to explain what needed to be explained, not only in the Bahamas but worldwide.

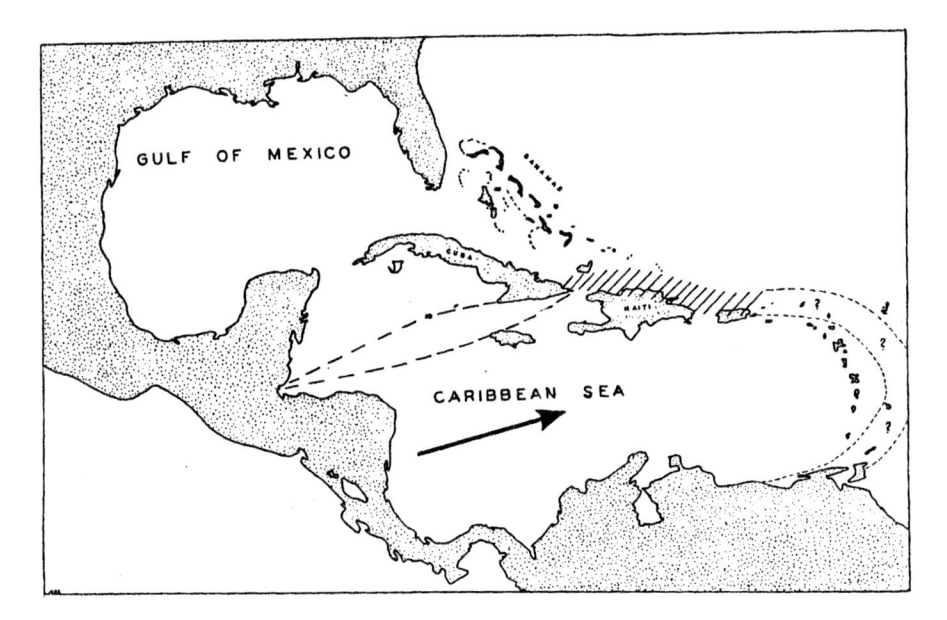

Figure 8.7. Movement of the Caribbean crust as indicated by gravity anomalies. In the official report of the S-48 expedition, in 1933, Harry Hess summarized the geological and gravity evidence and proposed that the belt of negative anomalies at the eastern edge of the Caribbean represented a zone of crustal compression resulting from an overall eastward movement of the crust underlying the Caribbean. Hess suggested that the Bartlett Deep was a pull-apart structure arising from transcurrent faulting along the edge of the moving block. The large arrow represents the "postulated direction of maximum compression." (From Field 1933, p. 33.)

Hess introduced the concept of a *tectogene*: a down-warping of crust at the site of regional compression. The tectogene could account for the formation of the Deeps and the trenches, the association of negative gravity anomalies with them, and ultimately, the deformation of the sediments that were accumulating in them (figure 8.9). Although Hess's illustration showed a symmetrical down-folding, he noted that in some cases, "an asymmetric arrangement of the anomalies" had been observed, which could be due to "an overturning of the down-buckled zone," as seen in nappes. Just as the Alpine overthrusts appeared to be detached from their underlying crust, Hess suggested—following Vening Meinesz—that a detachment occurred between the surface layer and the down-warping crust: "A thin upper layer a few miles thick does not partake of this downfolding, but folds or wrinkles upward as the underlying crust slides beneath it."[82] Whether symmetrical or not, the result was a thick root of material beneath the crust—precisely as postulated by the Airy model of isostasy.

An obvious question was the cause of these down-warpings. In a 1934 volume analyzing the results of his own worldwide gravity work, Vening Meinesz proposed that the crustal roots—tectogenes—developed at the site of down-welling convection currents.[83] In a section entitled "Convection Currents in the Sub-stratum," Vening Meinesz addressed the theoretical arguments of Harold Jeffreys, who claimed that the cooling history of the earth would have long since eradicated any significant thermal

ALTERNATIVE EXPLANATIONS OF THE ORIGIN

OF THE BAHAMA ISLANDS

General Questions

1. How may minus anomalies for the islands be rigidly interpreted as evidence of isostatic equilibrium if such is postulated?

2. To what extent may the different types of pendulum machines used on the sea and on the islands not give identical figures for the same conditions?

3. Might it be advisable to check some of the pendulum experiments with the seismograph and detonations?

4. To what depth may the pendulum be depended upon to faithfully determine the specific gravity of the rocks beneath it in such a region as the West Indies where there is bold relief and relatively closely spaced stations? Is it not true that specific gravities have to be __assumed__? Pendulum will only give average value for various masses times inverse of square of distance.

5. What specific gravities are assigned to rocks when reductions are made, and how are these rocks of different specific gravity assumed to be distributed?

6. For the purpose of __this__ expedition should the calculations be made by the Hayford-Bowie method?

Possible Structures

1. Consisting of __crystalline rocks__ (a continuation of the Piedmont region and of __Florida__) with but a __thin veneer__ of calcareous sediments, separated from the main land and from one another by grabens produced by a __lowering__ of narrow belts.

2. __Thick masses of reef-rock__ built up on a slowly __sinking but oscillating outer part__ of the Piedmont-Florida __crystalline basement__.

3. Belts of __rising mountain folds__ (of Southern and Central Rocky Mountain or of Alpine Type) consisting of an outer mantle of __thick sediments__ ranging from the modern sediments of the surface at least through the Tertiary and Cretaceous periods and a core of __crystalline rock__ (or at least metamorphosed paleozoics (such as those shown in the geanticline of Haiti) such as may form the base of the sediments in this region.

Figure 8.8. Richard Field's notes for the AGU Committee on Geophysics and Tectonics, December 1931. In these notes, Field outlined questions to be addressed by marine geophysical work and possible structural interpretations of the crust that would fit with available evidence. But within two years, Harry Hess had settled on a single hypothesis: the *tectogene*, a down-warping of crust at the site of regional compression. (RFP, Princeton University, Dec. 1931, Notes for AGU Committee on Geophysics and Tectonics.)

SYNOPSIS OF THE FOUR ALTERNATIVE EXPLANATIONS

No. referring to above list	Nature of Crystalline basement	Direction of displacement of basement	Nature of sedimentary mantle	Thickness of sediments
1.	Gneisses and schists and intrusive rocks (like Piedmont or substructure of Florida).	Sinking in "grabens" otherwise essentially none.	Little consolidated, late Tertiary to present.	Thin
2.	Same as above	Slow sinking over whole region	Same but extending possibly far back into Tertiary time.	Thick
3.	Same as above, with evidence, possibly, of folding of more recent date.	Relatively rapid orogenic rising in whole region.	Little consolidated near surface, more so downward; ranging in age from present to Cretaceous or perhaps further. (Perhaps even more recent folding.)	Thick
4.	Basic igneous rock, but little folded	Essentially none.	Little consolidated later Tertiary to present.	Thin
5				
6				

---

4. Consisting of a thick "pile" of largely basic igneous rocks with a thin veneer of calcareous sediments. Possible truncated basic volcanic cones composed of interrelated ashes and flows of dacitic or andesitic composition (such as are found on the South Eastern limit of the arc).

5. Granitic shell underlain by basaltic rock.
   - Alternating Upfolds and downfolds, produced by horizontal compression. Upfolds truncated by erosion. Downfolds form deeps.
     A. Folds have vertical axial planes.
     B. Folds are overturned.
     C. Upfolds overturned and overthrust onto downfolds.
   - Veneer of unconsolidated sediments on top of upfolds – thick terraces on flanks, with outer slopes at angle of repose. Practically no sediments on bottoms of synclinal downfolds.

6. Granitic shell overlain by thick layer of consolidated sedimentary rock, and underlain by basaltic rock.
   - Same as 5-A, 5-B and 5-C; in two phases.
     A. Truncated upfolds eroded completely through consolidated sediments along crests.
     B. Consolidated sediments only partially truncated along crests.
   - Same as in 5

7. Granitic shell; thinned by erosion on crests of anticlines; underlain by basaltic rock. Overlain by lithified sediments which are mantled by unconsolidated sediments.
   - Thrust faults (7-A) or folds (7-B and 7-C) in basement crystallines, and metamorphics. Flat-lying sedimentary rocks and unconsolidated sediments near surface.
   - Thin uniform veneer of unconsolidated sediments lying an thicker uniform series of lithified sediments resting on basement complex.

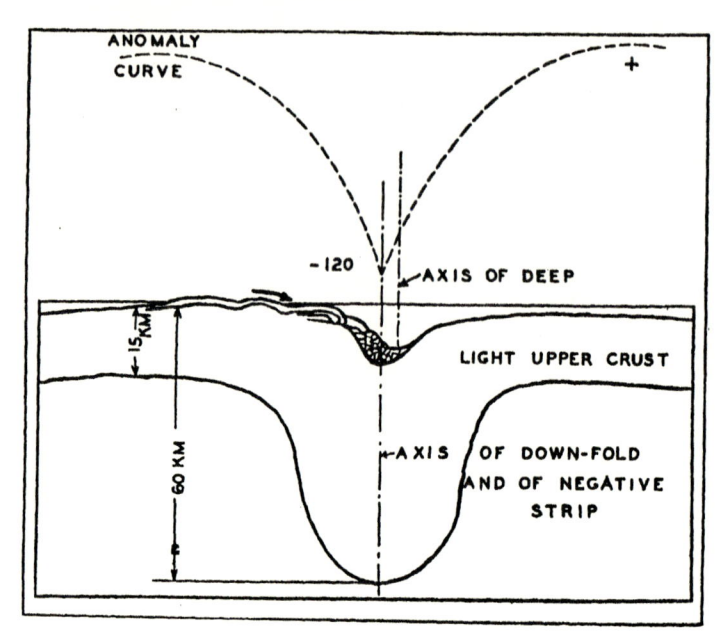

Figure 8.9. Harry Hess's tectogene concept, 1933. Hess proposed that the linear belts of negative gravity anomalies associated with ocean deeps, in both the Caribbean and the East Indies, could be attributed to a down-buckling of the crust to form a low-density root. Sediments accumulating in the depression would eventually be deformed and uplifted, which would explain the transformation of geosynclines into mountain belts. The dashed line at the top of the diagram represents an idealized gravity profile over the down-buckled crust. (Field 1933, p. 30.)

differentials. Vening Meinesz disagreed. Mathematically idealized earths might be homogeneous, but the real earth was heterogeneous, with an uneven distribution of radioactive constituents and thermal properties, and thus would convect: "Leaving aside whether convection would be likely when the distribution of temperature in the earth were perfectly regular . . . in the actual earth there can be no doubt that convection currents must develop."[84]

Vening Meinesz presented a mathematical treatment of the deformation potential of the crust, from which he concluded that there were three possible causes of large regional gravity anomalies: strong horizontal compression, recent disturbances like sedimentation or ice removal, and convection currents in the substratum. Ice removal could be dismissed as a factor in the Java Trench and West Indies, and sedimentation was restricted to coastal zones. This left horizontal compression and convection currents. But these were actually the *same* explanation, as the latter was the most likely cause of the former. Referring to the negative anomaly associated with the Indonesian archipelago, he explained:

> We must assume that the mass defect [that causes the anomaly] is narrowly related to the tectonic activity in the archipelago. Now all the geological evidence points towards great lateral compression in the Earth's crust in this area and this conclusion is likewise borne out by the submarine morphology . . . There are two possible

ways of explaining an excess of gravity over such an extensive area, lateral compression in the earth's crust or descending currents in the substratum, and as this latter assumption likewise involves lateral compression of the crust, both explanations concur in corroborating the above result. The fact that descending currents in general must imply compression of the crust, is made clear by realizing that such a current must be fed by converging horizontal currents below the crust and these currents must exert a viscous drag on the crust which will lead to compression. . . . The tectonic phenomena in the archipelago is a giving way of the crust to this lateral compression.[85]

There were only two possible explanations and each one implied the other. Large anomalies were the result of horizontal compression caused by the convergence of two convective cells, which in turn created the archipelago of Indonesia with its earthquakes and volcanoes. Similar explanations could apply throughout the globe.[86]

For Hess, as for Wright, a key aspect of these data was that they were actualistic. That is, they reflected processes actually taking place at present, rather than being residual effects of processes long inoperative. "Up to now," Hess wrote, "the study of diastrophism has been limited largely to the examination of highly complex structures of mountain systems, deformed in some previous geologic period, and eroded deeply enough so that the "roots" are partly exposed. By means of gravity anomalies, it may now be possible to study orogenic belts in the earlier, largely uneroded formative stages." In addition, the patterns appeared to be global:

It is also significant that only one negative strip was found in the East Indies, and that this strip was continuous for the length surveyed, 5000 miles. *It implies that deformation of the earth's crust is taking place in a single and continuous narrow belt rather than in several parallel belts or in discontinuous en echelon lines.* The terminations of the negative strip were not reached in the regions surveyed. . . . Do they terminate against faults or shear zones? Or are they perhaps continuous around the world?[87]

As early as 1933, Harry Hess was discussing the possibility that crustal deformation occurred in distinct belts associated with deep ocean trenches, and that these belts might be somehow connected all across the globe.

Experimental Support

The tectogene concept soon received experimental support, first from Vening Meinesz's Dutch colleague, Philip H. Kuenen, and then from American geologist David Griggs. Kuenen was unusual among geologists in being a consummate experimentalist: he would later become famous for his experimental demonstration of submarine mud flows—turbidity currents—through a series of scale models that reproduced the features of natural marine deposits.[88] Having participated as geologist on board the Dutch *Snellius* expedition to the East Indies in 1929 and 1930, Kuenen helped supply the geological interpretation of Vening Meinesz's gravity data and in the course of this work was inspired to demonstrate how a crustal down-buckle might form.[89] He devised a simple experimental apparatus: a glass aquarium filled with various combinations of layers of paraffin, vaseline, mineral oil, and warm water, and equipped with beams at either end to generate horizontal compression. Kuenen illustrated the results in a series of photographs: depending upon the viscosity of the material layers and the rate of compression, one could create symmetrical or asymmetrical downbuckles, as proposed by Hess, or underthrusting of the crust[90] (figure 8.10).

Kuenen did not presume to have proved the existence of tectogenes, which could "hardly be either proved or disproved by an experiment," but he did feel that, "when experimental conditions are properly chosen they are of considerable value for checking the theoretical deductions and guiding further research."[91] David Griggs agreed, particularly about choosing the right experimental conditions. Griggs, a professor of

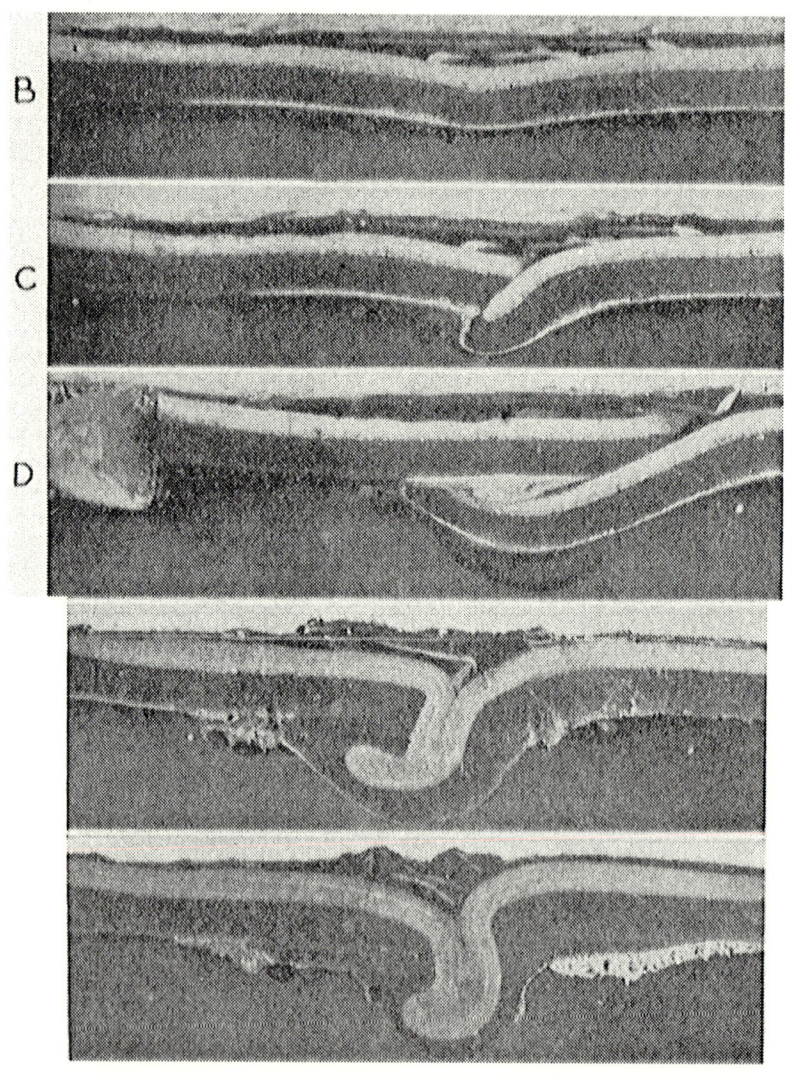

Figure 8.10. Philip Kuenen's scale model of a tectogene. Depending upon the viscosity of the material layers and the rate of compression, Kuenen could create symmetrical or asymmetrical down-warpings, or underthrusting of the crust. Kuenen postulated that as the root deepened, it would begin to melt. Also note the folding of the surface layers. (Kuenen 1936, p. 185.)

geology at the University of California at Los Angeles, was inspired by Kuenen's success in reproducing the postulated entities, but the question—as with all experiments—was whether his bench-scale success was indicative of earth-scale conditions. He decided to replicate Kuenen's experiments using Hubbert's theory to scale the viscosities of the experimental materials. And rather than using beams to compress the materials from the sides, he inserted into the system a pair of rotating drums to simulate the effects of viscous drag from below.[92]

In later years, Griggs recalled the motivation for his work: he had been positively inspired by Hess, Vening Meinesz, Reginald Daly, and Arthur Holmes, while Jeffreys's "static earth concept and thermal contraction theory of mountain building served as convenient targets for attack."[93] Like Daly, Griggs noted that American geologists had given far less consideration to these questions than their European counterparts and suggested that, given the inadequacy of existing orogenic theory, one had to take seriously the idea of convection currents as a driving force of crustal motion. Like Holmes, Griggs saw his work as a unification of recent discoveries and conceptual developments: Vening Meinesz's "discovery of the great bands of gravity deficiency in the East and West Indies," Hess's tectogene idea and Kuenen's experimental demonstration of its possibility, and continental drift.[94] Also like Holmes, he noted the arguments against the polflucht force, the tidal force, and the coriolis force as explanations of crustal displacement. That left convection currents, which, he noted, a number of workers including Vening Meinesz had recently treated mathematically and shown to be at least a *possibility*. Thermal gradients within the earth appeared to be sufficient to cause convective overturn. From the geological point of view, the question was whether drag on the base of the crust caused by convection could produce the features observed in mountain belts. Griggs developed a "dynamically similar scale model" to try to find out.[95]

The scale model worked: it produced down-folds in the "crust" which were maintained so long as the simulated convection was maintained. Where two convective cells met, the down-foldings were symmetrical; where only one cell was operative, the resulting down-fold was asymmetrical. Moreover, the resulting forces moved the crustal material *laterally*. Although this was hard to portray "short of moving picture[s]," Griggs did his best to explain. Starting with a layer of crustal material over the entire setup and one rotating drum, he found that "as the current velocity is increased, the crust is markedly thinned above [the rotating drum], and transported into the thickened part of the crust [above the stationary drum]. Finally, the current sweeps all the crustal cover off and piles it up in a peripheral downfold"[96] (figure 8.11). Here was a suggestion as to the origin of ocean basins and the thickening of continental crust peripheral to them.

Although warning his readers that, "in contrast to the rest of the paper, this section is purely speculative," Griggs forged ahead with theoretical interpretation of his results. Not only did convection currents make sense to explain crustal deformation, but the scale on which the cells could work appeared to be very large, thus opening "the attractive possibility of a convection cell covering the whole Pacific Basin, comprising sinking peripheral currents localizing the circum-Pacific mountains and rising currents in the center. Such an interpretation would partially explain the sweeping of the Pacific basin clear of continental material, in the manner demonstrated by the model."[97] Moreover, the thickened crust could help to explain the location of deep-

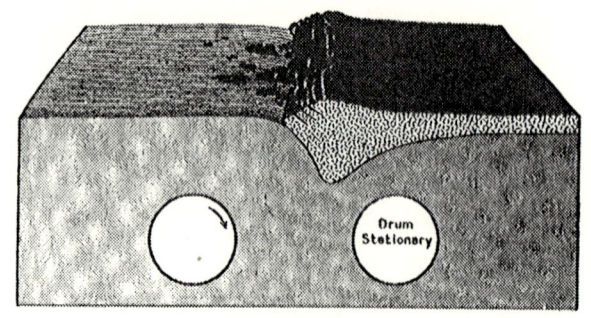

Fig. 15. Stereogram of Large Model with only One Drum Rotating, Showing Development of Peripheral Tectogene.

THE MOUNTAIN BUILDING CYCLE

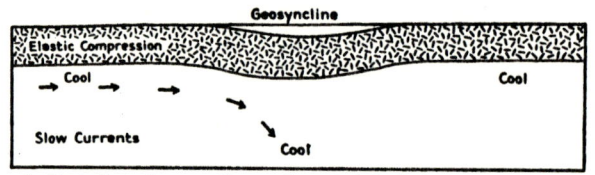

1. First stage in convection cycle – Period of slowly accelerating currents.

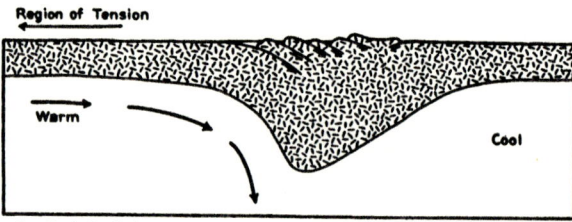

2. Period of fastest currents – Folding of geosynclinal region and formation of the mountain root.

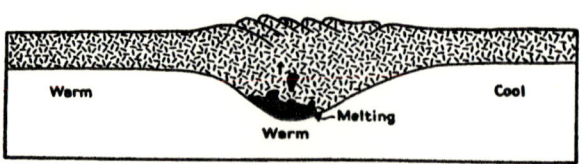

3. End of convection current cycle – Period of emergence. Buoyant rise of thickened crust aided by melting of mountain root.

Figure 8.11. David Griggs's 1939 experimental demonstration and theoretical interpretation of convection currents as the driving force of crustal tectonics. Using M. King Hubbert's theory of scale models, Griggs devised an experimental setup to demonstrate the plausibility of the ideas of Holmes, Vening Meinesz, Hess, and Kuenen. The results, he noted, "showed striking resemblance to structures developed in the earth," and Griggs suggested geologists should pursue research aimed at testing the theory of convection currents as the cause of crustal deformation. The upper portion of the diagram illustrates his experimental setup, in which a rotating drum simulated the effect of convection in a

focus earthquakes. Seismologists such as Gutenberg and Charles Richter "all agree that foci of deep earthquakes in the circum-Pacific region seem to lie on planes inclined at about 45° towards the continents." Griggs noted. "It might be possible that these quakes were caused by slipping along the convection-current surface."[98]

Griggs developed his experiments to "make more plausible the suggestion that convection currents may have operated in the earth's substratum to cause the development of our mountain systems," and Hess in turn used Griggs's work as supporting the plausibility of tectogenes:[99]

> Recently an important new concept concerning the origin of the negative strip in island arcs has been set forth by David Griggs. It is based on model-experiments in which the strength of the materials and rate of application of the forces are adjusted to the scale of the model. By means of horizontal rotating cylinders, convection currents were set up in a fluid layer beneath the "crust," and a convection cell was formed. A down-buckle in the crust, similar to that produced in Kuenen's experiments, was developed where two opposing currents meet and plunge downward. *So long as the currents are in operation the down-buckle is maintained.* . . . This concept has one great advantage, namely, that the down-buckle, and hence the negative strip, can be maintained for a long period of time. If the buckle were formed entirely by tangential compression in the upper crust, it would be so weak and unstable that it could last for only a very short period of time geologically. . . . [But] by means of convection-currents the difficulty of maintaining the tectogene can be removed.
>
> The currents which Griggs suggested would have velocities of one to ten cm per year.[100]

Griggs's theory was testable: it suggested a specific rate of crustal displacement. A few decades later his prediction would be confirmed; geologists today place crustal motions at 1 to 10 cm per year.

A New Global Synthesis?

The formal results of the *S-21*, *S-48*, *Barracuda*, and Dutch East Indies cruises were published in thick government documents, but Vening Meinesz, Hess, and Ewing also published concise scientific papers on their work. At the 1937 AGU meeting, Ewing, Hess, and Field reported on the *Barracuda* work. Ewing discussed technical aspects of the gravity measurement; Hess interpreted the geology of the Caribbean and added a twist: the strips of negative anomalies were also associated with serpentinized peridotite intrusions—altered, high-density, ultramafic rocks—which Hess interpreted to have been pushed upward from the substratum as the crustal root pushed downward.[101]

---

plastic substrate beneath a rigid crust, resulting in thickening of the crust below a region of surficial compression. The lower portion relates the experimental results to widely accepted phases of mountain-building: Stage 1, currents accelerate and a down-warping (geosyncline) develops in the crust. Stage 2, the crust thickens to form a root (tectogene) beneath a near-surface zone of compression. Stage 3, the root begins to melt, and magmas intrude the overlying mountain belt. (Griggs 1939b, pp. 643, 645.) Reproduced with permission of the American Journal of Science.

Similar serpentinite belts, Hess noted, had long been known in the Swiss Alps. The connection between ocean deformation and continental structures was getting stronger.

Perhaps the most important forum for this work in the United States was the 1937 symposium Geophysical Exploration of the Ocean Bottom, organized by Field under the auspices of the American Geophysical Union and the American Philosophical Society. Field introduced the symposium with a paper on the general project of developing submarine geophysics, Ewing spoke on gravity instrumentation, and Hess presented his tectogene concept.[102] Hess had redrawn his tectogene diagram to show how it could account for geological features observed in mountain belts, such as the Alpine nappes (figure 8.12). Papers were also presented by Charles Piggot of the Carnegie Institution's Geophysical Laboratory, who described a new technique of obtaining core samples from the ocean floor, John Fleming of the Carnegie's Department of Terrestrial Magnetism, who spoke on the application of magnetic studies to the structure of the ocean basins, and N. H. Heck of the Coast and Geodetic Survey's Division of Terrestrial Magnetism and Seismology, who spoke on implications of recent seismological work for the structure of the crust.

In his opening paper, Field framed current knowledge in terms of the two "opposed major geological hypotheses" of continental drift and permanency of ocean basins and suggested that permanency, and with it, geosynclines, was nearing the end of its useful life:

> We know now that the structure of the Appalachians not only suggests that the thrusting came from the oceanic area, but also that we can see only the western side of what was probably a more symmetrical and widespread orogeny which was not terminated at the continental margin, unless it has been torn away from complementary, transatlantic structures. This suggestion . . . is diametrically opposed to the earlier and simpler concept of the . . . Appalachian geosyncline.[103]

This knowledge was scarcely new—Barrell had made the same point twenty years before—but Field implied that the problem was now coming to a head, particularly in terms of the Atlantic, "the stratigraphic and structural matching of whose continental margins has been the chief stimulus to the several hypotheses which attempt to explain the major patterns of the face of the Earth."[104]

While Field, a geologist, was most moved by evidence speaking against the geological theory of permanence of ocean basins, Heck, a geophysicist, was moved by the evidence against the geodetic theory of isostasy as a driving force of crustal motion. He began his paper with a discussion of the interpretation of submarine seismicity in terms of isostasy. In Bowie's view, earthquakes were the result of isostatic adjustment to mass transfer in regions of erosion or sedimentation; therefore submarine earthquakes were expected, but only in coastal regions. It was now clear that this was not the case. Earthquakes were abundant on the "great rises or ridges in the Pacific, Indian, and Atlantic Ocean," Heck explained, and these were submerged features where "no transfer of material by erosion and deposition could take place."[105] Heck presented a map illustrating the correlation between the mid-Atlantic ridge and the distribution of earthquakes in the north and south Atlantic, and he suggested that the ridge might link up to similar features in the Arctic Ocean (figure 8.13).

Besides refuting isostasy as a cause of crustal deformation, seismic studies were also revealing the structure of the crust. Once again, the results were at odds with the

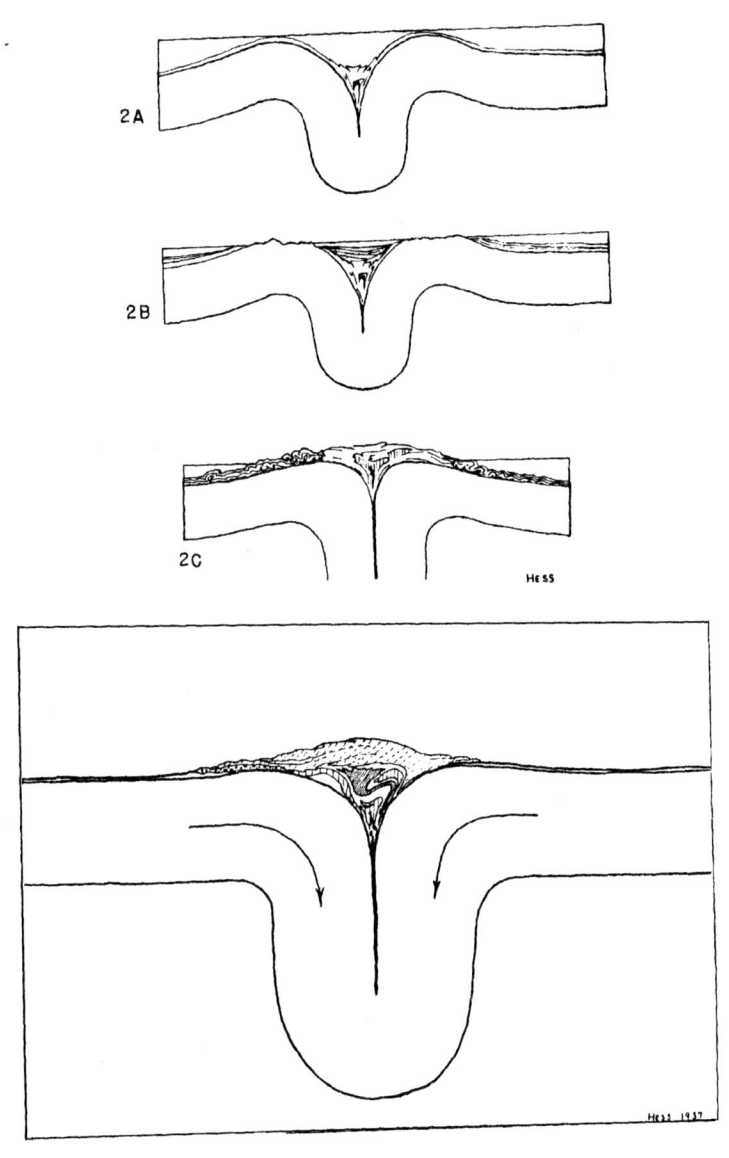

Figure 8.12. Hess's 1938 development of the tectogene concept. In this series of diagrams, from Hess's presentation at the 1937 AGU symposium on the ocean bottom, Hess explained how the tectogene concept could explain the origin of features observed in mountain belts, including the formation of nappes in the Swiss Alps. The top of the diagram illustrates progressive compression of the tectogene in response to crustal shortening; the bottom illustrates Hess's conception of the deep structure of the Swiss Alps. Alpine nappes are, in this model, the surface expression of the large-scale compression that causes the (much larger) tectogene. (Hess 1938, pp. 78–79.)

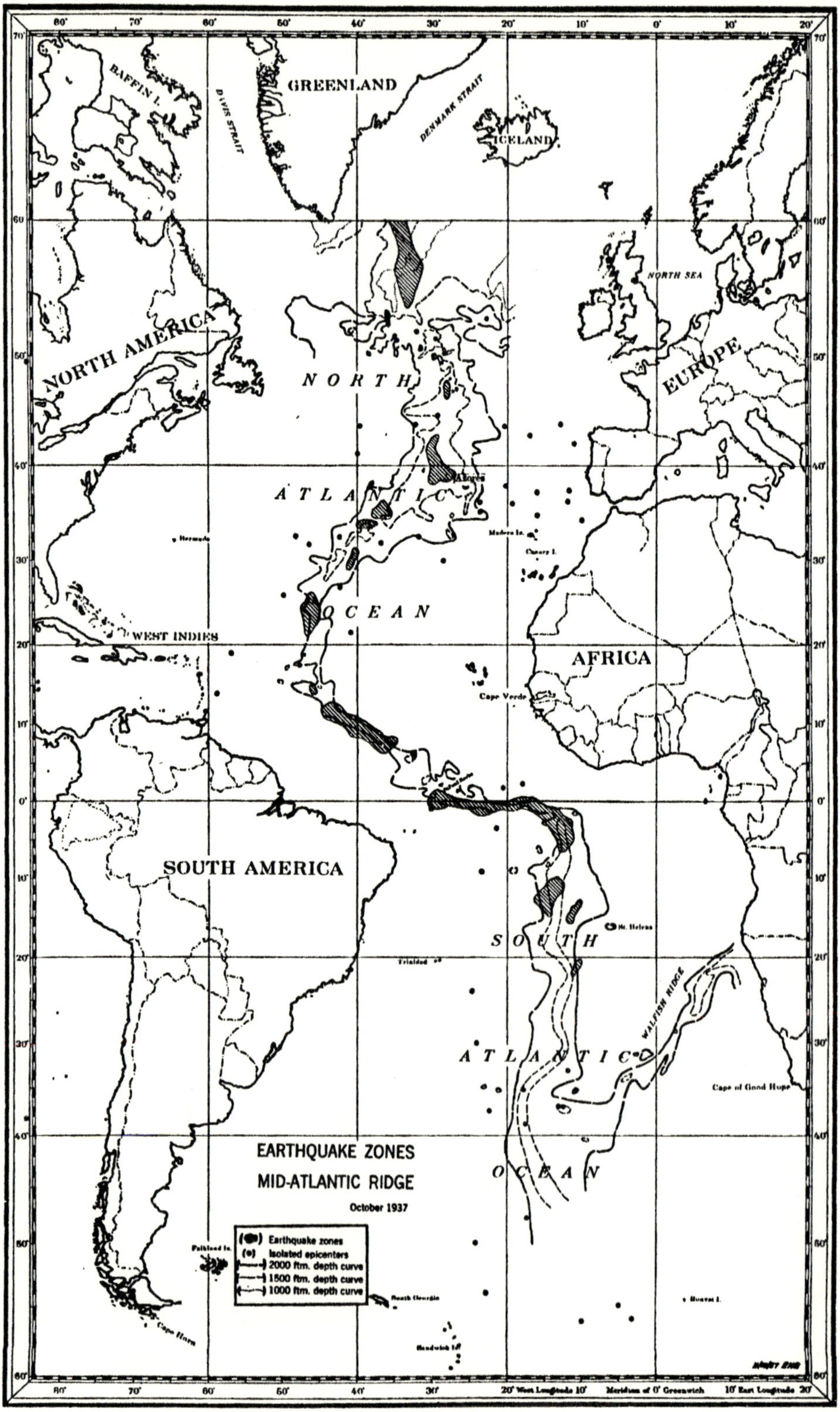

GREENLAND
BAFFIN I.
DAVIS STRAIT
DENMARK STRAIT
ICELAND
NORTH SEA
NORTH AMERICA
EUROPE
NORTH
ATLANTIC
OCEAN
Azores
WEST INDIES
AFRICA
Cape Verde
SOUTH AMERICA
SOUTH
ATLANTIC
OCEAN
WALFISH RIDGE
Cape of Good Hope
Cape Horn
EARTHQUAKE ZONES
MID-ATLANTIC RIDGE
October 1937
Earthquake zones
Isolated epicenters
2000 ftm. depth curve
1500 ftm. depth curve
1000 ftm. depth curve
20° West Longitude 10°
Meridian of 0° Greenwich
10° East Longitude 20°

Pratt–Hayford model. "The thickness of the continental crust varies from place to place," Heck wrote bluntly.[106] The depth to the base of the crust was not uniform. Heck summarized the recent work of Gutenberg at Caltech and that of several European and Japanese seismologists analyzing seismic wave travel paths and velocities, as well as preliminary studies that Ewing had undertaken to study the refraction of seismic waves from explosions. The continents were composed of thick layers of light, low-velocity material, to depths of perhaps 30 to 50 km. The Pacific basin contained no continental material at all; the Atlantic, according to Gutenberg's studies, seemed to contain a thin layer of sialic crust beneath the basaltic surface layer.[107] (figure 8.14).

John Fleming summarized recent work done by Carnegie scientists in terrestrial magnetism. The Department of Terrestrial Magnetism had been studying the earth's magnetic field for a long time, but recently a new phenomenon had come to light. In South Africa, rocks had been found that were strongly magnetized in a direction opposite to the earth's present-day magnetic field. Although he proposed no specific hypothesis to test, Fleming suggested that studies of both secular variation of the earth's field and remnant magnetism in rocks might shed light on the structure and forces active within the earth, including "crustal or subcrustal movements."[108] Piggot's apparatus to obtain core samples at sea might be used to determine the remnant magnetism of ocean bottom materials, a "fossil" record of the earth's magnetic field at the time the rocks formed. In a best-case scenario, one day rocks might be dated on the basis of their fossil magnetism just as geologists had long dated rocks on the basis of their fossil biology. In conclusion, Fleming noted the physical coincidence between Vening Meinesz's gravity anomalies, Gutenberg's recently documented zones of deep-focus earthquakes, and the DTM's maps of positive magnetic anomalies in the Pacific basin, and he suggested that all three were related to motions in the earth's interior, perhaps, as Vening Meinesz suggested, to "current systems in the Earth's interior, downward below the oceans and upward below the continents . . . May not then the anomalies of gravity and of terrestrial magnetism and the deep earthquakes have a common origin in current-systems in the inner Earth?"[109]

For the second time in just over a decade, at least some geologists thought that a conceptual breakthrough was at hand. Hess certainly did. In the opening words of his paper at the ocean bottom symposium, he declared:

> Meinesz's discovery of huge negative anomalies in the vicinity of island arcs is probably the most important contribution to knowledge of the nature of mountain building made in this century. . . . When the significance of the huge anomalies and the structures which they indicate become appreciated, a radical change in some of our ideas on mechanics and processes of mountain building becomes necessary.[110]

Figure 8.13. N. H. Heck's map of seismicity along the mid-Atlantic ridge. At the 1937 symposium where Hess presented his tectogene concept, Heck presented a map illustrating the relationship of seismicity in the Atlantic basin to the mid-Atlantic ridge. Heck argued that the occurrence of seismicity in regions that were far from sources of sedimentation and too deep to experience erosion was conclusive evidence against isostatic adjustment as the primary cause. He also discussed seismic evidence pointing to a variable depth to the base of the crust. (Heck 1938, opposite p. 98.) Reproduced with permission of the American Philosophical Society.

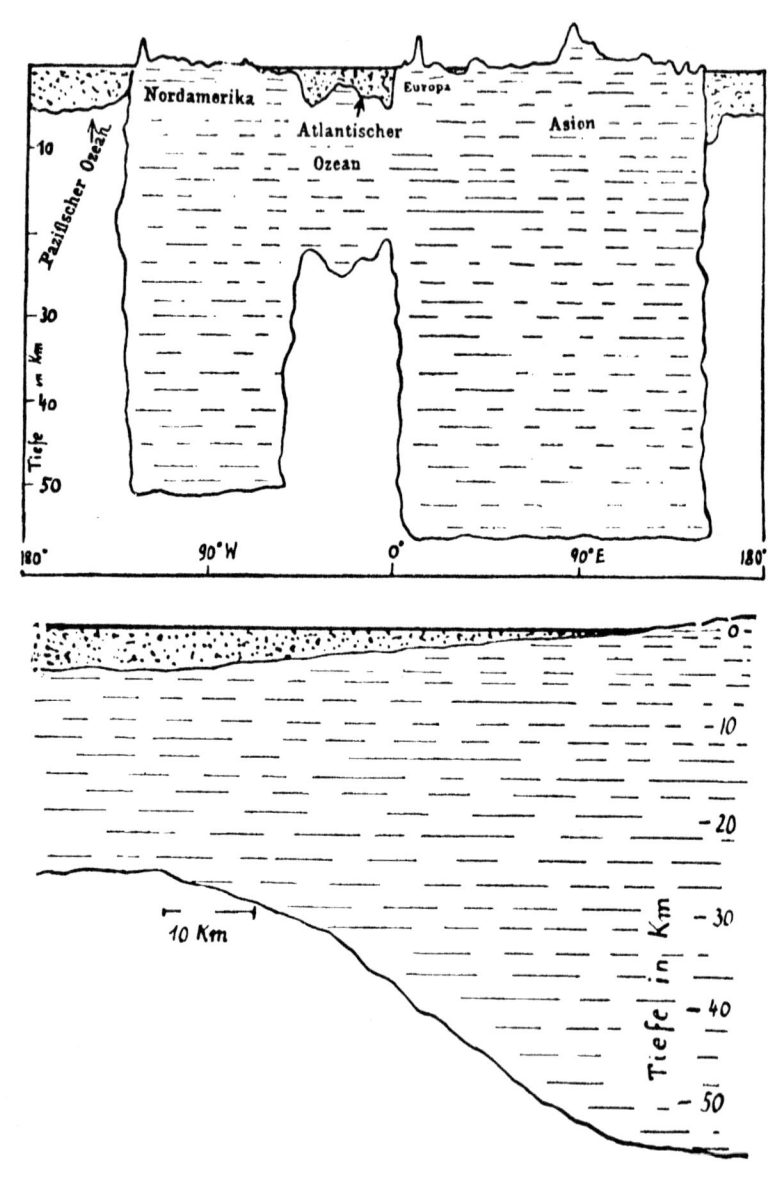

Figure 8.14. The structure of the continental and oceanic crust, circa 1937. By the mid-1930s, seismic evidence was refuting the Pratt–Hayford–Bowie model of the crust. In this diagram, based on the work of Beno Gutenberg, the sialic continental crust is shown to be at least fifty kilometers thick in contrast with a simatic oceanic crust only 5 km thick over most of its area. Top: Overview of crustal structure. Bottom: Detail of the continental–Atlantic ocean transition. Note that the model contains elements of both Pratt and Airy hypotheses and that the Atlantic Ocean was thought to possess a layer of sialic material beneath its basaltic floor. (Heck 1938, pp. 104, 107.) Reproduced with permission of the American Philosophical Society.

The evidence was clear: the ocean basins were not static, major horizontal compressive stresses were present in the crust, and these stresses were materially involved in crustal down-warpings that caused earthquakes and vulcanism. And there was a theoretical model available to account for it all: convection currents in the substrate, along which the crust was dragged.

Just as Holmes's convection model was not precisely the same as modern conceptions, neither was Hess's. Hess, Vening Menesz, Kuenen, and Griggs all focused on the development of a crustal root rather than on the possibility of underthrusting (which would have been more consonant with current understanding). Some historians have puzzled over this, but in the context of gravity measurement the reasoning is obvious: the formation of crustal roots is needed to account for gravity anomalies according to the Airy model.[111] Vening Meinesz and Hess had demonstrated that the Pratt model was wrong: there were large gravity anomalies that could not be accounted for by density differences, and their work was increasingly supported by seismic evidence, which was showing that the depth of the Mohorovičić discontinuity—the sudden increase in seismic velocities at the base of the crust discovered in 1909 by Croatian seismologist Andrija Mohorovičić—was highly variable. The crust was thicker under the mountains than under the oceans, just as the advocates of Airy isostasy held. These two lines of evidence, gravitational and seismic, together indicated that the depth of isostatic compensation was not uniform, the depth of the base to the crust was not uniform, and there were crustal roots. But if there were roots, then the question was, as Bowie had argued all along, what force created and sustained them? Bowie had rejected continental drift on the ground that horizontal compression would create crustal roots; now that it appeared that there were such roots, did this mean that continents were drifting?

While Hess did not explicitly make the link to continental drift, Richard Field did. In January of 1937, Field gave an address to the Philosophical Society of Washington entitled "Structure of Continents and Ocean Basins," in which he tried to summarize the present state of knowledge. Things were in a state of flux, he suggested, and much of it had to do with the basic question posed by Wegener as to whether or not continents moved:

> With the possible exception of isostasy and the postulate of the geosyncline, no hypothesis has so served to stimulate further and more varied geological and geophysical investigation, as has Alfred Wegener's suggestion of drifting continents. In my own case, at least, I can affirm that the hypothesis has strongly influenced my motives in attempting to obtain, through geophysical means, original data on the structure of the submerged continental margins.[112]

The gravity data were pointing to Airy isostasy, and from there to continental drift.

Besides Field, the man who saw the significance of these developments most clearly was William Bowie. Even before the 1937 symposium, Bowie realized that the demise of the Pratt model was at hand. In fact, everything he had believed was being challenged. At the 1936 AGU General Assembly, Bowie—now aged sixty-four—gave a talk entitled "Recent Trends in Geophysical Research." His entire scientific career was linked to Pratt isostasy, but he now acknowledged that the Pratt model was at best uncertain, and he retreated to an agnostic position and the reality of isostasy:

> We have not yet been able to determine whether the crustal material is limited by a level surface at some unknown depth below sea level [Pratt model] or whether the lower limit of the crust is an undulating surface [Airy model]. But in any event, it is quite definitely established that for at least the major features of the Earth's surface, such as oceans and continents, there is a compensating excess or deficiency of matter extending to a limited depth below sea-level. ... We can now quite safely say that isostasy, in its general aspects at least, can be accepted as a scientific principle.[113]

What an admission for the man who had written so lucidly—and so extensively—on the "fact" of complete and local isostasy compensated at a uniform depth of 113.7 km! From complete and local equilibrium, Bowie now retreated to isostasy in its "general aspects." There was more. Gravity was sounding the death knell for permanent ocean basins, too. Referring obliquely to the gravity work that he had helped to bring about, Bowie continued:

> It is most probable that there is no such thing as any part of the land surface being stable. ... The troughs of the ocean are apparently not permanent features. How long they have existed and what will be their future, no one, of course, can tell. ... [But] since there have been great changes in the elevation of portions of the Earth's surface, and perhaps all of it, we must conclude that those changes are due to a change in density of crustal material or to a regional compressive force or to a combination of the two.[114]

The demonstration of isostasy had led Bowie to the belief that the continents and oceans were formed and stabilized very early in earth history and therefore that continental drift was impossible. But if isostasy were neither complete nor local, then everything changed. The truth might lie with the Airy model and regional compressive forces, or perhaps with a combination of Pratt and Airy models: density differences *and* compressive forces.

Within twenty years, geologists and geophysicists would show that the structure of the earth's crust did in fact contain elements of both the Pratt and the Airy models: the oceanic crust is denser than the continental crust, and the continents do have deep roots. But for now, all Bowie knew was that what he had "proved" twenty-five years before was now being unproved. He continued bravely:

> I think we can ignore the question for the present of whether we have the Pratt or the Airy isostasy. If the Airy, then the continents are underlain by so-called roots that extend down into what would normally be subcrustal space. For the Pratt idea, we should have more or less uniform depth of crustal material. ... There are changes now going on which depress crustal material under areas of sedimentation and elevate crustal material that lies under areas of erosion. This would mean that for certain areas some crustal material is depressed into subcrustal space and for other areas the subcrustal material enters crustal space.[115]

Bowie had not ignored the question whether isostasy was Pratt or Airy; he had answered it. Pratt isostasy, as he and Hayford had conceived and used it, was dead.[116] Bowie clearly saw the implications, and he concluded his paper by raising once again the question of the Wegener hypothesis:

> The Wegener hypothesis has received a great deal of attention in recent years and deservedly so. It is based upon the idea [of] isostasy. ... Many students of the Earth's crust feel that the Wegener hypothesis does violence to certain mechanical prin-

ciples, but, in any event, it is something that should be looked into. The geodesists have cooperated with the astronomers in two great world-wide longitude campaigns which were designed to furnish the basis for determining whether continents and islands are wandering over the face of the Earth. Nothing has been found up to this time that would substantiate the Wegener hypothesis, but we cannot tell what the future may disclose.[117]

Ten years after he had called continental drift impossible, William Bowie now suggested that geodesists would be the ones to prove it.

9

An Evidentiary and Epistemic Shift

World War II dashed the hopes and dreams of many, among them Richard Field. As President of the American Geophysical Union, Field was to host the International Geological Congress when it met in Washington, D.C. at the end of the summer of 1939, an event Field was greatly anticipating. Scientists were expected from around the globe, and the permanence—or nonpermanence—of ocean basins was high on Field's list of topics for discussion.[1] But as the meeting was about to begin, Adolf Hitler's armies invaded Poland; many delegates turned around mid-voyage, and others who had already arrived in Washington quickly returned home. The following year William Bowie died; two years later Charles Schuchert died at the age of eighty-four, and Field was involved in a near-fatal car crash which effectively ended his scientific career.

Bowie's and Field's scientific goals would be realized, however, albeit not by them. Together, they had assembled an advisory committee for gravity studies that included five subsections—navigation, geophysics, tectonics, oceanology [sic], and marine microbiology—with prominent members from academia, and industry in the United States and abroad and from the U.S. Navy.[2] Their ambitions went beyond gravity: Bowie and Field hoped to foster a global science of geophysics and oceanography to explore the three-fifths of the earth that scientists had scarcely visited by enlisting the material and financial support of the U.S. Navy and other navies.[3] The U.S. Navy, for its part, was making plans to enlist geophysicists, both figuratively and literally.

When war broke out, Harry Hess joined the naval reserve, as did many other young and aspiring geophysicists and oceanographers. In 1941, Hess was called to active duty and became the captain of an assault transport, the USS *Cape Johnson*. Among the ship's tasks was the echo sounding of the Pacific basin; Hess subsequently became famous for the discovery of flat-topped sea-mounts, which he named guyots after the first professor of geology at Princeton, Arnold Guyot.[4] Hess published this discovery, but much geophysical work done during the war, and after, was classified.

In the war years and beyond, the U.S. Navy supported a huge array of geophysical and oceanographic work deemed to be of military relevance, particularly with respect to submarine and antisubmarine warfare. American scientists fanned out over (and under) the world's oceans. There were gravity surveys, magnetic surveys, temperature–depth studies, acoustic studies, bathymetry, studies of marine plankton, and studies of snapping shrimp. After the war, much of this work was supported by the newly created Office of Naval Research (ONR), and nearly all of it was classified. Many reports were written, but it is an open question how much and how widely this unpublished material was discussed. Some documents were declassified or downgraded within a short period of time; others remained classified for decades.

While much has been written on the contributions of physicists to the development of nuclear weaponry and radar, the military contributions of American earth scientists, and the impact of their relationship with the U.S. Navy on the development of their science, is scarcely known. One thing that is known, however, is that while American geophysicists worked largely under conditions of secrecy, British geophysicists were openly pursuing research areas that would reopen the question of continental drift. One of these areas was remanent magnetism in rocks.

From Gravity to Magnetism

The phenomenon of remanent magnetism was first described scientifically by Pierre Curie, who found that rocks heated above a certain temperature—later named the Curie point—lost their magnetic properties, and that when they were cooled again they took on the magnetism of the surrounding field. Under normal geological conditions, this is the earth's prevailing field; ordinary rocks typically contain a weak magnetic signature aligned with the earth's polarity. But, as John Fleming noted in 1937, some rocks are polarized *opposite* to the earth's field. In addition, some rocks have the same polarity as the earth as a whole, but indicate varying positions for the locations of the poles. During the war years, scientists at the Carnegie Institution's Department of Terrestrial Magnetism worked on the development of sensitive magnetometers to detect weak magnetic fields in earth materials; physicists soon realized that studying magnetism in rocks might be a means to understanding the nature and origins of the earth's magnetic field.[5]

One such was P. M. S. Blackett, the 1948 Nobel Laureate for his work in nuclear physics and cosmic rays. During the war, Blackett helped found the science of operations research; he also worked with geophyscist Edward Bullard, professor of geophysics in the newly created department of geodesy and geophysics at Cambridge University, on magnetic minesweeping devices. Blackett was a man of profound intelligence and insight. A member of the British postwar Advisory Committee on Atomic Energy, his 1948 best-seller, *Fear, War and the Bomb: Military and Political Consequences of Atomic Energy*, challenged the assumptions of atomic diplomacy and foresaw many aspects of postwar U.S.–Soviet relations.[6] After the war, Blackett turned away from military work and took up more peaceful questions associated with magnetism. In 1947, working at the University of Manchester, he proposed a theory in which magnetism emerged as a fundamental property of rotating matter. To test the theory, he designed an "astatic" magnetometer—a highly sensitive device using an array of magnets to mitigate extraneous effects—in which he would rotate a massive object in an attempt to generate a detectable magnetic field. Drawing on his rich political connections, Blackett arranged to borrow 17 kg of gold from the Royal Mint, which he rotated at high speed to simulate the effects of the more massive earth rotating at lower speed.[7]

The experiment failed—no magnetic field was detected—but it led Blackett into the study of magnetism in rocks. The Carnegie workers had suggested that, if Blackett's theory were correct, then magnetism in rocks should record a stable field with little directional variation. If, on the other hand, the earth's magnetic field arose from transient factors such as internal convection currents, as suggested by Bullard and others, then the remanent magnetism in rocks would be quite variable. Bullard also suggested another test. If the earth's magnetic field were a distributed property, emanating as Blackett proposed from the mass of the earth, then magnetic intensity should decrease with depth (like gravity). This suggestion was pursued by Stanley Keith Runcorn, a student of Bullard at Cambridge and then a lecturer in Blackett's department at Manchester. Runcorn began taking magnetometers down the shafts of coal mines in Lancashire and elsewhere. But Runcorn found no change, and by 1951 his measurements had made it clear that Blackett's theory was incorrect. The three men turned their attention to magnetism in rocks.

In 1937, Bullard had visited Princeton where he spent time in the field with Richard Field, and on the *Barracuda* with Hess and Ewing; he also worked in the same department as Harold Jeffreys and was thus familiar with the geological and geophysical questions surrounding crustal mobility.[8] Blackett knew less of such matters. But in 1953 he moved to Imperial College, where he encountered H. H. Read, the geology professor who had inspired Arthur Holmes. During the war years, with few students to teach, Holmes had written a comprehensive textbook, *Principles of Physical Geology*, which by the early 1950s was being widely used in university courses. The entire last third of the book, "Internal Processes and Their Effects," dealt with questions surrounding mountain-building, earthquakes, vulcanicity, and crustal motions. The final chapter was devoted to continental drift. Holmes presented a lucid summary of Wegener's argument, as well as du Toit's more recent evidence in support of it, and he reprinted both Griggs's and his own illustrations of a convection-driven model.

Holmes was well aware of the continued North American opposition to drift, and on his own side of the Atlantic Harold Jeffreys had hardened rather than softened his negative views. Holmes therefore closed his book with words of politic disclaimer—that his ideas were "purely speculative," and that it would be a long time before they could be adequately tested.[9] But the bulk of Holmes's text was far less ambiguous and allowed little doubt as to his convictions: they were the same in 1945 as they had been in 1925. Continental drift was the only theory that really addressed the geological conundrums that first Wegener and then du Toit had so effectively laid out.[10]

Of all the various aspects of the problem, Holmes most liked the drift solution to the problem of Permian glaciation, which he considered the strongest display of the theory's explanatory power: "The problem [of Permian glaciation] remains an insoluble enigma, unless the straightforward inference is accepted that all the continents except Antarctica lay well to the south of the present positions and that the southern continents were grouped together around the South Pole."[11] As for the mechanism objection, Holmes, like Bailey Willis before him, noted that the same objection could be raised to almost any tectonic theory, including Willis's own land bridges: Willis and Schuchert had provided no account of how land bridges were raised in the first place. "The continental drift solution has the advantage," Holmes therefore concluded "that it reduces two baffling problems [the cause of land bridges and the cause of glaciation] to one while at the same time it removes many other less intractable difficulties."[12] Moreover, and more important, Holmes reminded his readers that he had already provided a mechanism, one that had been seconded by Griggs and Vening Meinesz:

> We have already seen that the peripheral orogenic belts probably mark the regions where opposing systems of sub-crustal currents came together and turned downwards. The movements required to account for the mountain structures are in the same directions as those required for continental drift, and it thus appears that . . . sub-crustal convection currents . . . may provide the sort of mechanism for which we are looking.[13]

Years later at Imperial College, it was said that when Blackett turned to Read to learn about rocks, Read sent Blackett to the library to read Arthur Holmes.[14] Whatever the truth of the matter, by the mid-1950s, Blackett and his students at Imperial, as well as Runcorn and Bullard, had found overwhelming evidence that rocks had not remained in the same position relative to the poles throughout geological history.[15] The earth's magnetic poles as indicated in rocks were not only different in the past, they were different in different geological epochs.

There were two evident interpretations of these data: that the earth's poles had moved relative to the land masses, or that the land masses had moved relative to the poles. If the land masses *had* moved relative to the poles, then they might have moved separately or together. Paleomagnetic interpretation was complicated, however, by disputes about the exact causes of and controls on remanent effects; some physicists suggested that rocks were not necessarily reliable recorders of prevailing magnetic fields. Perhaps the signatures were affected by later oxidation, alteration, or even by laboratory analysis itself. The fact that some rocks were entirely reversed in polarity seemed to support such a view. John Graham at the Carnegie Institution, arguably the first researcher to suggest that paleomagnetism might help answer "questions regarding large-scale movements of the crust, such as continental drift and polar wandering," ironically gave up his own inquiries in 1950 for just this reason: he concluded that reversed polarity in rocks was a consequence of their physical and chemical constitution and turned away from the question of crustal displacements.[16]

Runcorn saw a means to answer this complaint. If rocks were not reliable recorders of prevailing magnetic fields, then data from different places would be random. If, on the other hand, the data were spatially coherent, this would be prima facie evidence for their reliability. Construction of "apparent polar wandering paths" based on remanent magnetism in sequences of rocks in particular geographic locations could help resolve the question. Runcorn, now back at Cambridge and working with Edward (Ted) Irving and Kenneth Creer, began compiling paleomagnetic data from rocks in the United Kingdom. They began with the Torridonian sandstones, Precambrian red beds from the northwest of Scotland that were rich in iron oxides and therefore comparatively easy to analyze. The data they obtained were internally consistent and pointed to a fossil magnetic field some 30° different from the prevailing one. In 1954, Creer, Irving, and Runcorn produced the first apparent polar wandering path from rocks in Great Britain, of ages ranging from Precambrian to recent (figure 9.1).[17]

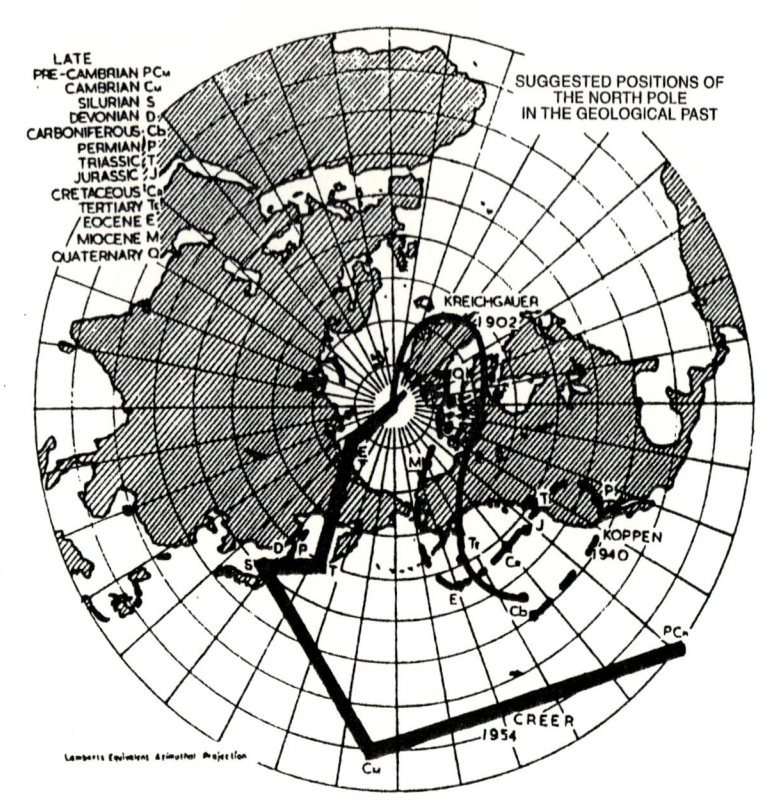

Figure 9.1. Apparent polar wandering path of Great Britain. (Creer et al., 1954.) Reproduced with permission of Terra Scientific Publishing.

Blackett, meanwhile, was analyzing rocks from the Deccan Traps: the heart of Gondwanaland. The Cretaceous Deccan lava flows corroborated the Precambrian results and more: in the lowermost flows, the polarity pointed inward toward the earth, and as one moved upward through the sequence, the polarity direction leveled out to horizontal and finally to a position consistent with the present-day field. Blackett, almost certainly influenced by reading Holmes, began to interpret these results in terms of continental drift. Runcorn, by contrast, leaned toward the interpretation that the poles had moved. Runcorn was perhaps skeptical of Blackett's theoretical instincts, given his highly publicized mistake on the source of the earth's magnetic field. (Bullard's alternative interpretation—that magnetic fields were generated by motions in the earth's iron core—was by this point gaining widespread acceptance.) Or perhaps Runcorn was influenced by Jeffreys at Cambridge. But Runcorn soon changed his mind.

Runcorn and Irving saw a solution to the question of whether the poles or the continents had moved: compare the apparent polar wandering paths from different continents. If they all showed the same pattern, then the poles had moved (or the entire crust had moved en masse). If they revealed divergent patterns, then the continents had moved autonomously. Irving left Cambridge for a position at the newly established

Department of Geophysics at the Australian National University in Canberra, where he began to collect paleomagnetic data and to compare apparent polar wandering paths for Australia, India, North America, and Europe. The result? The paths were distinctly different between continents. And they were consistent with drift positions suggested by Wegener on paleoclimatic grounds. By 1956 both Irving and Blackett were explicitly arguing for paleomagnetic data as evidence of continental drift. Runcorn, pursuaded by Irving, soon accepted their views.[18]

Runcorn, the Cambridge-trained geophysicist, now echoed the argument that geological supporters of drift had been making since the 1920s: whether or not one understood the mechanism by which it happened, continents had *moved*.[19] He noted the contradiction in geophysicists' own epistemological stance:

> Probably most geologists and geophysicists feel reluctant to admit the possibility of relative displacements of the continental masses in the recent history of the earth. It is often stated that a sound reason for such skepticism is the absence of any adequate theory of the mechanism by which such continental displacement could have taken place. This is an argument which should not be given much weight. Not until the last few years has there been an adequate theory for the existence of the geomagnetic field, but scientists did not previously disbelieve in the existence of the field for this reason.[20]

The same could be said for the magnetic reversals that earth scientists were about to uncover.

Convection Revisited: Seafloor Spreading

In 1960, inspired by these developments, Harry Hess wrote an article—an essay in geopoetry, he called it—in which he proposed a theory that would account for a variety of features of the seafloor now known from Navy-sponsored work and that would simultaneously provide a mechanism for continental drift. Hess suggested that convective overturn in the earth both created the crust initially and continued to drive its migrations. His "History of Ocean Basins" was published in a 1962 Geological Society of America Festschrift honoring Princeton petrologist A. F. Buddington, and even before publication Hess widely circulated a preprint of it. The argument centered on recent discoveries regarding the mid-ocean ridges. William Menard, now at the Scripps Institution of Oceanography, had mapped the ridges in detail and shown that they occurred along the medial lines of the oceans; Bullard, John Maxwell, Roger Revelle and others had confirmed earlier suggestions that the ridges were sites of exceptionally high heat flow; Bruce Heezen and Marie Tharp at Lamont had demonstrated the existence of medial grabens along the ridge crests and suggested that the crust was extending at right angles to the trend of the ridges. All these features were consistent with the view that the ridges were the sites of upwelling convective currents and that the oceanic crust migrated laterally on the limbs of these currents at a rate of approximately 1 cm/yr.[21]

Hess noted that submarine mapping had revealed very few features older than about 100 million years and that seismic studies suggested scant thickness of sediments on the ocean floor. These observations suggested that geological features were swept away as the ocean floor migrated laterally and was finally destroyed as it sank back into the mantle at sites of descending currents, marked by belts of crustal deformation and vulcanism. Connecting this to Runcorn's and Irving's studies, which "strongly suggest that the continents have moved by large amounts in geologically comparatively recent times," Hess concluded:

> One may quibble over the details, but the general picture of paleomagnetism is sufficiently compelling that it is much more reasonable to accept it than to disregard it . . . It strongly indicates independent movement in direction and amount of large portions of the Earth's surface with respect to the rotational axis. This could be most easily accomplished by a convecting mantle system which involves actual movement of the Earth's surface passively riding on the upper part of the convecting cell[22] (figure 9.2).

The sites of crustal consumption, such as "the circum-Pacific belt of deformation and volcanism," were the same places in which Hess had previously placed his tectogenes, except that now instead of being down-foldings of the continental crust, either transitory or long-lived, they were areas where the oceanic crust permanently sank down into the substrate. Hess made no reference, however, to his own earlier concept; he cited none of his own work prior to 1946, nor any of Vening Meinesz's 1930s or 1940s-era work on convection. Hess mentioned Holmes only in passing. "Long ago Holmes suggested convection currents in the mantle to account for deformation of the crust," Hess wrote, citing Vening Meinesz, Griggs, and "many others" (but with no specific reference to any papers by Holmes).[23] What Hess did not say—be it deliberately or inadvertently—was that Holmes (as well as Griggs) had explicitly argued for convection as the driving force of continental drift.[24] Instead, he implied that convection as the force behind drift was a novel idea. His own theory, Hess wrote in the abstract of his paper, provided a "more acceptable mechanism . . . for continental drift whereby continents ride passively on convecting mantle instead of having to plow through the oceanic crust."[25] And in the body of his text: "This [theory] is not exactly the same as continental drift. The continents do not plow through the oceanic crust impelled by unknown forces; rather they ride passively on mantle material as it comes to the surface at the crest of the ridge and then moves laterally away from it."[26]

In retrospect, many geologists would point to Hess's 1962 paper as the first "shot" fired in the plate tectonics "revolution," and there is little doubt that Hess's paper marked the beginning of the change in many people's views.[27] But to view Hess's paper as revolutionary rather than evolutionary is to ignore the work that Holmes, Vening Meinesz, and Hess himself did in the 1930s and 1940s, work that was widely read and cited.[28] Hess was right when he claimed that his theory was not exactly the same as continental drift, if by that he meant the theory proposed by Wegener in 1915. But he was wrong about the novelty of his mechanism: it *was* the same as that proposed by Arthur Holmes.

On the penultimate page of *Principles of Physical Geology*—published in 1945 and by the late 1950s widely read—Holmes wrote: "To sum up: during large-scale convective circulation the basaltic layer becomes a kind of endless travelling belt on the top of which a continent can be carried along, until it comes to rest (relative to the belt) when its advancing front reaches the place where the belt turns downwards and disappears into the earth."[29] This declaration was not buried deeply in obscure text: it was the conclusion of a chapter entitled "Continental Drift" and a section entitled "The Search for a Mechanism." In Holmes's 1945 theory, continents rode passively

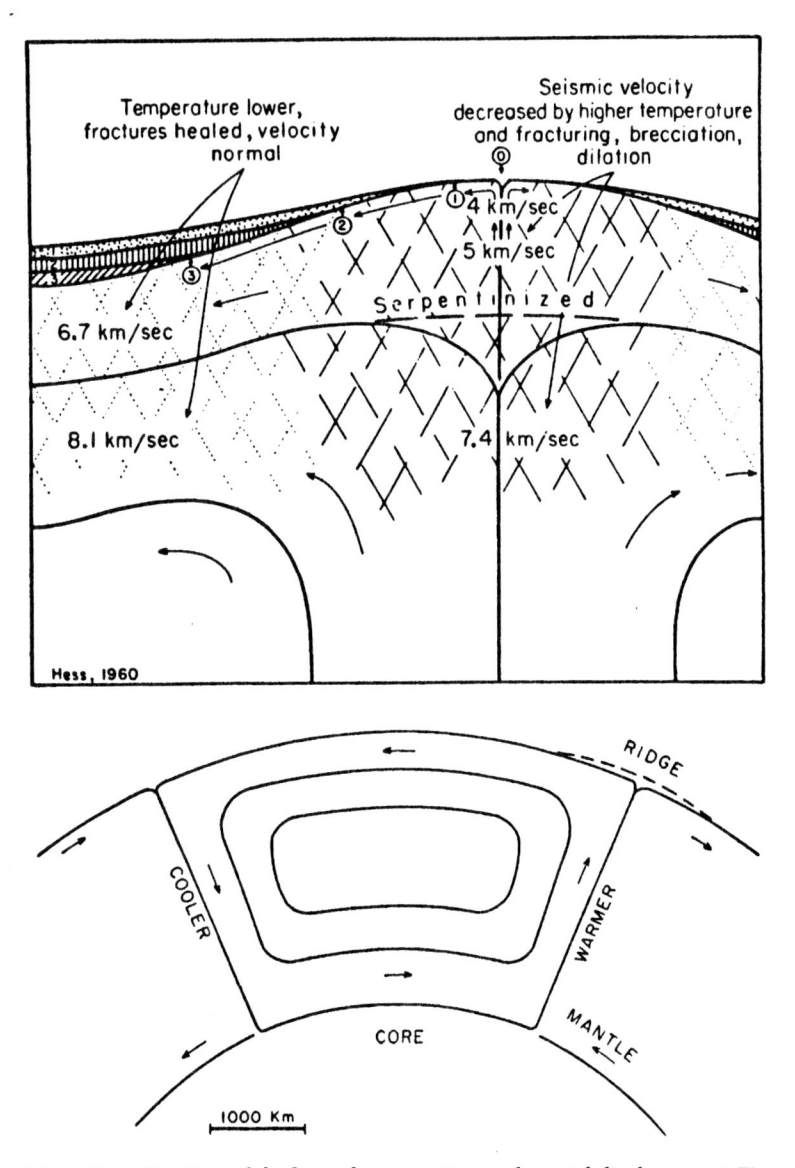

Figure 9.2. Harry Hess's model of mantle convection and crustal displacement. Top: Hess proposes that the ocean floor is actually exposed mantle, and that its low-seismic-velocity surface layer is a zone of serpentinization, formed where hot mantle rocks have interacted with cold ocean water. The thickening of the sedimentary cover away from the mid-ocean ridge (points 0, 1, 2, and 3) is a by-product of the mantle moving laterally away from the ridge crest: as one moves farther from the ridge, the exposed rocks are older and have had more time to accumulate sedimentary cover. Bottom: Hess suggests that the driving force is convective heat transfer within the mantle. The mid-ocean ridges represent the sites of up-welling convection currents. (From Hess 1962, p. 612. Reproduced with permission of the publisher, the Geological Society of America, Boulder, Colorado, U.S.A. Copyright © 1962, Geological Society of America.)

on the tops of convection currents, just as they did in Hess's 1962 theory. By comparing his idea not to Holmes's comparatively recent work, but to Wegener's much older (and by now anachronistic) proposals, Hess seemed to imply that he was solving the mechanism problem for the first time. Lest it be suggested that Holmes's model was somehow misunderstood, recall Menard's response to Daly's 1948 exam question on the requirements for an adequate orogenic theory. The best causal explanation, Menard wrote, was "sub-crustal flow resulting from thermal convection currents. . . . These currents rise to the bottom of the crust, drag along the bottom of the crust for an indefinite distance and descend to depth for reheating."[30] And this idea had been discussed before the war by Gutenberg. At Harvard and Caltech at least, students had been taught in the 1940s about a theory in which continents rode passively on the tops of convection currents.

When geologist Arthur Meyerhoff raised the issue of Holmes's priority in 1968, Robert Dietz tried to defend Hess by differentiating between the two men's theories. Having given Hess's theory its name—"seafloor spreading"—Dietz argued that Holmes's idea was "not really sea-floor spreading." He claimed: "[Holmes] resorts to thinning the crust rather than creating new oceanic rind at the mid-ocean swell and destroying it in trenches in the conveyor belt fashion of sea-floor spreading . . . Holmes thus follows Wegener."[31] But Holmes did *not* follow Wegener, except to agree with him that continents moved. Nor did he "resort to thinning" the oceanic crust "rather than creating . . . and destroying" it. On the contrary, Holmes explicitly argued that oceanic crust would be consumed by sinking back into the mantle at the site of descending currents. He suggested that this sinking would be facilitated by the density contrast created when basalt was transformed to eclogite under pressure (a view consistent with current knowledge):

> When rocks like basalt or gabbro (density 2.9 or 3.0) are subjected to intense dynamic metamorphism they are transformed into . . . a highly compressed form of rock called eclogite, the density of which is about 3.4. . . . [A] heavy root formed of eclogite would continue to develop downwards until it merged into and became part of the descending current, so gradually sinking out of the way, and providing room for the crust on either side to be drawn inwards by the horizontal currents beneath them.[32]

Holmes also argued that new oceanic crust would be created at the mid-ocean ridges by basalt intrusion and submarine lava flows. Although this may not be clear from the illustration in his 1931 paper "Radioactivity and Earth Movements" (which Meyerhoff reproduced in 1968) it is perfectly clear from Holmes's 1945 text:

> Basaltic magma would naturally rise with the ascending currents of the main convectional systems until it reached the torn and outstretched crust of the disruptive basins left behind the advancing continents or in the heart of the Pacific. There it would escape through innumerable fissures, spreading out as sheet-like intrusions within the crust, and as submarine lava flows over its surface.[33]

Holmes's theory is not vague, it is specific. The seafloor is not stretched and thinned, it is torn apart by submarine magmatism.[34] Arthur Holmes *had* proposed a mechanism for creating and destroying oceanic crust, and it was the same mechanism as that later proposed by Harry Hess.

In denying Holmes credit for the work he had done, Dietz ironically credited Hess with work he had *not* done. Dietz credited Hess with having worked out the mechanism of oceanic crust generation, and most geologists now accept this accreditation. But in his 1962 paper Hess *rejected* the explanation that geologists soon came to accept (and has since been proven on the basis of submersible work): that oceanic crust is generated by submarine basalt eruptions. Hess argued that seismic data, which suggested a uniform thickness for the oceanic crust, belied the possibility that the crust was generated by a physical process. "It is *inconceivable*," he wrote, "that basalt flows poured out on the ocean floor could be so uniform in thickness. . . ."[35] Hess thus opted for a chemical interpretation instead, suggesting that the oceanic crust was a hydration rind—a zone of chemical alteration formed by the interaction of hot rocks and cold ocean water—at the surface of the mantle "starting at a level 5 km below the sea flow."[36] It was Dietz, in 1961, who argued for the mechanism now accepted today—the mechanism Holmes proposed in 1945—that oceanic crust is generated by suboceanic intrusion and submarine eruption of basaltic lava.[37]

The Confirmation of Seafloor Spreading

If Harry Hess did not provide a novel solution to the problem of the mechanism of continental drift, he did inspire the work that would lead earth scientists to accept it. While Hess was writing geopoetry, Allan Cox, a California graduate student who had moved into geology after taking an undergraduate degree in chemistry, was studying paleomagnetism in Tertiary basalts from the Snake River Plain.[38] In an enormous thickness of seemingly homogeneous basalts, Cox found reversals everywhere within certain layers and their adjacent, baked country rocks, regardless of whether the rocks were oxidized, while other layers were entirely normal. Cox became convinced that the reverse magnetizations were real and reflected polarity reversals in the earth's magnetic field.[39] From this and the work of other geophysicists, it soon became apparent that the earth's field had reversed itself many times in the recent geological past. Cox and his coworkers Richard Doell and Brent Dalrymple published these results in the form of a geomagnetic time scale in 1963.[40] John Fleming's prediction—that fossil magnetism might one day be used to date rocks—had come true.

Cox's reversals and Hess's poetry added up to a testable hypothesis: if the seafloor had spread, and if the earth had repeatedly reversed its magnetic field orientation, then the ocean floor should record these events in a series of parallel "stripes" of normally and abnormally magnetized basalts. And this hypothesis could be tested using the large volume of magnetic data that had been collected during the war and post-war years. The hypothesis of seafloor magnetic stripes was proposed independently by a Canadian geophysicist, Lawrence Morley, and by two Cambridge researchers, Frederick Vine and Drummond Matthews.[41] Although Morley's paper was rejected by the scientific journal to which he submitted it, Vine and Matthews's proposal was published in *Nature*, and by 1966 the existence of seafloor magnetic stripes had been independently confirmed by Vine and Matthews and by Walter Pitman and J. R. Heirtzler at the Lamont-Doherty Geological Observatory in New York.[42] At Lamont, Neil Opdyke showed that marine sediments recorded the same magnetic events as terrestrial basalts, thus linking the continents and oceans.[43]

At this point, a flood of work emerged supporting the concept of moving continents. Among the most influential was that of J. Tuzo Wilson, a geophysicist at the University of Toronto and former student of Richard Field. Wilson had been studying the Hawaiian Islands, where the age of the islands increased as one moved northwest away from the focus of active vulcanism. Wilson suggested that this relationship could be explained if the islands were formed as the oceanic crust moved northwestward over a "hot spot" of volcanic activity. This seemed to be evidence in support of moving continents.

Wilson then turned his attention to the mid-ocean ridges, the subject of several recent papers by Menard, Victor Vacquier and coworkers at Scripps, and by Bruce Heezen and Marie Tharp at Lamont.[44] It had become apparent that the oceanic ridges formed a network that was essentially continuous around the globe, and the ridges were periodically offset by huge fracture zones running transversely to their long direction. Wilson had an ingenious insight about these offsets. Normally, geologists looking at a feature offset along a transcurrent (or strike-slip) fault can determine the direction of offset by displacement of that feature: across a right lateral fault, features are displaced to the right, across a left lateral fault, they are displaced to the left. But if the ridges were *rifts*, with the ocean floor splitting apart along them, then the slip directions on the tranverse offsets would be the opposite of what they would be along conventional strike-slip faults (figure 9.3). In 1967, Lynn Sykes at Lamont demonstrated using seismic evidence that the slip directions on ridge-to-ridge transform faults agreed with Wilson's interpretation.[45] The offsets were not normal transcurrent faults at all. They were, in Wilson's new terminology, *transform* faults—where a mid-oceanic ridge was locally *transformed* into a zone of crustal sliding, and then back again into another rifting ridge segment. There now seemed no doubt that the oceans were splitting apart.

Conceptual unification and scientific unity soon followed. In 1967 and 1968, Dan MacKenzie at Cambridge and Jason Morgan at Princeton independently succeeded in analyzing the known displacements of the seafloor magnetic strips as rigid body rotations on a sphere about a pole of rotation. The motions were mutually consistent: the displacements of each sector could be fitted geometrically together. Meanwhile, Jack Oliver and Brian Isacks of Lamont used seismic evidence to demonstrate the consumption of crust by subduction in oceanic trenches: the slip directions of deep-focus earthquakes in oceanic trenches were consistent with the overlap of one crustal plate onto another, with the lower one slipping downward. These data were consistent with the long-established geological and geodetic evidence for deformation at these sites. As Wilson summarized it in 1968, in language reminiscent of 1938, "The surface expression of this overlap takes the form of deep ocean trenches, overthrusts, and folded mountains."[46] That same year, Oliver and Isacks's Lamont colleague Xavier LePichon visually summarized these data in a map of the world divided into rigid lithospheric plates, labeled with calculated displacement rates[47] (figure 9.4).

Joseph Barrell's terminology, developed in the early 1900s, was now being used in full force. As Dietz had noted in 1961, "For considerations of convective creep and tectonic yielding, we must refer to a lithosphere and an asthenosphere."[48] Crustal plates were pieces of a rigid lithosphere that rode along the top of a plastic asthenosphere in motion. By the end of the decade, there were few holdouts against the new theory of plate tectonics. There had been a revolution in acceptance.

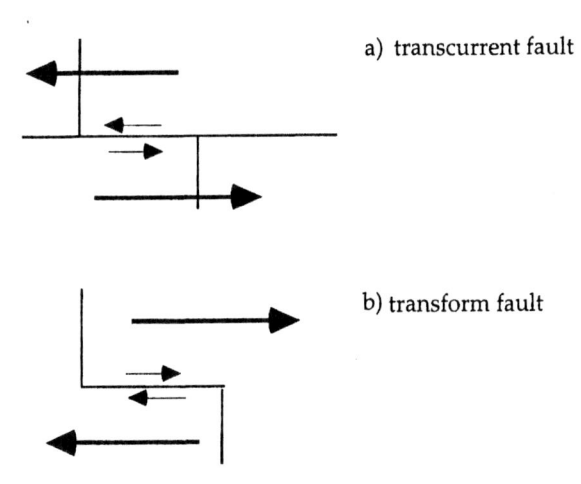

Figure 9.3. Tuzo Wilson's differentiation between transcurrent and transform faults. (a) In a transcurrent (or strike-slip) fault, the slip direction can be discerned from the offset of a feature intersecting the fault. Shown here is a left-lateral fault. The north side of the fault moves westward while the south side moves eastward, and the fault may continue indefinitely in either direction. (b) At mid-ocean ridges, the crust is splitting apart, with the east side moving east and the west side moving west. In this case, the motion along the orthogonal segment of the ridge is the opposite of case (a).

Two Types of Evidence

The evidence of plate tectonics was different from the evidence of continental drift. In support of plate tectonics in the 1960s, geophysicists presented the following:

- Land-based paleomagnetic measurements, which revealed divergent polar wandering paths on different continents
- Marine paleomagnetic data, which confirmed the hypothesis of seafloor spreading
- Seismological resolution of earthquake focal mechanisms, which differentiated transform faults from transcurrent faults and showed that the oceanic lithosphere splits apart at midocean ridges and slides down into the asthenosphere at ocean trenches
- Geometrical solution of plate motions as rigid body rotations around fixed poles, which showed that the postulated motions could be reconciled one with another

The evidence that Wegener had used in his attempt to substantiate continental drift was different, consisting of the following:

- Paleontological parallels suggesting extensive interchange of fossil species between continents during certain geological periods
- Stratigraphic parallels suggesting prior physical proximity of the continents
- Geological indicators of paleoclimatic change, change that could not be explained by simple warming or cooling trends
- The jigsaw-puzzle fit of the continents

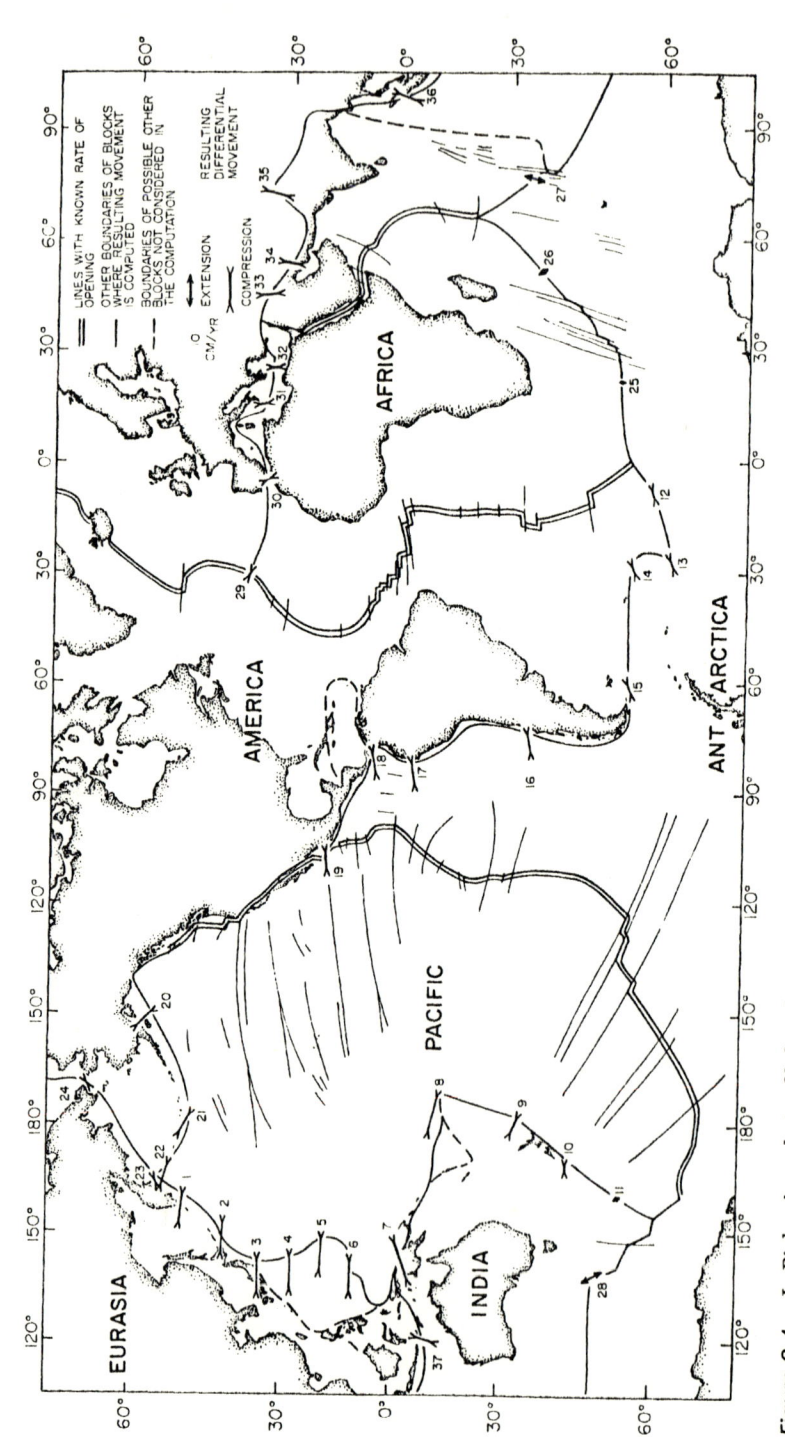

Figure 9.4. LePichon's analysis of lithospheric plates. Recalling John Joly's earlier work (figure 4.9), Xavier Le Pichon noted that plate boundaries account for the sites of most of the world's earthquake activity. (LePichon 1968, p. 3675.)

- Longitudinal measurements from the Danish Greenland expeditions, which indicated a present-day westward displacement of Greenland

And, in the late 1920s:

- Radio wave transmission time measurements.

With the exception of the Greenland data, which were widely held to be erroneous, and the radio wave data, which were inconclusive, all of this evidence consisted of *geological homologies*: similarities of patterns and forms based on direct observation of rocks in the field. In contrast, the data that instantiated plate tectonics were *geophysical*: instrumental determinations and calculations of physical properties of the earth. The earlier evidence consisted of observations described in words or pictures, the later of numerical data; the tools of the continental drift debate were hammers, hand-lenses, and field notebooks, those of the plate tectonics revolution seismographs, magnetometers, and computers. The advocates of plate tectonics seldom invoked the older evidence in support of the new theory, although many were aware of it.[49] The stratigraphic similaries, the fossil resemblances, the paleoclimatic data, played no important role in the establishment of plate tectonics. Rather, the old homologies were used ex post facto as a demonstration of the far-reaching explanatory power of the new theory.[50]

If a concept is rejected in the face of one set of data and later accepted in the face of another, one might conclude that the early data were poor and the later data were good. On a sociological level, such an account would be indisputable: the evidence of plate tectonics was efficacious in convincing earth scientists of moving continents in a way that the evidence of continental drift was not. But on an epistemological level, there is a problem: *plate tectonics validated Wegener's evidence*. Geologists in the 1970s used stratigraphical and paleontological evidence to reconstruct earth history in just the manner Wegener proposed, and the jigsaw puzzle fit of the continents was attributed to rifting and drifting. In the context of plate tectonics, Wegener's evidence is seen as substantively correct. More important, it was seen that way *at the time*. The evidence that Wegener used was borrowed from the lifetime work of prominent stratigraphers, paleontologists, and field geologists. In the nineteenth century, field-based pattern recognition had produced the geological time scale, the stratigraphy of the British Isles, and the unraveling of the structure of the Swiss Alps. In the early twentieth century it led first to the idea of Gondwana, then to Barrell's theory of fragmentation, and then to continental drift. Wegener's evidence came from the mainstream of geology, and leading geologists sought explanations for it.[51] Few disputed that there had been some kind of intercontinental connections; the issue was always how these connections were forged. Wegener's data were not poor data. On the contrary, they were rich.

Historian Henry Frankel has argued that seafloor spreading was accepted by the geological community because it predicted novel facts that turned out to be true.[52] No doubt. But continental drift also made predictions, beside the obvious one that continental motions might be detected. The theory predicted that matched fossil assemblages in reconstructed continents would tend to end abruptly, rather than gradually, that paleoclimatic patterns could be linked in the unified lands, and that mountains within continents would be older than those on their margins. In some cases, these predictions could have led to research programs, such as a geochronological study of mountain belts, a reconstruction of the earth's climatic history, or, as Arthur Holmes suggested, a study of the temporal relations of tensional and compressive motions on opposite ends of crustal blocks. Even William Bowie, who generally showed little affinity for fossils, imagined a test of the theory on the basis of paleontological evidence. In a 1924 paper on plans for the worldwide longitude operation, Bowie wondered about other possible tests and thought of one. Given that climatic conditions are generally milder along coastlines and harsher inland, he suggested that one might examine the paleobiology of regions now coastal, but according to drift theory previously interior. If the theory were true, then the fossil assemblages from appropriate strata in those locations would reflect harsher conditions. Conversely, one might look at assemblages now inland—such as the Himalayan sequences—but previously thought to be coastal.[53]

In some cases, the predictions of drift *did* lead to research programs: Alexander du Toit's comparative field study was undertaken as a test of continental drift; the *S-21* expedition was undertaken to test competing theories of crustal mobility; and some Dutch geologists, inspired by G. A. F. Molengraaf, used drift theory to help formulate and attack research problems in the Indonesian archipelago.[54] Emile Argand used drift to unravel European geological structure. And both du Toit's study and the *S-21* expedition produced novel facts. But even where the facts needed to test the predictions were already known, this should have made the confirmation no less compelling: one of the great successes of the theory of general relativity was its ability to explain the anomalous advance of the perihelion of Mercury, a problem that had been identified by the mid-nineteenth century.[55] Darwin's theory of natural selection rested largely on facts and observations that had been known for some time. It also predicted that fossil similarities should *not* occur, since isolated communities should evolve in different ways. Previously these similarities were explained by sunken continents, but with the failure of that theory, they became once again anomalous. One needed an explanation.[56]

Evidential Asymmetries and Underdetermination

The facts that Alfred Wegener compiled in support of continental drift were anomalous within existing explanatory frameworks, but they were not the only facts with which earth scientists had to contend. Geologists also grappled with the fact of isostasy, which, as interpreted by the leaders of North American geodesy, was incompatible with horizontal drift and indeed with any large-scale horizontal motion. Willis and Schuchert proposed isthmian links in an attempt to resolve these seemingly conflicting facts. On the face of it, the coexistence of the theories of continental drift and isthmian links may be interpreted as a case of underdetermination: two different theoretical explanations for the same empirical information. One could explain faunal homologies either by continental drift or isthmian links. Isostasy was similarly underdetermined: one could account for isostatic compensation with either the Pratt or the Airy model. John Hayford and William Bowie chose the Pratt because it was simpler—simple in this case being defined by ease of calculation. Under different circumstances they could have made a different choice.

But were these theoretical choices truly underdetermined? Not quite. Consider the case of isostasy. Having inherited a choice made on pragmatic grounds and then developed into a methodological program, Bowie became bound to it intellectually as well. He thus judged continental drift not so much on the evidentiary basis but on the basis of its compatibility with this prior commitment. Thus he had to ignore the paleontological and stratigraphic evidence of intercontinental connections and to focus instead on the geodetic evidence of equilibrium. The question of Pratt versus Airy isostasy *was* underdetermined when viewed solely in the context of geodetic data, but not when viewed against a broader evidential base. Bowie's problem in the late 1920s was not too few data but too many. To sustain his theoretical views — and preserve his methodological, intellectual, and institutional commitments — he had to ignore large bodies of data from ancillary fields. The same could be said for many scientists. Bowie focused on the data to which he was closest — intellectually, professionally, personally. Charles Schuchert did the same.[57]

In a footnote in his 1958 textbook, *Elementary Seismology*, Charles F. Richter (for whom the earthquake magnitude scale is named) spoke to the issue of how scientists ignore or attend to different forms of evidence. Richter recalled an anecdote about a visit to Berkeley in 1906 by the famous Dutch botanist and geneticist Hugo De Vries:

> He [De Vries] was taken into the field to see some of the effects of the great earthquake of that year. Being told in advance of displacements ranging from 15 to 20 feet, he listened with polite interest but obvious scientific skepticism. The first few localities to which he was conducted appeared not to convince him; he was clearly not familiar with geological matters. The party finally arrived at [a] locality . . . where the road was offset 21 feet. The fault trace here was close [by] . . . small trees and bushes were uprooted, some were torn and some killed. "Ah, here is *botanical evidence!*" cried De Vries, and he fell to examining these details with close attention, until his companions insisted on going on.[58]

Richter concluded: "We are all best impressed by evidence of the type with which we are most familiar."[59] Or, as William Menard put it some years later, "Some earth scientists believe in God and some in Country, but all believe that their own field data are without equal and they adjust other data to fit them."[60]

Given this, it is not surprising that Bowie, a geodesist, would prefer geodetic to geological evidence. It is consistent with Richter's and Menard's comments, and with the arguments for epistemological affinity presented earlier. What is surprising, however, is that *geologists* would prefer geodetic to geological evidence. For there is a central asymmetry in this story. While William Bowie felt free to argue for a theoretical position that ignored historical geological data, Charles Schuchert did *not* feel able to ignore geodetic data. Nor was he alone. Bailey Willis was fundamentally a field geologist, but he felt compelled to attend to geophysical constraints. So did Reginald Daly. So did most American geologists involved in the drift debate. In Schuchert's case, the compulsion to respect isostasy led to his decades-long quest for an alternative to sunken continents, a quest that ended in the adoption of isthmian links. But did isthmian links explain all the facts that drift did? No, it did not. Isthmian links left several evidential issues unresolved. Chief among them in Schuchert's mind was the stratigraphic evidence for the felsic nature and wide expanse of the Atlantic land bridge. Perhaps more profoundly (although it seems not to have bothered Schuchert), the theory ignored the stratigraphic homologies — one of the central arguments for continental drift. Isthmian links accounted for the intercontinental dispersal of plants and animals but did little to explain the homologous structural and stratigraphic patterns across oceanic divides.

Du Toit saw this clearly in May of 1933. While others praised Schuchert, du Toit felt a rising frustration to see his own work so patently ignored. "[Your papers] involve a great deal of research and form very valuable contributions to the subject," he wrote to Schuchert as diplomatically as he could, "but if I may be permitted to voice a little criticism ... the real problem, that of the Gondwana continent, its shape, linkages, etc. remains hardly touched."[61] The structural and stratigraphic correlations had not been explained so much as ignored. Willis had spoken faintly to the issue of structural homology by suggesting that compressive forces emanating from the ocean basins might equally affect all sides of those basins. But this was a thin reed on which to hang an explanation, and it failed entirely to account for the stratigraphic patterns. How could Willis (a structural geologist) and Schuchert (a paleontologist) supply an explanation that failed to account for the evidence from their own science? How could American geologists be content with it?

From the history recounted here, and from their own correspondence, it is clear that the constraints of geophysics were overriding. As Willis wrote to Schuchert early in 1932, "I recognize the validity of your paleontological evidence and its imperative demand for routes of migration and barriers to migration, [while you recognize] that the solution to the problem must be in conformity to the requirements of geophysical principles."[62] Schuchert's entire theoretical life orbited around this need: to explain his paleontological data in a manner consistent with the demands of his geophysical colleagues. Edward Berry, also a paleontologist, reflected a similar stance when he wrote that isthmian links would "stand inspection." Inspection by geophysicists, that is. But why? Why should the constraints of geophysics be deemed more important than the constraints of stratigraphy? Why was isostasy more important to an adequate theory than homologous strata? Why did geophysics have the last word? To answer these questions, we need to consider the disciplinary traditions from which these constraints emerged.

Evidentiary and Epistemic Traditions

Throughout the late eighteenth and nineteeth centuries, there were two distinctly different ways of thinking about the earth — two different evidentiary and epistemic traditions. Scientists like Buffon and Elie de Beaumont in France and Hopkins and Kelvin in the United Kingdom tried to understand the history of the earth primarily in terms of the laws of physics and chemistry. Their science was mathematical and deductive, and it was closely aligned to physics, astronomy, mathematics and, later, chemistry. With some exceptions, they spent little time in the field; to the degree that they made empirical observations they were likely to be indoors rather than out. Historians such as Nathan Reingold and Daniel Kevles have called this tradition *geophysical*, although at the time these men mostly thought of themselves as mathematicians, astronomers, or physicists. (William Hopkins was an exception: he called himself a "physical geologist," but with a meaning distinct from how the term came later to be understood.) In contrast, men like Werner in Germany, Cuvier in France, and Lyell in England tried to elucidate earth history primarily from physical evidence contained in the rock record. Their science was observational and inductive, and was, to a far greater degree than their counterparts', intellectually and institutionally autonomous from physics and chemistry. With some exceptions, they spent little time in the laboratory or at the blackboard; the rock record was to be found outside. By the early nineteenth century, students of the rock record called themselves *geologists*.

The identification of these two distinct traditions is not to imply that they were necessarily mutually exclusive or insulated from each other. In some institutions, for example Cambridge University, they existed under the same roof.[63] And quite a few individuals attempted to transcend the gap between them or to argue for advancement by unification. Henry De la Beche did so in the first half of the nineteenth century, Osmond Fisher in the second, John Joly in the early twentieth. But these men argued for a middle ground that few were able to occupy successfully; for the most part, when geologists and geophysicists addressed themselves to the same question, the result was not rapprochement but rancor.

The dispute over the age of the Earth is the most famous historical clash between geology and geophysics, but it was neither the first nor the last. In the 1850s, John Forbes argued with William Hopkins (and later John Tyndall) over the mechanisms of glacier motion. Forbes, together with Louis Agassiz, argued on the basis of field observation that glaciers flowed like a stream, with some parts moving faster than others and the whole mass deforming internally. Although they looked solid, glaciers were really fluid. Hopkins argued instead on theoretical grounds that glaciers slid downhill as solid objects lubricated by a melted layer at the base.[64] The dispute over the rheology of glaciers closely paralleled the later argument over the rheology of the crustal substrate. Proponents of a fluid substrate argued for its necessity on phenomenological grounds; opponents argued the impossibility on theoretical grounds. Osmond Fisher was unusual in pleading for his geophysical colleagues to take seriously the geological evidence of subcrustal fluidity. John Joly likewise argued for geologists and geophysicists to work toward reconciliation of their competing claims. But few did. Most geophysicists assumed that geologists were simply mistaken, while geologists either ignored geophysics or lived uncomfortably with the contradictions. As time went on, each side became increasingly frustrated with the other.

Throughout these debates, each side affirmed the superiority of its methods and denied the claims of the other: geophysicists argued for the greater rigor of mathematical analysis and dismissed phenomenological counterarguments, while geologists defended the accuracy of their observations and not infrequently ignored theoretical arguments that might call their conclusions into doubt. The same issues arose with respect to continental drift. Advocates like Alexander du Toit argued that drift *had* occurred; opponents like Harold Jeffreys and William Bowie argued that it *couldn't have*. The question is, Why should American geologists have taken more seriously the claims of Bowie than those of du Toit? Nationalism no doubt played a part, but du Toit's research was supported by an American institution, and he was seconded by Reginald Daly at Harvard and Gutenberg at Caltech. Schuchert is particularly perplexing: one might reasonably expect him, as a historical geologist, to be more inclined toward du Toit's evidence than toward Bowie's, and as an American to be more inclined toward empirically driven rather than toward theoretically driven arguments. But the opposite was the case. Why did historical geologists seemingly discredit their own data?

Searching for Certainty Through Measurement

When the drift debate languished in the 1940s, Chester Longwell argued that what drift really needed was some "incontrovertible evidence in its favor, not . . . reiteration of diffuse and qualitative arguments."[65] Longwell had earlier suggested that "incontrovertible evidence" might come from "the geophysicist," who could tell finally whether "it is or is not possible for continents to leave their moorings and drift widely." Incontrovertible evidence would be "at least roughly quantitative."[66] Longwell was not the only one to express this view: it was shared by the French geologist, Pierre Termier, who spoke to an American audience in the *Annual Report of the Smithsonian Institution* in 1924. Termier found continental drift too "convenient," in that it appealed "to occult forces, perhaps inexistent, and which afford us the formidable power of disposing at will of the continents and oceans."[67] The problem, Termier concluded, was that there was no direct evidence that the required forces existed. Like Longwell, Termier looked to geophysicists to resolve the question. On the subject of drift, he concluded, "we can trust to the geophysicists."[68]

The desire for "incontrovertible" evidence was embedded in the worldwide longitude operation, and Termier reflected on it: "Of the present drifting, supposing it exists, we will soon be advised by the . . . measures of longitude with a precision heretofore unknown." Termier was skeptical, however, not because he disdained the idea of measurement, but on the contrary because he realized the difficulty of it:

> It is to be feared that the relative movement, if it exists, may be an extremely slow one, and that a century may be necessary to surely establish its existence. We shall then be condemned to die without knowing whether the Atlantic is advancing or receding, and on that account many of us will find difficulty in consoling ourselves.[69]

But if American earth scientists could not console themselves with the direct measurement of continental drift, they could console themselves with other kinds of measurements, and they had been doing so for some time.

Throughout the nineteenth century, the U.S. Coast and Geodetic Survey had provided an institutional and intellectual base for scientists pursuing geophysical questions, largely through instrumental measurement.[70] As Nathan Reingold has pointed out, the Coast Survey was for many years the largest employer of physical scientists and mathematicians in America. In the mid-nineteenth century, American geodesists were competing on nearly equal terms with their German counterparts; the Survey's pioneering use of telegraphy in determining longitudes, for example, led Europeans to call it the "American method."[71] At the end of the nineteenth century, Germans were still considered the world's leaders, but by the end of World War I Americans held that distinction. So although the term *earth science* may be anachronistic, the concept is not: geology and geodesy grew up side by side in the United States and each influenced the other.

Geodesy was different from the theoretically based geophysics developed by men like Hopkins and Kelvin in Great Britain, in that it was heavily field based. Whereas theoretical geophysicists built their arguments on mathematical analysis, geodesists built their science on measurement. The determination of the Hayford spheroid, while necessarily relying on mathematics, was not fundamentally *based* on mathematics. It was based on instrumental measurement. Geodesy, like geology, was an empirical science, but its empirica were described in numbers rather than in words. What geodesy shared with other forms of geophysics—what made it reasonable to consider it a branch of geophysics, as Hayford and Bowie did—was the application of physical principles of the study of the earth. And the success of American geodesists in applying physical principles to empirical science was paralleled by efforts on the part of many American geologists to incorporate physical and chemical reasoning into their work.

As early as the mid 1800s, some American geologists looked for a fuller incorporation of physics and chemistry into their work. Prominent among them was James Dana. After returning from the Wilkes expedition, Dana spent the rest of his life ensconced at Yale. Field work played little role for him.[72] His system of mineralogy was organized along chemical principles; his permanence theory was grounded in the physics of materials. And his rivalry with James Hall, as we have seen, was in part a clash over epistemic and evidential preferences: although both men drew on a variety of kinds of evidence and arguments, Hall leaned heavily towards the field-empirical, Dana toward the laboratory-theoretical. The near-equal historical standing of these two major figures might suggest an argument for intellectual parity between their approaches (one was the first president of the Geological Society of America, the other the second), but it was Dana who held a university chair, Dana who edited the nation's most prestigious scientific journal, and Dana who more clearly left his mark upon subsequent generations of geologists.[73]

By the 1870s, reasoning from physics and chemistry was evident in the work of many of America's most important geologists. Clarence Dutton, Clarence King, T. C. Chamberlin, and G. K. Gilbert all emphasized the application of physics and chemistry to understanding earth processes and structures. Chamberlin was one of the earliest scientists to consider the role of atmospheric chemistry in climate change, and his cosmological theory was arguably more influential among astronomers than among geologists.[74] Writing in 1872, Berkeley professor of geology and natural history Joseph LeConte explicitly argued for greater integration of physics into geology. Having proposed a version of contraction theory based on a solid earth largely in response to physicists' arguments against a fluid interior, LeConte suggested that, be his own theory faulty, it was nevertheless based on correct *methods*. Physically based reasoning, he argued, was a needed antidote to the vagueness that arose from reliance on observational evidence alone. His theory might be wrong, but at least it was not uncertain:

> [My] theory . . . rests upon an insufficient knowledge of the structure of the earth. It is possible that the state of knowledge is not yet such as to warrant any attempt at a general theory. . . . [But] even if entirely wrong, it is at least a little more definite than anything we have. It is at least something tangible which may be attacked and overthrown by facts and by physical reasoning. We have had enough of vague theorizing in geology; of vague shadows through which the trenchant sword of science passes with no effect. It is time that the more perfect methods of physics were applied to geology.[75]

The grounding of geology in the principles of physics and chemistry was also the goal of Charles van Hise, a pioneer of chemical and physical analysis of rocks at the turn of the century, later president of the University of Wisconsin. In "The Problems of Geology," an address to the International Congress of Arts and Science in St. Louis in 1904, van Hise laid out a prescription for the future of his science: large doses of physics and chemistry. Geology, he argued, should be no more or less than "the science of the physics and chemistry of the earth." Van Hise allowed that traditional geologists might object that "this definition is inadequate to cover descriptive and historical geology," but he felt they were missing the point. History without explanation was of little use or interest:

> It may be said that the history of events, as shown by the rocks and fossils, does not necessarily require physical or chemical treatment. There is some truth in these statements, but on the other side it may be held that the facts are the results accomplished by physical and chemical work. These facts become important and significant mainly as they are interpreted in physical and chemical terms. The objects of the earth—the complex results of chemical and physical work—if described without reference to the manner in which the results came about, have comparatively little interest.[76]

Van Hise wanted a sea change in geology, from a tradition in which it was sufficient to recount the history of the earth to a science in which it was not. Geology "did not consciously begin as the science of the physics and chemistry of the earth," he acknowledged; "the phenomena of the earth were studied as objects, and thus geology was at first an observational study." This was fine as a start, but the next step, he argued, "a revolutionary one, was to explain the observed phenomena in terms of physical and chemical processes." Van Hise thus introduced the notion of "process-oriented geology" and suggested that the goal of geologists should be to understand geological products in terms of the chemical processes and physical forces that created them.[77] "Phenomena may be, and often are, observed and described in advance of their physical-chemical interpretation," he wrote. "But the naming or even the description of phenomena of the earth without reference to energy or agent is very unsatisfactory."[78] Why it was unsatisfactory he did not say; he evidently assumed this was obvious. And if it was obvious, then one *had* to turn to physics and chemistry because this is where energy and agents were to be found.

Like most American geologists in the early twentieth century, van Hise followed Chamberlin's and Gilbert's methodological prescriptions; he particularly agreed with Gilbert's argument that understanding emerged by viewing a problem simultaneously from multiple perspectives. But whereas Gilbert had spoken primarily in terms of *individual* understanding—using the analogy of the field geologist locating himself by triangulation—van Hise extended the metaphor to the disciplinary community as a whole. Because the historical perspective was already well established in geology, van Hise argued, the physical and chemical perspectives needed to be brought on par. Scientists needed to triangulate conceptually among geology, physics, and chemistry to gain a clear picture of the earth. Chamberlin's new textbook—which one reviewer thought went too far toward causes—was in van Hise's opinion exemplary, because Chamberlin brought together geology, geochemistry, physics, and astronomy to bear on causal questions. Chamberlin's first volume, appropriately entitled *Processes and Their Results*, was a "great step in advance."[79]

Admittedly, some geologists interpreted the method of multiple working hypotheses as interdicting strong theoretical stands, or perhaps any stand at all. Some geologists supposed they ought only to lay out the various interpretive options.[80] (Mining engineers thus quipped that the best geologist was the one with only one hand and therefore unable to say "But on the other hand . . ."). Van Hise found this absurd, particularly given Chamberlin's own theoretical convictions, and he argued that

Chamberlin's method not only encouraged theoretical resolution but also required one to strive toward quantification. For if Chamberlin were right that "almost every complex geological phenomenon has not a simple, but a composite, explanation," then one had to turn to mathematics to evaluate the relative importance of contributory causes. "As soon as it is appreciated that to explain a complex phenomenon several causes are usual, if not invariable, rather than exceptional," he argued, "it becomes plain that their relative importance should be determined, and this can be done only by quantitative methods."[81] The very complexity of geological phenomena that was sometimes used as an argument against quantification was in van Hise's view the most potent demonstration of the need for it.

"The consideration of the processes of geology by quantititive methods is superlatively difficult," van Hise admitted in closing, but it "must be undertaken if the science ever approximates certainty of conclusions." To apply quantitative methods, it would be necessary for mathematicians and geologists to work together. For if van Hise was bored by purely descriptive geology, he was equally annoyed by fallacious quantification:

> Those geologists who have made the attempt to combine mathematical with their geological reasoning usually have shown marked deficiency in their mathematics. Upon the other hand, those mathematicians who have attempted to handle the problems of geology mathematically have usually been so deficient in a knowledge of geology that their work has been of comparatively little value.... [T]he time has come for co-operation between geologists and mathematicians in the advancement of the science of geology to a quantitative basis.[82]

Van Hise's aspirations were in part fulfilled by the establishment of the Geophysical Laboratory of the Carnegie Institution of Washington.[83] The establishment of the laboratory went against the Carnegie grain of supporting "exceptional individuals"—indeed, Andrew Carnegie himself was opposed to the idea—but the advisory committee, which included physicists A. A. Michelson and Carl Barus and geodesist Robert Woodward, as well as Van Hise, Chamberlin, and U.S. Geological Survey Director Charles Doolittle Walcott, successfully lobbied the Trustees for the creation of a laboratory that would advance geophysics as a whole. In 1904 Woodward succeeded Daniel Coit Gilman to become the Carnegie's second president, and with his support the Geophysical Lab, founded in 1907, became a major center for research in the physical and chemical aspects of earth science.

The recommendation of the advisory committee to the Trustees was, in the words of Nathan Reingold, that geology being "in good shape, the future was in geophysics."[84] But by any historical standard geophysics was in at least as good shape in the United States as was geology; the desire on the part of the committee to promote geophysics in a broad sense (for the laboratory in the end pursued geochemistry more than geophysics) reflected not the problems of geophysics in America but its success.[85] As Reingold has shown, the Carnegie Trustees were not so much interested in creating new fields of study as they were in solidifying and enriching successful existing ones, and "concern with the physical and chemical properties of the earth's constituents was ... an intrinsic part of a widespread, successful research enterprise in geology in nineteenth century America."[86] The Geophysical Lab was in fact an extension of work already being done at the U.S. Geological Survey: its first director, Arthur Day, came from the Survey's Washington offices only a few blocks away. Laboratory work had been supported by Clarence King, who in the 1880s had hired Carl Barus to undertake experimental studies of the physical properties of rocks. Experimental work fared less well under the Survey's second director, the brilliant but embattled John Wesley Powell who under budgetary and political pressure fired Barus in 1894, but it was revived by Walcott, the Survey's third director. Walcott created a division of chemistry and physics under the auspices of geophysicist George F. Becker, who in turn hired Day. But funding for Becker's division remained difficult to justify; by transferring experimental work to the Carnegie Institution, Walcott hoped not to end it, but to secure its future.[87]

From Theory to Fact?

If the creation of a Geophysical Laboratory reflected a desire for greater exactitude among those who studied the earth, it also reflected a desire to find means to settle debates that often seemed interminable. Woodward had described the issue of the earth's interior as a "vexed" question that had "lingered on the battlefields of scientific opinion" to be brought up and fought all "o'er again." The same could be said for many other issues, and the Carnegie Committee hoped that physics and chemistry, by providing firm constraints, might help slay certain warhorses permanently. But, as the earlier comments of Joseph LeConte suggest, for some scientists the role of physics and chemistry lay not only in providing factual or conceptual constraints but also in suggesting "more perfect methods." For some, improvement in the science of geology meant a shift to a more theory-driven stance, a more deductive mode of reasoning.

American earth scientists in the early twentieth century were skeptical of theory-driven science, skeptical of hypotheticodeductive methods. As an alternative, Chamberlin and Gilbert championed a pluralistic methodology; Daly and others followed it; John Merriam, Woodward's successor, enforced it.[88] But by the late 1920s, strains of dissent were evident. Indeed, Merriam would not have had to enforce his preferred rules of scientific engagement had Fred Wright not broken them. Wright was no doubt influenced by his graduate education in Heidelberg, but he was not unique. Those who sought greater certainty and exactitude in geology frequently expressed an ancillary desire to make geology more deductive. After all, physics and chemistry were lawlike; perhaps this was whence their power derived. Make geology more nomological, and perhaps it would be more powerful, too.

Van Hise thought so. He explicitly linked the application of physics and chemistry to earth science with a hypotheticodeductive methodology. Anticipating later philosophers of science, van Hise even doubted the *possibility* of genuinely inductive science:

> No man has ever stated more than a small part of the facts with reference to any area. The geologist must select facts which he regards of sufficient note to record and describe. But such selection implies theories of their importance and significance. In a given case the problem is therefore reduced to selecting facts for record, with a broad and deep comprehension of the principles involved, a definite understanding of the rules of the game . . .[89]

For van Hise—as for most who would follow in the later twentieth century—theory came first.

Echoing the sentiments of LeConte, van Hise expressed vexation at the vagueness of geological theorizing and the lack of clear constraints for interpretation. He wanted definite "rules of the game." Uniformitarianism was one such rule; this is in part why geologists were so protective of it. But it was only one rule, and it could take geologists only so far. Uniformitarianism gave geologists grounds on which to interpret the geological record in terms of presently observable processes, but it did little to illuminate the forces behind those processes. For this, van Hise argued, one needed a deductive science based on the principles of physics and chemistry: "the problem of geology," he concluded, "is the reduction of the science to order under the principles of physics and chemistry."[90] In 1904 van Hise's position was anomalous. By the 1930s, it was on the way to becoming mainstream.

The desire for a nomological science of geology—and the confusions and contradictions that this desire generated—is most clearly seen in the work of Walter Bucher (1889–1965). Bucher was born in the United States, but his family moved to Germany, where he was educated in Heidelberg and received his Ph.D in geology in 1911. Appointed at the University of Cinncinati, Bucher taught there for twenty-seven years until he moved to Columbia University in 1940, where he served as professor of structual geology and later chair of the department until his retirement in 1956. A member of the National Academy of Sciences, he was awarded an honorary doctorate from Princeton in 1947, served as president of the Geological Society of America in 1954, and earned the Bowie medal in 1955 and the Penrose medal in 1960.

Scientifically, Bucher was best known for his 1933 book *The Deformation of the Earth's Crust* (reprinted in 1957) in which he attempted to organize "all essential geological facts of a general nature that bear on the problem of crustal deformation and to derive from them inductively a hypothetical picture of the mechanics of diastrophism."[91] Stated this way, his methodology seems no different from most other American geologists: his goals were theoretical and his strategy was inductive. His subtitle, *An Inductive Approach to the Problems of Diastrophism*, seems to confirm this. But, like Frederick Wright, Bucher was influenced by his German education toward a more theory-driven epistemic stance in tension with the traditional values of American field geology. Despite the implied methodological clarity of his subtitle, Bucher's book is contradictory in ways that reflect this tension, a tension increasingly being felt by others as well.

The body of Bucher's text was organized around the articulation of forty-six laws of crustal deformation, recapitulated in full in a final appendix. But these laws were not laws in any sense that scientists and philosophers would normally understand. For one thing, they are not quantitative. Bucher's laws consist of verbal summaries rather than mathematical equations—akin to the geological "law of superposition" (which holds that younger rocks lie atop older rocks) rather than the laws of motion or thermodynamics. More important, they are not simple: some of his laws are paragraphs long; others include various a's and b's. Nor are they statements of invariant relationships. They describe observed relationships, but it is implicitly understood that there are exceptions.

Bucher's first law, for example, reads, "Aside from the effects of erosion and deposition, the earth's surface deviates from the relatively simple form of an ellipsoid, to which it may be referred, through outward (upward) and inward (downward) deflections."[92] Or consider Law 3: "In ground plan, the forms of crustal elevations and depressions represent two types. First, elevations and depressions which are essentially equidimensional (swells and basins); and second, others that show a distinct linear development with one horizontal dimension decidedly greater than the other."[93] Or Law 17a: "The modern crustal folds, on one side of the globe, radiate outward from a focal point in the Hindu Kush and Pamirs. Strong multiple branches extend out toward the west, to the southeast, and to the northeast, and weaker branches run northward across the Eurasian continent and southward into the Indian Ocean. On the other side of the globe, the crustal folds encircle the Pacific Ocean. Offshoots exist in both parts."[94]

Using almost any reasonable definition of a scientific law—a simple and general statement about entities or processes, their causes or interrelationships—Bucher's laws are not laws at all.[95] They are summaries of geological observations—geological patterns—and he admitted as much. In his opening paragraph, he explained that his goal was to "assemble all essential geological facts" into "carefully worded generalizations which are designated as 'laws'."[96] But then why call them laws? Bucher gave two reasons. One was that he wished to distinguish them from "personal interpretation."[97] Although his laws might not have been universal or invariant, he did believe that they were accurate accounts of observable relations as distinguished from inference and interpretation. Bucher noted that he had been advised by one reviewer "to use the more cautious term 'theses' instead of 'laws'," but he rejected this advice: "In common language a thesis is too much a matter of individual conviction to make the word servicable here. Too much of the geotectonic literature of the past was concerned with the theses of men rather than with the laws of nature."[98] Like other geological colleagues, Bucher was striving to generate reasonably unfiltered accounts of geological phenomena, and to clearly demarcate these accounts from their causal interpretation.

Interspersed with his laws, which he printed in boldface, Bucher gave his "opinions" in italics. In a discussion of isostasy and the rheology of the substrate, for example, Bucher praised Barrell's concept of lithosphere and asthenosphere but concluded that the subject was still uncertain. Therefore he summarized with "opinions":

> Opinion 3. The "crust" is the outermost shell of the earth, which on the whole possesses sufficient strength to offer resistance to deformation and to transmit long-continued stresses within certain limits.
>
> Opinion 4. As a whole, the crust rests in isostatic equilibrium on the weak asthenosphere. When the vertical stresses set up by local excess loads exceed the limits of crustal strength, vertical movement accompanied by horizontal transfer of matter in the subcrustal asthenosphere restores equilibrium.[99]

In hindsight, more than a few of Bucher's opinions have held up better than his laws!

Bucher's second reason for presenting his empirical generalizations as laws was to stimulate thinking. He acknowledged that many of his laws might "prove untenable," but that was the point. He wanted his colleagues to test and even disprove them. "Although yet of uncertain value," he wrote, the generalizations "are cast here into the form of 'laws' in order to stimulate critical penetration of the vast body of information that is being accumulated. Not until we have succeeded in the formulation of a body of specific laws, which are recognized as valid by all, can we expect to arrive at an intrinsically satisfactory solution of the problems of geotectonics." Bucher's laws were meant to play the role of hypotheses in organizing information and stimulating testing of ideas. He wanted to foster an atmosphere in which the idea of having laws would seem to be as natural in geology as it was in physics. He chose deliberately to describe his statements as laws as a kind of provocation—to stimulate his colleagues into a stance of consciously entertaining and testing specific, well-articulated theoretical ideas.

Just as van Hise wanted geology to operate according to "definite rules" and Schuchert defended uniformitarianism as one of those rules, Bucher wanted to move geology away from the rule of men to the rule of law—or what he perceived as the rule of law. He wanted his science to be, as philosophers would soon describe it, deductive and nomological. In a rare example of an American geologist quoting a philosopher of science, Bucher drew on the words of Alfred North Whitehead: "Laws . . . keep alive that 'atmosphere of excitement,' which 'arising from imaginative consideration, transforms knowledge.' In which 'a fact is no longer a bare fact; it is invested with all its possibilities. It is no longer a burden on our memory; it is energizing as the poet of our dreams, and as the architect of our purposes.'"[100]

Bucher's treatise was not exactly a success—his laws and opinions were more forgotten than refuted—but his impulses are enormously revealing. For his work speaks eloquently to the tension many geologists felt: on the one hand committed to a field-based enterprise grounded in experience and observation of the natural world, and on the other hand feeling that science should be made of firmer stuff.

10

The Depersonalization of Geology

Some historians have concluded that plate tectonics caused a change in the standards of the geological community, but the shift in standards of the American scientific community was not so much the result of the development of plate tectonics as it was a larger trend that helped to cause it.[1] Geologists consciously chose to move their discipline away from observational field studies and an inductive epistemic stance toward instrumental and laboratory measurements and a more deductive stance.[2] This shift helps to explain why geologists felt compelled to attend to the demands of geodesists even at the expense of their own data: it was the geodesists' data, rather than their own, that seemed to be in the vanguard of their science.

From the Field to the Laboratory

Geologists at the start of the twentieth century had high hopes for their discipline, and they were not disappointed. The Carnegie Institution's Geophysical Laboratory became one of the world's leading locales for laboratory investigations of geological processes, and work done there inspired scientists at other American institutions.[3] At Harvard, for example, Reginald Daly joined forces with Percy Bridgman to raise funds for a high pressure laboratory to determine the physical properties of rocks under conditions prevailing deep within the earth.[4] The application of physics and chemistry to the earth was also advanced at the Carnegie's Department of Terrestrial Magnetism, where scientists pursued geomagnetism, isotopic dating, and explosion seismology.[5] By mid-century, the origins of igneous and metamorphic rocks had been explained, the age of the earth accurately determined, the behavior of rocks under pressure elucidated, and the nature of isostatic compensation resolved, largely through the application of instrumental and laboratory methods.[6] Similar advances occurred in geophysics and oceanography. The work that Bowie and Field instigated in cooperation with the U.S. Navy, and that scientists at places like Wood's Hole and the Scripps Institution of Oceanography greatly furthered, had grown by the 1950s into a fully fledged science of marine geophysics and oceanography with abundant financial and logistical backing.[7] This work—in gravity, magnetics, bathymetry, acoustics, seismology—relied on instrumentation, much of it borrowed from physics.

Van Hise and his colleagues had never argued for laboratory methods as a *replacement* for field geology. They saw it as a complement to it. Their successors, however, increasingly viewed the matter in terms of competing alternatives, of new methods replacing the old. The result was that, over the course of the mid twentieth century, there was an uneven yet unmistakable decrease in the perceived centrality of field work in the geological sciences. Perhaps the strongest claim in this regard is the title of the 1984 memoir by Francis J. Pettijohn (1904–), *Memoirs of an Unrepentant Field Geologist*. A sedimentologist at the University of Chicago and later at Johns Hopkins University, and for several years the editor of the *Journal of Geology*, Pettijohn in 1949 wrote the most widely used textbook of sedimentology of the postwar period. An undergraduate at the University of Minnesota and a graduate student at U.C. Berkeley, Pettijohn had done his Ph.D. on Precambrian sedimentary rocks, a project that was "largely mapping."[8] But when he joined the Chicago faculty in 1929, he found a department in which comparatively little fieldwork was being done. Despite being "meagerly equipped," the focus was on laboratory studies. The geology faculty even wore lab coats![9] Field geology was taught only in the summer, rather than as an integrated part of regular classes, and, after World War II, not taught at all. Mapping, field work— "outcrop geology" in general—these were considered "something we are trying to get away from."[10]

Pettijohn had been hired on the strength of his Ph.D. work, which was theoretically significant in demonstrating the existence of sedimentary processes in earliest earth history, showing therefore that the surface was neither too hot nor too cold to sustain liquid water. But he came to feel defensive about a project so largely based on mapping. He also felt sheepish about having found his thesis topic "by chance"—i.e., by observation—rather than having articulated a problem first and then finding a way to pursue it.[11] At Chicago, science was to be done another way. He thus turned to the laboratory: "Chicago forced me to turn in new directions. One of these was the laboratory analysis of sediments . . . The illusion that to quantify the subject was to make it a science spurred us on."[12]

Although Pettijohn interpreted his experience as an encounter with the ideology of quantification, the larger historical context suggests that events at Chicago were part of a broader move in the earth sciences to the laboratory from the field, with the concomitant values of exactitude and control that laboratory work suggests.[13] There are aspects of field work that can be quantified, but Chicago geologists were not striving to make their work in the field more quantitative, they were striving to remove it from the field altogether.

Pettijohn moved from Chicago to Johns Hopkins in part to find an atmosphere more congenial to his methodological preferences, and in retrospect he emphasized that not all efforts at quantification are productive (recall Elie de Beaumont's *réseau pentagonal!*). "The innumerable grain-size analyses by sieving and other methods," he wrote in 1984, "have not . . . led to unambiguous identification of either the agent or the environment of deposition, despite some lingering faith that they can."[14] And he bemoaned what he now saw as his own minority position. Geological mapping has "fallen out of favor," he concluded ruefully. "It is regarded as a routine activity, not oriented to the solution of some problem or the answer to a significant question. It is just 'coloring a piece of paper.'"[15]

The work of historians Martin Rudwick, James Secord, and David Oldroyd has left no doubt that mapping is a theoretical as well as an observational activity, far from "just coloring a piece of paper." Geologists themselves know this: it becomes self-evident when one pauses to consider the theoretical results that emerged from mapping the structure of the Swiss Alps or the Himalayas, the stratigraphy of the earliest Precambrian, or the alteration sequences of porphyry copper deposits, to cite a few examples. Nevertheless, Pettijohn was not alone in his experience. In a 1964 lecture, "Is geologic field work obsolete?," noted field geologist Charles Park discussed a survey conducted by the American Geological Institute in which respondents were asked to discuss the nature and future of geology. To Park's chagrin, many believed that geology in the future would be almost entirely a laboratory science. "Field work," one suggested, "is an escape mechanism used by geologists to avoid serious scholarship."[16] While this may have been an extreme view, many American geologists tell anecdotes about the decline of fieldwork. One professor at Stanford recalled that, until the late 1960s or so, it was easy to find research topics for graduate students: you sent them out to "map a quadrangle," confident that interesting theoretical issues would arise.[17] But by the 1970s, students were being advised not to pursue "outcrop geology"—meaning geology in which problems were defined by the outcrop, rather than by the scientist. Others noted changing epistemological preferences as reflected in dress: at national meetings, some said, it used to be that nearly everyone wore boots, jeans and flannel shirts, as if they had come straight from the field. Whether or not geologists actually did field work, they dressed as if they did. By the 1980s, such pretenses no longer seemed necessary.

None of this is to imply that fieldwork ceased altogether. To be sure, many geologists continued to engage in field studies; what changed was the communitarian valuation and centrality of those studies in comparison to other methodological options. Nor should one imagine that field geologists did not defend their methods. Of course they did. The pejorative term "outcrop geology" was matched by the term "box-of-rocks geology" applied to the work of laboratory geologists. Those who affirmed the primacy of field experience frequently disparaged laboratory work as *disconnected*, because it can be done by scientists receiving samples in the mail, out of their natural context. Field evidence, some argued, was *in* context, and therefore more likely to be understood. As in the nineteenth century, each group affirmed the values and strengths of its chosen methodological approach and implicitly or explicitly denied the values and strengths of others. Laboratory scientists promoted exactitude, precision, and control; field geologists promoted authenticity, accuracy, and completeness. But, as compared with the nineteenth century, the balance had tipped. Field work, once unquestionably the backbone of geology, no longer was.

Contemporary attitudes toward field mapping are perhaps best illustrated by the comments of an anonymous reviewer who was asked to judge a proposal submitted to the National Science Foundation by a distinguished geologist in the mid-1980s. The geologist, famous for sedulous studies of ore deposits, had requested funds for a large project that would involve both field mapping and laboratory components. The reviewers had few complaints about the laboratory work, but one reacted viscerally to the mapping. "There is considerable question in my mind whether NSF should be providing funds for the basic mapping of a large mine," the reviewer wrote. "This is *hardly* research."[18]

The Discountenance of du Toit

A shift in methodological standards accounts for a remaining conundrum in the history of continental drift: the discountenance of the work of Alexander du Toit. Throughout the 1930s and into the 1940s, du Toit continued to promote the stratigraphical evidence of continental drift—to show how continental drift could make sense of the continents. Yet, his arguments had scant impact on American thinking, even after the deaths of early opponents like Bowie and Schuchert. And unlike the *S-21* expedition, which fostered an entire research tradition, the work that du Toit completed in South America was not followed by further studies of the same type.

After publication of his South American monograph, du Toit's reputation as South Africa's leading geologist became firmly established. He served as President of the Geological Society of South Africa (1927), President of the South African Geographical Society (1932), and President of the South African Association for the Advancement of Science (1934). He also earned wide international recognition. Beside the support of the Carnegie Institution, he received the Murchison Medal of the Geological Society of London (1933) and served as one of the principal hosts when the International Geological Congress met in Pretoria in 1929. He also attended the IGC when it met four years later in Washington, D.C. His 1926 *Geology of South Africa* became the standard reference on the subject; Reginald Daly, reviewing it for the *Journal of Geology*, called it a "perfect . . . survey of one of the most instructive parts of the planet."[19] Yet, in spite of du Toit's widening reputation, he remained nearly alone in pursuing the comparative geology of the world in terms of continental drift.

In 1937, du Toit published what he considered to be his magnum opus. *Our Wandering Continents* attempted to present a comprehensive, coherent, and well-documented account of the evidence and explanations of continental drift. He reminded his readers that drift stood in opposition not to an existing unified theory, but to a confusion of competing and frequently conflicting ideas. "What has to be termed 'current theory,'" he wrote, "is no general doctrine, but embraces the most diverse of views":[20]

> As individually interpreted, geosynclines and rift valleys are ascribed alternatively to tension or to compression; fold-ranges to shrinkage of the earth, to isostatic adjustment or to plutonic intrusion; some regard the crust as weak, others as having surprising strength; some picture the subcrust as fluid, others as plastic or solid; some view the land-masses as relatively fixed, others admit appreciable intra- and intercontinental movement; some postulate wide land-bridges, others narrow ones, and so on. Indeed on every vital problem in geophysics there are . . . fundamental differences of viewpoint.[21]

Like Wegener before him, du Toit suggested that drift theory could reconcile many of these differences. Why scientists continued to resist it was a vexing question. The answer, he suggested, was complex. In part, it was a matter of "conservatism," the general tendency of ideas to become "entrenched."[22] But he also saw substantive issues.

The lack of a mechanism for drift was one, but du Toit argued that if geologists had accepted global glaciation on "inductive grounds"—that is, without a causal explanation—they could do the same for drift.[23] Another issue was uniformitarianism, which he suggested had exerted a "cramping influence" on both geologists and geophysicists. It was time to dismiss the shibboleth of the present being the key to the past and acknowledge that the earth today was not the same as it once was, nor did it need to be:

> Having passed through various revolutions [he wrote, following Joly and Holmes], the present Earth is at one particular stage in this, the latest, cycle, yet seldom is it recognised that the conditions today within and below the crust may not be constant but experiencing related cyclic changes also. During various past epochs such things as internal temperature, geothermic gradient, depths of zones of discontinuity, degree of isostatic adjustment, state of magmatic activity, etc., might have been somewhat different to those of the present, just as is admittedly the case with climate as well as the rates of erosion and sedimentation.[24]

Drift, du Toit argued, needed to be evaluated on the basis of the evidence. Wegener had made mistakes in his presentation of the data, but more and better evidence was now available, in large part thanks to du Toit's own work. "Wegener was much criticised for the incorrectness of his facts," du Toit wrote privately to a friend, explaining his strategy, "& I have set out to produce stacks of them for orthodox minds to explain or refute."[25] If only the evidence were set out fully, clearly, and without error, perhaps people would become convinced.

The bulk of du Toit's book was thus dedicated to a detailed recounting of the geological evidence for the prior existence of Gondwana and Laurentia (a northern supercontinent that embraced Europe and North America). This comprised over one hundred pages. The remainder of the work attempted to demonstrate how the theory of drift explained these data, and a final chapter discussed possible causes. While arguing for drift on phenomenological grounds, du Toit nevertheless acknowledged that many colleagues would more readily accept it if an adequate mechanism existed. Here he followed Holmes. The most likely explanation, he wrote, was that "under differential heating a *system of magmatic streaming* must develop."[26] Du Toit republished Holmes's 1933 illustration of convection-driven drift and suggested that convection currents, supplemented by density changes due to melting and recrystallization and isostatic effects arising from sedimentation and erosion, could drive the motions of the continents (figure 10.1). In conclusion he suggested presciently that the most important impact of theory would ultimately be economic, as geologists one day would use it to guide exploration for minerals and oil in the displaced fragments of once unified land masses.

In South Africa, reviews of *Our Wandering Continents* were highly favorable, and the book was frequently cited by supporters of drift in the 1940s and 1950s.[27] T. W. Gevers, Professor of Geology at the University of the Witwatersrand, called it a "brilliant exposition" by a geologist possessed of the "spark of genius."[28] Du Toit received many laudatory letters from European colleagues, such as Philip Kuenen, who—while still somewhat skeptical—felt that du Toit's "marshalling of the points in favour of drift makes it even more probable to me that it has actually occurred."[29] Du Toit soon became a one-man clearinghouse for geologists around the world who found support for the theory in their own work. In Africa, India, Australia, and South America, field

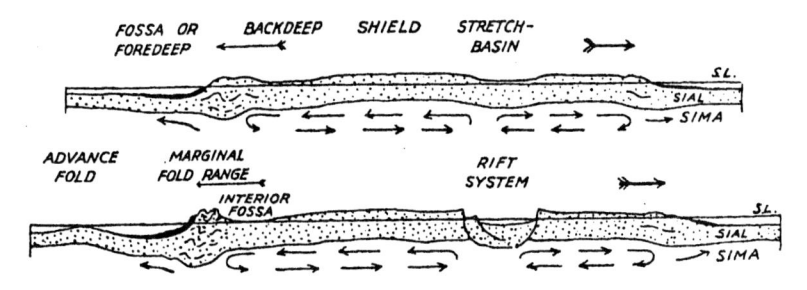

Figure 10.1. Du Toit's model of continental drift. Arrows represent the direction of crustal motion: rift valleys form over areas of ascending convection currents and crustal extension; mountains form over areas of descending convection currents and crustal compression. (From du Toit 1937, p. 325.)

geologists and paleontologists, men working for oil and mining companies and for national and local surveys, sent du Toit letters, photographs, even rock and fossil samples, in support of continental drift. British colleagues expressed support as well.

There were also copious reprints. One example is an article entitled "The Origin of Forces Responsible for Disruption of Continents, Mountain Building, and Continental Drift: Origin and Permanence of Ocean Basins and Distribution of Land and Sea," by Dr. H. L. Chhibber, published in the *Bulletin of the National Geographical Society of India, Benares*, in May 1947. The article began, and du Toit underscored, "It is generally familiar that the shape of the continents has been changing.... It is also true that the land-masses on breaking up underwent *drift*." Chhibber related the Deccan Traps to an extensional event associated with the breakup of Gondwana and argued that compressional forces arising from continental drift had their complement in tensional forces manifested in flood basalts. Du Toit marked the following passage: "Just as the continental masses ... have not occupied the same position, likewise the oceans have never been permanent. For example, the distribution of land and sea during the Gondwana times was absolutely different from that of today.... The permanence of ocean basins ... appears a definite myth."[30]

But only a few letters came from America. Daly wrote to congratulate on "all sides"; Gutenberg wrote to "agree in general" with du Toit's exposition.[31] Howard Baker echoed Du Toit's sentiments about uniformitarianism, complaining about American "uniformitarian penuriousness" and Schuchert and Willis's "almost missing links."[32] Baker wrote: "To get right up and call Lyellian uniformitarianism in question certainly takes something that few geologists in this country possess. I guess they all like their jobs too well.... Lyell's ghost still haunts."[33] Frank Taylor, writing only a few months before his death, expressed gratitude for the "full measure of recognition" du Toit had given him.[34] But these letters from America were exceptions; most of du Toit's correspondence came from elsewhere. From his perspective, as he received letters of support from at least three of the four corners of the globe, it began to seem as if *only* Americans still opposed the theory.

Bailey Willis had been in correspondence with du Toit since meeting him in 1929 at the International Geological Congress in Pretoria, and the two renewed their acquaintance when du Toit visited the United States for the next IGC in 1933. After the congress, du Toit made a grand tour of the United States, including a visit to Stanford University. Willis at this point apparently imagined du Toit to be a follower of Suess, for shortly before du Toit arrived, Willis wrote him a letter in which he reiterated that it was "impossible that any continental areas should have sunk to become the bottom of an ocean."[35] Du Toit did not record the substance of their discussions when they met in Palo Alto on August 9th, but they surely ranged far and wide; at the end of the day, du Toit wrote simply in his diary, a "memorable day!"[36] But neither man changed the other's mind. Three years later, they were still at an impasse. Drift, du Toit wrote to Willis in September of 1936, "affords the simplest and most logical interpretation of the evidence. . . . The hypothesis provides the only intelligent picture of earth evolution that has yet been formulated. One can predict so much & by searching the literature find fact upon fact in support thereof."[37] But Willis made it clear that his mind was made up. In November of 1937, after receiving a copy of *Our Wandering Continents*, he replied by way of closing off further discussion:

> It is not surprising that you have difficulty accepting notions founded upon . . . a permanent continent, since you have long been convinced of the instability of continents. I recognize, of course, that each of us is the victim of circumstance, and that our convictions are more or less imposed upon us by bent of mind and experience. I therefore shall not dispute with you the question of whether or not continents can drift but would assure you that I appreciate the courage with which you have set forth the state of the problem. You certainly have succeeded in putting the best possible face upon your solution of it.[38]

Charles Schuchert replied to *Our Wandering Continents* rather similarly. "You surprised me greatly by presenting me with a copy of your interesting book . . . and I thank you for it," he wrote in December of 1937. "I have now finished reading it from cover to cover, but after your kindness I can not write a rejoinder. You know where I stand." Schuchert instead found what there was for him to praise in du Toit's work. "I like your enthusiasm," he privately allowed, "and surely you have presented the drift theory in its best form. I like your honesty, because you now tell us where you stand and why you prefer to stand on a sliding base. . . . Your unlocking of the drift theory permits us to see your view of how the continents of Gondwana have flown away from Darkest Africa."[39] But du Toit's view was not Schuchert's view. Like Willis, Schuchert was unmoved.

Du Toit was frustrated by what he saw as American intransigence, particularly when antipodean colleagues increasingly spoke of drift as if it were established fact. What irked him especially was the suggestion that he was somehow lacking in necessary detachment while his opponents were not. Du Toit's American supporters had always defended his scientific *equanimity*: Wright had assured Merriam that du Toit did not "go overboard;" Daly, in reviewing *The Geological Comparison of South America with South Africa*, insisted that du Toit showed "a commendable restraint and impartiality of judgment."[40] But more typical was Schuchert's earlier public accusation that du Toit, like Wegener, was guilty of "unbounded enthusiasm and . . . special pleading."[41] Similar was the complaint of Chester Longwell on reviewing *Our Wandering Continents* in 1938 that "du Toit's presentation . . . betrays somewhat too clearly the viewpoint of the zealous advocate."[42] Why, du Toit wondered, could one be adamant against drift but not adamant for it?

In 1943 and 1944, the issue of American attitudes surfaced again. These were war years; Hess was at sea, Veining Meinesz was underground with the Dutch resistance, and many scientists were engaged in war work. Nevertheless, some academic research went on, and after Schuchert's death in 1942 the paleontological argument against drift was taken up by American paleontologist George Gaylord Simpson. In 1943, Simpson published an article in the *American Journal of Science* entitled "Mammals and the nature of the continents."[43] An expert on Cenozoic mammals at the American Museum of Natural History, Simpson argued that evidence from mammalian evolution did not support drift. He had performed a quantitative analysis of species distributions and concluded that they were consistent with the present continental configuration. Minor overlaps could be explained by occasional "rafting"—chance dispersal of species, perhaps on floating logs.

Du Toit was indignant. Simpson had framed his response on the now anachronistic 1924 Skerl translation of Wegener and ignored du Toit's more recent first-hand work. Moreover, there was a fundamental and obvious flaw in Simpson's logic. By all accounts, mammals evolved *after* the breakup of the continents. Both supporters and opponents of drift agreed that the Gondwana connections were severed by the end of the Cretaceous period. The Laurentian connections were somewhat less certain, but they appeared to have been severed early in the Tertiary. It was hardly surprising, du Toit argued in a published reply in 1944, that mammalian distributions were consistent with the present configuration of the continents: mammals *evolved* under this configuration. "The bitter fact," he concluded, no longer able to contain his frustration, "is that current Geology, sublimely unconscious of its impotence, is wholly unable with the continents remaining fixed to account for [global] tectonics ... in a physical, logical, and convincing fashion!"[44]

Simpson by this point had been inducted into the U.S. military, and, with Schuchert gone, Chester Longwell decided to reply to du Toit "for the good of geological science."[45] Despite his earlier enthusiasm for drift, Longwell had by this point calmed down considerably and advised du Toit to do the same. While calling du Toit his "good friend," Longwell nevertheless criticized his "open zeal," which he contrasted unfavorably with Simpson's "cool objectivity."[46] "The concept of drifting continents will be strengthened only by establishing a body of incontrovertible evidence in its favor," Longwell wrote patronizingly, "not by reiteration of diffuse and qualitative arguments."[47]

The alternative to diffuse qualitative arguments was sharp quantitative ones, and Longwell singled out for praise Simpson's mathematical analysis of the degree of similarity of Triassic reptilian fauna in South Africa and South America. "Simpson, reducing the available information to a quantitative basis, shows that [only] 43 per cent of families and 8 per cent of genera are common to the two regions, with no identical species."[48] This, he argued, gave one pause about the supposed homological relations: perhaps paleontologists had become overly focused on a few arresting resemblances that were not statistically significant. Following his call for quantitative measures, Longwell noted the "special appeal and importance" of the telegraphic longitude measurements, which, alas, had produced ambiguous results.[49] Longwell thus counseled patience. "I cannot believe," he concluded, "we have arrived at a stage that permits discarding of the method of multiple hypotheses. In this study the several concepts at our disposal are essential tools, which should be sharpened continuously with the abrasive of objective criticism."[50]

Du Toit could scarcely bear this. Why was *his* criticism considered partial and Simpson's not? "Longwell . . . takes me to task, largely, for my lack of 'objectiveness,'" he complained to his South African colleague, Lester C. King, Professor of Geology at the Natal University College in Pietermaritzburg, but why was there was no comparable criticism of the other side? "'Orthodox' persons have had their minds so moulded by 'Conformity,'" he bemoaned, "that they continue to regard the sponsor of every 'Unorthodox' theory as not *impartial*."[51] Du Toit's criticism was not ungrounded: that very month in the *Journal*, Bailey Willis dismissed continental drift as a "fairy tale," yet Longwell would have nothing to say about his lack of detachment. King shared du Toit's frustration with Americans, a frustration he had commented on in earlier correspondence and blamed on American parochialism. "The Americans are about the toughest isolationists in existence," he concluded, "geologically as well as politically."[52]

Quantifying Quality

Did Americans equate objectivity with quantification? It seemed to du Toit that they did. His initial reaction was to suggest that the opponents of drift were dazzled by mathematics. In a response to Longwell published in July 1945, du Toit noted that mathematical analysis was at the heart of the mechanism objection, yet these analyses were based on a large number of unverifiable assumptions, one or several of which might be wrong. "Too much has been made of the impossibility of [crustal] sliding because adequate forces have not yet been disclosed by mathematical analysis. All such calculations have been based on assumptions that, considering the complexity of the problem, are . . . in certain cases extremely doubtful."[53]

But while arguing that mathematical analyses *might* be flawed, du Toit could not demonstrate that they *were* flawed, except by reiterating the phenomenological evidence. He thus tried to remind Longwell of how much evidence there was—more than was even contained in the now copious literature on the subject. "Keeping in close touch with progress in other lands," du Toit wrote to Longwell in September 1944, "especially those of the S[outhern] hemisphere, I could fill pages with data on the 'Comparative Geology of the Earth' that are not only new but in the main favor drift." But he now saw that this would have no effect. "With regret I have come to realise that such laborious syntheses are unwanted, and the labor involved seemingly futile."[54] Apparently no amount of geological data—no matter how copious or well-documented—would break the impasse.

Du Toit decided to try another approach. He would take on the opponents of drift on their own terms. His first impulse was to attempt to prove that convection currents could drive continental motions. For years he had closely followed the work of Joly, Holmes, Hess, and Vening Meinesz, as well as that of various German scientists who had written on convection, heat, and gravity, and he kept a thick file of notes, reprints, and letters on these subjects. He had also met Vening Meinesz when the latter had stopped in Johannesburg on one of his submarine circumnavigations, and the two had corresponded throughout the 1930s. In 1936, Vening Meinesz told du Toit that he considered convection currents the "best explanation" for gravity anomalies, but he admitted that they were nevertheless speculative.[55] Kuenan expressed similar feelings. But Holmes had argued that convection was more than a speculation: it was a necessary consequence of the earth's internal heat. Was there some way to show this? Du

Toit considered the idea, but in the end acknowledged that there was little he could do on the geophysical side of the problem. He could, however, address the geological side. Du Toit decided to try to quantify the qualitative evidence of drift.

Supporters of drift had always emphasized the *unlikelihood* that close homological relations could arise by chance, and du Toit began to play with the idea of using probability theory to quantify this unlikelihood. Wegener himself had suggested something along these lines in the third edition of his book. Discussing the correspondences of the South Atlantic coast about which du Toit would later become expert, Wegener argued that their sum total might "yield a proof," even if any one individual observation would not. Continuing his analogy of a newspaper torn in half, in which the lines of print can be accurately matched, he wrote: "If but a single line rendered a control possible, we should have already shown the great possibility of the correctness of our combination. But if we have n rows, then this probability is raised to the nth power."[56]

Could one quantify the quality of resemblance? Du Toit floated the idea past Daly in May 1943. "I feel that a mathematical study of the hypothesis, based on the theory of probabilities, using some of the tabulated similarities in my book, would bring out a remarkably high ratio in favour of the theory."[57] Daly, now emeritus at Harvard, was dubious. Distracted by the illness of his wife and demoralized by the war, he doubted that any such project was possible. The Harvard department, he wrote, was "all shot to pieces. Graduate students number 4 or 5, perhaps, and none is reported to have brains enough to attack such a problem as cont[intental] drift on a quantitative basis." Having failed at his own attempt to convince geologists of the possibility of drift, he had lost some of his earlier enthusiasm. "All the paleontological evidence in the world will not convince until the geophysical problem is solved," he lamented, "if this be possible at all."[58]

But du Toit was not deterred, and in any event the task would be a distraction from the tedious war work on asbestos he was now doing and the distressing lack of correspondence from colleagues in Europe. In late 1944 or 1945, he began work on a manuscript entitled "On the mathematical probability of continental drift." The crux of his thinking was as follows: other things being equal, one expects plants and animals to be most similar to their nearest neighbors and most different from those far away. Cast mathematically, one might say $S = f(D)$, where S is the degree of similarity between forms and D is the distance between them. If one could somehow quantify the degree of similarity among forms, and compare this to the distance between them, homologies resulting from continental drift would appear as mathematical anomalies. How to do this, however, was not obvious. "It is far from easy," he wrote in his notes, "instituting a scheme of mathematical comparison (probability) out of relative amounts of likeness or unlikeness a) of the opposed shorelines and b) of their hinterlands [and of] the relative weighting that should be allotted to large and small points of resemblance." If it could be done, however, "it should go far to dispel much doubt."[59]

Du Toit considered various options: a digital scheme, where assemblages would be rated +2 if homologous and –2 if not; a graduated scale from –3 for no resemblance to +3 for extremely close resemblance; and a multivariate scheme, in which he would evaluate resemblances in five categories: tectonics, stratigraphy, climate, volcanism, and paleontology, on either the digital or the graduated scale. These measures of similarity would then be weighted by a factor: l^2/d, where l was the present distance between locales and d the average transverse distance between outcrops in reassembly.

The farther the distance today, the less likely the resemblance arose by chance. The larger this number, therefore, the greater the significance of the homologies.

Du Toit settled on the digital scheme (2 points if homological, –2 if not), applied to all five categories of potential resemblances. For any given locale and time period, one could tabulate the number of homological relations and the number of non-homological relations of all types. If the former be labeled x and the latter y, then the degree of resemblance might be defined as $2^x/2^y = 2^{(x-y)}$. Multiply this number by the calculated length factor to obtain a number representing the *unlikelihood* that the observed similarities have arisen by chance. In the margins of the penultimate page of his own copy of *Our Wandering Continents*, du Toit penned "mathematical probability of C[ontinental] D[rift] . . . For example, on no. of relations Ch VI = $2^x/2^y = 2^{(x-y)}$ which is so huge that [it] gives practical certainty. Argue this point by examples."[60] Du Toit was evidently planning a revised edition of his book, in which he would make this argument. But it was not to be. Before finishing the project, du Toit died at his home in Pinelands, near Cape Town, in 1948.

Objectivity and Distance

Alexander du Toit's attempt to quantify the qualitative evidence of drift was not successful, but like Bucher's effort at nomothesis it speaks to feelings of need. Geology had long been a *personal* science, in which individuals experienced the natural world with their own eyes, hands, and feet. One walked the ground, one collected samples, one examined and observed. One developed geological intuitions through the unconscious analyses of experience. Geological evidence was hardly ever mathematical, it was almost always circumstantial.

Arthur Holmes argued in 1929 that the "circumstantial evidence of geology is not likely to lead us far astray, so long as we read it right," but others were not so sanguine.[61] How did one know if evidence was being read right? The traditional answer was to "read" it for oneself. Geologists felt sure of the things they had *seen*, and a common tactic for discrediting opponents was to accuse them of not having seen enough—of being armchair geologists. Accordingly, professors advised their students to be greedy for experience: H. H. Read famously professed that the best geologist was the one who had seen the most rocks; Charles Lyell was frequently quoted as having advocated "travel, travel, travel."[62] Du Toit was logically the last geologist in the world who should have been accused of familiarity with only furniture, but when Bailey Willis wished to discredit him in the 1940s, he found a way: he accused du Toit of being a disciple of an armchair geologist, "Eduard Suess . . . a charming, genial German, who never traveled far."[63]

The logic of seeing for believing was clear enough, but the practicalities were another matter. How could a science be built in which everyone had to see everything? How could such a science progress? Historians Steven Shapin and Simon Schaffer have lucidly described experimental demonstrations in the early years of the Royal Society, in which the gentlemanly members served as witnesses to attest to the purported phenomena. In time, actual witnessing was increasingly replaced by "virtual" witnessing: scientific reports were written with sufficient detail of description so as to give the reader the impression that he or she was there. As in legal affairs, the more witnesses (albeit of appropriate stature) the more likely the events attested to were true.[64]

But in geology, witnessing was a more complicated affair: outcrops could not be brought into the halls of a learned society. One could bring back specimens, and geologists always did (usually crates full), but to remove samples from the field was to remove them from the context that gave them much of their meaning.

Martin Rudwick has described how, in nineteenth-century Britain, when Henry De la Beche and Roderick Murchison clashed on the interpretation of field evidence, members of the Geological Society of London repaired to Devon to examine the disputed strata.[65] Such field excursions were common then and are significant even today. But as geology grew as a science, and particularly as it grew in North America where scientists and sites were widely scattered, such approaches became impractical.[66] Although geologists went to great lengths to visit type localities and famous sites of past contestation, it simply was not possible for everyone to see everything. There were other resources for conveying field information, to be sure—maps and cross sections existed precisely for this purpose. But how did one know if a map was correctly drawn? How did one know if a cross section was right?

Even if one did go to the field to see for oneself, single outcrops were rarely revealing. Geological interpretations were built on the amalgamation of pieces of evidence—widely scattered and typically observed over many field seasons—and they were not infrequently undermined when new evidence became available. Thus by the early to mid-twentieth century, many earth scientists expressed some degree of dissatisfaction with reliance on field evidence. Writing in his examination papers at Harvard, William Menard concluded that "field evidence is seldom sufficient to make any statement wholly acceptable, and yet seldom sufficient to make it wholly unacceptable."[67] Even van Waterschoot van der Gracht, America's most vocal proponent of the empirical evidence of drift, was reluctant to rely too heavily on the fossil evidence, conceding that "there are few subjects where there exists a greater diversity of opinion regarding practically everything than in paleontology."[68]

Geophysics offered an alternative, one that challenged traditional understandings of "observation." On some level, no one observes geophysical data, or, conversely, everyone does, but from a remote position. This quality of remoteness may help to explain the sense expressed by many earth scientists that data generated with geophysical instruments were somehow more objective. William Bowie explicitly looked to quantification and instrumentation as a way to avoid the subjectivity and ambiguity of traditional field geological evidence. Speaking at the 1937 AGU symposium devoted to the ocean bottom, he

> wondered at the slowness with which the geologists have employed mathematics, physics, chemistry and mechanics in attacks on some of their baffling problems. It is very seldom that students in dynamics and structural geology agree. The reason for this would seem to be that they have not used enough geophysical technique to get the data that they should have in order to interpret what has really gone on in the earth.

Walter Bucher, in his copy of the symposium proceedings, underscored this passage.[69]

"What has really gone on in the earth"—this is what all earth scientists wanted to know, and many, like Bowie, were increasingly trusting geophysical data to tell them. Earth scientists of all stripes came to favor evidence collected with instruments over evidence collected with hands and feet and eyes. Perhaps it was because of the remoteness of such evidence: few could claim—at least in the gross physical aspect—to have a privileged *view*. Or perhaps it was because of the sense that the data were not actually observed at all; they were *collected*. Bowie introduced an epistemological reversal of the roles of nature and laboratory—a reversal that by now has become so commonplace among earth scientists that they scarcely question it: that the earth be viewed as a "natural laboratory." When considering the earth, he suggested, "one is observing the working of the greatest laboratory on earth, with nature as the operator."[70] Laboratories were once viewed as places where men tried to recapitulate the operations of nature. Now the operations of nature were being cast as recapitulations of the work of men. And if the experiments had been performed by nature, then men had merely to collect the results. In principle anyone could do this.[71]

On one level, the appeal to instrumentation makes a certain sense: there is clearly a notion of objectivity that involves distance and detachment, what the historian Lorraine Daston has called the "escape from perspective."[72] By removing the observer one degree further from the thing observed, geophysical instruments seemed to obviate that quality of irreproducible expertise that made geology feel so subjective. As the historian Theodore Porter has put it, measurement contributes to self-effacement.[73] But on another level, the belief or hope that instruments would somehow eliminate the subjectivity of interpretation is highly ironic, for not only is geophysical evidence no less demanding of interpretation than any other kind of scientific data, it is arguably *more* demanding of it. All data collection evokes problems of accessibility, and of linguistic and physical interpretation, but geophysical instrumentation involves at least two additional steps not present in traditional geological observation: instrument design and construction, and signal processing. The introduction of instrumentation does not eliminate the need for interpretation. On the contrary, it increases it. It is one of the ironies of twentieth century earth science that, in the hope of decreasing the subjectivity of their science, earth scientists turned to data that required even more layers of interpretation.

A Culture of Distance?

Earth scientists were not alone in their impulses; historians have documented the transformation that occurred in biology in the early twentieth century, a movement away from morphological and toward experimental traditions, away from whole organisms and toward cells and organelles.[74] As geology increasingly moved from the field to the laboratory, so did biology. The reasons for this transformation are complex, but the patterns of opinion and activity among biologists in the early twentieth century resonate with those of geologists. The desire for greater standardization, the hope that new instruments would lead to new insights and new potential for power and control, and the reductionist impulse to trade generality for specificity and comprehensiveness for precision—all these were implicated in the changes that occurred in biology in the early twentieth century.

Historian Philip Pauly has described the mind-set of the new biology as informed by an engineering ideal, in which scientists asked not so much "What is this?" or "How does it work?" but rather "How can I make this work?" To make things work increasingly required scientists to take an interventionist role toward their objects of study, and this implied the need for greater instrumentation and measurement and therefore a greater overall commitment to the laboratory at the expense of the field. Biologists, like geologists, increasingly sought to study their world under conditions defined by men rather than by nature.[75] Indeed, throughout American natural and social science in the 1920s and 1930s, measurement, mathematization, and mechanization, often hand in glove, were becoming increasingly widespread. On the day in 1925 when Bowie told the readers of the *New York Times* about Vening Meinesz's advances in gravity measurement, the adjacent article soberly informed educated readers of the attempt by a French mathematician to use delicate instrumentation to measure the human soul (figure 10.2).

While soul measurement did not spawn a major research tradition, measurement in other fields did. Philosopher Ian Hacking has referred to the early twentieth century as possessed by a "fetish of measurement" as the techniques of the physical sciences spread not only to the life sciences but to the social sciences as well.[76] Reflecting on this trend, Norton Wise has described numbers as canonical of modernity.[77] Although measurement and particularly mathematization were viewed with skepticism and even dismay by many, they were also seemingly inexorable; in some fields it became nearly impossible not to do quantitative work. Economics particularly embraced quantification in the interest of prediction and control; at the University of Chicago, measurement became a mantra, and Lord Kelvin's famous complaint, that unless one could measure a thing, one could hardly be said to know anything about it, was enshrined as a motto on the facade of the social sciences building.[78]

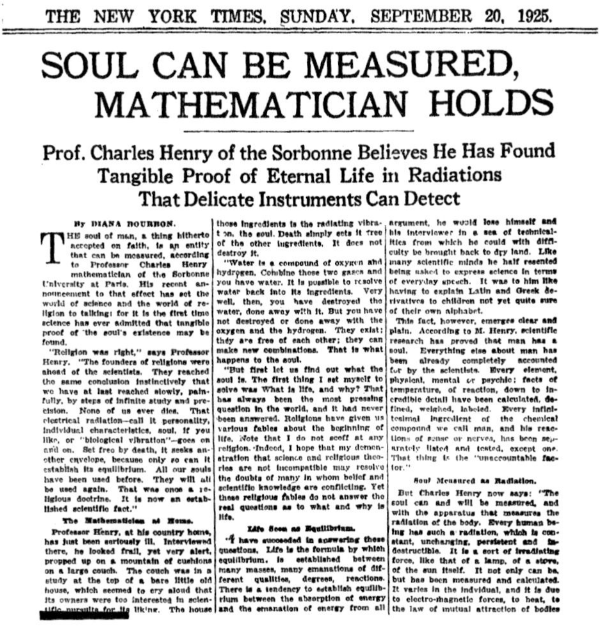

Figure 10.2. "SOUL CAN BE MEASURED." (*New York Times*, Sunday, Sept. 20, 1925. Copyright © 1925 by The New York Times Co. Reprinted by permission.)

These changes reflected the desires of many scientists to emulate physicists and chemists, and thereby, they hoped, to share in their successes; but internal disciplinary impulses were not the only cause of these transformations.[79] Historian John Servos has shown how improvements in mathematical education in American schools and universities circa 1910 both reflected and enabled increasing mathematical prowess throughout American society. Better mathematical education, he has argued, was in part the result of a campaign by the well-financed Society for the Promotion of Engineering Education, which represented the interests of industries that increasingly needed numerate engineers. This led in turn to increasing mathematization of American physics and engineering, as young people entered these disciplines prepared to tackle problems mathematically. Once so trained, American physicists were for the first time able to engage in advanced theory—and they transformed their science.[80] Biologists and social scientists similarly responded to the perceived or expressed needs of the culture at large: the development of standardized tools and techniques in bacteriology was in part a response to the context of public health; the development of standardized techniques in sociology was in part a response to the context of immigration and population growth.[81] Thus, although reflecting the goals and aspirations of scientists to give a reliable and impersonal account of the world, the desire for objectified data was also fostered by the needs of the larger society. Indeed, it is this context that helps to explain why these changes occurred just at this time—even when the techniques being applied had for the most part already been in existence for some time.[82]

Just as the desire of American manufacturers for a better educated workforce had an enabling effect on theoretical physics, and the demands of urban public health had a motivating effect in bacteriology, so the Navy's interest in probing the ocean depth helped to empower instrumental earth science. American geophysicists wanted to go to sea to answer scientific questions, but they could not have done so without the Navy's material and logistical support. But the U.S. military did more than merely foot the bill for scientific curiosity. Proximity to military matters affected how scientists thought about their work, and the desire on the part of American scientists to remove the "personal equation" was surely reinforced by similar concerns in the military.

American earth scientists throughout the late nineteenth and early twentieth centuries had close relations with the U.S. military—geophysicists with the Navy through the Coast Survey and the Hydrographic Office, geologists with the Army through the surveys of the West—and the elimination of the exercise of personal judgment was a pervasive desideratum of the military. Authority is, by definition, achieved at the expense of the will and discretion of the *unauthorized*, and no military organization can afford to have large numbers of its members exercising unauthorized personal judgment. But military concern with judgment and subjectivity went beyond issues of discipline and command; the military also needed to be assured of the reliability of the information on which command decisions were based. Scientists who worked alongside military officers might not have been concerned with discipline per se, but, like their military colleagues, they were concerned about the reliability of scientific knowledge. If the data they were generating were of interest and possible use to the military, then that concern would only have been strengthened.

The phrase "personal equation" appears frequently in publications of the Hydrographic Office and the Coast and Geodetic Survey from the early twentieth century.

A 1904 publication on telegraphic longitude determinations, for example, contains a section by that title. The largest problem in determining the accuracy of longitude measurements, the authors reported, was variation caused by individual observers. But "no correction for personal equation [could be] introduced," the authors explained, because "the observers had no practical method of obtaining a value for this changeable quantity." Lacking a way to calculate the "personal equation," they concluded it was "better to omit it altogether rather than put in an arbitrary value."[83] But ignoring the problem did not make it go away.

When Vening Meinesz and Wright began gravity measurements on the USS S-21, they developed a military-like discipline in which measurement error would be avoided by following a simple but structured protocol. "In the design of the apparatus," they explained, "the effort has been made to reduce to a minimum the number of operations necessary to start the apparatus and to have these operations follow a definite sequence, so the procedure becomes one of routine. This is necessary because on a submarine the mind is not alert and mistakes are likely to be made if anything beyond routine and purely mechanical details is demanded. . . . The operation of the gravity apparatus," therefore, "is made as simple as possible, so that the observer makes each measurement almost mechanically . . . thus eliminating the personal factor in the method."[84]

Mistakes, of course, are hardly confined to submarines, and the desire for protocols, instruments, and techniques to remove the personal equation was increasingly being found in many kinds of scientific work—indeed, many kinds of work in general.[85] Mass marketing, assembly lines, the growth of the consumer economy—these developments all relied upon the capability to produce standard products. As American culture and economy became increasingly standardized, so did scientists strive to produce increasingly standardized knowledge. Instrumentation was a means to the goal of standardization.

Induction and Deduction

If the move toward depersonalized knowledge was sustained by both scientific values and cultural patterns and interactions, so was the move from induction to deduction. Like the shift toward objectification, the move toward deductive methodology was not confined to geology. At least since August Comte articulated his Positive Philosophy in 1842, many if not most philosophers of science assumed that knowledge developed inductively. The early twentieth century was a period of particularly optimistic convictions about the power of observation, indeed, the power of human capabilities in general. Philosophically, it saw the flourishing of logical positivism, the ambitious philosophy that sought to ground all knowledge in verifiable observation statements. Rudolf Carnap, the leader of the "Vienna Circle," was held by the American philosopher W. V. O. Quine to be "the dominant figure in philosophy from the 1930s onwards"; Albert Einstein claimed to have been inspired by positivism in his development of special relativity.[86] Historians, too, were influenced by positivism: in his encyclopedic *Introduction to the History of Science*, George Sarton declared that "my appointed task, the task which justifies my existence, is to prove inductively the essential unity of knowledge and the essential unity of mankind."[87]

By the 1940s, however, philosophers and historians saw a different justification for their existence. Most famously, Karl Popper argued against the notion of inductive science in general and verification in particular; he argued that all science begins with an idea—that theories are the "searchlights" under which scientists view empirical information—and the purpose of empirical evidence is not so much to generate truth as to refute falsehood. Carl Hempel similarly promoted the idea that the true method of science was hypothetical and deductive, rather than descriptive and inductive. By the 1960s, philosophers like Gilbert Hanson and Thomas Kuhn were denying even the possibility of observation without prior theoretical commitment. All observation, they claimed, was "theory-laden."[88] Whether the changed views of philosophers were a cause or an effect of the changed views among scientists themselves, or whether both were the result of still larger cultural patterns, is a question beyond our reach here. The point is that attitudes *did* change, both in philosophy of science and in science itself. What seemed not only right but necessary in the 1920s seemed unnecessary and mistaken in the 1960s.

The Self-Validating Power of Success

The changes that occurred in American science in the first half of the twentieth century help to explain why certain kinds of evidence were pursued and others were not. American earth scientists showed little interest in pursuing further studies in comparative geology, but great interest in gravity and magnetic studies. The latter, but not the former, were consistent with their desire for greater objectification in their science, and they were able to pursue this desire in large part because the U.S. military made it logistically and financially possible. The resulting work substantiated plate tectonics.

Du Toit's effort to mathematize field relations throws into relief the contrast between the traditional methodologies of geology and the desire for objectified data. By objectified, however, one should not necessarily read "objective," for, as emphasized above, the preferred data were by no means immune to interpretation. Rather, they were depersonalized. This is the practice that historian Gerald Holton has described as "a step basic to all scientific work . . . but rarely discussed, . . . the process of removing the discourse from the personal level . . . to arrive at statements which are invariant with respect to the individual observer."[89] Anyone who has ever done geological field work or attempted to describe a rock knows how difficult it is to arrive at a description that is "invariant with respect to the observer."[90] The appeal to instruments is an appeal to a perceived objectivity—to *statements that appear to be observer-independent*. The step of removing the discourse from the personal level occurs in the choice and construction of the instrument.[91] The rejection of continental drift was an implicit rejection of observer-dependent evidence despite the strength of that evidence. The acceptance of plate tectonics was an affirmation of a depersonalized alternative.

Perhaps the most telling example of the impulse toward objectification comes in the early 1960s work of Edward Bullard and colleagues in demonstrating the so-called "Bullard fit" of the continents. As we have seen, few geologists or geophysicists had disputed the observation that the outlines of the continents seem to fit together like the pieces of a jigsaw puzzle, but they had argued mightily over the meaning of that fit. American doubters of drift in the 1920s had claimed that the matches were not as good as Wegener claimed: at the New York AAPG meeting, Rollin Chamberlin pronounced that America and Europe fitted "very badly indeed"; Charles Schuchert cut and pasted a paper globe to show that Wegener had taken "liberties with the earth's ... crust."[92] Those who thought drift plausible faced the criticism of Bailey Willis, who argued that the fits were *too* good, because faulting during breakup and subsequent coastal erosion would surely have modified the continental outlines![93] If the similarity of outline held, then there could have been no significant normal faulting and no movement of one continent away from the other. Schuchert, having first argued that the fits were not good enough, later argued the opposite in this light. "Wegener," he wrote, "wants us to believe that the original fracture lines have practically retained their original geographic shape during 120 million years. Is there a geologist anywhere who will subscribe to this startling assumption?"[94]

The meaning of the jigsaw-puzzle fits was heavily contested, not least of all because there was little agreement on what kind of data were needed. One person looked for similarities of outlines, another looked for differences. Put epistemologically, the issue was worse than whether or not the theory fulfilled its predictions, the issue was what those predictions were. And yet, in the early 1960s, these issues suddenly dropped away, when Bullard and coworkers revisited the question. Responding to Harold Jeffreys's continued insistence that the jigsaw-puzzle fits were really very poor, Bullard turned to computer modeling. Using Euler's theorem—which permits description of motion on a sphere as rotation around a pole to that sphere—Bullard calculated the best fit of the coastlines of North America, South America, Africa, and Europe. The result, published in 1965 (figure 10.3), was considered by many to be a compelling piece of evidence, and it helped turn the tide of geological opinion.[95] But, as Homer Le Grand has noted, the Bullard fit had its problems; the most obvious one in retrospect is the complete absence of Central America! Nor had the theoretical issues gone away: the mechanisms of rifting were scarcely better understood in 1965 than they were in 1925. Yet few earth scientists complained. As Le Grand has put it, Bullard's work had an "aura of objectivity."[96] The data had been objectified, and opposition quickly dissipated.

It is noteworthy that Wegener, too, attempted to objectify the homological evidence with a chart graphically displaying the consensus of expert opinion on paleontological evidence for land bridges throughout geological time (figure 10.4). Beside the paleontological opinion poll, Wegener's later editions also contained a score of testimonials—quotations from geologists whose fieldwork supported his theory, who could not explain *their* rocks without it. By including these testimonials, Wegener revealed that he missed the point. The point, as Bucher labored so valiantly to show, was neither to count nor count on the opinions of men. The point was to try to transcend them.

In hindsight, it might seem that earth scientists made the right choice. Some readers may see here a tale of progress: geologists replaced their hoary but vague field data with the exactitude of instrumental and laboratory measurements, generated reams of new information, and the result was plate tectonics. Earth scientists have in fact read their history this way, and in retrospect they have discredited much of the work that preceded the development of the new global theory. An example of this emerges from the work of John Stewart, a sociologist who interviewed earth scientists in the wake of plate tectonics. As a result of these interviews, Stewart developed a highly pejorative if not disdainful view of geology prior to the 1960s. Work before this, he held, was "trivial"

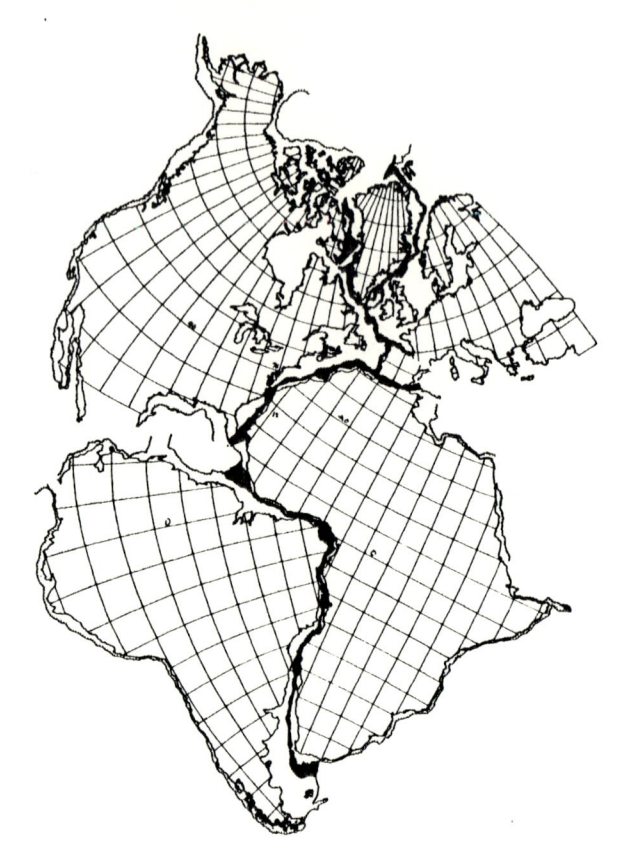

Figure 10.3. Objectification of the evidence of continental drift: the "Bullard fit." Using Euler's theorem, which permits description of motion on a sphere as rotation around a pole to that sphere, Edward Bullard and coworkers calculated the best fit of the coastlines of North America, South America, Africa and Europe. The result, published in 1965, was considered by many to be a compelling piece of evidence and helped turn the tide of geological opinion. (From Blackett, P. M. S., E. Bullard, and S. K. Runcorn, eds. A symposium on continental drift. London: Royal Society, 1965, opposite p. 49.)

and "narrow." But on what basis was such a claim made? It was made on the basis of the retrospective and self-congratulatory comments of his participant-informants. One geophysicist, for example, argued that prior to plate tectonics "there was a retreat . . . into data collection and analysis [instead of theorizing]." But he elaborated: "[But] I didn't see it as a retreat, and other people didn't see it as a retreat at the time."[97] Science that met the standards of the community at that time, that earned awards and prizes and tenure, was retrospectively dismissed as a "retreat." Or consider this exchange, between Stewart, William Menard, and geologist Kevin Burke, on geosyncline theory and the role it played prior to plate tectonics:

> MENARD: That [geosyncline theory] was a model for how you got thick wedges of sediments, not for how you produced mountains.
>
> STEWART: I guess I had a mistaken impression about it.

MENARD: Well, I think everyone, including me, at the time had an impression that it was a way of producing mountains.

BURKE: We had a vague feeling that subsequently these became mountains.[98]

What Menard accepted for twenty years as an orogenic theory—what according to his teachers in graduate school *was* an orogenic theory—he now realized was no such thing! Burke went further, concluding retrospectively, "a lot of what we were doing was fairly unscientific."[99]

There is a sharp irony here. For if the instrumental data of the 1960s are right and continents do move (and no credible scientist today denies it), then the homological data of the 1920s were *also* right. The oceanographic and geophysical work of the 1940s and 1950s generated tremendous amounts of new information, to be sure, but in the end both data sets—the geophysical and the geological—pointed to the same result. In fact, the geological data pointed there *first*. Sociologically, the geophysical data proved to be more powerful in moving the minds of men, but epistemically they proved to be equivalent with respect to moving continents.[100] Moreover, the phenomenological evidence of geology has proved right more than once. Field data may have been vague, descriptive, and imprecise, but in the end they were more often than not *accurate*. By the standards of contemporary knowledge, geologists were right when they insisted that the earth was older than Lord Kelvin's calculations allowed, they were right when they held that the substrate behaved in a fluid manner, and as for those who insisted on the

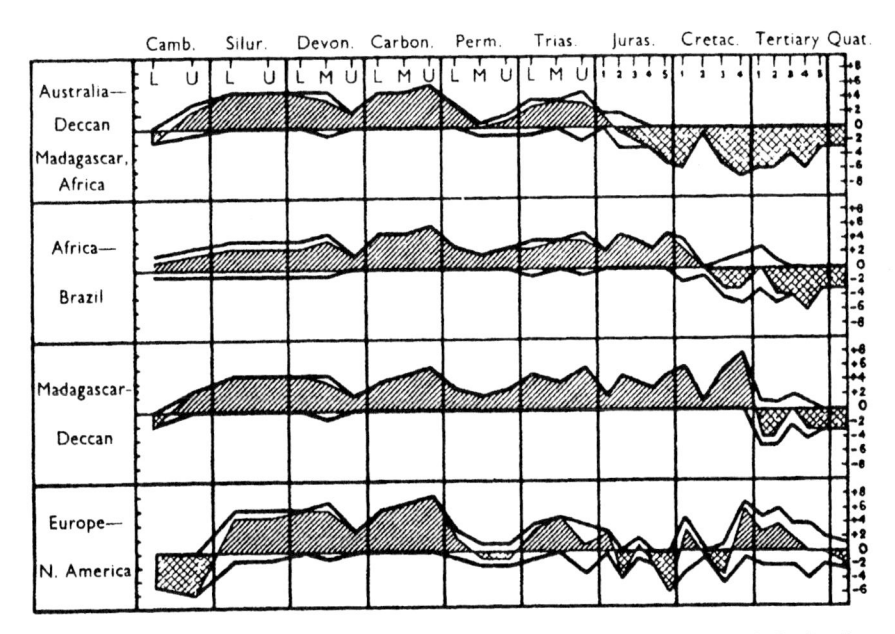

Figure 10.4. Wegener's graphical representation of paleontologists' demands for land bridges from the Cambrian period to the present. The upper curve represents the number of paleontologists advocating the specified bridge during that period of geological time; the lower curve represents those opposing the said bridge. The majority opinion is shown by the hatching: single hatching if in favor, cross-hatched if opposed. (From Wegener [1929] 1966.) Reproduced with permission of Dover Publications, Inc.

reality of continental movement in the face of geophysical opposition—who, like Daly, invoked Galileo's memory and muttered *E pur si muove!*—"nevertheless, it does move"—they were right, too.

But these are victories in retrospect only, for at the time geologists felt browbeaten and defeated. Geologists' conflicts with geophysicists led them not to a greater confidence in their own methods but to an increasing sense that they had no choice but to adopt the strategies and tactics of their opponents. Robert Woodward was right: the thrice slain combatants did rise up again. Only this time not merely to fight but to rule the kingdom.

To What End?

There is an alternative reading to the progress tale. It is the reverse: American scientists went to great lengths, spending large sums of money, huge amounts of effort, and even a few lives, to prove to themselves something they might have known all along. The desire for precision and objectification cost them forty years of scientific advance. So we might well ask, what did earth scientists achieve through the methodological transformation of their science? To what purpose were the changes for which they so mightily strove? Such questions are impossible to answer without venturing into the realm of counterfactual history. We cannot know what might have happened; we only know what did happen: A new era in geology was ushered in not by the elucidation of the ultimate cause of crustal motions, nor even by the mechanism by which they occur, but by the availability of a new kind of evidence. Perhaps the important point is not so much what *kind* of evidence it was but simply the fact that it was of a new kind.

Why did Americans react so differently to two data sets that implied the same theoretical result? One reason has already been made clear: the instrumental data of the 1950s satisfied the desire for objectified data while the field data of the 1920s did not. But there is another, related yet distinct, reason. Geologists in the 1930s, 1940s and 1950s were aware of the homological evidence of continental drift but believed that it was *by nature* inconclusive. Schuchert and Willis had shown that the homological evidence of drift could also be explained by isthmian links. The data did not uniquely entail the theory of continental drift and thus were powerless to compel it. From this position, it was a small step to the conclusion that they were powerless to compel *any* particular theory. Hence Menard's comment about fieldwork, cited above, a comment that could be readily reinforced by any one of the several (in)famous historical disputes triggered by divergent interpretations of field evidence. Even as Menard wrote, another controversy was raging, this one over the origins of granite. As had been the case so many times before, field evidence was invoked on both sides of the debate—to show unequivocally that granite had intruded its surrounding in liquid form, and to show equally unequivocally that it had formed in situ by the selective transformation of particular rock layers.[101]

No matter how much more homological evidence du Toit or others compiled, it had little effect. Once drift had been rejected on the basis of these data, the data themselves seemed powerless to reopen the case. A whole category of data had been rendered incompetent to contribute further to the debate. Du Toit himself ironically contributed to this perception. In writing his South American monograph, he had focused primarily on the structural and stratigraphic parallels between South America and South Africa, on the ground that the fossils could be explained by other theories. In fact they *had* been previously explained by the idea of sunken continents. Only the structural and stratigraphical relations, he argued, were uniquely explained by continental drift. When viewed from the perspective of paleontology, the problem of continental drift was underdetermined, but when viewed from the perspective of structure and stratigraphy it was not. Therefore the latter, but not the former, could be the basis of a strong argument for drift.

By focusing geologists' attention on the aspect of the problem he believed was not underdetermined, du Toit hoped to further the cause of continental drift. But he did the opposite: his position had the effect of helping to discredit data which, albeit imperfect, were a large part of the overall case for drift. For du Toit himself—a supporter of the theory—had judged these data incompetent. Alfred Wegener complained about this in the foreword to the final edition of his book. "*All* earth sciences," he pleaded, "must contribute evidence towards unveiling the state of our planet in earlier times, and the truth of the matter can only be reached by combining all this evidence."[102] But it was too late. Although du Toit's intention had been to convey that paleontological data were incompetent *alone* to resolve the question of drift, geologists leapt to the conclusion they were incompetent *altogether*. By showing how one aspect of the problem was underdetermined, du Toit inadvertently led others to conclude that the whole problem was underdetermined, that the available data were essentially meaningless. Thus it was that Keith Runcorn wrote in 1959 that the evidence from structural geology and paleontology "by its *nature*, could hardly appear decisive to scientists in other fields."[103]

Walter Bucher argued this point in the proceedings of a 1952 symposium on land bridges organized by biologist Ernst Mayr. Claiming that the homological evidence was unable to compel any particular interpretation, he insisted on a distinction between "suggestive" and "decisive" evidence. "Facts that follow by necessity from a hypothesis constitute decisive evidence," he wrote. "The vast geological literature on continental drift deals largely with suggestive evidence."[104] Bucher's logic was flawed: homologies *did* follow by necessity from drift. If the continents had been connected, then necessarily fauna and flora would have been able to migrate, and homological patterns would result. What he evidently meant is that homologies did not necessitate drift. There were other explanations that could accommodate these data, which therefore remained merely "suggestive." Even those who sympathized with the idea of drift in the 1950s felt they had to apologize for the qualitative nature of the evidence.[105] Kenneth Caster, a geologist who had worked in Brazil, where he became convinced of du Toit's correlations, argued in the same symposium that antagonism toward the theory had caused geologists wrongly to discount the data.[106] Like du Toit, Caster sought to mathematize his data, which he presented in the form of a Venn diagram illustrating the degree of faunal affinity between various areas, in the hope that this would somehow show that the data were better than most people thought (figure 10.5).

Gravity and magnetic data were different, not because they did not require interpretation—of course they did—but because they were an entirely new *kind* of data. And because they were new, their "nature" was an open question. Marine gravity data did not exist until the 1920s, and continental gravity data had never been used in a major geological debate. Magnetics were newer still, only first collected in the 1930s

BOREAL REALM

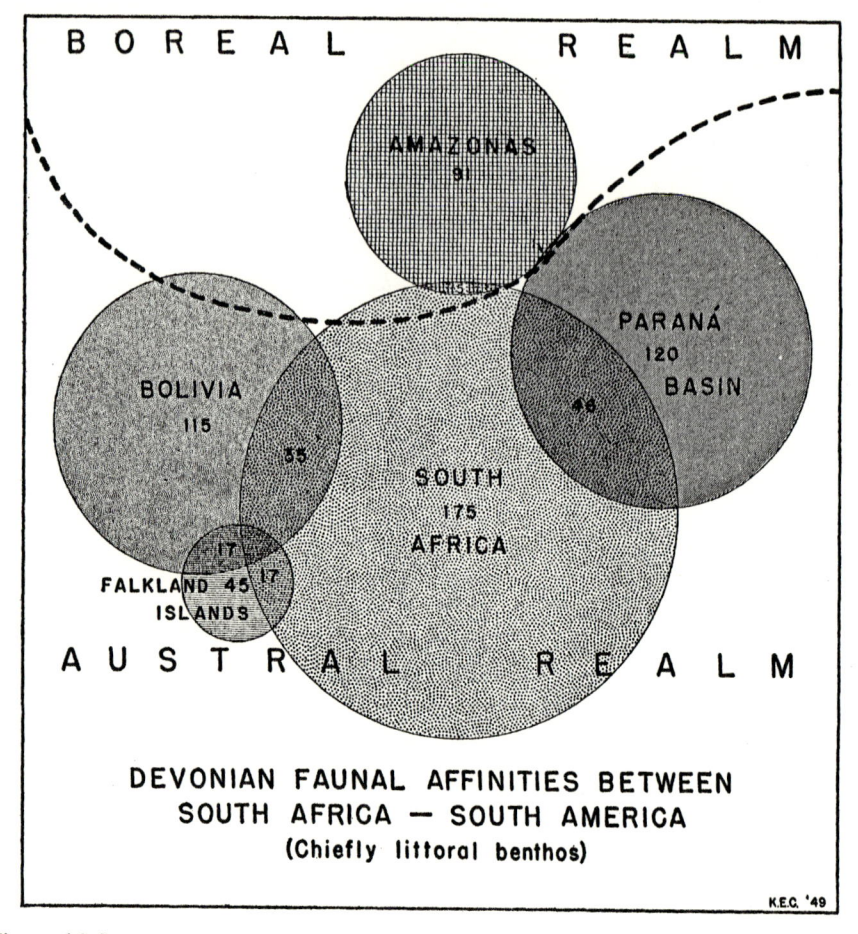

AMAZONAS
91

PARANÁ
120
BASIN

BOLIVIA
115

35

48

SOUTH
175
AFRICA

17

FALKLAND 45 17
ISLANDS

AUSTRAL REALM

DEVONIAN FAUNAL AFFINITIES BETWEEN
SOUTH AFRICA — SOUTH AMERICA
(Chiefly littoral benthos)

K.E.C. '49

Figure 10.5. Caster's 1952 Venn diagram illustrating the degree of overlap between Gondwanaland faunal assemblages. In response to arguments that the evidence of continental drift was merely qualitative and suggestive, paleontologist Kenneth Caster attempted to quantify the degree of overlap between faunal provinces postulated to have been contiguous in the Devonian period. The radii of the circles are drawn proportional to the number of species in each faunal province. (From Caster 1952, p. 116. Courtesy of the American Museum of Natural History.)

and soon thereafter largely sequestered by the military. The potency of paleomagnetic data to resolve the question of migrating continents was entirely to be determined in the future; their relationship to theoretical questions was yet to be established. Scientists could thus approach these data without preconceptions as to their power or lack of power to settle a debate, and without preconceptions as to which side of the debate these data would support. For it was not that *any* of these data exactly compelled plate tectonics. As Patrick M. Hurley, a professor of geology at MIT, pointed out in *Scientific American* in 1968, the new data did "not unequivocally call for continental drift."[107] Indeed, when evidence of seafloor spreading first became widely discussed, the initial impulse of some scientists was to assume that it did *not* entail drifting continents: the ocean floor could circulate in place, like a treadmill, without necessarily pushing continents anywhere. Except for the preexisting homological evidence, one might have had seafloor spreading without continental drift.[108]

Novel Data for Novel Theories?

The history of continental drift and plate tectonics suggests a crucial yet perhaps counterintuitive reason for the valorization of instrumentation in scientific work: new instruments create new forms of data that lack established relations to existing theoretical belief. And this makes it possible to look again at important theoretical questions. This conclusion—that geologists were more receptive to the new geophysical data *because* they were new—in effect unsullied by earlier arguments—goes against the grain of recent historical and philosophical wisdom. In his detailed studies of twentieth century physics, Peter Galison has argued for the role of well-established instruments and techniques in helping to bridge theoretical divides. When faced with contested theoretical claims, physicists can turn to well-trod evidential paths, to machines and tools whose reliability has already been established in other contexts. Continuous experimental traditions may span the gulfs across theoretical breaks and, conversely, new machines and tools can be tested against well-established theories.[109]

Galison's arguments make sense, yet this is not what earth scientists did. They did the opposite. They rejected well-established evidential traditions in favor of a new tradition whose reliability could have been considered as suspect as the theory it was used it to support. How could they do such a seemingly reckless thing? There are several things to note here. Although the instruments used to produce the data of plate tectonics were callow, the theory behind them was not. Some individuals did dispute the meaning of paleomagnetic data, for example, but no one disputed the existence of magnetism or the fact, established by Pierre Curie at the turn of the century, that rocks preserved the effects of ambient fields. Geophysical instruments borrowed on the success of the theories behind them. They also borrowed on the prestige of the discipline from which those theories arose. For although geologists were right more often than they were wrong in their disputes with geophysicists in the nineteenth century, by the middle of the twentieth century there was no disputing which discipline held the position of greater prestige. Geologists in the twentieth century did not live in a vacuum; they lived in a scientific world increasingly inhabited by physicists. Whatever successes they had or should have had in their debates with physicists were beside the point. Instruments borrowed from physics traded on the overall success of physics, and that success was read as implicit validation of the instruments even before they were tested in geological realms.

The new data were also consistent with the old. Harry Hess scarcely ever mentioned fossils in connection with plate tectonics, yet he knew about them and their earlier connection with continental drift. So did William Menard. So did many others. But how to reopen the debate? Hess cited the evidence of remanent magnetism as the motivation for his development of the theory of seafloor spreading. Had the new data been inconsistent with the old, perhaps he would have studied these data more closely before venturing an account of them. Perhaps geologists would have raised more questions about technique. Perhaps plate tectonics would not have been accepted so rapidly. But the new data *were* consistent with the old, and geologists embraced them rapidly. After a forty-year wait, scientific opinion on the question of moving continents turned around in just a few short years. Perhaps this is why: after forty years of nervous joking and discussions on back staircases, earth scientists found a way to start fresh.[110]

This is the point to be made here. Scientists evaluate evidence in light of networks of prior belief, including beliefs about the nature of evidence itself. When evidence is invoked in highly contested issues, scientists may come to conclusions not only about the issue being contested but also about the evidence used to contest it. Scientists make value judgments about the nature of evidence, even though the alliances between particular kinds of evidence and particular theories may be fortuitous. But when new forms of evidence are created, such a priori value judgments are avoided, and a greater degree of conceptual flexibility may be achieved. It need not be instruments that create new forms of data; it may be new methods of any kind. Perhaps, implicitly, geologists understood this when they turned, in the early twentieth century, to laboratory and instrumental methods. When new instruments or methods are developed, or are extended from one discipline to another, the data produced have no established relation to the theoretical issues at hand. New methods and machines provide rare opportunities for conceptual flexibility, because they produce categories of data that lack an established relation to existing theoretical belief.

Epilogue

Utility and Truth

> While everyone well knows himself to be fallible, very few think it necessary to take precautions against their own fallibility.
>
> *John Stuart Mill*, On Liberty

In 1928 Rollin Chamberlin complained that if continental drift were true, geologists would have to "forget everything which has been learned in the last 70 years and start all over again."[1] And that is what they did. Between 1928 and 1968, many things changed in American earth science. The inductive and pluralistic stance that geologists defended against Wegener's theory-first approach had by mid-century given way to the very stance they had earlier criticized. The Pratt model of isostasy, which suggested that continental drift was physically impossible, gave way in the face of evidence from gravity and seismology. And uniformitarianism, the principle that informed the interpretation of field data, gradually became less methodologically significant as fieldwork lost its position of centrality.[2] In three distinct domains, the methods of earth science made it difficult for scientists to accept the evidence of continental drift—and the evidence became discredited. When these methods changed, it became not only possible but easy for earth scientists to accept the new evidence of plate tectonics—and the new evidence was heralded for its power.

Throughout the history recounted here, earth scientists were constantly making choices: about the questions they pursued, about the methods they used to answer them, about the ways in which they interpreted and presented their results. Scientists are always making choices. But on what grounds are such choices made? Many are possible—disciplines have internal cultures that condition collective choices, and individuals have idiosyncratic reasons as well—but when it comes to the choices made by the scientists in this study, one criterion stands out. American earth scientists in the early twentieth century did *what worked*.

The clearest example of this is the choice of the Pratt model for calculating isostatic compensation. The Pratt method enabled geodesists to make better, more accurate maps, and to make them faster and cheaper. This is utility in its most obvious sense. But utility played a role in a more subtle, epistemological sense as well. Geologists subscribed to the principle of uniformitarianism because it was *enabling*. It enabled them to interpret field evidence in a consistent and logical way. It enabled them to build a science of the past that would otherwise have remained logically inaccessible. It enabled them to demarcate scientific from theological explanation. Uniformitarianism helped geologists to unify their science.

Why geologists chose induction and the method of multiple working hypotheses is a more difficult question. How people and cultures come to their broadest and deepest convictions is the most difficult of questions and can scarcely be resolved here. Nor can the more narrowly relevant question of why geologists (and others in the mid-twentieth century) changed their minds about science's version of the chicken and the egg problem—which comes first, theory or observation? What may be said is that induction made sense in a field-based science where theoretical questions arose out of experience and observation, and an ideology of tolerance and pluralism seemed a reasonable response to the question of how to forge a common discipline among individuals with diverse backgrounds, education, and experience—just as it did for American culture at large. Conflicting beliefs can quickly turn into personal antagonisms, and in the history of geology they often had.[3] The method of multiple working hypotheses helped American geologists avoid such antagonisms by accustoming them to stances of receptivity and inclusion rather than advocacy and exclusion.

Thus, on several levels—economic, epistemic, social—the choices that American geologists made were enabling. Their choices helped them to do their science. Their choices worked. And because they worked, scientists came to believe that they were true, and to deduce consequences from them. William Bowie came to believe that the depth of isostatic compensation really was uniform, and that the base of the crust was, too. And if so, then large-scale horizontal compression was impossible. The same may be said for uniformitarianism. Because it worked so well as a heuristic principle, geologists came to believe that the present really *was* the key to the past—that the present and the past were, in effect, no different. And if so, Charles Schuchert reasoned, then continents could not have circumnavigated the globe.

But utility is not truth.[4] Ideas can work and still be wrong. And this is where scientific knowledge becomes problematic. There have been many scientific theories that were eminently successful in their day—in making accurate predictions, in enabling people to do things they wanted to do—but which we no longer accept. The Ptolemaic system of the planets, the caloric theory of heat, classical mechanics—these were all extremely powerful theories. Classical mechanics can be partly rescued from the scrap heap of discarded knowledge by arguing that it remains approximately true, that relativistic mechanics reduces to classical mechanics at speeds far less than light; many physicists make this claim.[5] But Ptolemaic astronomy? It is difficult to see how to make the geocentric universe a limiting case of a heliocentric one. And today only historians ever think about caloric.

How can wrong theories make right predictions? One answer, which philosophers have stressed, is that more than one theory can predict the same result. Our predictions may be right, but not for the reasons we think. The history of continental drift illustrates this a fortiori: American geodesists chose the Pratt model for pragmatic reasons—and it worked—but they could have made a different choice. And had they made a different choice, they might well have come to different conclusions about continental drift. Theories are not uniquely confirmed by their predictions.

But there is another reason, far more obvious, why false theories can work: Theories can be right in some respects and wrong in others. The Pratt model of isostasy was correct with respect to the existence of isostatic compensation, but not with respect to how that compensation was achieved. Uniformitarianism was correct insofar as it underscored the idea that the earth had a natural history, and that geologists could look to the present to interpret it, but it was wrong when it was taken to imply that monumental changes could never occur. And so for the method of multiple working hypotheses. Open-mindedness is a virtue — one that is often overlooked — but not all theories deserve equal time, and open-mindedness can degrade into dithering and vacillation. In short, to paraphrase John Stuart Mill in his famous discussion of political polemics, each of these ideas was right in what it affirmed, wrong in what it denied.

All this may seem to imply that the task of scientists is clear: to recognize that utility is not truth, and, as Mill put it, to take precautions against our own fallibility. Scientists would argue that they do take precautions — in peer review, in open discussion at national and international meetings, in replication of observations and experiments. Indeed, self-skepticism and collective surveillance have been taken to be the defining norms of science.[6] But the difficulty is not so easily removed. For why else do we believe in the truth of scientific theories except that they work? The most common defense of science — among scientists and philosophers — the most common justification for the belief that science gives us an objectively true account of the world — is its *success*. By success, the aficionados of science almost always mean success in *doing things*. Using scientific knowledge, humans have created new materials, built nuclear weapons, predicted collisions of comets, found oil and mineral deposits, invented VCRs, and saved lives — how could we do these things if the knowledge upon which these discoveries were made were not true?[7] The theories of science must be true — some would argue — the entities that those theories postulate really exist. In the words of the philosopher Ian Hacking, speaking of electrons, "If you can spray them, they are real."[8]

The conflation of utility and truth is deeply rooted in the scientific psyche. Sir Francis Bacon promoted the advancement of science as a means to the advancement of men, but he also argued for the tangible fruits of scientific knowledge as the best evidence of its veracity. Other philosophical systems had borne little fruit, he argued, but his philosophy — later known as science — would demonstrate its bona fides through its tangible results. "Truth therefore," Bacon wrote in the *New Organon*, "and utility are here the very same things, and works themselves are of greater value as pledges of truth than as contributing to the comforts of life."[9]

Bacon chose his words with consummate care: "works" are just what he said they are — pledges of truth, but not guarantors of it. And this is why the objective truth of science cannot be taken for granted. Not because of politics or ideology or vested interests — although these factors assuredly affect science more than scientists wish them to — but because the way we judge the legitimacy of scientific knowledge is insufficient. We judge scientific theories primarily by their ability to predict natural phenomena and enable humans to do things in the world.[10] But false theories can make true predictions. False theories can work. Ptolemaic astronomy did, contraction theory did, and the Pratt model of isostasy did. The Pratt model did not correspond to the actual structure of the earth, and geodesists' belief that it did impeded the development of geological knowledge. Geology would have been better off had William Bowie simply accepted the Pratt model for what it was—what *all* scientific theories are—an illuminating and productive way of thinking about the world.

The Unmaking of Scientific Knowledge

American earth scientists in the early twentieth century were not irrationally fixated on old ideas; they were committed to principles and practices that had served them in good stead. The conclusions they came to were not predetermined, but they were conditioned by what had gone before. Earlier methodological choices affected later theoretical outcomes; indeed, in some cases they *became* theoretical outcomes. The method of multiple working hypotheses was developed by American geologists in the late nineteenth century in reaction to the perceived ills of European science, but in the twentieth century it lived on as an intellectual legacy—long after the concerns that inspired it were only faintly remembered—and made the methods and rhetoric of Alfred Wegener seem unscientific. Uniformitarianism was accepted by geologists in the nineteenth century for epistemological and sociological reasons, but what was accepted initially as a heuristic principle gradually lost its flexibility, and in the twentieth century it became a default assumption under which available conceptual options were narrowed. And so for Pratt isostasy. Geodesists chose the Pratt model in the 1900s for reasons of practicality and parsimony; in the 1920s adherence to that model delimited the possibilities for what could come next.

In judging new ideas in terms of existing theoretical and methodological commitments, geologists were implicitly applying another standard of truth, in addition to the standard of utility, namely coherence. Coherence is a common way people judge the truth of interrelated ideas. It is also a way in which philosophers have sometimes defined truth. But as anyone who has studied logic or mathematics knows, systems of thought can be internally consistent yet bear no resemblance to physical reality. Incoherence may be a sign of falsehood, but coherence is no guarantor of truth. Yet, as historian David Oldroyd has demonstrated in his work on nineteenth-century field geology, and as this study also shows, scientists do judge new ideas by coherence with the old.[11] Thus if new knowledge is to be made, old knowledge has to be *unmade*.

For earth scientists to accept the idea of moving continents, many things had to change. And many things did change. They changed in the light of empirical evidence, they changed in the face of a changing social and intellectual context, and they changed under the positive pressure of patronage and the negative pressure of disciplinary competition. But none of these changes occurred overnight. The unmaking of scientific knowledge does not happen quickly. It is not—contra the widely held views of Thomas Kuhn—a gestalt shift. At least in the case of continental drift, an edifice of scientific knowledge was not smashed and rebuilt, but rather an interwoven fabric of methods and belief was gradually unwoven and restitched. As Kuhn himself showed in his own early work, the unmaking of Ptolemaic astronomy took one hundred fifty years.[12] By that standard, geologists moved fast!

Historicity and Self-Ratification

Philosophers and critics of science have made much of the fact that our access to the natural world is incomplete. And so it is. But the history recounted here suggests that, at least in field-based sciences, the problem for scientists is not too few data but too many. Scientists are often unable to reconcile all the information at their disposal, and so they make choices. To forge a reliable representation of the natural world out of the infinite sea of sensory perception, one must, to paraphrase the famous Dr. Doolittle, choose what to put in and what to leave out. Scientists make choices based on epistemological affinities, disciplinary affiliations, evidential and methodological preferences—and all of these are, at least in part, constituted in advance of any given theoretical debate. Theoretical debate may alter affinities and affiliations, but it nevertheless encounters a preexisting set of them. For this reason, if for no other, the development of scientific knowledge is inescapably *historical*.[13]

Cherishing the method of multiple working hypotheses as an indigenous alternative to European methodological stances, American geologists defined themselves in juxtaposition to Europeans just as intellectuals did in other arenas of American cultural life. Uniformitarianism was in part a defensive posture framed in reaction to intellectual competition from physics and chemistry on the one hand and from theology on the other. Pratt isostasy was adopted in response to the exigencies of patronage. Even the move toward instrumentation and quantification—of all the methodological commitments described here the one most arguably driven by the aspirations of the scientific community—was motivated in large part by geologists' desire to use the methods of other sciences that had lately come to seem more potent than their own. In each of these cases, social and historical context played a significant—even determinative—role in the choices that scientists made. And these choices affected their theoretical beliefs. Their realms of reasonable belief were delimited by their communal standards and practices, and these standards and practices were forged in response to a unique historical and cultural context.

The choices these scientists made, moreover, were self-ratifying. American earth scientists chose not to pursue the field-based observational evidence relevant to the question of continental drift; instead they solicited the patronage of private institutions and military bureaucracies in support of instrument-driven science. Not surprisingly, then, little new observational evidence in support of the theory was gathered, while reams of instrumental data were. And when these instrumental data were made public and their support of moving continents became evident, earth scientists were satisfied that they had made the right choice. Yet had the Navy not been interested in supporting marine geophysics—had submarine warfare not existed—earth scientists would by necessity have taken a different route, and perhaps been well satisfied with that, too.

In closing we may note a striking contrast between the two realms in our culture held most responsible for determinations of truth. In the legal realm, juries are impaneled on the basis of impartiality: anyone with prior connection to the case at hand is dismissed. The establishment of truth is taken to require an absence of background knowledge. In science, the opposite is the case: we hold those most knowledgeable about matters to be the ones most qualified to judge. In doing so, we recognize the importance of scientific expertise and experience in judging matters of evidence and plausibility; implicitly we are recognizing the value of background knowledge and of intuition that accrues from experience. Yet, ironically, in doing so we are also affirming—willingly or not—the role of the subjective in establishing the objective. We are acknowledging the role of the unique and particular in establishing the general and universal. And we are placing responsibility for making new knowledge in the hands of those who have the most old knowledge to unmake. The recognition of scientific expertise—the very stuff that enables scientists to build on prior results—that enables science to progress—at the same time makes scientific judgments inescapably personal and historical, undermining our deepest wishes for knowledge that might somehow be transcendent.

NOTES

Introduction

1. Many would distinguish, with some justice, between the observations of science and the theories deduced from those observations. Although the latter obviously change, the former, many scientists would argue, do not, and therefore the "perishability" of science applies only to theories, not data. The problem with this distinction is at least twofold. Observations *do* perish; there are plenty of examples of corroborated data in science that were later rejected as mistaken—but for obvious reasons these examples are usually forgotten! Scientific data do not accumulate *pyramidically*; they accumulate *sequentially*: good data replace the bad, and this is a good thing. But even when observations do endure, their meaning may be radically reinterpreted in the light of new theories; this is the critical point that has been emphasized by philosophers and historians of science. Insofar as most people would agree that what matters about science is its *meaning*, the transformation of meaning in observational facts is a profound matter.

2. Exceptions include some American philosopher-scientists near the turn of the century, such as C. S. Peirce and Percy Bridgman.

3. Quoted in Holton (1973, p. 17). Reichenbach was specifically distinguishing between epistemology and psychology—how ideas are justified logically rather than how they are arrived at creatively (Reichenbach 1938, pp. 5–7). Historians nowadays interpret the "context of discovery" to include all the factors—psychological, sociological, political, economic—affecting the development of scientific knowledge.

4. Rosenberg (1988). The move toward the context of discovery is generally credited to Thomas Kuhn (1962), because he emphasized the community nature of scientific work and the role of extraevidential considerations in theory choice, but Kuhn said little about the larger social context of scientific work.

5. An explicit example of this in the philosophy of earth science is found in von Engelhardt and Zimmerman (1988, p. 313): "The change in theory initiated by Wegener and the researchers of the 1950s can be rationally understood and reconstructed in accordance with the [philosophical] model of sophisticated falsificationism." The phrase "rational reconstruction" was apparently coined by Carnap in *Der Logische Aufbau der Welt*; see (Reichenbach 1938) note 1, p. 5. Two widely cited post-Kuhnian defenders of scientific rationality are Imre Lakatos (Lakatos and Musgrave 1970) and Larry Laudan (1977, 1984).

6. A widely cited example is Paul Forman's (1971) argument that the abandonment of causality in modern physics was a response to the unhappy social climate of Weimar Germany, but his account is not fully convincing because the social forces Forman invokes were not applicable to all the most important advocates of acausality, particularly Niels Bohr (Hendry 1980). For entry into the literature on the sociology of scientific knowledge, see Barnes, Bloor, and Henry (1996), Collins ([1985], 1992), and Bloor (1976).

7. The success of science is also used as the basis for realist arguments in philosophy of science (Boyd 1991). The most notable recent defense of science in response to a perceived attack on it from academic quarters is Gross and Leavitt (1994); see also Holton (1993).

8. Some examples relevant to the history of earth sciences include Rudwick (1985), Secord (1986), Smith and Wise (1989), and Oldroyd (1990).

9. For example, see J. Z. Buchwald's (1991) review of Smith and Wise's *Energy and Empire*.

10. Meretricious because the ideology of objectivity and context-independence permits scientists to deny responsibility for the negative consequences and applications of their work (Harding 1986, 1991).

11. Exceptions include Cartwright (1983), Franklin (1986), Galison (1987), and Hacking (1983). See also Achinstein and Hannaway (1985), Gooding, Pinch, and Schaffer (1989), Lenoir (1988), and Warwick (1995).

12. I acknowledge that my use of the term "American" to mean the United States is chauvinistic, but I follow what is, after all, common if regrettable usage. I use the term in the same sense as the *American Geophysical Union* or *Geological Society of America*. I have included Canadian geologists in my purview, where relevant, most notably Canadian-born Reginald Daly. Mexican geology was only beginning to develop during the time period covered by this study. Some readers may find fault with my focus on North America, perhaps desiring a broader study including the Germans, French, or others. Such a study would certainly be desirable; my self-justification is that it was in America that the rejection of continental drift was most pronounced and hostile, and that this hostility invites close examination. On the European response to continental drift, see Carozzi (1985) and Marvin (1985). For broad overviews, see Hallam (1973), Marvin (1973), Le Grand (1988), and Stewart (1990).

13. Some will argue that continental drift is not the same as plate tectonics, and that the evidence used to support plate tectonics was different from the evidence used to support drift. These claims are true, and they are addressed in chapters 4 and 8. For the purposes of the question being posed here, however, they are not germane. Whether or not the precise contours of the theory that was finally accepted match those of continental drift is beside the point. Scientists commonly reserve judgment on theories pending better evidence or fuller causal explanations, but the idea of moving continents was not merely put on hold; it was rejected as false. The theory of plate tectonics has itself been modified since its original formulation, but none of these necessary modifications stood as obstructions to acceptance of the theory in the 1960s. Both theories shared the claim that continents are not fixed, just as we might say that Galileo and Kepler shared the claim that the earth was not fixed.

14. Of course there have been modifications since the original idea, but in broad outlines, the arguments made by Wegener and Holmes have been upheld by the work of the past three decades.

Chapter 1

1. Cox (1973).
2. Gass, Smith, and Wilson (1972), p. 9.
3. Wegener ([1929]1966).

4. In the 1980s, it was commonly reported that geologists in the Soviet Union did not accept plate tectonics, but this may no longer be the case. In the Anglophone world, a rare dissenter is S. Warren Carey, who for the past thirty years has advocated whole-earth expansion as the cause of continental displacements (Carey 1958, 1983, 1988). Objections to plate tectonics in the 1970s were summarized by Meyerhoff and Meyerhoff (1972); rare examples of holdouts against the theory are found in Wilde (1990) and Keith (1993); see also Saull (1986). Contemporary resurgence of interest to alternatives to plate tectonics appears in Chatterjee and Hotton (1992). The editors and many of the authors of this latter work are not Americans; once they accepted plate tectonics Americans became, and in general remain, its staunchest defenders.

5. Greene (1982), esp. pp. 39, 76.

6. Suess (1904–1924). For background to Suess's contributions, see Greene (1982), pp. 167–191.

7. Zittel (1901), pp. 457–459; also see Leviton and Aldrich (1991).

8. Zittel (1901), pp. 316, 320–321. Emphasis in original.

9. De la Beche (1834), p. 121, quoted in Greene (1982), p. 99. A slightly different version of this same point is found in the first edition of De la Beche (1832), p. 501. De la Beche took pains to point out his general agreement with Elie de Beaumont's emphasis on secular cooling as a mechanism of orogenesis, but not necessarily requiring that the resulting dislocations were simultaneous worldwide.

10. Hooykaas (1975), pp. 336–337; Greene (1982), pp. 93–121; see also Rudwick (1990).

11. Whewell (1858).

12. Whewell (1858), p. 592. Emphasis in original. Whewell also argued (p. 594) that the nebular hypothesis, if correct, proved that contra Hutton the universe did have a beginning.

13. See Greene (1982); Rudwick (1985, 1990). Charles Darwin helped Lyell to become famous and overshadow De la Beche; Darwin's son George ironically became a leading contractionist in the late nineteenth century.

14. Greene (1982), p. 99. The law of rational intercepts states that the disposition of the external faces of crystals can be described according to integral number intercepts on a Cartesian coordinate system. This was (and is) construed by many scientists as a compelling example of the intrinsic mathematical order to be found in nature.

15. Elie de Beaumont's attempts to explain the entire world in terms of geometry is a case study in the mathematization of nature in the nineteenth century. His spectacular failure is a significant point: not all mathematization in science has been successful or productive.

16. Greene (1982), pp. 88–89; also Hooykaas (1975) and Zittel (1901), pp. 300–301. On Cuvier's influence on continental geology, see Zittel (1901), pp. 135–141. Cuvier estimated the time since the last major revolution as 5,000–6,000 years, which is where geologists now place the end of the Flandrian transgression in Eurasia.

17. Hooykaas (1975); Toulmin and Goodfield (1965).

18. Kant (1755); Laplace (1796). Kant's work was not well known until it was later revived by Humboldt. On the importance of Laplace for nineteenth century geology, see Zittel (1901), p. 79. In the United States, the nebular hypothesis would later be greatly criticized by T. C. Chamberlin (1916a, 1916b), who argued that the earth's atmosphere and hydrosphere were incompatible with the nebular hypothesis. See Numbers (1977) and chapter 5.

19. Greene (1982), p. 90.

20. Greene (1982), Laudan (1987).

21. On Silliman and Dana, see Gilman (1899), Struik (1957), Wilson (1979), Bruce (1987), and Brown (1989).

22. Hurlbut and Klein (1977).

23. The other was T. C. Chamberlin, editor of the *American Journal*'s main competitor, the *Journal of Geology*; see note 18 above and chapter 5.

24. See Zittel (1901), pp. 161–163. G. K. Gilbert and Arthur Day argued that the lunar craters were impact structures, but most geologists interpreted them as volcanic. This debate is documented in the papers of Bailey Willis in the Huntington Library, and discussed in Doel (1996).

25. Dana (1846), p. 353. The geological implications of differential melting points had been earlier discussed by De la Beche (1834); see Zittel (1901), pp. 167–168.

26. Dana (1846), p. 353.

27. Dana (1846), p. 353.

28. Dana (1846), p. 355. Also see Dana (1847a, 1847b, 1847c, 1873a, 1873b).

29. In retrospect, this does not entirely make sense; logically one might expect *tension* along the continental margins as they were pulled down by the adjacent subsiding basin. James Hall would later accommodate this by placing geosynclines there.

30. Willis (1910), p. 243. Emphasis in original.

31. Greene (1982), pp. 123–124.

32. Rogers and Rogers (1843), p. 396; see also Adams (1954).

33. Reade (1886), p. 1, quoted by Greene (1982), p. 122.

34. Hall beat out Dana for the first presidency of the GSA (Dana served as the second president, in 1890), but Dana preceded Hall as a president of the AAAS.

35. On the history and development of geosyncline theory, see the excellent articles by Robert Dott, Jr. (1974, 1979, 1985) and the Ph.D. thesis of Mayo (1985).

36. Dana (1873a), on p. 347; also Dana (1873b). Hall claimed to reject the kind of broad deductive theorizing in which Dana engaged, although given his own theoretical work, Hall's rejection seems more rhetorical than actual; see Greene (1982), p. 135, and chapter 3. The two men were at odds throughout most of their professional careers, but Hall—an arrogant man and a virulent racist—was at odds with many of his professional colleagues.

37. Adams (1954), p. 397. See also Greene (1982), especially p. 140; and Mayo (1985).

38. Dana (1875), pp. 241–242.

Chapter 2

1. Zittel (1901), p. 321. Greene (1982), p. 192, suggests that Suess was so influential largely because he outlived his adversaries.

2. Trümpy (1991); Greene (1982), p. 195.

3. Bertrand was also the first to interpret the history of European orogeny in terms of three distinct episodes, the Caledonian, the Hercynian and the Alpine; the first to recognize the progression of sedimentation in the Alps from deep marine cherts and shales to shallow marine flysch to nonmarine molasse; and the first to recognize the continuity of the Calendonian and Hercynian events across the Atlantic Ocean to North America (Greene 1982, p. 203).

4. Greene (1982); Rogers and Rogers (1843).

5. Oldroyd (1990); Greene (1982), pp. 201–204 and 213–216, and Trümpy (1991), pp. 397–398.

6. Greene (1982), p. 224. Ironically, Suess himself had been responsible for some of the Alpine field work that led to these conclusions, and he eventually endorsed nappe theory. In the context of Alpine geology, Suess and Heim were seen as fighting on the same side, against earlier European theories that emphasized the role of plutonism in uplift; see Greene (1982), pp. 219–200; Trümpy (1991), p. 398.

7. Dawson (1894), p. 104. Dawson, however, thought this could be reconciled with contraction theory: in a typically American manner, his approach was to to combine contraction, expansion, and isostasy so as to avail oneself of "all possible causes of elevation." Although his American colleagues were highly sympathetic to the notion of multiple causes (see chapter 5), the version of contraction they were thinking of was Dana's.

8. Markham (1878); Kushner (1990).

9. Pratt (1855, 1859a, 1859b). See also Fisher (1886) and Heiskanen and Vening Meinesz (1958), pp. 126–127. The phenomenon of "missing mass" had been earlier recognized by Bouguer (1749).

10. Airy (1855).

11. Pratt (1871), p. 200; see also Greene (1982), pp. 242–243.

12. Pratt also believed in a fully solid earth, as advocated by his Cambridge teacher William Hopkins, which Airy's model seemed to contradict. See Kushner (1990).

13. The differences between these two models, and the very different methodological strengths and theoretical implications bound up with them, played a major role in later discussions of drift—the subject of chapter 6.

14. On the mathematical tradition in physical geology—later geophysics—at Cambridge, see Smith (1985), Wise and Smith (1987), Smith and Wise (1989), Hevly (1996), and Kushner (1990, 1993).

15. Fisher (1882b); see also Kushner (1990).

16. Many geologists had interpreted volcanic phenomena as evidence of a fluid interior, but William Hopkins argued on physical principles that the solid crust had to be at least 1/4 to 1/5 the radius of the earth. Thomson augmented Hopkins's calculations to show that the whole earth must be solid (Thomson 1862a; also see Smith and Wise 1989, especially pp. 556–559, and Kushner 1990). This argument was seconded in the United States by Joseph LeConte (1872, p. 352), who argued that, given Hopkins's and Thomson's work, "the whole foundation of theoretic geology must be reconstructed on the basis of a solid earth." See chapter 3.

17. Fisher (1874, 1878, 1879).

18. Kushner (1990).

19. Fisher (1881).

20. Fisher (1881), p. 270. Pratt (1871) had made the same point in his discussion of the deflections of the plumb bob and the evidence for excess mass over the oceans.

21. I take a small historical liberty and refer to Thomson as Kelvin, although he had not yet, at this point in the story, been elevated to peerage. I do this for the sake of my readers, most of whom will know Kelvin by his lordly moniker.

22. Smith and Wise (1989), pp. 573–578.

23. Melting by pressure release had been earlier proposed by the American Clarence King (1878).

24. Fisher (1881), p. 275. This idea harks back to Herschel, although Fisher credited his ideas to Airy.

25. Fisher (1881), p. 83–84, original emphasis.

26. Pratt (1871), p. 236, quoted by Fisher (1881), p. 76.

27. Fisher (1881), pp. 277–278.

28. Darwin (1879); Fisher (1882a).

29. Joly (1924), p. 1.

30. Quoted by Fisher (1882b), p 77.

31. Adams (1954) p. 397; Hayford (1909), p. 68; Barrell (1918a), p .190.

32. Geikie (1880), p. 325.

33. Dupree (1986), pp. 198–202, quote on p. 200. See also Smith (1918), p. 203; Bartlett (1962); Goetzmann (1966); and Bruce (1987), esp. pp. 320–321.

34. Dutton (1874). Although Fisher's first publication on contraction preceded Dutton's, Dutton was the first to reject contraction; in 1874 Fisher was still trying to salvage it.

35. Thomson (1862b). On Forbes's experiments, see Smith and Wise (1989), pp. 559–560. In later years, Kelvin revised his estimate downward to 20 million.

36. Dutton (1874), p. 120.

37. Dutton (1874), p. 121.
38. Dutton (1876a, 1876b); Dutton (1889), p. 58; Greene (1982), pp. 108 and 238. Mayo (1985, p. 5) notes that Hall and Dutton cite the same passage from Babbage's Bridgewater treatise, thus demonstrating that Hall was drawing on the idea of isostatic adjustment to explain the accumulation of sediments in geosynclines.
39. Dutton (1889), p. 58.
40. Dutton (1889), p. 60. Plications are, literally, fanlike folds, and the term was used to describe any large-scale systematic folding but especially that which resulted in repetitions of sedimentary sequences.
41. Dutton (1889), pp. 61–62.
42. Dutton (1889), pp. 56–60.
43. Smith (1918); Bruce (1987), pp. 322–323; Mendenhall (1920); Yochelson (1980); Pyne (1980).
44. Gilbert (1886, 1890); also see Smith (1918), Mabey (1980), and Sack (1989). Because Gilbert's physically oriented approach to geomorphology has since become the foundation of that science, he is retrospectively credited with being one of its founders, but at the time William Morris Davis's evolutionary approach was much more widely utilized (Sack, 1989, 1992; Yochelson 1980).
45. Gilbert (1893).
46. Gilbert (1893), pp. 186–187.
47. Comments on Dutton (1889); see also Greene (1982), p. 253.
48. Willis (1893), p. 280, see also Greene (1982), p. 253. An insightful but ironic comment, given Willis's later objection to continental drift in part on these grounds! Willis later accepted isostasy; see chapters 6 and 7.
49. Fisher (1886), p. 2.
50. Dutton (1889), p. 57.
51. Gilbert (1893), p. 183.
52. Fisher (1886).
53. Gilbert (1895a), p. 333.
54. Gilbert (1889), p. 24.
55. For example, see discussion of Gilbert (1889), pp. 25–27.
56. Dupree (1986), especially pp. 202–203, 215–231, and 377; Reingold (1991). For background on the Coast Survey, see Reingold (1964), Manning (1988), and Slotten (1994).
57. Gilbert (1895b); Putnam (1922); Mabey (1980). Gilbert was still an employee of the U.S. Geological Survey; cooperative work was not too uncommon.
58. Gilbert (1859b) referred to the use of variable densities as his "geological correction"—since it involved attention to the geology of the district (see also Mabey 1980). In effect, he was assuming regional compensation and comparing the degree of local compensation of one portion of the crust relative to another.
59. Gilbert (1895b), p. 70.
60. Putnam (1896); see also Putnam (1922).
61. The Faye correction was named after M. Faye, a French geodesist who pioneered gravity reductions in the late nineteenth century (Faye 1880, 1883, 1895). Gilbert claimed that his adjustment to the mean plain was the same as Putnam's use of the Faye correction, but this was not strictly true: the latter took no account of local variation in rock density, and Putnam used a 100 km rather than a 30 km radius. Furthermore, Gilbert subtracted the effect of the calculated average slab to reduce the gravity stations to the level of the interior plain (i.e., a negative correction), but Putnam used the average elevation correction as a positive correction to be added back after making a Bouguer reduction. Putnam: "The procedure was equivalent to subtracting the effect of topography, as in the Bouguer reduction, and then adding as a compensation term the effect of a plate of average surface density, whose thickness is the

average elevation of the surrounding region" (Putnam 1922, p. 293). What he ended up with was essentially an average free-air reduction.

62. Putnam (1922), p. 293.
63. Putnam (1922), p. 296, quoting Putnam (1896), p. 192.
64. Hayford (1909), p. 9. For biographical background see Burger (1931).
65. Hayford (1909), pp. 171, 173; Hayford (1906), p. 29.
66. Woodward (1889), p. 341.
67. Hayford (1909), p. 20. More limited studies of the deflection of the vertical and of gravity had been made by Bonsdorff, in Spitsbergen, and of gravity by Schiotz, during Nansen's polar expedition. Both concluded that isostasy applied, but their studies were far smaller and less detailed than Hayford's. See Heiskanen and Vening Meinesz (1958), p. 130.
68. On female computers at government and private observatories, see Rossiter (1982), Lankford (1994), and Oreskes (1996). Working in a government agency whose basic function involved massive amounts of computation, Hayford had access to skilled (or semiskilled) scientific workers in a way that few others had. For an example of female computers outside the United States, see Bruck (1995).
69. Hayford (1909), p. 130.
70. This may seem trivial in retrospect, but prior to this vertical deflections were generally calculated only for very local effects.
71. Hayford (1909), p. 65.
72. Hayford (1909), p. 66.
73. Hayford (1909), p. 9.
74. Hayford (1909), p. 173. Hayford also made the point that the measurements were better because the stations were connected by continuous triangulation, rather than along an arc. Constructing a theoretical spheroid from measurements along meridianal arcs was equivalent to trying to construct a physical model from a series of small pieces of bent wire, whereas Hayford's measurements connected by triangulation were equivalent to a large piece of properly bent sheet metal. Clearly, it would be easier to complete the physical model from the sheet, for "such a bent sheet essentially includes within itself the bent wires, and moreover, [they] are now held rigidly in their proper relative positions" (Hayford 1909), pp. 169–170.
75. Burger (1931).
76. Hayford (1909), p. 175.
77. For biographical background see Fleming (1951), pp. 208–232.
78. Hayford and Bowie (1912); Bowie (1922).
79. Making gravity measurement was laborious, so data reduction and plotting was the surveyor's bottleneck; see RUSCGS, Entry 33, Box 775, Annual Office Reports, 1926–1929.
80. Bowie (1922), pp. 616–617; see also Hayford (1911).
81. In at least one case, he explicitly claimed the truth of isostasy: "Hayford's researches indicated that the theory of isostasy is substantially true" (Bowie 1924b, p. 26).
82. Gilbert (1914), pp. 29 and 37; also see Mabey (1980).
83. Becker (1915), p. 179.
84. Bowie (1924b), p. 44.
85. Joly (1924). Original emphasis and parentheses.
86. Hayford (1909), p. 68.
87. Willis (1910), p. 243.
88. Pratt (1871), pp. 201 and 204.
89. On Curie, see McGrayne (1993) and Quinn (1995). On Rutherford, see Romer (1964), Bunge and Shea (1979), and Badash (1979).
90. Quoted in Smith and Wise (1989), p. 603. Also see Burchfield (1974) and Brush (1996b).
91. Discussed further in chapter 5.

92. Chamberlin (1899a, 1899b); Mather and Mason (1939), p. 612.

93. Mather and Mason ([1939], 1967), p. 615.

94. James Orr to Kelvin, January 19, 1906, cited in Smith and Wise (1989), p. 550.

95. Joly (1903), p. 526. Ironically, the modifications ended up being far greater than Joly, who had calculated the age of the earth based on the salt content of the oceans, wanted. The quote concludes, "The hundred million years which the doctine of uniformity requires may, in fact, yet be gladly accepted by the physicist." Famous last words.

96. Strutt (1905), Boltwood (1904a, 1904b, 1905a, 1905b, 1905c, 1907a, 1907b).

97. Holmes (1913c) gave the following dates: Carboniferous 340 million, Silurian or Ordovcian 430, Precambrian 1025 or greater; see also Holmes (1911a, 1911b, 1913a, 1914, 1926a, 1926b). He did this work before Nier's invention of the mass spectrometer and without knowledge of the isotopic complexities of uranium and lead. Later in life, so equipped, he was able to refine his work and provide a maximum estimate of the age of rocks at approximately 3 billion years (Holmes 1946).

98. Geologists generally interpret the story as a vindication of observational methods over the hobgoblin of fallacious quantification. However, as Joe D. Burchfield points out, many geologists in fact capitulated to Kelvin's estimates or responded with fallacious quantification of their own. Indeed, the more prominent they were, the more likely geologists were to have felt the need to accommodate results from physics, leading to the ironic situation that in the early twentieth century many distinguished geologists argued *against* the new radiogenic dates. Others, having been once forced by physicists to reduce their estimates of geological time well below what they felt geology indicated, now refused to increase them above it. See Burchfield (1974), Smith and Wise (1989), Greene (1982), Gould (1983), and Kushner (1990).

99. Joly (1903). See also Eve (1939), pp. 393–394.

100. Joly pursued this topic intensely for over a decade, trying to use the haloes as a dating method. Later, he helped to establish a radium institute in Dublin for the promotion of medical applications of radioactivity; see chapter 4. For his own retrospective, see Joly (1923), pp. 205–216.

101. Joly (1923), p. 69.

102. Joly (1923), p. ix.

103. Rutherford (1913); see Becker (1915), p. 195.

104. Willis (1942b), p. 172.

105. In principle, one might have wondered if the earth had passed through a peak of radioactive heating and was now cooling again, but so far as I know, no one made this argument.

106. Joly (1909), p. 131.

107. Quoted in Barrell (1918a), p. 182.

108. Le Grand (1988), especially p. 97, footnote 97. Le Grand notes that sociological "interests" are a version of the same problem.

Chapter 3

1. See preface to Wegener ([1929] 1966) and Greene (forthcoming). In writing this chapter, I have had to confront an awkward issue. The edition of Wegener's book that was most widely read in the United States was the English translation of the third edition (1924). However, this edition is now difficult to come by; I have had to weigh accuracy against accessibility. I have opted for a compromise. In cases where the substance of Wegener's arguments changed from the third to the fourth edition (chapter 4), or where the exact structure and wording matters historically (chapter 5), I have relied on and refer to the historically correct third edition. However, for the sake of my readers who may wish to revisit Wegener's arguments, the majority of my quotations are taken from the widely available Dover reprint (1966) of the fourth (1929) edition. I believe my choice is historically as well as pragmati-

cally defensible, for two reasons. One, the overall structure of Wegener's argument and the bulk of the evidence presented are the same in the two editions. Thus for the purposes of this chapter, which deals with the broad outlines of the situation, the use of the more accessible edition does not misrepresent the historical situation. Two, my argument here—that the standard explanation for the rejection of continental drift is historically anachronistic and false—holds true irrespective of which edition of *The Origin of Continents and Oceans* one considers.

2. Taylor (1910, 1925); Baker (1914); Pickering (1924). An even earlier precursor is Snider-Pelligrini (1858)—see discussions in Hallam (1973) and Carozzi (1970). As far as I can discern, geologists in the 1920s were not aware of Snider-Pelligrini's work. They were aware, however, of the work of Taylor, Baker, and Pickering.

3. Although Wegener's theory found its most abundant support in paleontology and stratigraphy, *he* saw his concept as a theory of *tectonics*—i.e., of movements of the earth's crust, including mountain-building. This is shown by the title of his book and of his original Frankfurt lecture, "The geophysical basis of the evolution of the large-scale features of the earth's crust," as well as by the structure of his arguments, which present drift as an alternative to the failed theory of contraction.

4. Wegener ([1929] 1966), pp. 5–6.
5. Wegener ([1929] 1966), p. 9.
6. Wegener ([1929] 1966), p. 16.
7. Wegener ([1929] 1966), pp. 17, 21.
8. Marvin (1973), pp. 54–56.
9. Frankel (1976, 1984, 1985).
10. Wegener ([1929] 1966), p. 110.
11. Wegener ([1929] 1966), p. 1. Mott Greene suggests that drift may have developed by analogy with meteorological phenomena, as Wegener pondered the coalescence and fragmentation of clouds; see Greene (forthcoming).
12. Hallam (1973) discusses early discussions of the jigsaw fit, pp. 1–6. The morphological match was also the motivation for the Darwin–Fisher hypothesis of fissiparturition; see chapters 2 and 6.
13. Wegener ([1929] 1966), p. 76.
14. Wegener ([1929] 1966), pp. 76–77. Readers may note the curiosity of this comment, because if the probabilities are less than one, then raising them to the *n*th power *decreases* their values, hardly bolstering Wegener's position! Presumably what he meant was that the probability of these homologies having arisen by *chance* becomes vanishingly small the more of them there are. In any case, the whole point is metaphorical, because no one had the slightest idea of how to determine the probability of a homology arising by chance, although Alexander du Toit would later try; see chapter 10.
15. Wegener (1911). After developing drift theory, Wegener coauthored with his father-in-law a book on past geological climate (Koppen and Wegener 1924).
16. Schuchert (1918), pp. 60–121.
17. Willis (1910), p. 250.
18. Wegener ([1929] 1966), p. 121.
19. Wegener ([1929] 1966), p. 23, recognizing the uniformitarian commitments of many geologists.
20. Wegener ([1929] 1966), p. 23.
21. Schwarzbach (1980); Hallam (1983).
22. Wegener ([1929] 1966), p. 26.
23. Wegener ([1929] 1966), p. 29.
24. Lambert (1922), quoted by Wegener (1929), p. 34.
25. Greene (forthcoming).

26. This is not to minimize evidentiary arguments that were made by Wegener's critics, but only to say that there was a great deal of evidence presented in the 1920s and that much of it was collected by well-respected scientists. Drift was not a "speculation" without supporting data. The issues of how good the data really were and how much evidence is enough are discussed at length in chapters 7 through 10.

27. Press and Siever (1974), p. 701. Similar or related views can be found in many places, *viz*: Wilson (1971, pp. 2, 44), Cox (1973, p. 14), Marvin (1973, p. 179), Hallam (1973, pp. 68–76), Hallam (1983, p. 171), and Menard (1986, p. 28). Walter Sullivan (1964, p. 15) gave a more temperate interpretation of the same problem but came to the same conclusion. Recent work tends to be less pejorative than the triumphalist accounts written in the early days of plate tectonics, for example Allegre (1988), discussed below.

28. Uyeda (1978), p. 22, his emphasis. Uyeda continues: "Even if Wegener's theory of continental drift did provide lucid explanations for many geological effects, such as ancient glaciation and mountain building, it could scarcely be regarded as scientific unless it could also explain what had originally caused the continental movements." Uyeda thus reveals his epistemic standards for a scientific theory, but this by no means proves that these were the standards of the geologists who rejected drift forty years before.

29. Hallam (1973), p. 151.

30. Alexander du Toit made this point in 1929 in response to a critical review of his work by Charles Schuchert, who had demanded to know what the mechanism of continental drift was. Du Toit replied, "The mechanism of continental sliding, being extraneous to the description, was not discussed any more than I felt myself compelled to explain the refrigerations of the Paleozoic, a phenomenon which Schuchert accepts, despite the fact that the most eminent scientists have hitherto failed to furnish any convincing explanation thereof" (du Toit 1929, p. 182). On the acceptance of the ice ages without causal explanation, see Oldroyd (1996, pp. 150–157). In 1959, Keith Runcorn made the same point with respect to paleomagnetic reversals, the existence of which was critical to the establishment of plate tectonics, yet the cause of which remains unknown (see Glen 1982, and chapter 8).

31. I heard this quip first in the mid-1980s from Allan Cox, the late dean of Earth Sciences at the School of Earth Sciences at Stanford University. Robert Dott Jr., Professor of Geology Emeritus at the University of Wisconsin, heard it many years earlier from Kay.

32. Rudwick (1971).

33. See Mayo (1985), Dott (1985), and Greene (1982), especially pp. 126, 132, 136–140.

34. With the advent of geophysical instrumentation in the twentieth century, the traditional understanding of "observation" was radically challenged; see chapters 8 and 9.

35. Robert Dott Jr. (personal communication, 1996). Even if Hall's use of this distinction was self-serving, the point remains that he perceived the demarcation as one that would resonate with his colleagues. Indeed, the word "speculation" was often preceded in American geological literature by the word "unbridled."

36. R. Laudan (1977, 1987). The Geological Society of London defined geology as "knowledge of the system of our Earth"; Laudan contrasts the geological with the mineralogical, the latter being Werner's primary domain.

37. Ospovat (1959), R. Laudan (1987). It is ironic that Werner is now mostly known for his discredited "geology"—neptunism—when in his own life he considered his "geognosy" to be far more important.

38. Toulmin and Goodfield (1965), pp. 150–152. Rachel Laudan (1987) has argued that geologists in the eighteenth and nineteenth centuries can be classified into causal and historical camps. Like most dichotomies, this one is too sharply drawn: Henry De la Beche was both causal and historical in his work, and the American James Hall, known in retrospect primarily for geosyncline theory, dedicated most of his life to the classification of fossils and

claimed to eschew grand theory. Nevertheless the point is well taken that it was *possible* to do historical geology and remain agnostic about causes, and many geologists, particularly in the United States and the United Kingdom, did.

39. Toulmin and Goodfield (1965) remind us of the argument of Bertrand Russell (among others) that the acceptance of the fact of earth history is a leap of faith: it *is* conceivable that God made the earth in a moment, exactly as it appears today, with all its rocks, fossils, historical relics, etc.—but then we would be dealing with a very ironic Creator.

40. Greene (1982), p. 39. While Werner promoted *geognosy* in order to produce "certain knowledge of the framework of the Earth," he ultimately abandoned this commitment and produced the most famously discredited theory in the history of geology. For many geologists, the term *geognost* thus came to take on the reverse of its original meaning; see Geikie (1897), p. 240, and Adams (1954), pp. 216–217.

41. Joly (1909), pp. 143–144.

42. Holmes (1927), p. 15, also quoted in the 2d edition, p. 215.

43. The proliferation of explanation was made a virtue by American geologists later in the century; see chapter 5.

44. Longwell (1928), pp. 145–157, on p. 152, original emphasis. In private, Longwell concluded that the evidence *was* compelling—see chapter 4—but in public he took a decidedly more cautious stand. In the 1940s and 1950s, Longwell was one of the few Americans who maintained a neutral stance with respect to drift. The larger question is *what* makes evidence compelling to some and not to others; this is the subject of the latter two-thirds of this book.

45. Marvin (1973), p. 179.

46. Allegre (1988), p. 216, original emphasis.

47. Bonatti (1994), p. 44.

48. Davis (1992), p. 493.

49. Saull (1986), p. 536.

50. Gould (1979), p. 161. The same argument can be found in Marvin (1973, p. 179) and Cox (1973, p. 14).

51. Laudan (1978), p. 230, original emphasis.

52. Wegener ([1929] 1966), p. 13. There is a complicated point here. Some readers will argue, and geologists writing retrospectively in the 1970s did argue, that a yielding *substrate* is no help to horizontal drift, because it was the ocean *floor* that had to yield. However, it is clear from his writing (see third edition, pp. 31–37, figures 6 and 7, and even more explicitly in the fourth edition, pp. 35–45, especially figures 9 and 10) that Wegener interpreted the theory of isostasy to require yielding of the ocean floor as well as the underlying substrate, for how could the continental crust experience vertical oscillation unless the oceanic crust to which it was attached also yielded? This point is developed further in chapter 4.

53. LeConte (1872). A liquid interior was implicit in von Buch's craters of elevation theory and many other global theories of the nineteenth century.

54. Thomson (1862a, 1862b, 1863, 1876). This point was essential to Kelvin's premise that the earth's cooling history could be adequately addressed by Fourier's theorem. The irony, as Wegener notes, is that steel and glass are *not* rigid when exposed to small stresses over geological time periods. George Darwin later developed a partial refutation of Kelvin's argument from earth tides in order to allow for a greater age of the earth and thus protect his father's theory of natural selection from refutation; Osmond Fisher seconded Darwin's arguments. For contemporaneous discussions of Kelvin's position, see LeConte (1872) and Fisher (1874, 1878, 1895); for recent historical treatment, see Smith and Wise (1989, pp. 552–578 and 600–602).

55. LeConte (1872), p. 352.

56. LeConte (1872), p. 472.

57. Fisher (1878), p. 293.

58. Fisher (1881), pp. 83–84, original italics. Fisher also emphasized that a plastic substrate was necessary to explain large-scale horizontal movement, as argued by Heim (1878).

59. Airy (1855), p. 101.

60. Airy (1855), pp. 244–246, on p. 244.

61. Fisher (1882c), p. 526. The impasse between Kelvin and Fisher was especially ironic because the advocates of isostasy were among the most quantitatively oriented of late-nineteenth-century earth scientists.

62. Dutton (1889), p. 60.

63. Dutton (1889), p. 60.

64. Gilbert (1886), p. 298. See chapter 5.

65. Woodward (1889), pp. 343–344.

66. Woodward (1889), p. 16. Woodward may have inspired Hayford with this insight. Note that Woodward is basing his discussion on the Pratt (rather than Airy) model of isostasy (see his p. 79). This raises an important point to be discussed further in chapter 6: Airy was an astronomer who wrote a paper about an idea; Pratt was a practicing geodesist whose treatise on the figure of the earth became a disciplinary standard (see Woodward 1889, pp. 85–86). Thus, even before he "proved" isostasy using the Pratt model, Hayford was working in the tradition of Pratt.

67. Woodward (1889), p. 352. The stanza is: "Sooth'd with the sound, the king grew vain/Fought all his battles o'er again/And thrice he routed all his foes/and thrice he slew the slain" (John Dryden, *Alexander's Feast*, line 66)—Kelvin's metaphor was thus unintentionally ironic. See Thomson (1876), p. 429, and the discussion in Fisher (1878).

68. Chamberlin (1898), p. 455. Chamberlin's planetesimal hypothesis is discussed in chapter 5.

69. Chamberlin and Salisbury (1909), vol. 2, pp. 233–234; original emphasis.

70. Chamberlin and Salisbury (1909), vol. 2, p. 235.

71. T. C. Chamberlin to Bailey Willis, March 16, 1894, BWP 11:32.

72. Hayford (1906), p. 38. This undertow was in the reverse direction to what Wegener required; nonetheless, notice the similarity between Hayford's and Wegener's language.

73. Dutton, comment on Hayford (1906), p. 39. On isostatic undertow and its discussion by H. F. Reid, originator of elastic rebound theory, see Wood (1915), p. 221 and footnote 14, and Barrell (1918a), p. 187.

74. Gilbert (1914), p. 34. Also see discussion in Mabey (1980), p. 66.

75. Gilbert (1914), p. 35.

76. Gilbert (1914), p. 35.

77. Adams (1912). A related, albeit later, example of experiments aimed at elucidating the rheology of the crust and subcrust is MacCarthy (1928).

78. Becker (1915), pp. 188–189. Bailey Willis also discussed the "undertow"; in early papers he advocated it (Willis 1907, 1910), but later he abandoned it (Willis 1922).

79. In Chamberlin's view, the weight of the continents would force the substrate outward in all directions, producing no net lateral motion. Hayford, in contrast, focused on the isostatic effects of sedimentation on continental margins to produce an undertow from the oceans toward the continents.

80. Lawson (1924).

81. Hall (1859), p. 81. Dwight Mayo has shown that Hall had Herschel's mechanism in mind when developing his theory (Mayo 1968, pp. 36, 50; and 1985). It is generally accepted that Herschel first proposed his ideas in a letter to Lyell, subsequently published as part of Babbage's ninth Bridgewater treatise (Babbage 1837, p. 207). Babbage (1847) used the concept of a mobile substrate subject to thermal expansion and contraction to explain the subsidence and reelevation of the Temple of Serapis, near Naples, attributing fluidity to partial

melting caused by the differential effect of the Earth's internal heat on strata with different melting points.

82. McGee (1896).
83. LeConte (1872), p. 468.
84. Dutton (1889), pp. 54–56.
85. Barrell (1914a), p. 36. Hall's theory only required isostasy in zones of sedimentation, but Hayford's work was widely interpreted as showing that isostasy was ubiquitous.
86. Wegener discussed the relation between geosynclinals and drift explicitly in the third (1924) edition of *The Origin*, pp. 163–165, but less clearly in the fourth ([1929] 1966) edition, e.g., p. 93.
87. Lyell (1850), p. 502.
88. Lyell (1850), p. 504.
89. Jamieson (1865, 1882), pp. 400, 457. See also Shaler (1874), p. 291, and the discussion in Daly (1940), pp. 310–326. Dana (1875) also suggested glacial ice as a cause of subsidence; see Mayo (1985), p. 9.
90. Fisher (1882c).
91. Wegener ([1929] 1966), pp. 43–45, and Daly (1940), pp. 313–315, drawing on De Geer (1888).
92. Nansen (1922); see also Nansen (1928).
93. Wegener ([1929] 1966), p. 44.
94. Nor was this consensus only on the level of specialists. By 1933, plastic flow was taught in Geology 1 at Dartmouth College as part of the overall knowledge of the structure of the earth. An exam paper in February of that year, for example, posed the question, "What reasons are there for the view that: (a) the earth has a core very different from the outer portion? (b) there is a thin surface shell, mostly granite, very different from the material beneath it? (c) rocks at depths of twenty miles can flow? (Geology 1 Course Notes, February 2, 1933, Dartmouth College, Papers of Lincoln Page, Courtesy of the family of Lincoln Page); see chapter 9.
95. Wegener ([1929] 1966), pp. 43–45.
96. Ruse (1978), p. 252. Today, geologists say that solid lumps of rock *do* move through other lumps of rock over time: this is what occurs when a lithospheric slab subducts into the asthenosphere. The study of the flow of solid materials is the stuff of rheology; questions of large scale movements are the stuff of tectonophysics.
97. Wegener ([1929] 1966), p. 43.
98. Wegener (1924), p. 40.
99. And there *was* evidence of deformation in the ocean floor, which Bailey Willis would later make use of; see chapter 7.
100. Wegener (1924), p. 131.
101. Wegener (1924), p. 131.
102. Wegener (1924), p. 130; see also Wegener ([1929] 1966), pp. 54–59.
103. Wegener ([1929] 1966), p. 45.

Chapter 4

Part of chapter 4 first appeared as "The Rejection of Continental Drift," *Historical Studies in the Physical and Biological Sciences*, vol. 18, part 2, pp. 311–348. Reprinted with permission of the Regents of the University of California.

1. Wegener (1924).
2. Taylor and Leverett (1915).
3. Taylor's early works are difficult to obtain; see Michele Aldrich's entry in the *Dictionary of Scientific Biography*.

4. Taylor (1910).

5. Taylor (1910); also see Taylor (1925, 1926) and Pickering 1907, 1924). On Taylor's and Pickering's contributions as indicative of widespread demand for theoretical innovation at this time, see Laudan (1985).

6. Wegener ([1929] 1966), pp. 3–4.

7. Wegener (1924), p. 190.

8. Wegener (1924), p. 200.

9. Wegener (1924), p. 205.

10. Wood (1915), p. 214. Wood's invocation of "true and sufficient" causes harks back to Isaac Newton (although he did not note this), suggesting that American geologists were working within an implicit legacy of British empiricism; see chapter 5. For a discussion of Newton's influence on nineteenth century British geology, see Laudan (1987), pp. 13–14.

11. van der Gracht (1928b), pp. 4–5.

12. Lambert (1921), p. 135.

13. Lambert (1921), p. 137. The geologist was Daly.

14. Lambert (1921), pp. 136–137.

15. Lambert (1921), p. 138.

16. Wright (1923), p. 31.

17. *The Earth: Its Origin, History, and Constitution* went through seven editions, the last of these in 1976. The sixth edition contained a discussion on the impossibility of plate tectonics; see Nur (1976) and "Continental drift loses a dedicated opponent" (1989). Besides his obstinacy regarding plate tectonics, Jeffreys held the record for the longest continuous tenure of a fellowship at an Oxford or Cambridge college—over 74 years at St. John's. For a recent paper contra plate tectonics, see Keith (1993). Keith's position is interesting but anomalous; earth scientists acknowledge various small difficulties with plate tectonics—as biologists acknowledge small difficulties in the evolutionary synthesis and physicists in the standard model—but there is no question that the overwhelming majority fully endorse plate tectonics.

18. Two example are Cox (1973) and Wilson (1971); many geophysicists propose this reason in conversation. For a contemporaneous summary of Jeffreys's views on contraction, see Gutenberg (1939), p. 189.

19. Nur (1976). See also Kozenko (1991).

20. If Jeffreys were so influential, then why did his objections hold so little sway in the 1960s when he was more famous and powerful than he was in the 1920s? Scientists who accepted plate tectonics in the 1960s did so over Jeffreys's objections; those who were sympathetic to drift did likewise in the 1920s. Those who understood Jeffreys best, like Walter Lambert and Arthur Holmes, were among his strongest resisters. Another irony, missed by those who have exaggerated Jeffreys's influence, is that he was British. If his views had been as influential as is often held, one might expect British scientists to have been more resistant to drift than Americans, but the reverse was the case. In fact, geophysicists at Cambridge in the 1960s played seminal roles in the development of plate tectonics in spite of Jeffreys's continued presence and opposition.

21. To reconcile the conflict between the geodetic and seismological perspectives, some scientists suggested that the substrate was heterogeneous, perhaps analogous to a sponge: pools of plastic or even molten rock sufficient to impart an overall flexibility to the substrate within a solid framework or skeleton that transmitted seismic waves.

22. Holmes to Bailey Willis, August 31, 1931, BWP 15: 45.

23. Willis to R. A. Daly, January 17, 1929, BWP 26: 43.

24. Willis to Margaret D. Willis, December 12, 1928, BWP 39: 7. On Jeffreys and Jeans vs. Chamberlin and Moulton, and Moulton's complaints against Jeffreys, see Brush (1978b), especially pp. 82–83 and 86–88, and Willis (1929) p. 361.

25. Willis to Margaret D. Willis, April 8, 1929, BWP 39: 9.

26. Willis to R. A. Daly, January 17, 1929, BWP 26: 43. Willis divided the world into those who supported Chamberlin and those who did not. Writing to Arthur Holmes prior to the latter's planned visit to Harvard to deliver the Lowell lectures, Willis wrote, "At Harvard, in association with Daly you will find a congenial atmosphere [in support of continental drift], but elsewhere agnosticism is the prevailing attitude toward the speculations of Joly, Jeffreys, and the European school that follows Wegener" (Willis to Arthur Holmes, October 1, 1931, BWP 28: 19). The following year, Willis blocked Jeffreys's nomination as a foreign correspondent of the Geological Society of America. In a letter to Charles Schuchert, he wrote, "Jeffreys is not a geologist, either by training or habit of thought. He is a mathematical astronomer, whose disregard for geologic facts where they conflict with his theories I have had occasion to learn in conversation as well as from his writings" (Willis to Schuchert, October 4, 1930, CSP 38 Book 2, 1929–1932).

27. Wegener ([1929] 1966), p. 167.

28. Burchfield (1974, p. 144) makes this point in the context of the origins of the sun's energy and the age-of-the-earth debate. Evoking J. B. S. Haldane, he writes, "the universe is not limited by the scientist's conception of it."

29. Daly (1912); Birch (1960).

30. Bowen (1956). Daly was one of the first geologists to look systematically at the composition of igneous rocks, but he attributed the observed differences to density segregation or immiscibility. Bowen, in *The Evolution of Igneous Rocks* (1956), outlined the sequence of crystallization of minerals in magmas (now known as the Bowen reaction series) and showed how it could explain the formation of igneous rocks of various compositions, thus laying the foundations of igneous petrology.

31. Birch (1960). Louise Daly, who graduated *summa cum laude* in history, was an author in her own right and edited a volume of her father's Civil War letters, *Alexander Cheeves Haskell: The Portrait of a Man* (1934). Bailey Willis's letters home to Margaret Willis include frequent instructions for substantive points and/or references to be added to manuscripts in progress. She may also have contributed illustrations. Charles Schuchert, who was unmarried, had an assistant, Clara de Vene, who did the illustrations for his works and was his co-author on some articles.

32. Longwell (1927), p. 525.

33. "Geology." Entry by Reginald Daly to Harvard College Guide to Undergraduate Fields of Concentration, RADP 3304.227.

34. Daly (1923).

35. Daly to Walter Lambert, Jan. 18, 1921, RADP 4315.7: 1920 [sic]. Also see Daly to Lambert, Jan. 19, 1921, RADP 4315.7: 1920 [sic].

36. Daly to Walter Lambert, Jan. 18, 1921, RADP 4315.7: 1920 [sic].

37. Walter Lambert to Daly, Jan. 29, 1921, RADP 4315.7: 1921.

38. CIWGF, Papers of the Geophysical Laboratory: F. E. Wright folder. On Wright, see Schairer (1954). Wright was primarily known for optical work on experimental melts; see Shepard and Rankin (1911) and Rankin (1925).

39. RADP field notebooks; especially Folder South Africa IV 3551–3710, p. 86. January to June were spent in South Africa, July to September working their way up the rift to Cairo.

40. Frederick Eugene Wright, letters from Africa, courtesy of Helen Wright Greuter. I am extremely grateful to Mrs. Greuter for lending me copies of her father's letters from Africa, which provide an eloquent account of the food, transportation, and social relations they encountered. His description of his first experience of a nectarine is blissful. The geological aspects are recorded in detail in RADP HUG 4315.55, field notebooks, and notes 1909–1922, Box 2 of 2, South Africa, especially Folder South Africa III 3369–3550 and IV 3551–3710, which includes notebooks relating field observations to his theoretical ideas about the origins of igneous rocks—discussed further in chapter 5.

41. A. L. Hall, *Bushveld Igneous Complex of Central Transvaal*, cited in Birch (1960); also see Daly and Molengraaf (1924).

42. Daly traveled separately to South Africa.

43. Carozzi (1985).

44. Molengraaf (1898). The Permian glaciation of the Southern Hemisphere is now accepted by geologists.

45. Molengraaf (1916), on p. 621. See also Molengraaf (1912, 1915a, 1915b).

46. Evans (1921), p. 119.

47. Pronounced "dew toy."

48. Letters from F. E. Wright to his wife Kathleen, June 12, 1922, and June 18, 1922; courtesy of Helen Wright Greuter.

49. RADP Folder South Africa IV 3551–3710.

50. RADP 4315.55 Field notebooks and notes 1909–1922, RADP Box 2 of 2, South Africa, Folder II, Note book South Africa II 3180–3368, Entry April 10, "at Pretoria, Eng train to Jo'burg Goldfields Hotel, where Mol'f lectures on Wegener" (p. 57).

51. Daly (1923), pp. 349–350. Also see Daly to Walter Lambert, Jan. 18, 1921, RADP 4315.7: 1920 [sic].

52. The flaw in this reasoning, viewed in retrospect, is that Daly neglected the effect of time: volcanic glasses freeze because they cool so quickly. Still, the point was to demonstrate the possibility of properties that would allow drift, and this Daly did.

53. Daly (1923), pp. 358–361.

54. Daly (1923), pp. 361–362.

55. Daly (1923), p. 364.

56. This contrasts with "ridge-push," whereby slabs are pushed forward by the force of igneous intrusions at the mid-ocean ridges. At present, ridge-push models are more widely accepted, but slab-pull models have been taken seriously and modeled by geophysicists. In plate tectonic models, the slab consists of the crust plus the rigid upper portion of the mantle, which are together referred to as the lithosphere.

57. Daly (1923), p. 365.

58. Daly (1923), p. 366.

59. Daly (1923), pp. 368–371.

60. Daly (1926), p. 260. The epigram was purportedly muttered by Galileo under his breath, even as he acquiesced to Pope Urban's demand that he teach Copernican astronomy only instrumentally.

61. Bowen's *Evolution of Igneous Rocks* was not published until 1928, but Daly was aware of Bowen's experiments and early articles.

62. Daly (1926), especially pp. 263, 268.

63. For example, Willis (1928) and Longwell (1928), pp. 150–151. This argument is often cited by geologists today, e.g., Laporte (1985).

64. Daly's model bears resemblances to a theory developed earlier in France by Eduard Reyer as a mechanism of generating crustal folding (Reyer 1892; see discussions in Zittel 1901, pp. 322–323, and Greene 1982, pp. 254–256).

65. Chamberlin (1928), p. 87.

66. On the planetesimal hypothesis, see Brush (1996c), and chapter 5.

67. Willis to M. D. Willis, Sept. 10, 1928, BWP 39: 7. This letter and comments like it elsewhere belie Robert Newman's recent claim (1995) that American loyalty to the planetesimal hypothesis was the root cause of the rejection of continental drift. By 1928, Willis was nearly alone in his loyalty to T. C. Chamberlin—except perhaps for the company of Rollin, son of T. C. (see chapter 5, esp. note 48).

68. Longwell (1927), p. 525.

69. Schuchert (1928a), p. 141.

70. Longwell (1928), p. 151.

71. On Joly's astounding range of activities, see TCD/Geol/Joly/0037, Dixon (1934), and Nudds (1986). On Joly's 1894 patent for the first single-image method for creating color plates and transparenices, see Coonan (1991). On his work on maritime navigation, see TCD/Geol/Joly/0025. Joly is also sometimes credited as the first person to suggest the existence of two isotopes of uranium, arising from his work on pleochroic haloes (see TCD/Geol/Joly/0037); I have been unable to confirm this suggestion. The best brief summary of Joly's life is Jackson (1992).

72. John Joly, August 28, 1886, "Oldhamia," quoted in Dixon (1934), and reprinted in *Upon Sweet Mountains*, a limited edition pamphlet published in Dublin by Trinity Closet Press in 1983, ed. J. Nudds.

73. JJP-TCDMS 3475, Diary, 1882.

74. His most important paper from this period is Joly (1889). For a full bibliography, see Dixon (1934).

75. Dixon (1957) and JJP-TCDMS 2313/12, MS 2313/14, and MS 2313/21.

76. Nudds (1986) and Coonan (1991).

77. Nudds (1986), p. 86.

78. Joly (1899), p. 23. Steven Jay Gould (1965) made this point many years later.

79. Joly (1899), p. 29. Joly neglected to factor in the processes that remove salt from the ocean—evaporite formation, adsorption onto clays, uptake into the shells of micro- and macrofauna—moreover weathering rates may not be constant through geological time.

80. Joly (1903). See also Brush (1996b), p. 68.

81. In his 1909 book *Radioactivity and Geology*, Joly increased his estimate to 150 million years, in deference to Bertram Boltwood's preliminary radiometric dates. Joly concluded, "With an interest almost amounting to anxiety, geologists will watch the development of [radiometric] researches . . ." (Joly 1909, pp. 250–251). On the history of radioactive dating, see Badash (1968) and Brush (1996b).

82. Joly and Rutherford (1913). Joly also founded the Dublin Radium Institute in 1914, where he helped to pioneer the use of hollow needles to deliver radium directly to cancerous areas of the body (Nudds 1986).

83. E. Rutherford to Joly, January 2, 1920, JJP-TCDMS 2312/362.

84. JJP-TCDMS 2297a-ND, p. 11 (sometime between 1903 and 1908). Later he changed this to "Efforts of this kind are healthy and in the cause of advance"!

85. JJP-TCDMS 2297a-ND, p. 11. Joly never entirely gave up trying to salvage an age in the hundreds of millions of years, but he accepted that his own 90 million was too low. See Holmes (1926b).

86. Joly (1909), p. 9.
87. Joly (1909), p. 94.
88. Joly (1909), p. 95.
89. Joly (1909), p. 148.
90. Joly (1909), p. 96.
91. Joly (1909), p. 111.
92. Joly (1909), p. 172. One cannot help but note the possible parallel with Irish nationalism, as many in Ireland were at this time imagining the precolonial past and looking forward to a postcolonial future. I have found no evidence of involvement on Joly's part in Irish nationalism—indeed, given his aristocratic heritage and professional links to England, he may well have been a loyalist—yet it is hard not to imagine at least some connection when reading these words.

93. Joly (1909), p. 101.
94. Joly (1909), pp. 113–144; see also pp. 127–131 for further discussion of the permanence of the overall configuration of the continents.

95. Joly (1909), p. 131.
96. Joly (1909), pp. 184–185.
97. Joly (1924, 1925, 1926a, 1926b, 1928a); Joly and Poole (1927).
98. Halley lecture, May 28, 1924, JJP-TCDMS 2302/6, published in Joly (1924), p. 39. Joly was not alone in being impressed by flood basalts; in 1872, the American Joseph LeConte listed the "lava floods" of the western United States as among the most "extraordinary" geological phenomena in the world and suggested they formed from cyclic melting leading to fissure eruptions (LeConte 1872, p. 470).
99. Joly (1924), p. 7; see also Joly (1925), Preface.
100. Joly (1925), pp. 87, 173.
101. Essay, "Radium and geology." Address to the University of Dublin Experimental Science Assocation, JJP-TCDMS 2297b (undated).
102. Joly (1925), p. 175.
103. Joly (1925), p. 177.
104. JJP-TCDMS 2295, "Notes for Ch. VI," and Joly (1925), pp. 96–106.
105. JJP-TCDMS 2297b (undated), "Radium and Geology, Dublin Address."
106. Joly (1925), p. 94. On volume changes associated with melting, see p. 115.
107. Joly (1925), p. 272.
108. Joly (1925), p. 125.
109. Joly (1925), p. 172.
110. Joly (1925), p. 180.
111. Joly (1925), p. 184. Now referred to as the Tertiary volcanic province.
112. Joly (1925), pp. 186–187.
113. See Dixon (1957), p. 16.
114. W. J. Sollas to Joly, February 6, 1923, JJP-TCDMS 2312/383.
115. November 25, 1925, signature illegible, Geological Survey of India, JJP-TCDMS 2312/683.
116. J. H. Jeans was Secretary of the Royal Society 1919–1929.
117. J. H. Jeans to Joly, March 15, 1923, TCD/Geol/Joly/0042/1.
118. J. H. Jeans to Joly, March 15, 1923, TCD/Geol/Joly/0042/2. For Jeffreys's published objections and Joly's replies, see Jeffreys (1926) and Joly (1926a, 1927).
119. Note, March 15, 1923, TCD/Geol/Joly/0031. It is not clear whether this note was formally communicated to Jeans; it seems likely that it was.
120. Joly (1928c), p. 88.
121. Joly (1928c), p. 89.
122. Joly ([1925] 1930), 2d ed., p. 149.
123. Dixon (1957), p. 16.
124. Dunham (1966); see also Allwardt (1990).
125. Dunham (1966).
126. See Brush (1996b), pp. 7–9 and pp. 66–86; Burchfield (1974), pp. 171–179; and Badash (1968). Boltwood's numbers were not reliable, because he measured lead from both uranium and thorium—a problem that was only later resolved after Soddy's 1913 discovery of isotopes and Nier's 1936 invention of the mass spectrometer.
127. Holmes (1911a, 1911b, 1913a, 1913b, 1913c, 1914). See also Holmes (1931) and (1946).
128. For three perspectives on the resolution of the debate over the age of the earth, see Burchfield (1974), Gould (1983), and Brush (1996b).
129. Holmes (1913a), p. 398.
130. Holmes continued to defer to Joly all his life. After publishing his seminal contribution to the National Research Council Volume *The Age of the Earth* Holmes wrote to Joly

to solicit comments: "It is so easy to get off the rails, that I would appreciate it if you could tell me quite frankly where you think I may have done so" (A. Holmes to Joly, August 31, 1931, JJP-TCDMS 2312/128). Throughout his private correspondence Holmes oscillated between being rather bold and rather obsequious.

131. Holmes (1926c), p. 306.
132. Holmes (1926c), p. 307.
133. Holmes (1926c), p. 307. Argand was a supporter of continental drift at least since 1916 but for obvious reasons said relatively little about it during the war years; see Argand ([1922] 1977).
134. Holmes (1928), pp. 237–238.
135. Holmes (1925a, 1925b, 1926c, 1929a, 1929b, 1931, 1933).
136. Holmes (1929a), p. 346.
137. Holmes (1929a), p. 287.
138. Holmes (1929a), p. 345.
139. Holmes (1926c), p. 308. Viz., "We do not claim to reduce tectonics to physics; that is the business of the future."
140. Chamberlin and Salisbury (1909), vol. 1, p. 560; also see Brush (1996b).
141. LeConte (1872), p. 351.
142. Fisher (1878), p. 293.
143. Fisher (1878), p. 295.
144. Fisher (1895), p. 246.
145. Fisher (1878), pp. 295–296.
146. Kushner (1990, 1993); Brush (1996a, 1996c).
147. Rothpletz is discussed in Zittel (1901) and Bucher (1933); German advocates of convection, including Ampferer, are discussed in Wegener ([1929] 1966), pp. 56–50, and in Gutenberg (1939), pp. 184–185. Ampferer's theory of "understreaming currents" was very close to Holmes's, but he did not specifically link these currents to heat transfer (Ampferer 1925a, 1925b; Gutenberg 1939). See also Bull (1921, 1927, 1929, 1931).
148. Griggs (1936, 1938, 1939a, 1939b); see also Lawson (1939). For a recent historical treatment, see Allwardt (1990).
149. Griggs (1974).
150. It was further propounded in Holmes's famous textbook, *Principles of Physical Geology*, which he wrote when he had few students during the war years. This textbook, used by more than a generation of aspiring geologists from grade to graduate school, was described by Sir Kingsley Dunham, Holmes's eulogizer, as "deserv[ing] to rank with Lyell's *Principles* as one of the most successful textbooks of geology ever written" (Dunham 1966). Several geophysicists involved in the development of plate tectonics, including P. M. S. Blackett, were said to have first heard of the idea of drift from this book; see chapter 9.
151. This point is emphasized especially by Rudwick (1985). For a recent review of the state of geophysical knowledge on mantle convection, see Bonatti (1994).
152. Longwell to Schuchert, January 31, 1926, CSP 20: 173.

Chapter 5

1. Hallam (1983), p. 152; van Andel (1985), pp. 85–86.
2. R. A. Daly to Walter Lambert, January 18, 1921, RADP 4315.7: 1920 [sic], and Longwell (1928), p. 145.
3. W. Bowie to C. Schuchert, June 17, 1927, CSP 19: 166. Rachel Laudan (1985) also makes this point: the proliferation of tectonic theories in the early twentieth century, including Frank Bursey Taylor's version of continental drift, is prima facie evidence that many

geologists, including rank-and-file ones, recognized the limitations of existing theory. That Taylor had so little impact is evidence, she argues, of how hard it is to change scientific thinking even when the need for change is widely recognized.

4. Lake (1926), pp. 181-182.
5. Van Andel (1984), pp. 337-343.
6. B. D. (1931), p. 861.
7. Cohen (1985, p. 453) blamed Wegener's nationality for the failure of his theory, but even if the ellipsis of Wegener did reflect anti-German bias, it should not have extended to the corollary work of British colleagues. With regard to the issue of disciplinary affiliation, Thomas Kuhn (1962) famously emphasized the positive role of outsiders. For a postwar complaint about German exclusion, see Brennecke (1927).
8. Schuchert made this point repeatedly in conversation, perhaps to make clear (or to reassure himself?) that nationalism had nothing to do with his objections. For example, to the Viennese geologist Carl Diener in 1921, Schuchert wrote, "Wegener's theory is a crazy one. It originated in America...." Schuchert to Carl Diener, November 30, 1921, CSP 37: Book I (1921-23).
9. Wright (1923), pp. 30-31.
10. Evans (1923), p. 393.
11. Wright (1923), p. 30.
12. Gregory (1925), p. 256.
13. Wright (1923), p. 30; Evans (1923), p. 393.
14. Meyrick (1925), p. 834.
15. Lake (1922), p. 338.
16. Lake (1922), p. 346.
17. Gregory (1925), p. 257.
18. Willis (1928), p. 76.
19. Berry (1928), p. 194. In the 1920s, autointoxication was a euphemism for constipation; Berry's implications are thus not subtle.
20. Burchfield (1974), p. x. Reingold (1991, p. 68) calls this the Turner thesis of American historiography of science—that the frontier mentality controlled American scientific perspectives until the twentieth century.
21. Le Grand (1988).
22. Stewart (1990). A variety of views have been expressed on this issue by other authors. For example, Rachel Laudan (1987) has argued that causal and historical geology should rightly be viewed as two different disciplinary programs that have coexisted throughout much of the history of geology; it is only the proponents of the latter who eschewed causal accounts. Thus, continental drift may have transgressed the aims of a significant *sector* but by no means all of the scientific community. However, David Kitts (1977) argued that the very nature of geology as a science privileges historical reconstruction over causal explanation. Kitts's work is a good example of the perils of making ahistorical claims about scientific methodology; twenty years ago, many geologists would probably have agreed with his statement; today, most would probably not.
23. Wilson (1971), p. 30.
24. Cox (1973), p. 5. Cox blames the lack of an adequate causal mechanism for the rejection of drift, but if earth scientists prior to plate tectonics were mere classifiers, then why should the lack of a causal mechanism have been a concern to them?
25. The role of history in the plate tectonics revolution has been discussed by Laudan (1983).
26. Wood (1985), p. 223.
27. A number of historians of science have valiantly tried to slay this particular historiographic dragon, from George Daniels (1972) to Julie Newell (1993), but the trope of naive

empiricists seems to be unfailingly satisfying for those who insist on having a boy (or girl) to whip.

28. Struik (1957), pp. 47, 153. Similar views were propounded by the eminent historians Richard Shyrock and I. Bernard Cohen and by the sociologist Robert Merton (see Reingold 1991, pp. 57–59). I wrongly accepted this characterization of American geology in my early work (Oreskes 1988, p. 344).

29. Dupree (1986), p. 9.

30. Bruce (1987), p. 65.

31. Kevles (1971) attributed the change in American physics largely to the influx of European emigrés fleeing Nazi Germany; others have given more credit to internal developments in American physics (e.g., Sopka 1988), but most agree that American physics was theoretically impoverished until the post–World War I period.

32. Kevles (1971), p. 37.

33. Reingold (1991), pp. 54, 63–64.

34. An obvious example of this is Hayford's work, the theoretical implications of which are almost impenetrable without first reading other people's work.

35. Reingold (1991), p. 55.

36. Reingold (1991), p. 61. Also see Miller (1972), Rosenberg (1972), and other articles in Daniels (1972).

37. Hannaway (1976). On the development of the American research university, see Veysey (1970) and Marsden (1994); on the development of chemistry in America, see Servos (1990).

38. Greene (1982); also see Newell (1993), her chapter 4. Newell notes that historians have mostly failed to come to terms with the complex positions Americans took vis-à-vis theory and often read twentieth-century meanings into nineteenth-century vocabulary. Too right!

39. Dutton (1880), pp. xxiv, xvii.

40. Taylor (1910, 1925); Baker (1914); Pickering (1907, 1924); Daly (1926). W. H. Pickering played an important role in the refutation of the nebular hypothesis in his discovery of the retrograde motion of Phoebe, the ninth satellite of Saturn; see Brush (1978a), p. 19. His version of continental drift, however, was less successful.

41. Chamberlin (1897b); see also Fleming (1992).

42. On Chamberlin, see the series of eulogies in the May-June 1929 edition of the *Journal of Geology*, and Schuchert (1929a); also see Brush (1978a, 1978b) and Greene (1982), pp. 258–275.

43. Some historians and geologists have suggested that Chamberlin created the *Journal of Geology* specifically to promote ideas that were opposed by agonists in control of the *American Journal of Science*; see, for example, Schultz (1976).

44. Brush (1978a, 1978b, 1996c); Smith (1982).

45. Willis (1929); also see Van Andel (1985), p. 365.

46. Hovey (1901). Stephen Brush has suggested that the planetesimal hypothesis had more impact on astronomers than on geologists: although few fully accepted it, it led many to abandon the nebular hypothesis and consider other theories of the origin of the solar system. See Brush (1978a, 1978b, 1996c) for a complete bibliography and account of the planetestimal hypothesis.

47. Chamberlin (1897b), p. 676. See also Chamberlin (1899e, 1899f).

48. Rollin T. Chamberlin (1928) argued that this provided Americans with an answer to the objections to contraction theory and that therefore there was no need for continental drift; but in so doing, he was ignoring all of the homological evidence. But Rollin was no genius — Charles Schuchert referred to him as a "very small chip" off his father's block, invited to the 1922 Ann Arbor meeting only for political reasons (Schuchert to Arthur Keith, January 31, 1922, CSP 37: Book I (1921–23)).

49. Schuchert (1929a), p. 196.
50. Chamberlin (1906).
51. Chamberlin (1906).
52. This point is also made by Brush (1978a), pp. 14–15.
53. Chamberlin (1906). Chamberlin's father was an evangelical minister; Bailey Willis (1929) claimed that Chamberlin had found in science an alternative to the inflexible theology in which he had been raised. But it would be wrong to conclude that Chamberlin rejected religion altogether; rather, like many early twentieth-century geologists, he sought a religious perspective compatible with his science. On this point, see Winnik (1970), Spanagel (1996), and Bork (1994).
54. Yochelson (1994).
55. Walcott (1902), p. 115.
56. Reingold and Rothenberg (1981), p. 3; see also Servos (1984), Reingold (1991), and Kohler (1991).
57. CIWGL: F. E. Wright folder.
58. Schairer (1954).
59. CIWGL: F. E. Wright folder.
60. Wright (1937), pp. 1–4. Wright's experience was not unusual; for a discussion of how wartime work strengthened the links between academic and applied chemistry, see Servos (1990), pp. 215–220.
61. Wright (1937).
62. This network of institutional support contrasts strongly with physics, which was not directly associated with any particular federal agency at this time (Kevles 1971).
63. Oreskes (1994); see also Manning (1967). This is not to say that geology did not already blend theory and practice: the importance of coal, limestone, iron, and other mineral resources made geological work inescapably practical, as illustrated by the work of men like William Smith who surveyed the lands of English gentry for their economic value, and by the growth of continental mining academies like Werner's famous one in Freiburg. The growth of state and federal surveys as a primary source of geological employment in America thus institutionalized a pattern that already existed in other places. What was distinctive about America, however, is how central a role these surveys played.
64. Yochelson (1994). One of the greatest advocates of the unity of theory and practice was John Wesley Powell, whose proposal to use drainage basins as the basis of western settlement cost him the Directorship of the U.S. Geological Survey. On tensions between theory and practice in government service see Rosenberg (1972).
65. Daly (1912).
66. Until recently Werner's important contributions were overshadowed by accounts that emphasized his discredited theoretical views (e.g., Geikie 1897 and Adams 1954; see also the discussion in Ospovat 1976). The best recent accounts are Hallam (1983) and Laudan (1987). On the role of experimental evidence in the neptunist-plutonist debate, see Newcomb (1990).
67. Hallam (1983), Brown (1989), Wilson (1979). The greenstones are now recognized as altered diabase dykes, consistent with plutonist interpretation.
68. Spanagel (1996), p. 32.
69. Quoted in Schuchert (1918), p. 70; originally in *American Journal of Science*, 43 (1842): 225. Although Silliman leaned toward plutonism, he tried in his teaching to articulate the strengths of both perspectives—consistent with the idea of pluralism later advocated by Chamberlin; see Spanagel (1996), p. 34, and Brown (1989).
70. Quoted by Barrell (1918a), p. 160, from Edward Hitchcock, "Greenstone dykes in old red sandstone," *American Journal of Science* 6 (1823): 56–60.
71. Barrell (1918a), p. 158. Barrell may have been specifically referring to William MacClure, who published the first geological map of the United States, using Wernerian

nomenclature; see Ospovat (1959). Or his may have been a more general complaint; Schuchert agreed that initially Americans attempted to describe stratigraphic sequences on the Wernerian system (Schuchert 1918, pp. 74–75). The idea that Werner pushed speculative theory without empirical justification was particularly promoted by Archibald Geikie, in his *Founders of Geology*: "Never in the history of science did a stranger hallucination arise than that of Werner and his school, when they supposed themselves to discard theory and build on a foundation of accurately-ascertained fact. Never was a system devised in which theory was more rampant; theory too, unsupported by observation, and, as we now know, utterly erroneous" (Geikie 1897, p. 212). Geikie was the main source on Werner for most American geologists in the early twentieth century.

72. Barrell (1918a), p. 170.

73. There were elements of envy in American critiques of Europeans, to be sure, particularly at the start of the nineteenth century when most American scientists lacked professional positions (Newell 1993, Spanagel 1996), but this was far less so by the century's end. As for Barrell, there is little reason to suspect jealousy in his critiques: he was widely admired, highly successful, and, as a result of his work in the mining industry, very rich.

74. Woodward (1889), p. 354. Woodward's antiauthoritarian view was obviously congenial to Americans hoping to make important contributions in their own right.

75. Willis to Margaret D. Willis, April 23, 1929, BWP 39: 9.

76. Barrell (1918a), p. 161.

77. Dutton (1880), p. 113.

78. Dutton (1880), p. 142.

79. Willis (1929), p. 359.

80. This philosophy clearly presupposes a notion of observation that is at least quasi-autonomous from theory; this point is discussed at length below. Suffice it to say here that although Chamberlin was an empiricist, he was anything but naive. His epistemology, his methodology, and his pedagogy were acutely conscious.

81. Chamberlin (1897a), p. 843.

82. Chamberlin (1897a), p. 843.

83. Chamberlin (1897a), p. 848.

84. Chamberlin (1897a), p. 844.

85. Chamberlin (1897a), p. 837.

86. Chamberlin (1897a), pp. 838–839.

87. Chamberlin (1897a), pp. 840–841.

88. This was not purely a stereotype. Zittel (1901), p. 307, argued that the "supreme scientific success attained by the first volume of *Das Antlitz der Erde*" could be attributed to its "universal aspect" and the "endless reserve" of knowledge on which it was based. In a paper ostensibly about the problem of longitude variation, the German geodesist Brennecke explicitly quoted Goethe advocating wholism (the context was German reinclusion in international geodetic work): "Immer strebe zum Ganzen, und, kannst du selber kein Ganzes Werden, als dienendes Glied schließ an ein Ganzes dich an!"; roughly, "Always strive toward the whole, and if you yourself cannot become a whole, ally yourself to a whole as a serving member!" (Brennecke 1927). (See *Goethes Werke*, Tome I, Hamburg, Christian Wegner Verlag, p. 226, no. 127.) Conversely, Karl Mannheim criticized Anglo-American social sciences in the 1920s and 1930s for its "hollow" empiricism and general lack of theorizing; see Kaiser, forthcoming.

89. Chamberlin (1897a), p. 842.

90. Lyell (1830).

91. Rudwick (1970); Porter (1976); Ospovat (1976); Taylor (1992).

92. Rudwick (1963); R. Laudan (1977). Laudan has argued that the British antitheoretical attitude greatly hindered British scientific advance in the first half of the nineteenth century, but more recently Martin Rudwick (1985) has shown how theory became embedded in out-

wardly empirical work in the construction of the geological time scale, which was, of course, one of the greatest scientific accomplishments of the nineteenth century.

93. Geikie (1897); see also Oldroyd (1980) and Taylor (1992).
94. Chamberlin (1897a), p. 842.
95. Pyne (1978).
96. Chamberlin (1890, 1897a, 1905, 1919, 1965).
97. For the hold that this idea continues to have on American geologists, see Johnson (1990) and Railsback (1990). The latter referred to the method of multiple working hypothesis as "one of the fundamental underpinnings of not only geology, but of science itself." Rachel Laudan (1980b) discusses the application of the principle in the development of plate tectonics.
98. Pyne (1980). Gilbert is the most famous but not the only scientist of this time who expressed views similar to Chamberlin's. Writing in 1889, Robert S. Woodward argued that progress in geodesy had been retarded because of adherence to the erroneous spheroid model determined by the great German astronomer and mathematician Friedrich Wilhelm Bessel, Director and Professor of Astronomy at the Observatory at Königsberg: "The weight of Bessel's great name may have been too closely associated in the minds of his followers with the weights of his observations and results. The sanction of eminent authority, especially if there is added to it the stamp of an official seal, is sometimes a serious obstacle to real progress" (Woodward 1889, p. 340). Woodward insisted on a sharp distinction between geodetic "facts" and "inferences" and quoted Marcus Aurelius: "Remember that all is opinion" (Woodward 1889, p. 346).
99. Gilbert (1886), p. 285.
100. Gilbert (1886), pp. 285–286.
101. Gilbert (1886), p. 286.
102. Gilbert (1886), pp. 286–287.
103. Gilbert (1900), p. 1012.
104. Van Hise (1904), p. 600.
105. Barrell (1919d), p. 282.
106. Chamberlin (1897a), p. 839.
107. Willis (1907). On Willis as a lover of theory, see CIWGF 1929–1947, Bailey Willis folder; also see Blackwelder (1961).
108. Willis (1907), p. 117.
109. Willis (1907), p. 133.
110. Scientists certainly encounter the unexpected in the laboratory, but their experiments are also created in such as way as only to make some kinds of phenomena observable (Galison 1987; Cartwright 1983). On the question of a priori knowledge in experimental work, see Kaiser (1992).
111. This is evident from the several excellent books on nineteenth century field practice, e.g., Rudwick (1985), Secord (1986), and Oldroyd (1996). It was certainly possible to go into the field with well developed theoretical goals, but it was *also* possible to go into the field with minimally theoretical goals.
112. Laudan (1987). This is not to imply that theory operates only on the level of causal explanation; theory is also embedded into classificatory systems and observational techniques. But when Chamberlin, Gilbert, and others talked about theory, they almost always meant causal and explanatory theory: What causes the geological features and phenomena that we are observing? How do we explain what we see? This kind of theory *could* be separated from the phenomena being explained. Not only could, but was, because many geological phenomena had no theoretical account, or, conversely, had multiple competing accounts, which complicates any simplistic notion of theory-laden-ness even on a microtheoretical level.
113. RADP Folder II, note book South Africa II 3180–3368, p. 46.
114. RADP Folder II, note book South Africa II 3180–3368 , p. 35.

115. Daly made the same demarcation in his classroom teaching: William Menard's notes from Daly's 1946 seminar in tectonics are filled with pages in which topics are outlined "A. Facts. B. Inferences." See HWMP 33: 11.

116. Merriam to Wright, February 1, 1927, FWP Folder 1927.

117. Dutton (1889), pp. 54–56; also see LeConte (1872), p. 468, viz., "Mountain chains are composed of enormous masses of sediments. This fact forms the basis of Hall's sedimentary theory."

118. Schuchert to Holmes, June 2, 1926, CSP 38: Book 1, 1925–1929.

119. Chamberlin (1897a), p. 839 n.

120. Merriam to Wright, February 1, 1927, FWP Folder 1927. The idea of theories as stepping stones towards scientific truth through a process of elimination was shared by Merriam's predecessor, Robert S. Woodward. Discussing the issue of the nature of the earth's interior in 1889, he had written, "We may expect to go on theorizing adding to the long list of dead theories which mark the progress of scientific thought, with the hope of attaining the truth not so much by direct discovery as by the laborious process of eliminating error" (Woodward 1889, p. 352).

121. Bowie (1926). Gilbert (1914, p. 33) likewise called isostasy a theory, saying explicitly that "we need no longer call it [a] hypothesis" given the amount and quality of empirical evidence of it.

122. Bowie (1926). Bowie himself was inconsistent; elsewhere he called isostasy a "fact" or spoke of the "facts of isostasy."

123. Hayford (1911), p. 207.

124. And not just in science. At the Carnegie Institution, historian J. Franklin Jameson, who helped to establish the U.S. National Archives, went even further toward empiricism than his scientific colleagues; he resigned from the CIW in part because he felt John Merriam went too far in pushing him towards interpretation! See Singer (1994).

125. For example, see Longwell's introduction to Barrell (1927), and Schuchert (1936).

126. Daly to Lambert, January 18, 1921, RADP 4315.7: 1920 (sic) and elsewhere in this collection; Longwell (1927), p. 525.

127. Schuchert to Alexander du Toit, April 25, 1922, CSP 37: Book 1, 1921–23; Schuchert to E. Blackwelder, May 23, 1922, CSP 37: Book 1, 1921–1923. On a much earlier discussion of the same problem, see Duchesneau's (1982) discussion of Buffon, especially p. 121.

128. LeConte (1893), pp. 549–550.

129. Schuchert to Malcolm Bissell, June 3, 1921, CSP 37: Book 1, 1921–23.

130. Willis (1929), p. 360. Similar comments were made regarding G. K. Gilbert and John Hayford, two of the most brilliant of American geoscientists. For example, in an obituary by W. C. Mendenhall, Gilbert's mental powers were praised as "unsurpassed by any geologist of his time," but his accomplishments were assigned to his capacity for unprejudiced hard work: "Having gathered his facts, he weighed them and marshalled them in proper order. . . . In his writings Gilbert never seems to be, and in truth never is, supporting a theory. He puts all theories of which he can conceive to the test of fact, indifferent as to which, or as whether any, survive." In short, he was true to his own method, and an American hero, "the very antithesis of the brilliant, temperamental, erratic genius" (Mendenhall 1920, p. 42). Similarly, writing of Hayford after his death, William H. Burger claimed that "the foundations of Hayford's work were the 'observed facts'; he never allowed imagination to lead him astray [and always] presented his conclusions to the scientific world in the simple language of an engineer: "Here are the observed facts, and here are the conditions of the crust that explain them" (Burger 1931, p. 212).

131. J. B. (1906).

132. Hofstadter (1963), p. 43.
133. Reingold (1991).
134. Kevles (1971), p. 42.
135. Chamberlin (1890), p. 96.
136. Kevles (1971), p. 19. Similar thoughts were earlier voiced by Benjamin Silliman at Yale; see Greene (1979).
137. Hannaway (1976); see also Kohler (1990).
138. And in other branches of intellectual thought, most notably pragmatism. Social scientists also felt the tension between empiricist and theoreticist perspectives; see Novick (1988) and Ross (1991).
139. Schweber (1986). Operationalism was an antitheoretical program, focusing on instrumental measurements and the formulation of explicit operational protocols for making such measurements.
140. Maienschein (1991b), p. 24. On patterns of practice in biology, see also Maienschein (1986, 1991a), Rainger (1981, 1993), and Pauly (1991). Pauly emphasizes that American biologists were catholic in both their interests and their explanations, agreeing with Maienschein that "their only strong antagonism was to explanations that seemed to reduce a wide range of phenomena to a single unifying cause" (Pauly 1991, p. 137). Besides leaving explanation to the end of their papers, many Americans also left it toward the end of their lives; the major protagonists in American geological debates were generally very senior scientists.
141. Wegener ([1929] 1966), p. 9.
142. Wegener (1924), pp. 11–12. Wegener did not back off from this position in his fourth edition, where he wrote that "even today [it] possesses a degree of attractiveness" ([1929] 1966), p. 9).
143. Wegener (1924), pp. 1, 9.
144. Pursuing the "offspring analogy" perhaps too far, continental drift was thus a bastard theory.
145. Wegener (1924), p. 5.
146. Wegener (1924), pp. 44–45.
147. Wegener (1924), pp. 41, 73, and passim.
148. Wegener (1924), p. 26.
149. Wegener (1924), p. 73.
150. Wegener (1924), pp. 1, 17, 27.
151. Wegener (1924), p. 56.
152. Wegener (1924), p. 120.
153. Schuchert (1928a), p. 139.
154. Schuchert (1928a), p. 139. Compare this with Bacon: "[the scientific method] derives axioms from the sense and the particulars, rising by a gradual and unbroken ascent, so that it arrives at the most general axioms last of all" (*New Organon*, Book 1, Aphorism XIX). On Boyle's program of linguistic demarcation of fact and interpretation see Shapin and Schaffer (1985, esp. pp. 140, 147).
155. Chamberlin (1928), p. 83.
156. Longwell (1928), p. 146.
157. Schuchert (1928a) p. 139.
158. Mentioned in passing in the third edition and more explicitly in the introduction to the fourth (and final) edition of his book.

Chapter 6

1. Reid was the originator of the elastic rebound theory.
2. Reid (1922), p. 674.

3. Wright to Merriam, November 1, 1922, CIWGF, Alexander du Toit folder.
4. Wright to Merriam, November 1, 1922, CIWGF, Alexander du Toit folder.
5. Wright to Daly, November 1, 1922, FWP, Folder 1923 [sic].
6. Wright to du Toit, February 9, 1923, CIWGF, Alexander du Toit folder.
7. Wright to Merriam, February 2, 1923, CIWGF, Alexander du Toit folder.
8. Wright to Merriam, February 2, 1923, CIWGF, Alexander du Toit folder.
9. Wright was not unique in studying in Heidelberg; one of his compatriots was Florence Bascom, the first American woman to receive a Ph.D. in geology.
10. Wright to Merriam, February 2, 1923, CIWGF, Alexander du Toit folder.
11. Gilbert to du Toit, February 9, 1923, FWP, Folder 1923.
12. Merriam to du Toit, May 4, 1923, CIWGF, Alexander du Toit folder.
13. Du Toit to Merriam, June 6, 1923, CIWGF, Alexander du Toit folder.
14. Du Toit to Merriam, February 2, 1926, CIWGF, Alexander du Toit folder.
15. Merriam to du Toit, March 15, 1926, CIWGF, Alexander du Toit folder.
16. Wright to Merriam, January 31, 1927, 1926, CIWGF, Alexander du Toit folder.
17. Merriam to Wright, February 1, 1927, CIWGF, Alexander du Toit folder.
18. Du Toit (1927), p. 17.
19. Du Toit (1927), p. 97.
20. Du Toit (1927), p. 97.
21. Du Toit (1927), p. 102.
22. Du Toit (1927), p. 103.
23. Du Toit (1927), p. 108.
24. Du Toit (1927), p. 119.
25. Du Toit (1927), p. 109.
26. Du Toit (1927), p. 109.
27. Du Toit (1927), p. 110. Original emphasis.
28. One might also have argued that the facies evidence was novel, whereas the fossil data were not, but du Toit himself did not make this claim.
29. Du Toit (1927), p. 110.
30. Bowie to Schuchert, June 17, 1927, CSP 19: 166.
31. Bowie (1925a, 1925b), see also WBP.
32. Bowie (1929a), p. 19; RUSCGS Entry 33, Box 775, Annual Office Reports 1926–1929.
33. Hayford (1911), p. 201.
34. Bowie to Schuchert, June 17, 1927 CSP 19: 166.
35. Bowie (1929b).
36. Bowie (1927c), pp. v, 247.
37. Bowie (1927c), p. 20.
38. Bowie (1929b), p. 221. Emphasis added.
39. Bowie (1931).
40. Bowie (1929a).
41. Bowie (1929a), pp. 181–82.
42. Bowie (1925a). What Bowie considered the "strong-point" is what Bailey Willis considered the weak point! See chapter 8.
43. Bowie (1929a), p. 180.
44. Bowie (1935), p. 444.
45. Bowie to Schuchert, October 11, 1928, CSP 21: 181.
46. Bowie to Schuchert, June 2, 1927, CSP 19: 166. Emphasis added.
47. Bowie (1927b), p. 107.
48. Bowie to H. O. Wood, April 13, 1927, HWP 16.1
49. Willis (1922) and unpublished correspondence with Arthur Holmes, October 1, 1931,

BWP 28: 19; also see Lawson (1924, 1938, 1939). Wood also argued — and was perhaps one of the first American geologists to argue so — that the mechanism of isostatic compensation might involve a combination of the Pratt and Airy models (see Wood to Bowie, March 24, 1925, HWP 16. 2, and Wood to Bowie, March 30, 1927, HWP 16.1).

50. Mott Greene suggests that Europeans' preference for Airy isostasy was in part related to their greater advancement over Americans in seismology: early European seismic data, which was only just beginning to be discussed in the United States, supported the notion of a variably thick crust. Bowie discounted this evidence as too preliminary, until the mid 1930s, when American seismologists began to corroborate their results; see Geschwind (1996) and chapters 8 and 9. The preference for Airy isostasy among British scientists was no doubt also in part because of Airy's stature. On Joly's preference for Airy isostasy and its relation to his support for continental drift, see Bucher (1933), p. 148.

51. Harold Jeffreys, published discussion in Bowie (1924b).

52. Burger (1931), p. 212.

53. Hayford (1909), p. 147.

54. Hayford (1909), p. 20.

55. Hayford (1909), p. 46.

56. Servos (1986). Hayford was not the only one to feel this way. Discussing the fissiparturition theory, Frank Bursey Taylor wrote: "G. H. Darwin's fission hypothesis impresses me as about the most preposterous doctrine that had ever been seriously advocated by scientific men, and when I see such men as Jeans and Jeffreys devoting such labor to a mathematical expression of that idea, I am filled with despair at the lack of intuitive insight as to physical reality displayed by men of such high training and reputed ability." (F. B. Taylor to C. Schuchert, December 23, 1927, CSP 22: 191). Geophysicists for their part expressed annoyance at geologists who proffered explanations in blithe disregard of physical or chemical constraints. See chapter 9.

57. Hayford (1909), p. 47.

58. On golden events versus statistics in experimental demonstration, see Galison (1987).

59. Bowie did finally draw such a picture in the last chapter of the book, where he discusses other people's opinions. Here he draws a picture of Airy isostasy as used by the geodesist W. A. Heiskanen but makes it clear that he considers this view to be in error (Bowie 1927c, p. 224; see also Heiskanen and Vening Meinesz 1958).

60. Bowie (1927c), pp. 40–41.

61. Bowie (1927c), pp. 40–47.

62. Bowie (1927c), p. 49.

63. Bowie (1927c), p. 237.

Chapter 7

1. In later years, the two men had a falling out, and Ulrich tried to block Schuchert's nomination to the Presidency of the GSA; see Weiss (1992).

2. On Schuchert, see Knopf (1952).

3. For example, Schuchert quoted from Daly's book in which the latter admitted that geologists were "going slow" in accepting the idea of drift because neither Wegener nor Taylor had given an account of the driving forces (Schuchert 1928a, p. 108). This might seem to support the argument that Americans rejected drift for lack of a mechanism, but the point is that Daly's comment here introduces the chapter of his book where he *proposes* a mechanism; see chapter 4.

4. Schuchert (1928a), p. 117.

5. Schuchert (1928a), pp. 141–142.

6. Greene (1982), p. 270.
7. Gregory (1923).
8. Joseph Barrell (1869–1919) (1919), p. 251.
9. Gilbert (1889), quoted in Barrell (1914a), p. 34.
10. L. Pirsson, quoted in Schuchert (1919), p. 274.
11. Barrell (1919d), p. 282.
12. Schuchert to Holmes, June 2, 1926, CSP 38: Book I, 1925–29.
13. Barrell (1919d).
14. Barrell (1914b), p. 146.
15. Barrell (1914b), p. 147.
16. Barrell pointed out that the sum of the squares of the residuals was not particularly sensitive to the depth of compensation and that a value anywhere between 66 and 305 km was compatible with Hayford's results; see Barrell (1914b), p. 151. As we have already seen, Hayford acknowledged that his assumptions were conventional, but Bowie took Hayford's end result and ran with it; see chapter 6.
17. Barrell (1914b), p. 158.
18. Barrell (1914b), p. 158.
19. Barrell (1914b), p. 164.
20. Barrell (1914b), pp. 164–165.
21. Barrell (1914f), p. 657.
22. Barrell (1914c), p. 227.
23. Barrell (1914c), pp. 227–228.
24. Barrell (1914d), p. 291.
25. Barrell (1914d), p. 292.
26. Barrell (1914d), p. 313. Barrell's interpretation is much closer to presently accepted views.
27. Barrell (1914d), p. 297.
28. Barrell (1927).
29. Barrell (1927), p. 298.
30. Barrell (1927), p. 288.
31. Barrell (1927), p. 290.
32. Barrell (1927), p. 291.
33. Barrell (1927), p. 292.
34. Barrell (1927), p. 295.
35. Barrell (1927), p. 296.
36. Charles Schuchert to Arthur Holmes, June 2, 1926, CSP 38: Book 1, 1925–29.
37. Schuchert to Eliot Blackwelder, May 23, 1922, CSP 37: Book 1, 1921–23. A copy of the meeting program may be found in the John Merriam Papers (JCMP 78: Geol. Soc. Am.). According to Merriam's records, 110 members attended the meeting and 75 papers were presented, which gives a sense of the size of the active geological community in the United States at that time.
38. Schuchert to Holmes, June 2, 1926, CSP 38: Book 1, 1925–29.
39. Introduction by Chester Longwell to Barrell (1927), p. 284.
40. Barrell (1927).
41. Schuchert to Holmes, September 27, 1927, CSP 38: Book 1, 1925–29.
42. Schuchert to Holmes, May 31, 1927, CSP 38: Book 1, 1925–29.
43. Holmes to Schuchert, August 27, 1927, CSP 20: 171. Original emphasis.
44. Holmes to Schuchert, August 27, 1927, CSP 20: 171.
45. Schuchert to Holmes, September 27, 1927, CSP 38: Book 1, 1925–29.
46. Chester Longwell to Arthur Holmes, September 21, 1927, CSP 20:171.

47. Holmes to Schuchert, November 12, 1927, CSP 20: 171. Emphasis added.

48. Holmes to Schuchert, November 12, 1927, CSP 20: 171. Historically, this could be considered a retroduction in the sense that the geological events involved had already occurred, but epistemologically it was a prediction in the sense that Daly predicted that such phenomena could be found before they actually *were* found.

49. Schuchert to Holmes, November 25, 1927, CSP 38: Book 1, 1925–29.

50. Thus belying the claim that scientists never abandon an old theory until they have a new one with which to replace it. This also tends to undercut Robert Dott Jr.'s (1997) claim that loyalty to Dana was a principal reason for American resistance to continental drift.

51. Schuchert to Holmes, September 27, 1927, CSP 38: Book 1, 1925–29.

52. Schuchert to Holmes, June 2, 1926, CSP 38: Book 1, 1925–29. Emphasis added.

53. van Waterschoot van der Gracht to Schuchert, September 21, 1926. CSP 21: 179. Original emphasis.

54. An enormous amount of drift history has been based upon this one symposium, an obviously perilous situation. This is underscored by Schuchert's private correspondence surrounding the meeting. Even while Schuchert was preparing Barrell's paper for publication and "sliding towards drift" in correspondence with Holmes, he suppressed this inclination in his published remarks and instead proclaimed his allegiance to permanence theory, accepted by most of his American colleagues, although it was a theory that he himself had *never* subscribed to; see Schuchert (1928a), p. 140: "The battle over the theory of permanency of the earth's greater features introduced by James D. Dana has been fought and won by Americans long ago."

55. Schuchert to Holmes, June 2, 1926, CSP 38: Book 1, 1925–29, p. 2.

56. Schuchert to Holmes, June 2, 1926, CSP 38: Book 1, 1925–29, p. 2.

57. Schuchert to Holmes, June 2, 1926, CSP 38: Book 1, 1925–29, p. 2.

58. Schuchert to Holmes, June 2, 1926, CSP 38: Book 1, 1925–29 p. 3.

59. Taylor (1926).

60. Schuchert to Holmes, July 6, 1926, CSP 38: Book 1, 1925–29.

61. Bowie to Schuchert, June 2, 1927, CSP 19: 166.

62. Schuchert to Bowie, October 22, 1928, CSP 38: Book 1, 1925–29.

63. Schuchert to Holmes, November 5, 1928, CSP 38: Book 1, 1925–29.

64. Howard Baker to Schuchert, December 5, 1928, CSP 21: 181.

65. Schuchert to J. W. Gregory, July 9, 1929, CSP 38: Book 2, 1929–32.

66. Schuchert to Longwell, January 22, 1931, CSP 38: Book 2, 1929–32.

67. Longwell to Schuchert, January 22, 1931, CSP 23: 199.

68. Schuchert to E. Haarman, January 22, 1931, CSP 38: Book 2, 1929–32.

69. Schuchert to Holmes, January 24, 1931, CSP 38: Book 2, 1929–32.

70. Schuchert to Holmes, January 24, 1931, CSP 38: Book 2, 1929–32.

71. Schuchert to Holmes, January 24, 1931, CSP 38: Book 2, 1929–32. Emphasis added.

72. Gould (1965).

73. On interpretations of uniformitarianism, see Shea (1982).

74. Gould (1965); Shea (1982). For examples of scientists using methodological uniformitarianism before it had that name, see Laudan (1987) and Rudwick (1961, 1971, 1985, 1990). On the acceptance of the invariance of natural law in the early history of geology, see Hooykaas (1963, 1975), Rudwick (1971), Wilson (1980), and Rupke (1983).

75. As Rudwick (1971) points out, Lyell's theory opposed not so much catastrophism per se as the theory of a cooling earth. When the cooling earth was (seemingly) refuted in the late nineteenth century, Lyell's work seemed prescient.

76. Rudwick (1970, 1971, 1990).

77. Rudwick (1971), p. 225; also see Rupke (1983), pp. 207–208. Lyell's work is illustrative of a certain kind of historical process where an idea, although obviously flawed, neverthe-

less comes to be deeply influential because of its applicability in a particular, important, situation, or its usefulness in contemporaneous but only marginally related debates. In Lyell's case one may suppose that the importance of his methodological stance and the clarity with which he presented it, coupled with his profound influence on Charles Darwin, caused his theoretical positions to be given more weight than perhaps they deserved. Rupke (1983) suggests that Lyell's independence from clerical authority made him an attractive figure to those wishing to break the yoke of Anglican control of geology at Oxford and Cambridge. A qualification: I refer here primarily to the United Kingdom and the United States. Not much work has been done on the influence of Lyell in the rest of Europe or in the twentieth century.

78. One need only look at the recent debates over the meteorite-impact theory of dinosaur extinction, as well as debates in biology and paleontology over the idea of punctuated equilibrium, to see how pervasive is the commitment to gradualism and the suspicion of sudden change even today. It seems likely that historians will look back on the 1980s as the period in which geologists reversed their long-standing opposition to abruptness in explanation. On the dinosaur debate see Alvarez (1997) and Glen (1994); on the influence of catastrophism on debates in planetary science, see Doel (1997).

79. This is not to say that others did not view continental drift as antiuniformitarian in other respects. Rupke (1970) notes the historical association of drift, in the guise of fissiparturition, with catastrophism, and it is notable that the advocates of fissiparturition were mostly not geologists.

80. Schuchert (1928a), p. 141.

81. Schuchert to G. A. F. Molengraaf, November 24, 1923, CSP 37: Book 2, 1913–25.

82. Schuchert to F. B. Taylor, June 11, 1925, CSP 37: Book 2, 1913–25.

83. Schuchert to Erich Haarman, January 22, 1931, CSP 38: Book 2, 1929–32. Schuchert obviously made the same point to Daly two years later (see next note).

84. R. A. Daly to Schuchert, February 9, 1933, CSP 24: 207.

85. Schuchert to R. A. Daly, May 29, 1926. CSP 38: Book 1, 1925–29.

86. Schuchert to G. A. F. Molengraaf, November 24, 1923: CSP 37: Book 2, 1913–25.

87. Schuchert (1928a), p. 140.

88. There has been little historical analysis of rhythmicity as a geological program, but it is clear that Barrell and Chamberlin interpreted rhythmicity as a variation on the theme of uniformitarianism. This may have been one reason why Schuchert found Joly's theory congenial; see Barrell (1917).

89. Whewell (1858) is the most commonly cited reference, although Whewell first introduced the terms in 1832.

90. Rudwick (1961, 1970, 1971, 1990); Rupke (1983); see also Hallam (1983) and Hooykaas (1963, 1975). In the early nineteenth century, many geologists were both progressionists and catastrophists, but by the late nineteenth century this was no longer the case.

91. See Smith and Wise (1989), ch. 16.

92. The philosopher David Kitts considered geology the "paradigmatic historical science" insofar as it dealt with singular events, and that analogies were therefore of limited use (Kitts 1980; also see Kitts 1977). While this view might be defensible in principle, it clearly fails in practice: analogies are everywhere in the work of the geologists studied here.

93. Hallam (1983); Oldroyd (1990); Rudwick (1985); Secord (1986).

94. Some geologists went further to claim that uniformitarianism was the fundamental principle of *all* science and thus constituted geology's most important contribution to science in general. See, for example, W. J. McGee, "Cardinal principles of science," *Proceedings of the Washington Academy of Sciences* 2 (1900): 6, discussed in Manning (1967), p. 88.

95. Schuchert (1924); see also Numbers (1985, 1992).

96. Numbers (1992), on p. 89.

97. Schuchert (1924), on pp. 486–487. Schuchert may have been particularly sensitive to this issue. Given his own lack of formal training, what separated Schuchert from Price was apprenticeships and field experience.

98. The other was a man named George F. Wright (1838–1921), a more complicated and interesting figure than Price (Numbers 1992). Professor of the Harmony of Science and Revelation at Oberlin College from 1892–1907, Wright was raised as a "New School Calvinist" and became a respected glacial geologist who was active in many professional scientific organizations, did extensive field work (including an 1895 excursion to Greenland and a trip across the entirety of Asia) and was a friend to such luminaries as Charles Hitchcock, Asa Gray, and Clarence King. In the 1890s, Wright argued—against T. C. Chamberlin and John Wesley Powell—that humans lived in North America during Pleistocene time, a position that geologists now unequivocally affirm.

99. Schuchert to Holmes, June 2, 1926, CSP 38: Book 1, 1925–29, and Schuchert to Holmes, November 5, 1928, CSP 38: Book 1, 1925–29.

100. It is unclear how Schuchert felt about his own past, for he rarely referred to it and then only obliquely. Some colleagues, however, clearly resented his success. E. O. Ulrich, for example, turned against Schuchert later in his career (Weiss 1992).

101. Schuchert to E. W. Berry, December 20, 1928, CSP 38: Book 1, 1925–29.

102. Schuchert to Willis, April 18, 1931, CSP 38: Book 2, 1929–32.

103. Chamberlin (1924); also see Willis (1942b), esp. p. 168.

104. Gilbert (1900); also see Barrell (1917) and Willis (1910). For the 1932 National Research Council volume on oceanography, Schuchert entitled his paper "The periodicity of oceanic spreading, mountain-making, and paleogeography" and wrote that periodicity of crustal motions had now come "into recognition" as a "grand fact" (Schuchert 1932b, p. 538). In this paper Schuchert introduced the idea of periodic eustatic sea level changes—a precursor to the contemporary "Vail curve." Americans were not unique, however, in their "saner" interpretation of uniformitarianism: see Smith and Wise (1989), pp. 592–593, and esp. their footnote 41.

105. Holmes to Schuchert, February 9, 1931, CSP 24: 209.

106. Schuchert to Holmes, January 24, 1931, CSP 38: Book 2, 1929–32.

107. Ben Page, Stanford University, personal communication; also see Geschwind (1996).

108. It is striking that both men were so highly advanced in age, and Bowie nearly so. Philip Pauley notes a similar situation in American biology: Morgan, Conklin, Osborn, Ritter, and others were educated in the 1880s and still guiding their discipline in the 1920s and 1930s. As the first generation of American scientists to be able to take advanced degrees in American institutions, these men not only led but to some extent created American biology and lacked exemplars for transferring leadership.

109. Willis to Schuchert, March 2, 1931, CSP 23: 204.

110. Willis (1910). Original emphasis.

111. See chapter 5.

112. Willis to Schuchert, January 15, 1932, CSP 25: 217.

113. Willis to Schuchert, March 2, 1931, CSP 23: 204.

114. Schuchert to Bailey Willis, BWP April 14, 1931 21:8; see also CSP 38: Book 2, 1929–32.

115. Schuchert to Bailey Willis, BWP April 14, 1931 21: 8. This quote also serves to refute the argument that drift was rejected because it was physically impossible: if Schuchert was willing to defend sunken land bridges against a mechanical objection, he could have defended continental drift in the same manner.

116. Willis to Schuchert, April 21, 1931, CSP 25: 217.

117. Willis to Schuchert, April 21, 1931, CSP 25: 217, p. 2.

118. Willis to Schuchert, April 21, 1931, CSP 25: 217, p. 2.

119. Schuchert to Willis, June 1, 1931: CSP 38: Book 2, 1929–32.
120. Longwell to Schuchert, June 10, 1931, CSP 25: 217.
121. Willis to Schuchert, June 9, 1931, CSP 25: 217.
122. Willis to Holmes, December 20, 1931, BWP 28: 19.
123. Schuchert to Willis, June 1, 1931, CSP 38: Book 2, 1929–32. For Willis's views of periodicity as derived from T. C. Chamberlin's work on climate change, and his rejection of a strictly steady-state earth, see Willis (1910). For other discussions of periodicity, see Woodward (1907) and du Toit to S. J. Shand, December 30, 1938, ADTP 7: Folder Correspondences P–S, 1930's–1940's.
124. Willis to Schuchert, June 9, 1931, CSP 25: 217.
125. Willis to Schuchert, June 9, 1931, CSP 25: 217, p. 3.
126. Bowie to Schuchert, July 2, 1931, CSP 22: 193.
127. Willis to Schuchert, July 7, 1931, BWP 31: 22.
128. Schuchert to Willis, July 14, 1931, BWP 21: 8.
129. Schuchert to Willis, July 21, 1931, CSP 25: 217.
130. Willis to Schuchert, July 27, 1931, CSP 25: 217.
131. Schuchert (1932a) and Willis (1932).
132. Willis to Schuchert, January 15, 1932, CSP 25: 217.
133. Holmes to Willis, August 31, 1931, BWP 15: 45.
134. Willis to Schuchert, July 14, 1932, BWP 31: 22.
135. Schuchert to Willis, May 2, 1932, BWP 21: 8.
136. Willis to Schuchert, May 9, 1932, BWP 31: 22.
137. Schuchert to Willis, May 12, 1932, BWP 21: 8.
138. On Holmes's plans for the Lowell lectures, see Holmes (1931) and various materials in RADP.
139. E. W. Berry to Schuchert, February 13, 1933, CSP 24: 205. Isthmian links were later supported by George Gaylord Simpson's theory of chance dispersal of creatures on floating logs and rafted grasses (Marvin 1973, pp. 114–117; Laporte 1985).
140. At Dartmouth College in 1935, students were given a handout of Willis's map illustrating the Permian isolation of the southern oceans as the explanation for the Permian glaciations. (Papers of Lincoln Page, General Geology Course notes, 1935, Courtesy of the family of Lincoln Page.)
141. Stewart (1986, 1990).
142. Joseph Poland, U.S. Geological Survey (personal communication, 1986).
143. Gould (1979), p. 160; see also Marvin (1973) and Vine (1977). Australian geologists tell anecdotes of traveling to America in the 1950s only to be subjected to rude American hosts who ridiculed their entertainment of drift.
144. Howard Baker to Schuchert, January 19, 1933, CSP 24: 205. To Alexander du Toit, however, Baker called them "almost missing links." Baker wrote to du Toit on Christmas Day, 1934, "It is a little amusing to observe how those two (W + S) try to stack the cards. I think that between them they are very effectively preventing geological progress in this country. There are plenty of thinkers, but they don't care to clash with the fashions set by the few" (Baker to du Toit, December 25, 1934, ADTP 8: "Continental drift correspondence").

Chapter 8

Part of chapter 8 first appeared as "Weighing the Earth from a Submarine: The Gravity-Measuring Cruise of the U.S.S. S-21," in *The Earth, the Heavens, and the Carnegie Institution of Washington*, edited by Gregory A. Good, *History of Geophysics*, vol. 5, pp. 53–68. Reprinted by courtesy of the American Geophysical Union.

1. Gutenberg (1939).
2. Gutenberg (1939), p. 184. This brilliant essay includes a review of European advocates of subcrustal convection (pp. 184–185) and one of the earliest suggestions that the correct interpretation of isostasy was a combination of the Pratt and Airy models (p. 179).
3. Gutenberg later calculated the depth of the plastic zone based on seismic wave velocities at a depth of 100–200 kilometers.
4. HWMP 1938–1986, 33: 5.
5. HWMP 1938–1986, 33: 5.
6. HWMP 1938–1986, 33: 5.
7. HWMP 1938–1986, 33: 5.
8. Gutenberg (1939) suggests that thermal contraction may still be occurring but can only account for a small fraction of crustal deformation, and that one might look to subcrustal currents for much of the rest (see especially p. 212).
9. Hubbert (1937, 1945).
10. Hubbert (1945), p. 1630.
11. HWMP 1938–1986, 33: 5.
12. Hubbert (1945), p. 1653.
13. Hubbert (1945), p. 1653.
14. T. Waylan Vaughan to Schuchert, 11 May 1928, CSP 22: 191.
15. Schuchert (1932b).
16. HWMP 1938–1986, 33: 11.
17. For the mid-1930s–1940s, I located classroom notes from four universities: Caltech, Harvard, University of Colorado, and Dartmouth. All four included some course in which continental drift was discussed, usually tectonics, but only Daly assigned Wegener's book. In each case, continental drift is discussed as one theory among many—including Barrell's fragmentation hypothesis—and generally interpreted specifically in terms of Wegener's westward drift of the continents, uniquely associated with the breakup of Pangea. In response, one student, Lincoln Page, a former geologist with the U.S. Geological Survey (now deceased), recorded, "Why drift apart all at once?" For Harvard and Caltech, see RADP and HWMP; for Dartmouth and the University of Colorado, I consulted "Notes from the University of Colorado, 1939" and "Course Notes, Dartmouth College," courtesy of the family of Lincoln Page. These course notes are consistent with Homer Le Grand's finding that textbooks in the 1940s and 1950s mostly *did* mention drift but usually only briefly and in the context of other (better?) theories (Le Grand 1988, p. 121).
18. HWMP 1938–1986, 33: 11. These appear to be rough notes for a paper or homework assignment. Alas, Daly's comments on the submitted version are not to be found.
19. Birch was one of the first to perform high-pressure experimental studies of rocks, as an assistant at Harvard to Percy Bridgman, and believed that his results refuted claims for plasticity at depth (Birch and Bancroft 1938a, 1938b). This work was funded by the Rockefeller Foundation.
20. Wegener (1924), p. 112.
21. Wegener (1924), p. 115.
22. Wegener (1924), p. xi.
23. Wegener (1929), p. 2.
24. Bowie (1925b), p. 5.
25. Bowie (1925b), p. 5.
26. Drifting apart (1925), p. 6.
27. Reingold (1991), p. 113; see also Norris and Witherspoon (1904). The late Charles Whitten, formerly of the U.S. Coast and Geodetic Survey, recalled that Bowie and Helmert had discussed the idea of reconciling German and American measurements, shortly before the idea was quashed by World War I (Charles Whitten, personal communication, 1991).

28. Pyensen (1990).
29. Paper read at Second General Assembly... (1925); see also Paper read at Third General Assembly... (1928). The proceedings records are extremely abbreviated; it is not clear when it was first suggested that the operation could serve as a test of continental drift.
30. Paper read at Second General Assembly... (1925), p. 200.
31. Bowie (1925b), p. 5.
32. Bowie (1925b), p. 5; also see Paper read at Third General Assembly... (1928). The choice of 20–25 feet as the minimum detection limit suggests a much greater expectation of the magnitude of drift than would be confirmed today; this followed from Wegener's estimates of the Greenland displacement.
33. Bowie (1925c), p. 183.
34. Bowie (1925c), p. 183.
35. For a review of recent work on motives for experiments, see Franklin (1993b).
36. On entertainment and pursuit of new theories as applied to the problem of drift, see Le Grand (1988), who draws on Larry Laudan (1977). On the issue of experimental pursuit in physics, Franklin (1993a) emphasizes that it is not necessary for a scientist to believe in a theory to pursue it. But usually it helps!
37. Comments on the importance of geodesy, geophysics, and geology working hand in hand can be found in various letters of Bowie to Bailey Willis, BWP. On the need for more data in general, and for cooperation with navies, see Bowie, "Round table discussion," in Field (1938).
38. Hayford (1909), p. 9.
39. Bowie (1929e) and other reports in this same series, RUSCGS Entry 33, Box 775, Annual Office Reports 1926–1933.
40. Manning (1988).
41. Bowie (1925b), p. 5. In the end, sound waves proved more useful for the Navy than radio waves; see note 78 below.
42. Charles Whitten (personal communication, 1991). Bowie never published on these negative results—indeed, almost nothing was said about them in the scientific literature. This tends to refute any suggestion that Bowie's goal in organizing the experiment was refutation: if negative results had been his aim, he would surely have advertised the ones he got.
43. Littel and Hammond (1928), Littell et al. (1929), Watts et al. (1938). These publications present compilations of the longitude data; discernible differences were attributed to operator error. Also see Gutenberg (1939), pp. 204–205.
44. Bowie (1925c), p. 181.
45. Triangulation data, of course, could not be collected at sea.
46. F. E. Wright to J. Merriam, December 27, 1928, CIWGF: Miscellaneous file #2, 1908–1938.
47. William Bowie, RUSCGS Entry 30, General Correspondence, "Terrestrial Magnetism," 1908–1914.
48. Vening Meinesz and Wright (1930); Laudan (1980a).
49. Wright (1929); Vening Meinesz and Wright (1930); see also Laudan (1980a).
50. Vening Meinesz (1932); Vening Meinesz et al. (1934).
51. Vening Meinesz and Wright (1930), p. 1; Vening Meinesz (1932). This compares with Hayford's and Bowie's production of approximately 100 gravity stations in all of the United States, work that took a decade to complete.
52. Vening Meinesz and Wright (1930), pp. 1–2; CIWGL Misc., 1908–1935, File 1.
53. Fleming and Piggott (1956); Wright (1926); Wright and England (1938); Wright (1941); and CIWGL Misc., 1908–1935, File 1. Also see Doel (1996), ch. 5.
54. Curtis Wilbur to John Merriam, June 19, 1928, CIWGL Misc., 1908–35, File 1.
55. Bowie (1925a, 1927c).

56. See discussion by Evans in Bowie (1924b), pp. 43–44.
57. Bowie (1929f), p. 589.
58. There is another reason why Bowie may have been eager to measure gravity at sea: the possibility that it would help differentiate between the two versions of isostasy. In general, geodesists spoke of Airy and Pratt as being mathematically equivalent, but in 1924 Bowie wrote an article in which he questioned whether this was entirely true.

> [In areas of minimal topography] the computed effect of [isostatic] compensation is the same whether it is in or below the crust and whether it is distributed through a thin or a thick layer of crust. [However] the situation is entirely different in mountainous regions where there is great variation in the elevations. Here the effect of the compensation will vary considerably for different methods of distribution. There will be a noticeable difference between the computed values of gravity at stations near the margins of mountain systems when computed on the theory that the crust is of uniform thickness with compensation in the crust, and when computed on the theory that the compensation is in roots extending below the normal depth of the crust and proportional in length to the elevation of the surface.
>
> The same will be true for an area of limited width, say about 100 miles wide, along the margins of the continental shelves and over the margins of the great deeps of the ocean. The gravity observations at sea now available are not sufficiently accurate to test this phase of the question, but it is hoped that the splendid work of Dr. Felix A. Vening Meinesz . . . may result in having adequate data for the purpose in the near future. (Bowie 1924a)

It is not clear from this article precisely how Bowie thinks the gravity work would resolve this question, although depending upon the boundary conditions between areas of high and low elevation, one could imagine scenarios where the two models would give different results along the boundaries.

59. Wright (1929).
60. Lamson (1930).
61. Manning (1988), pp. 37–38.
62. Du Toit (1927); CIWGF Meinesz folder, Memorandum, October 1, 1928.
63. Vening Meinesz and Wright (1930), p. 11.
64. Vening Meinesz and Wright (1930).
65. Vening Meinesz and Wright (1930), p. 76.
66. Bowie (1924a), pp. 3–4.
67. Bowie (1924a), p. 5.
68. Vening Meinesz and Wright (1930), p. 76.
69. Vening Meinesz and Wright (1930), p. 76.
70. Vening Meinesz and Wright (1930), p. 53.
71. Vening Meinesz and Wright (1930), p. 4.
72. Vening Meinesz (1932), Vening Meinesz et al. (1934), Vening Meinesz (1941).
73. Bowie (1929b).
74. Freeman (1930).
75. Initially, AGU was a committee of the NRC, so Bowie and other heads were called "chairman." Field was the first president of the autonomous American Geophysical Union.
76. The original plan had been for the work to be taken over by the CIW, but this plan perished along with the ship *Carnegie* and her captain, J. P. Ault, in a tragic fire in November of 1929; see Forbush (1946) and Oreskes (1994). Wright continued to serve as an advisor to the project.
77. RFP, Magazine clipping, *Princeton Alumni Weekly* 32: 368.

78. Sound channeling is the phenomenon by which sound waves are refracted in the thermocline, resulting in a zone of highly focused waves that travel with little attenuation. Ewing and Worzel used this as the basis of a system of locating downed airmen (SOFAR: Sound Fixing and Ranging); later it was the foundation for the Navy's underwater sound surveillance system (SOSUS) (Ewing and Worzel 1945; Weir 1993). Virtually all of this work was classified at the time.

79. RFP, "1932 international expedition to the West Indies," Abstract for National Academy of Sciences Meeting.

80. 4,000 fathom depth found in Atlantic Ocean (1932). The interpretation of the Bartlett Deep was complicated. The initial negative anomalies discovered there turned out to be rather minor, and Hess concluded that the Deep was a transcurrent rift—a pull-apart basin in today's parlance—created as the Caribbean block slid eastward.

81. Field et al. (1933), p. 31.
82. Field et al. (1933), p. 31.
83. Vening Meinesz et al. (1934), pp. 117–125.
84. Vening Meinesz et al. (1934), p. 55.
85. Vening Meinesz et al. (1934), p. 118.
86. Vening Meinesz et al. (1934), pp. 126–139.
87. Field et al. (1933), p. 32. Original emphasis.
88. Geologists in the 1930s were also arguing about the origins of recently discovered submarine canyons. Most geologists, including Francis Parker Shepard, who helped discover the canyons, and Harry Hess, believed they were drowned river valleys formed when sea level was lowered during the Pleistocene glacial period. Reginald Daly disagreed, arguing that they were carved out by submarine landslides—later dubbed turbidity currents. Kuenen's work, first published in 1937 and later confirmed by Bruce Heezen, demonstrated that Daly was right (Kuenen 1937, 1938, 1948; Kuenen and Migliorini 1950; Kuenen 1951a, 1951b; Kuenen and Menard 1952). Daly's dispute with Shepard is documented in the papers of Francis Parker Shepard, Scripps Institution of Oceanography. On Heezen's confirmation of Kuenen's theory of turbidity currents, see Oreskes (1998).

89. Allwardt (1990, p. 69, footnote 70) credits Kuenan with supplying the term *tectogene*, but Kuenan says he borrowed it from the German geologist Erich Haarmann.

90. Kuenen (1936).
91. Kuenan (1936), pp. 170–171.
92. Griggs (1939b), pp. 611–650.
93. HWMP, 1930s, 69: 3, Retrospective by David Griggs. This item is undated, but Menard collected recollections in preparation for his book *The Ocean of Truth* (1986).
94. Griggs (1939b), pp. 616–620; quote on p. 616.
95. Griggs (1939b), p. 637.
96. Griggs (1939b), p. 643.
97. Griggs (1939b), p. 647.
98. Griggs (1939b), p. 648.
99. Griggs (1939b), p. 612.
100. Hess (n.d.), pp. 47–48. The reference to Griggs's "new concept" dates this paper as late 1930s.
101. Hess (1937); Ewing (1937); Field (1937b); see also Field (1936).
102. Ewing (1938); Hess (1938).
103. Field (1938), p. 5.
104. Field (1938), pp. 5–6.
105. Heck (1938), p. 97.
106. Heck (1938), p. 103.

107. This is no longer held; all ocean basins are now known to be underlain by basaltic ocean crust. Gutenberg was working with limited data and may have been misled because the Atlantic, with its slower spreading rate, contains older and therefore colder crust.

108. Fleming (1938), p. 120.

109. Fleming (1938), p. 125; also see Gutenberg and Richter (1939), who illustrate the distribution of deep-focus earthquakes along the margins of the Pacific Basin, which they attribute to shearing or faulting at depth—a mechanism no different from that supposed for near-surface earthquakes. They also suggest that the motions involve "each continental block as a whole" (p. 298).

110. Hess (1938), p. 71.

111. Cf. Allwardt (1990).

112. Field (1937a), p. 187.

113. Bowie (1936), on pp. 15–16.

114. Bowie (1936), p. 16.

115. Bowie (1936), p. 16.

116. It did not take long for others to realize in retrospect the conventional nature of Hayford's assumptions. For example, Daly wrote in his 1940 book *Strength and Structure of the Earth*, "Hayford['s] . . . type of compensation was a frankly artificial conception, adapted for the convenience while attempting to improve the formula for the earth's figure" (p. 340). Others had recognized this earlier. John Joly's copy of Hayford and Bowie's 1912 gravity treatise, which Joly received from Bowie in 1923, is intensely annotated and underlined, particularly in the places where Hayford provided pragmatic justification for his choice of the Pratt model. On page 11 of the text Hayford and Bowie wrote, "The authors believe that the assumptions on which the computations are based are a close approximation to the truth"; Joly underlined this in red and in the margins wrote "!!". Further down on the page, the authors wrote, "it is certain that the assumptions are nearly correct," and Joly's red pencil underscored the word "certain" six times; on the top of the next page, he concluded, *"This is the source of error."* On page 133, where Hayford and Bowie concluded in favor of a uniform depth of compensation in spite of some data to the contrary—the data that Barrell noted—Joly wrote in large letters at the top of the page, "This shows that the depth of compensation is *not* a unique depth—but varies from plains to mountains" (see JJP–TCD MS 2315 and MS 2316).

Bailey Willis was another who saw the distinction between isostasy in general and the Hayford–Bowie interpretation of it. In a letter to R. T. Chamberlin in October 1928, Willis wrote that it was "high time the geological objections to Bowie's extreme assumptions were stated." Calling himself a "Gilbertian isostasist"—because Gilbert insisted on some residual rigidity for the crust—Willis noted that "Isostasy . . . is not all Isostacy Bowieii" (Willis to R. T. Chamberlin, October 18, 1928, BWP 25: 72). A few weeks before this he had discussed the same matter in a letter to his wife, Margaret, after Arthur Day had visited Stanford and told him about the *S-21* work. "Day and I discussed the reduction of the gravity observation by the Hayford–Bowie method and agreed that the result obtained was an artificial product of mathematics rather than reality. The pendulum [sic] is beginning to swing back toward a recognition of partial rigidity of the earth's crust; i.e., towards Gilbert's view" (Willis to Margaret D. Willis, September 10, 1928, BWP 39: 7, p. 4). Despite these privately expressed reservations, little direct opposition was revealed in print, suggesting the strong hold that the model had on American scientific opinion until the late 1930s.

117. Bowie (1936), p. 20.

Chapter 9

1. RFP: Eulogy to Richard M. Field, Department of Geological Sciences, Princeton University.

2. For a complete list of the members, which included representatives from the petroleum industry, foreign universities, and the Navy, see RFP.

3. On Field's goals for a global science, see Field (1941).

4. Hess (1946).

5. Glen (1982); Le Grand (1988, 1994). DTM had a long history of supporting research on the earth's magnetic field as a basic scientific problem, which shifted in the late 1930s and 1940s toward rock magnetism. The exact reasons for these shifts have only recently begun to be explored by historians: the loss of the ship *Carnegie* was clearly one reason; Navy interest in sensitive magnetometers is clearly another. Merle Tuve, the postwar director of the DTM, was a physicist who had worked during the war on radar and on the proximity fuse.

6. Blackett (1948). It also led the FBI to keep a file on him (Nye 1996).

7. Blackett (1947, 1952); see also Wood (1985), p. 113; Le Grand (1988), pp. 142–143; and Nye (1996).

8. RFP, see also Wood (1985).

9. Holmes (1945), pp. 508–509.

10. Which is not to say that he did not have residual doubts, particularly in his later years as his health began to fail. Holmes occasionally shared his uncertainties with Daly, his closest confidant in the United States. Writing in 1943, Holmes sought Daly's advice on whether to include convection and continental drift in his book at all (Holmes to R. A. Daly, March 27, 1942, RADP 4315.7: Correspondence 1942). Under the influence of Bowen, Daly had abandoned his own idea of the glassy basalt substrate in favor of a peridotite layer, but he still believed in continental migrations. Indeed, he appears to have accepted Holmes's explanation of it; among his notes from the 1930s is an address to the Boston Geological Society entitled "Convection in the Earth's body." Daly outlined the history of thinking on convection and suggested that geologists in the past had wrongly "bowed to the authority of Kelvin and other leaders in geophysics who thought they had found in the high rigidity of the globe direct evidence of true solidity . . . and therefore systematically discredited the idea of convection." Now, with the work of Ampferer, Joly, Poole, Holmes, and others, he wrote, "the pendulum of thought has . . . swung back" (RADP Lecture notes, "Convection in the Earth's body," 4315.45: Manuscripts and speeches 1924–c1948). In his 1940 book, *Strength and Structure of the Crust*, which reviewed isostasy and its implications, Daly concluded that the hypothesis of convection should be "put into competition" (Daly 1940, p. 376).

In 1951, when Holmes expressed renewed skepticism, Daly—now eighty years old—shored him up: "I still hold to the possibility of continental migration, in spite of Wegener's poor presentation of the doctrine. But," he allowed, " I may be quite crazy!" (Daly to Holmes, September 30, 1951, RADP 4315.7: Correspondence 1951). Holmes may have felt conflicted precisely because of the depth of his understanding, appreciating as well as he did the arguments on both sides. As one postdoc later put it, "Holmes was an anachronism in our time, because he knew so much" (Allwardt 1990, p. 91).

11. Holmes (1945), p. 503. Holmes was not alone in this view. In the 1949 version of Schuchert's original *Historical Geology*, now in the hands of Schuchert's protégé Carl O. Dunbar, continental drift is mentioned but once: in connection with the Permian glaciations. "The most remarkable feature of the Permian glaciation," Dunbar wrote, "is its distribution. It was chiefly in the southern land masses and in regions which now lie within 20–25° of the equator. This circumstance, more than any other, has made attractive the belief in continental drift" (Dunbar 1949, p. 263).

12. Holmes (1945), p. 504.

13. Holmes (1945), p. 505.

14. This story was told to me as an undergraduate at Imperial in the late 1970s; it may not be literally true, but one way or another Blackett did apparently read Holmes's textbook and instruct those who worked with him to do so as well.

15. The best brief summaries of this work are given by Le Grand (1988) and Frankel (1979b; 1982; 1985); for a more detailed treatment based primarily on interviews with participants, see Glen (1982).

16. See Le Grand (1994), p. 175.

17. Creer, Irving, and Runcorn (1954).

18. Irving (1956), Blackett (1956), Runcorn (1959).

19. Runcorn (1959); see also Le Grand (1988).

20. Runcorn (1959), on p. 1011. Bullard later made the same argument; see Le Grand (1988), pp. 202–203.

21. Hess (1962); also see Dietz (1961). Although Hess's paper succeeded Dietz's by a year, a preprint had been circulated in 1960 (Meyerhoff 1968, Dietz 1968). On Heezen and Tharp, see Tharp (1982); Oreskes (1998).

22. Hess (1962), p. 608.

23. Hess (1962), p. 607.

24. The question has been raised as to what Vening Meinesz really believed about continental drift and convection currents, and how much Hess knew about early work on these questions (e.g., Allwardt 1990). Both the published and the archival record leave no doubt that Vening Meinesz believed in the existence of convection currents from the 1930s onwards, and that Hess was intimately acquainted with early work on this topic.

As early as 1932, Hess and Vening Meinesz discussed tectonics questions in correspondence. In September of 1935 Vening Meinesz explicitly suggested convection currents, along the lines proposed by Arthur Holmes, to explain gravity anomalies in the Java Trench. "I cannot see another reasonable explanation," he wrote, "than by admitting convection-currents in the substratum below the crust, rising below the continents and sinking below the oceans. (They may perhaps be caused by temperature differences because of differences in radioactivity, as e.g. Arthur Holmes assumes.)" (F. A. Vening Meinesz to H. H. Hess, September 22, 1935, HHP 21: Unlabeled folder). In 1939, the two men discussed David Griggs's experiments, which Vening Meinesz called "a very important result," and which he "hope[d] to be able to hear more about . . . in America" (F.A. Vening Meinesz to H. H. Hess, July 7, 1939, HHP 7: W1). Hess agreed, and suggested to seismologist Perry Byerly at U.C. Berkeley that Griggs offered a solution to the problem of how positive gravity anomalies could be maintained, namely, by "a downward convection current of a few cm per year" (H. H. Hess to P. Byerly, August 9, 1940, HHP 7: W1).

Did Vening Meinesz and Hess link convection to continental drift? Vening Meinesz's discussion of Holmes and Hess's discussion of Griggs strongly suggests that they did: the whole point of convection, for Holmes and Griggs, was to explain the cause of drift. Alexander du Toit also made the connection between gravity anomalies, convection, and drift, in a letter to Hess in 1939. "Thank you for your paper on "Gravity Anomalies in the West Indies," he wrote, "which sheds such a great deal of light on the tectonics of that region. . . . I have been much struck by the evidence therein submitted of big horizontal movements, such as would support the hypothesis of continental drift" (A. du Toit to H. H. Hess, March 16, 1939, HHP 7: W1). In the 1930s, at least, the connection was clear.

Vening Meinesz's later work is more ambiguous. His papers from the 1940s and 1950s clearly advocate a close tie between subcrustal drag and crustal compression, but they sidestep the question of large-scale continental migrations (Vening Meinesz 1941, 1952). In Daly's (1940) *Strength and Structure of the Earth*, he quotes from a passage in which Vening Meinesz seems to suggest convection currents as the *result* rather than the *cause* of crustal motions (Daly 1940, pp. 267–268). As for Hess, his course notes from the 1940s and 1950s show that, at minimum, he assigned readings from Daly, Holmes, du Toit, Vening Meinesz, and Griggs in his classes; in one course handout, he lists Wegener among famous geological luminaries!

The most plausible interpretation is that both men did initially link convection to large-scale crustal motions of some kind, but they lost the momentum of this work during the war years and later took more circumspect positions. Then, in the 1950s, with the emergence of the paleomagnetic evidence, Hess returned to these earlier ideas.

25. Hess (1962), p. 599.

26. Hess (1962), p. 609. Hess was not the only one to foster the notion that no serious prior work had been done on the mechanism question. When Walter Elsasser discussed "The mechanics of continental drift" for an American Philosophical Society Symposium in 1968, his paper contained not one reference (Elsasser 1968).

27. See, for example, Allwardt (1990) and Cox (1973) and references cited therein. For the dispute over this issue that occurred among earth scientists in 1968, see Meyerhoff (1968), Hess (1968), and Dietz (1968).

28. Daly reviewed this work in his (1940) *Strength and Structure of the Earth* (see esp. p. 374), but Hess did not cite this either.

29. Holmes (1945), pp. 507–508. John Stewart, to his credit, is one of the few historians to note this (although he gives the year of Holmes's book as 1944 rather than 1945). See Stewart (1990), pp. 250–251.

30. Examination Book, Geology 24-a, 1948, HWMP 33: 11. These appear to be rough notes for a paper or homework assignment. Alas, Daly's comments on the submitted version are not to be found.

31. Dietz (1968), p. 6567. Note the use of the word *rind*, referring to Hess's chemical rather than physical interpretation of the oceanic crust.

32. Holmes (1945), p. 507.

33. Holmes (1945), p. 508.

34. Chester Longwell understood this in 1944. Addressing an argument presented by Bailey Willis against continents "plowing" through the ocean floor, Longwell noted that Willis's objection "has no validity," because it had already been addressed by Holmes. See Holmes (1933) and Longwell (1944b), p. 514.

35. Hess (1962), pp. 602–603; emphasis added.

36. Hess (1962), p. 599.

37. Dietz (1961). Deitz also reintroduced Barrell's terminology, pointing out that "if convection currents are operating 'sub-crustally', as is commonly written, they would be expected to shear below the lithosphere and not beneath the 'crust' as this term is now used" (Dietz, 1961, p. 855). While this paper is little cited, in deference to Hess's temporal priority, it is far closer to contemporary understanding than Hess's. On the other hand, Dietz, like Hess, compares his model to Wegener rather than to Holmes and thus perpetuates the "mechanism myth."

38. Glen (1982), pp. 154–174.

39. Thus belying Cox's own argument about causal mechanisms: Cox himself accepted the reality of the reversals he observed, although he did not know what caused them.

40. Cox et al. (1963).

41. Morley (1986); Vine and Matthews (1963).

42. See Glen (1982) for a full account of this work. Some philosophers have interpreted the confirmation of the Vine–Matthews–Morley (VMM) hypothesis as a textbook example of the power of hypothetico-deductive reasoning in science. Henry Frankel and Rachel and Larry Laudan, for example, have presented arguments that earth scientists were rational in their acceptance of seafloor spreading because it predicted novel facts that turned out to be true (Frankel 1979b, 1982; Laudan and Laudan 1989). On an epistemological level, this claim is reasonable, and probably true viewed from the perspective of individuals. But on a historical level, there is a profound ambiguity as to what one means by "novel," and what lessons one

might therefore derive from this history. The paleomagnetic data used to test the VMM hypothesis were not *new*: they had already been collected by Navy-sponsored research, which is how Vine, Matthews, and Morley knew to think to use them to test Hess's theory. In 1961, Mason and Raff (1961) had already published their results from offshore California; presumably far more data were available in ONR reports or unpublished files. Philosophically, it may suffice that the data were new to the scientists seeing it for the first time, but historically the data were collected for reasons completely independent of the theory that they were eventually used to test.

43. Le Grand (1988).

44. On Menard and Vacquier, see Le Grand (1988) and Stewart (1990); on Heezen and Tharp, see Tharp (1982) and Oreskes (1998).

45. Sykes (1968).

46. Wilson (1968).

47. Le Pichon (1968).

48. Dietz (1961), p. 855.

49. Hess is one obvious example; another is Menard. A less obvious example is Drummond Matthews, who, in an interview with historian Henry Frankel, testified that his experience as a geologist in the Falkland Islands convinced him of the stratigraphic homologies between Africa and South America, but he never invoked these data to support seafloor spreading (Frankel 1982; Oreskes 1988).

50. In *Continents Adrift and Continents Aground*, a compendium by Wilson (1976) of papers from *Scientific American*, only two papers dealt with the question of faunal relationships and both were placed in the final section of the volume entitled "Some Applications of Plate Tectonics." What had been previously used as evidence of the theory was now being used as an application of it!

51. This is not to deny the point, emphasized by Henry Frankel and Homer Le Grand, that one tactic for refuting the theory was to take issue with the details of the evidence: no doubt many scientists did this (Frankel 1976, 1979a, 1984, 1985; Le Grand 1986, 1988). The point is that the major anomalies *were* recognized before Wegener drew attention to them; why else would Barrell, Schuchert, and Willis have spent so much time and effort trying to account for them? This refutes the argument of Lightman and Gingerich (1991) who claim that the jigsaw-puzzle fits were not "noticed" by geologists until Wegener drew attention to them. They were indeed noticed, but they were also the least of geologists' problems.

52. Frankel (1979b). There is also the issue of whether we mean epistemologically novel or temporally novel. Hess's predictions were retrodictions in the sense that the paleomagnetic data had already been collected; see n. 42 above. An obvious question is how much Hess knew about these data; the circumstantial evidence suggests that he knew quite a bit.

53. Bowie (1925c).

54. Carozzi (1985). So did some South Africans, such as Lester King, and some Australians, such as S. W. Carey, who developed an alternative theory of earth expansion (Carey 1958, 1976, 1983, 1988).

55. Cohen (1985).

56. By this point I hope that it is evident that I use "fact" to mean empirical observations that the scientists involved accepted as accurate.

57. To some extent, this is a question of trust: if data seem to conflict, which do you trust? On trust as a scientific and social phenomenon, see Shapin (1994).

58. Richter (1958), p. 82. Original emphasis. I thank Duncan Agnew for calling this gem to my attention.

59. Richter (1958), p. 82.

60. Menard (1986), p. 24.

61. Alexander du Toit to Charles Schuchert, May 15, 1933, CSP 24: 205.

62. Bailey Willis to Charles Schuchert, January 15, 1932, CSP 25: 217, p. 5.

63. George Darwin held a chair in astronomy; Cambridge later established three separate departments dealing with the earth sciences: geodesy and geophysics, mineralogy and petrology, and paleontology.

64. Hevly (1996). The argument was complicated by the entry into the debate of John Tyndall, who, while taking Hopkins's position, attempted to validate it by field observation. Thus the identification of these two "traditions" should not be construed to imply that they were mutually exclusive.

65. Longwell (1944a), p. 221. Geophysicists were of course collecting such data even as he spoke.

66. Longwell (1928), p. 146.

67. Termier (1924), p. 227.

68. Termier (1924), p. 231.

69. Termier (1924), pp. 235–236.

70. Wilson (1979), pp. 1–10, on p. 6. How this compares with the institutional bases for geological work is a tricky subject. Earlier historiographies tended to discount American science in the early nineteenth century, but Julie Newell's (1993) Ph.D. thesis provides an excellent corrective. As she has emphasized, the situation was complex. Although the U.S. Geological Survey was not founded until 1879, state and private surveys flourished in antebellum America, and it was possible to be a professional geologist in America as early as the 1830s. On the other hand, the security of positions in state surveys was highly variable: several state surveys disbanded during hard economic times or when they perceived that their work had been completed. A lack of infrastructure further impeded progress in stratigraphy until after the Civil War (Newell 1993, pp. 66–68). Overall, it is probably fair to say that geology did rather well compared to other sciences in antebellum America, but perhaps not as well as it did in Europe. By the end of the nineteenth century, with the establishment of the U.S. Geological Survey and graduate programs in many universities, geology in America came on par with both geodesy in America and geology elsewhere.

On the history of British field geology, see Woodward (1907), Oldroyd (1990), Rudwick (1985), and Secord (1986). On continental geology see Laudan (1987). On the origins of the U.S. Geological Survey, see Manning (1967), and on the theoretical strength of U.S. geology in the nineteenth century, see Greene (1982).

71. Reingold (1991), p. 113.

72. For a recent discussion of Dana, see Newell (1993).

73. Newell (1993).

74. Brush (1978a, 1978b).

75. LeConte (1872), p. 472.

76. van Hise (1904), pp. 590–591. For background on van Hise, see Servos (1990), pp. 227–229.

77. Van Hise (1904), pp. 592–593.

78. Van Hise (1904), p. 600.

79. Van Hise (1904), p. 601. This is an interesting point: some works considered exemplary in their day nevertheless fail to become paradigmatic.

80. Given Chamberlin's own work, this was clearly a misinterpretation of his intent, but nevertheless a common one.

81. Van Hise (1904), pp. 604–605.

82. Van Hise (1904), p. 603.

83. Van Hise (1904), p. 611; also see Carnegie Institution of Washington, Year-books No. 1 (1902) and No. 2 (1903). For secondary literature on the establishment of the Geophysical Laboratory, see Servos (1984); Reingold (1991), pp. 190–223; Yoder (1989, 1994); and Yochelson (1994).

84. Reingold (1991), p. 206; also see Servos (1984).

85. What we would now identify as geophysics was pursued at the Department of Terrestrial Magnetism. The Geophysical Laboratory under Day pursued primarily what would now be thought of as geochemical questions. John Servos (1984) argues that the move towards geochemistry reflected the influence of van Hise, who as a metamorphic petrologist was greatly influenced by the work of physical chemists.

86. Reingold (1991), p. 206.

87. Yoder (1994); Servos (1984).

88. Reingold (1991), p. 207.

89. Van Hise (1904), p. 612.

90. Van Hise (1904), p. 615.

91. Bucher (1933), p. vii.

92. Bucher (1933), p. 1.

93. Bucher (1933), p. 4.

94. Bucher (1933), p. 99.

95. It is interesting how little agreement there is on what scientific laws are—or even are supposed to be. Philosophers variously emphasize invariance, generality and/or universality, explanatory power or causality; scientists typically emphasize simplicity (particularly of the mathematical kind) or the reduction of such statements to the smallest possible number. In the main, I follow Cartwright (1983) in thinking of laws as idealizations of natural interrelationships—with or without causal explanations.

96. Bucher (1933), p. vii.

97. Bucher (1933), p. v.

98. Bucher (1933), p. vii. Beno Gutenberg also chose to use the word "law," albeit in quotation marks. "The books and papers dealing with hypotheses on the development of the earth's crust are as the sands of the sea," he wrote. Our "'laws' . . . are not laws of nature—which we do not know—but have been derived from observations and may be 'recalled' any time by scientists following the finding of new data or even a rediscussion of older ones" (Gutenberg 1939, pp. 177–178). Nevertheless, Gutenberg argued, along similar lines to Bucher, that it was heuristically beneficial to articulate them.

99. Bucher (1933), p. 39. This is not to say that Bucher was entirely consistent. For example, on p. 42 he stated unequivocally, labeled as neither law nor opinion, that "recent work on the melting points of acid and basic rocks has demonstrated beyond doubt that 'roots-of-mountains' cannot exist." Famous last words again.

100. Bucher (1933), pp. vii–viii, quoting A. N. Whitehead, "Universities and their function," *Atlantic Monthly*, May 1928, p. 639.

Chapter 10

1. Versions of this interpretation can be found, for example, in Le Grand (1988) and Stewart (1990).

2. A choice that was made possible by patronage, first by the Carnegie Institution and then more abundantly by the U.S. Navy.

3. The other important locale was Paris, where experimental work in petrology and crystallography had been going on for some time. There is very little secondary literature on this work; Lydie Touret (1997) has made a start. On the influence and importance of the Geophysical Laboratory, see Servos (1990) and Yoder (1989).

4. RFP, newspaper clipping, "Will attach puzzle of the earth's core," *New York Times*, March 18, 1932, p. 23; also discussed in RADP.

5. Good (1994) and other papers in that volume.

6. Le Grand (1994) and other papers in Good (1994). At Harvard, Daly helped Percy

Bridgman to receive funding from the Carnegie Institution for his studies of experimental deformation of materials.

7. It is impossible for me to track here the entire development of marine geophysics and oceanography in America. Suffice it to note that, although oceanography began to grow at Wood's Hole and Scripps, and later at Lamont, in ways that Bowie and Field never imagined, Bowie was a member of the original NAS committee on oceanography and Field was the man who first connected Harry Hess and Maurice Ewing with the Navy.

8. Pettijohn (1984), p. 122.

9. Pettijohn (1984), p. 166; also see Penrose (1929).

10. Pettijohn (1984), p. 207. Pettijohn exaggerates somewhat: the paleontologists at Chicago were primarily field-workers (Rainger 1993), and geologists at the other end of the spectrum—for example M. King Hubbert—complained that *their* brand of science was not fully appreciated either (R. Doel, personal communication.) It would not be surprising if Hubbard's intensely mathematical work was not fully understood by his colleagues, but the trend—unmistakable in retrospect—was toward the kind of work Hubbard was doing. It is also interesting to note that the only nonpaleontologist strongly associated with fieldwork in that department, J. Harlen Bretz, was vilified by colleagues in the Scablands dispute. Bretz argued in the 1930s that the scablands of Washington state had not formed by gradual erosion but by the bursting of an ice-dammed glacial lake. Decades later he was vindicated and received the GSA Penrose medal at the age of 97; see Pettijohn (1984), p. 149, and Doel (1996).

11. Pettijohn (1984), p. 117.

12. Pettijohn (1984), p. 240.

13. On the values of exactitude and control in the laboratory see Olesko (1991); on its extension beyond, see Wise (1995).

14. Pettijohn (1984), pp. 240–241.

15. Pettijohn (1984), p. 244; also see Mackin (1963), p. 135. Mackin was that rare individual who was able to see the benefits and deficits of both sides of an argument, and rightly saw that much measurement in geology was no less descriptive than traditional field data: the observations were merely being described in numbers rather than in words. He thus preferred to distinguish not between the qualitative and the quantitative, but between the "rational" and the "empirical." Mackin was also one of the first people, as far as I know, to comment on the neglect of geology in the history of science; see his footnote on p. 137. My epigraph about evidence may be found on his p. 159.

16. Charles Park, *Geological Society of America Bulletin* 59, no. 4 (1964), cited in Wallace (1992). Economic geologists continue to defend the importance of fieldwork—28 years later—because of its persistent utility in exploration, and in the generation of "occurrence models" for ore deposits. Although the theory of ore deposition is relatively well developed, it remains of little help in locating individual deposits; this is done primarily by heuristic, descriptive models, based mostly on field experience.

17. Ben Page, personal communication, 1986.

18. Anonymous review, NSF Program in Geochemistry and Petrology, 1985. I am grateful to Prof. Marco Einaudi for sharing this.

19. Daly (1927), p. 672.

20. Du Toit (1937), p. 2.

21. Du Toit (1937), p. 2.

22. Du Toit (1937), p. 1.

23. Du Toit (1937), p. 318.

24. Du Toit (1937), p. 7.

25. A. du Toit to S. J. Shand, December 30, 1938, ADTP 7: Correspondences P–S, 1930s–1940s.

26. Du Toit (1937), p. 321, original emphasis.

27. See, for example, Caster (1952), pp. 105–152.

28. Press clipping, Review of *Our Wandering Continents* by T. W. Gevers, Johannesburg Star, Wednesday, 29 December 1937, ADTP 8: "Correspondence on problem [continental drift]."

29. Kuenen (1938).

30. H. L. Chhibber, 1947. The origin of forces responsible for disruption of continents, mountain building, and continental drift: Origin and permanence of ocean basins and distribution of land and sea. *Bulletin of the National Geographical Society of India, Benares*, May 1947, in ADTP 5: "Urgent Correspondence 1947–1948."

31. R. A. Daly to du Toit, November 25, 1937; B. Gutenberg to du Toit, January 5, 1938, ADTP 8: Correspondence on Problem [Continental Drift].

32. H. Baker to du Toit, April 30, 1939, ADTP 6: Correspondence A-B (1930's–1940's) and ibid., December 25, 1934, ADTP 8: Continental Drift—Correspondence.

33. H. Baker to du Toit, November 26, 1937, ADTP 6: Correspondence A–B c. 1930s–1940s.

34. Frank B. Taylor to du Toit, January 24, 1938, ADTP 7: Correspondence T–V 1930s, 1940s. Taylor died the following June.

35. Bailey Willis to du Toit, July 11, 1933, BWP 26: 76.

36. ADTP 3: Diaries 1928–1944, 1933.

37. Du Toit to Bailey Willis, September 25, 1936, BWP 12: 87.

38. Bailey Willis to du Toit, November 24, 1937, BWP 26: 76.

39. Schuchert to du Toit, December 20, 1937, ADTP 7: Correspondence P–S, 1930s, 1940s.

40. Rastall (1928), p. 140.

41. Schuchert (1928b), p. 272.

42. Longwell (1938).

43. Simpson (1943).

44. Du Toit (1944), p. 153.

45. Longwell (1944b), p. 218.

46. Longwell (1944b), p. 219.

47. Longwell (1944b), p. 221.

48. Longwell (1944b), p. 221.

49. Longwell (1944b), p. 226; also see Longwell (1944c).

50. Longwell (1944b), p. 231.

51. Du Toit to L. C. King, September 4, 1944, ADTP 5: Correspondence 1937–1947.

52. L. C. King to du Toit, October 9, 1940, ADTP 5: Correspondence 1937–1947. King tried to persuade du Toit to write a textbook of geology in the framework of continental drift, a text which, he argued, would be the "first *real* geology textbook."

53. Du Toit (1945), p. 407.

54. Du Toit to C. Longwell, September 2, 1944, ADTP 6: Longwell Correspondence 1939–1944.

55. F. A. Vening Meinesz to du Toit, April 8, 1936, ADTP 8: Paramorphic zone of the earth's crust.

56. Wegener (1924), p. 56.

57. Du Toit to R. A. Daly, May 22, 1943, ADTP 6: Correspondence C–E (1930s–1940s).

58. R. A. Daly to du Toit, September 6, 1943, ADTP 6: Correspondence C–E (1930s–1940s).

59. "On the mathematical probability of drift," undated ms. circa 1944–1945, ADTP 14.

60. Du Toit's personal copy of *Our Wandering Continents*, courtesy of Bob du Toit, grandson of Alexander du Toit, Johannesburg, South Africa.

61. Holmes (1929a), p. 347.

62. The actual quote is from a letter to Roderick Murchison, dated Jan. 12, 1829. The full sentence is: "We must preach up travelling, as Demosthenes did 'delivery,' as the first, second, and third requisites for a modern geologist, in the present adolescent state of the science." *Life, Letters and Journals of Sir Charles Lyell*, edited by Katherine Lyell, vol. 1, pp. 233–234. Note the final phrase: it is not clear that Lyell would have still been preaching travel in the twentieth century.

63. Willis (1944), p. 510. The irony of this can scarcely be overstated. As Homer Le Grand notes, Lester King used this as an argument against those who opposed drift on theoretical grounds: "The hypothesis of continental drift is not to be proved by idle, arm-chair theorizing, but stands or falls on one thing—hard work in the field" (L. C. King, 1953, quoted in Le Grand 1988, p. 83). King was arguing in defense of field geology against geophysics, but this argument had general resonance in America where it was also used as a defense of professional geology against amateurism, especially creationist amateurism (see chapter 7). On the emphasis of seeing for oneself in American geology, see Newell (1993), esp. p. 31.

64. Shapin and Schaffer (1985), especially pp. 55–61. On the credentials required for witnesses, see Shapin (1994).

65. Rudwick (1985).

66. Newell (1993), esp. pp. 151–156.

67. HWMP 33: 11.

68. van der Gracht (1928a), p. 224.

69. Bowie (1938), p. 138. Bucher's annotated copy of this symposium proceedings was given to me by the late Charles Drake.

70. Bowie (1925a).

71. The idea that instrumental data are independent of the theories they test is as faulty as it is widespread. Allan Franklin (1993b, p. 131) quotes from Ackerman (1985): "Instruments create an invariant relationship between their operations and the world, at least when we abstract from the expertise involved in their correct use." As Franklin notes, Ackerman's "at least" clause is precisely the rub, because determinations of correct use—and correct construction—often require invoking the very theory the instrument is supposed to test, and we are still left with a logical circularity: there is no way to know for sure if a magnetometer is working without a theory of remanent magnetism, which is precisely the thing we are trying to measure.

72. Daston (1992); Daston and Galison (1992).

73. Porter (1995a); also see Porter (1995b) and Latour and Woolgar (1979).

74. The literature on the history of biology has grown enormously in recent years, but see especially Allen (1978), Benson et al. (1991), Maienschein (1986, 1991a), Pauly (1987), Rainger (1981), and Rainger et al. (1991).

75. Pauly (1987). This is a complex topic. Feminist historians, especially Evelyn Fox Keller (1985), have emphasized the importance of themes of power and control throughout the history of science, which might undermine the claim that something distinctive happened in biology at the turn of the century. There are two points to note about this. Historically, tropes of power have competed with tropes of knowledge, and scientists have (disingenuously or not) generally argued for knowledge as their "real" mission. (Even Bacon argued for power as *evidence* of veracity, i.e., as a testament to truth—see Epilogue.) The willing embrace of an engineering ideal by scientists in any discipline *is* notable, as Smith and Wise (1989) have shown in their study of Lord Kelvin. Second, illustrations of the theme of power and control in the history of science have typically been drawn from the natural philosophic rather than the natural historical traditions. In biology and geology, metaphors and motifs of control are fairly scarce until the twentieth century. Their use in the twentieth century *is* a signifier of change, and is it not coincidental that just when biologists embrace a control-oriented approach to their discipline, the term "natural history" is dropping away?

Control, of course, functions on a variety of levels. Olesko (1991) has emphasized the role of quantification of error as an internal disciplinary standard; Porter (1995b) focuses on quantification as a means to establish authority in the face of external skepticism; Kay (1997) discusses the goal of social control in the Rockefeller Foundation support of molecular biology; and Wise (1995) discusses the engineering ideal in American physics in relation to commercial enterprise. What I am emphasizing here is the conscious adoption of methods of control in the hope of generating more certain knowledge. In this vein, it is noteworthy that the Rockefeller Foundation—so central to the reductionist transformation of modern biology—mostly declined to fund geology because program directors Warren Weaver and Max Mason considered it insufficiently "fundamental" (R. Doel, personal communication; also see Doel, 1997), but they did fund geophysics, including Percy Bridgman's studies of the properties of earth materials.

76. Hacking (1983), esp. pp. 242–245, and Ross (1991).

77. Wise (1995).

78. Porter (1995b), p. 72; Wise (1995), p. 5. Although measurement and mathematization are not the same thing, as Norton Wise emphasizes, in the twentieth century they are hard to separate. Sociology embraced measurement, economics embraced mathematization, but both attempted both. Philosophers, too, adopted a scientistic attitude at this time, turning first to symbolic logic and then to linguistics (Wilson 1990).

79. On quantification and its motivations in the eighteenth and nineteenth centuries, see Frangsmyr et al. (1990) and various papers in Wise (1995).

80. Servos (1986); also see Servos (1990), esp. his chapter 5.

81. Gossel (1992); Ross (1991).

82. Frederic Holmes (1992) makes the point that there are always more problems than scientists have time to address, so it is not surprising to find delays between the availability of a tool and its first use. So if a particular tool suddenly gets picked up broadly across disciplines, then it seems reasonable to ask what made that tool seem so right at that time to so many people? Put another way, given the relatively long history of the hammer of quantification and measurement, how is it that in the 1920s the whole world finally started to look like nails?

83. Norris and Witherspoon (1904), p. 18.

84. Vening Meinesz and Wright (1930), p. 16.

85. On deskilling of labor in the twentieth century and its relation to industrial capitalism, see Braverman (1974). To consider objectified science as one example of a larger pattern of deskilled labor is an extremely suggestive idea, but it is well beyond the scope of this study.

86. Creath (1990), esp. 463–466; Holton (1973), pp. 219–259.

87. Sarton, *Introduction to the History of Science*, quoted in Singer (1994), p. vi. Perhaps not coincidentally, Sarton was funded by the CIW.

88. Popper (1963); Hempel (1966); Hanson (1958); Kuhn (1962); see also discussion in Boyd (1991).

89. Holton (1973), p. 101.

90. Which is not to say that geologists have not tried. On the contrary, the various grain-scale and rock classification protocols attest to the effort and energy people have dedicated to finding ways to standardize geological observations. These methods work up to a point: geologists generally can agree on the *name* of a rock, but what they still cannot do is provide observer-invariant *descriptions*.

91. Latour and Woolgar (1979); Oreskes (1988); Daston (1992); Daston and Galison (1992).

92. Chamberlin (1928), p. 85; Schuchert (1928b), p. 111.

93. Willis (1928), p. 80.

94. Schuchert (1928b), p. 114. At least one geodesist did, and went further: William Bowie ascribed the fit to the breakup of the earth's primordial crust at the very beginning of geological time.

95. See, for example, Maxwell (1968) and the discussion in Suppe (1993).

96. Le Grand (1988), p. 204. Echoing Walter Bucher, Wilson (1968) proposed that the new, objectified science should be renamed "geonomy"—the study of the laws of the earth.

97. Stewart (1990), p. 139.

98. Stewart (1990), p. 138.

99. Stewart (1990), p. 140. There are many more examples, both of scientists making such claims and of historians believing them. Alan Cox, for example, claimed that "it was only after the introduction of the new geometrical concepts [that global] tectonics became a modern science characterized by an interplay of theory and observation" (quoted in Allwardt 1990, p. 90). The absurdity of this claim should by now be apparent. Even more absurdly, historian Stephen Pyne (1979, p. 176) claimed that, prior to plate tectonics, "geology had simply run out of significant questions."

100. Several of the most prominent supporters of drift in the 1920s—John Joly, Arthur Holmes, Reginald Daly—were men whose interests and training encompassed both field geology and laboratory science.

101. Read (1957).

102. Wegener ([1929] 1966), foreword. Emphasis added.

103. Runcorn (1959). Menard (1986) also makes this point—that geologists had effectively discredited their own evidence.

104. Bucher (1952), p. 93.

105. Mayr (1952), p. 85.

106. Caster (1952), p. 105.

107. Hurley (1968), p. 7.

108. Tjeerd van Andel, personal communication; Maurice Ewing apparently held this view (Heezen and Tharp 1965; Tharp 1982). In retrospect this point surely would have been pressed more if not for the preexisting evidence of drift. Another alternative was earth expansionism, which Bruce Heezen and S. Warren Carey promoted; see Heezen and Tharp (1965) and discussions in Le Grand (1988) and Nunan (1988). It was also possible to accept convection currents and not seafloor spreading; this was William Menard's position as late as 1964; see Menard (1964) and discussion in Stewart (1990), p. 79.

109. Galison (1987, 1997); see also discussion in Franklin (1993b).

110. When J. Tuzo Wilson documented the history of plate tectonics, he buried the earlier evidence even while acknowledging its historical importance. In 1976, when reediting a 1971 *Scientific American* volume on plate tectonics, *Continents Adrift and Continents Aground,* Wilson decided to omit "all five articles that constituted the first section of *Continents Adrift.* They dealt with observations that preceded the acceptance of the theory of continental drift. Their chief interest is now historical, for they paved the way for recognizing the validity of the theory." T. S. Kuhn famously noted this revisionist feature about scientific revolutions, and Wilson cited Kuhn here and elsewhere. In thus rewriting his own disciplinary history, Wilson intimated that he was doing what he was *supposed* to do. Life imitating philosophy? See Wilson (1971, 1976).

Epilogue

1. Chamberlin (1928), p. 87.

2. In 1967, M. King Hubbert asked rhetorically whether "the Principle of Uniformity, having played a strategic role in the development of a valid history of the earth, has by now largely lost its usefulness?" (Hubbert 1967, p. 29). Steven Jay Gould (1965) asked similarly, "Is uniformitarianism necessary?" The answer is obvious from the fact of their having posed the question: although geologists throughout the 1950s and 1960s continued to tout uniformitarianism as the guiding principle of modern geology, it had lost much of its substantive

meaning. Sedimentologists especially came to emphasize the role of rare large events in depositional environments—the hundred-year flood, the sporadic turbidite; see, for example, Miller (1940). The recent acceptance of a catastrophist interpretation of dinosaur extinction suggests that the uniformitarian period in geology has come to a close.

3. Julie Newell (1993) discusses personal antagonisms in the early history of American geology; see esp. pp. 131–132.

4. Pragmatist philosophy notwithstanding. It is not coincidental that C. S. Peirce, credited as the founder of pragmatism, was the son of Benjamin Peirce, Director of the U.S. Coast and Geodetic Survey, and worked for a time as a survey employee engaged in pendulum gravity measurements. Peirce was deeply influenced by scientific practice as a model for knowledge, as were his followers. William James (1907, p. 35), in his attempt to define what pragmatism means, wrote that "Dewey, Schiller and their allies, have only followed the example of geologists, biologists, and philologists." (Note geologists come first!) The best introduction to pragmatism is Hacking (1983); on Peirce, see Moore (1993) and Houser and Kloesel (1992).

5. Philosophers, on the other hand, are skeptical, because relativity implies fundamentally different conceptions of space and time than did classical mechanics. Either space and time are absolute or they are not; these are not epistemologically reconcilable positions.

6. Most famously by Merton (1942, pp. 267–278) and Conant (1947, pp. 7–8). For two recent discussions of collective scrutiny in science, see Traweek (1988) and Longino (1990). For insight into problems of contemporary peer review, see "Methodical progress," *Economist*, September 27, 1977, p. 89.

7. Carl Hempel (1966, p. 2), in his *Philosophy of Natural Science* wrote, "The high prestige that science enjoys today is no doubt attributable in large measure to the striking successes and the rapidly expanding reach of its applications." More recently, Carl Sagan (1996), in his last lectures and penultimate book, made the same argument: people are fooled into believing in all kinds of nonsense, but none of these things actually work. Perhaps the declining prestige of science that Sagan lamented may be accounted for in a parallel vein: an increasing sense of the ways in which science has *not* worked has led to disenchantment.

8. Hacking does not however conclude that our *theories* of electrons are necessarily correct, only that electrons themselves exist (Hacking 1983, esp. pp. 262–275). For an account of the success of science that skirts realism—either about theories or entities—see Laudan (1984); for a discussion of the criterion of manipulation as justification for realist views of experimental entities, see Franklin (1993b). Admittedly, I am conflating what some would consider two different senses of success—predictive and utilitarian. Philosophers have typically emphasized the former—viz., the "no miracles argument": if our theories were not true, how could they make such accurate predictions? (for example, Boyd 1991). But prediction *is* a kind of use, and when we use scientific theory to predict a phenomenon—such as the collision of comet Shoemaker-Levy with Jupiter—we are *doing* something on the basis of scientific knowledge no less than when we build a VCR or design a nuclear waste disposal site.

9. Sir Francis Bacon, *New Organon*, Book One, Aphorism 124; also see Cohen (1985, pp. 150–151). In the political realm, Mill sagely notes the reverse: "The truth of an opinion is part of its utility." See Mill (1859, Collini edition p. 25).

10. Not that other factors do not come into it—symmetry, elegance, parsimony; philosophers have made much of such extra-evidential considerations. But most scientists, most of the time, judge theories mostly by how accurately they make predictions, how well they explain known phenomena, and how effectively they can be used to solve problems. At least this is my belief.

11. Oldroyd (1990); also see Oldroyd (1987) and (1996), pp. 246–247, 281, and 308–310. For a recent attempt to enshrine coherence as a central explanatory resource in philosophy of science, see Thagard (1992).

12. Kuhn (1957), esp. pp. 227–228. At times it is hard to fathom what people mean by scientific revolutions: Steven Brush (1979, p. 149) says that "the scientific revolution" took place from 1500 to 1800, followed by a second revolution beginning around 1800 and lasting at least until the 1950s. In this light, we might as well call Darwin's work the theory of revolution by natural selection! For a recent review and critique of Kuhn's historiography, see Westman (1994).

13. Yet scientists are notoriously oblivious, if not hostile, to their own history. Science looks to the future, not the past. Why should one be interested in a history of mistakes? As a student of mine once put it, "Why should we learn these old ideas that aren't even true, anyway?" Why, indeed.

BIBLIOGRAPHY

Archival Sources

ADTP   Alexander du Toit Papers, University of Cape Town Archives, Cape Town, Republic of South Africa. Courtesy of the University of Cape Town.

BWP   Bailey Willis Papers, Huntington Library, San Marino, Calif. All items reproduced or cited by permission of the Huntington Library, San Marino, California.

CIWGF   Carnegie Institution of Washington, General Files, Carnegie Institution of Washington, Washington, D.C. Courtesy of the Carnegie Institution of Washington.

CIWGL   Carnegie Institution of Washington, Papers of the Geophysical Laboratory, Carnegie Institution of Washington, Washington, D.C. Courtesy of the Carnegie Institution of Washington.

CSP   Charles Schuchert Papers, Manuscripts and Archives, Yale University Library, New Haven, Conn. By permission of Yale University Library.

FWP   Frederick Eugene Wright Papers, Carnegie Institution of Washington, Washington, D.C. Courtesy of the Carnegie Institution of Washington, Helen Wright Greuter, and Finley Wright.

HHP   Harry Hess Papers, Manuscripts Division, Department of Rare Books and Special Collections, Princeton University, Princeton, N.J. Published with permission of the Princeton University Library and Mr. George B. Hess.

HWMP   Henry William Menard Papers, Scripps Institution of Oceanography Archives, University of California, San Diego, La Jolla, Calif. By permission of the University of California Regents per deed of gift.

HWP   Harry O. Wood Papers, California Institute of Technology Archives, Pasadena, Calif. Courtesy of the Archives, California Institute of Technology.

JCMP   John C. Merriam Papers, United States Library of Congress, Washington, D.C.

JJP–TCD   John Joly Papers, Trinity College Dublin Archives, Dublin, Ireland. By permission for brief quotation by the Archives of Trinity College, Dublin; access to the papers by permission of The Board of Trinity College, Dublin.

TCD/Geol/Joly   John Joly Papers, Geology Department, Trinity College, Dublin, Ireland. Cited with permission of the Geology Department, Trinity College, Ireland.
RADP   Reginald A. Daly Papers, Harvard University Archives, Cambridge, Mass. Courtesy of the Harvard University Archives.
RFP   Richard Field Papers, Department of Geological Sciences, Guyot Hall, Princeton University, Princeton, N.J. Courtesy of the Department of Geological Sciences, Princeton University, and the American Geophysical Union.
RUSCGS   Records of the U.S. Coast and Geodetic Survey RG23, United States National Archives and Records Administration, formerly Washington D.C., now located in College Park, Md.
WBP   William Bowie Papers, National Oceanographic and Atmospheric Agency/ National Oceanic Survey and National Geodetic Survey, Rockville, Md.

List of Abbreviations

AAAS   American Association for the Advancement of Science
AGU   American Geophysical Union
CIW   Carnegie Institution of Washington
DTM   Department of Terrestrial Magnetism, Carnegie Institution of Washington
GSA   Geological Society of America
NAS   National Academy of Sciences
NRC   National Research Council
ONR   Office of Naval Research, U.S. Navy
SIO   Scripps Institution of Oceanography
USGS   United States Geological Survey
USCGS   United States Coast and Geodetic Survey

Published Sources

Achinstein, Peter, and Owen Hannaway, eds., 1985. *Observation, Experiment, and Hypothesis in Modern Physical Science: Studies from the Johns Hopkins Center for the History and Philosophy of Science*. Cambridge: MIT Press.

Ackerman, Robert, 1985. *Data, Instruments, and Theory*. Princeton, N.J.: Princeton University Press.

Adams, Frank Dawson, 1912. An experimental contribution to the question of the depth of the zone of flow in the earth's crust. *Journal of Geology* 20, no. 2 (Feb.–Mar.): 97–118.

Adams, Frank Dawson, 1954. *The Birth and Development of the Geological Sciences*. New York: Dover.

Airy, G. B., 1855. On the computation of the effect of attraction of mountain masses as disturbing the apparent astronomical latitude of stations in geodetic surveys. *Philosophical Transactions of the Royal Society of London* 145: 101–104.

Alessio, A., 1924. Doubts and suggestions on terrestrial isostasy. *Geographical Journal* 13: 33–45.

Allegre, Claude, 1988. *The Behavior of the Earth*. Translated by Deborah Kurmes van Dam. Cambridge: Harvard University Press.

Allen, E. T., and Arthur Day, L., 1928. Natural steam power in California. *Nature* 122: 17–28.

Allen, Garland, 1978. *Life Science in the Twentieth Century*. Cambridge: Cambridge University Press.

Bibliography

Allwardt, A. S., 1990. *The roles of Arthur Holmes and Harry Hess in the development of modern global tectonics*. Ph.D diss., University of California, Santa Cruz.
Alvarez, Walter, 1997. *T. Rex and the Crater of Doom*. Princeton, N.J.: Princeton University Press.
Ampferer, Otto, 1925a. The origin of continents (Review). *Nature* 116 (2917): 481.
Ampferer, Otto, 1925b. Über Kontinentverschiebungen. *Naturwissenschaften* 13: 639–620.
Argand, Emile, 1977. *Tectonics of Asia*. Translated by Albert V. Carozzi. New York: Hafner Press. Originally published 1922.
B. D., 1931. Obituary: Prof. Alfred Wegener. *Nature* 127: 861.
Babbage, Charles, 1837. *The Ninth Bridgewater Treatise*. London: John Murray.
Babbage, Charles, 1847. Observations on the Temple of Serapis, at Pozzouli, near Naples, with remarks on certain causes which may produce geological cycles of great extent. *Quarterly Journal of the Geological Society of London* 3: 186–213.
Badash, Lawrence, 1968. Rutherford, Boltwood, and the age of the Earth: The origin of radioactive dating techniques. *Proceedings of the American Philosophical Society* 112: 157–169.
Badash, Lawrence, 1979. *Radioactivity in America*. Baltimore: Johns Hopkins University Press.
Baker, Howard B., 1914. The origin of continental forms. *Annual Report of the Michigan Academy of Sciences* 99–103.
Barnes, Barry, David Bloor, and John Henry, 1996. *Scientific Knowledge: A Sociological Analysis*. Chicago: University of Chicago Press.
Barrell, Joseph, 1914a. The strength of the Earth's crust: Part I. Geologic tests of the limits of strength. *Journal of Geology* 22: 28–48.
Barrell, Joseph, 1914b. The strength of the Earth's crust: Part II. Regional distribution of isostatic compensation. *Journal of Geology* 22: 145–165.
Barrell, Joseph, 1914c. The strength of the Earth's crust: Part III. Influence of variable rate of isostatic compensation. *Journal of Geology* 22: 209–236.
Barrell, Joseph, 1914d. The strength of the Earth's crust: Part IV. Heterogeneity and rigidity of the crust as measured by departures from isostasy. *Journal of Geology* 22(4): 289–314.
Barrell, Joseph, 1914e. The strength of the Earth's crust: Part V. The depth of masses producing gravity anomalies and deflection residuals. *Journal of Geology* 22: 441–468.
Barrell, Joseph, 1914f. The strength of the Earth's crust: Part VI. Relations of the isostatic movements to a sphere of weakness—the asthenosphere. *Journal of Geology* 22: 655–683.
Barrell, Joseph, 1914g. The status of hypotheses of polar wanderings. *Science* 40(1027): 333–340.
Barrell, Joseph, 1917. Rhythms and the measurements of geologic time. *Bulletin of the Geological Society of America* 28: 745–904.
Barrell, Joseph, 1918a. A century of geology—The growth of knowledge of Earth structure. In *A Century of Science in America, with Special Reference to the American Journal of Science, 1818–1918*, edited by E. S. Dana. New Haven: Yale University Press, pp. 153–192.
Barrell, Joseph, 1918b. The growth of knowledge of Earth structure. *American Journal of Science* 196(46): 133–170.
Barrell, Joseph, 1919a. Grove Karl Gilbert: An appreciation. *Sierra Club Bulletin* 10: 397–399.
Barrell, Joseph, 1919b. The status of the theory of isostasy. *American Journal of Science* 198: 291–338.
Barrell, Joseph, 1919c. Sources and tendencies in American geology. *Scientific Monthly* 8(March): 193–206.
Barrell, Joseph, 1919d. The nature and bearings of isostasy. *American Journal of Science* 198: 281–290.

Barrell, Joseph, 1927. On continental fragmentation and the geologic bearing of the Moon's surface features. *American Journal of Science* 213: 283–314.

Bartlett, Richard A., 1962. *Great Surveys of the American West*. Norman: University of Oklahoma Press.

Becker, George F., 1915. Isostasy and radioactivity. *Bulletin of the Geological Society of America* 26: 171–204.

Benson, Keith R., Jane Maienschein, and Ronald Ranger, eds., 1991. *The Expansion of American Biology*. New Brunswick, N.J.: Rutgers University Press.

Berry, E. W., 1928. Comments on the Wegener hypothesis. In *The Theory of Continental Drift: A Symposium*, edited by W. A. J. M. van Waterschoot van der Gracht. London: Murray, pp. 194–196.

Birch, Francis, 1960. Reginald Aldworth Daly. *National Academy of Sciences Biographical Memoirs* 34: 31–64.

Birch, Francis, and Dennison Bancroft, 1938a. The effect of pressure on the rigidity of rocks. *Journal of Geology* 46(1): 59–87.

Birch, Francis, and Dennison Bancroft, 1938b. The effect of pressure on the rigidity of rocks: Part 2. *Journal of Geology* 46: 113–141.

Blackett, P. M. S., 1947. The magnetic field of massive rotating bodies. *Nature* 159: 658–666.

Blackett, P. M. S., 1948. *Fear, War and the Bomb: Military and Political Consequences of Atomic Energy*. New York: McGraw-Hill.

Blackett, P. M. S., 1952. A negative experiment relating to magnetism and the earth's rotation. *Transactions of the Royal Society* A245: 309–370.

Blackett, P. M. S., 1956. *Lectures on Rock Magnetism*. Jerusalem: Weizmann Science Books.

Blackett, P. M. S., E. Bullard, and S. K. Runcorn, eds. 1965. *A Symposium on Continental Drift*. London: Royal Society.

Blackwelder, Eliot, 1961. Bailey Willis, May 31, 1857–February 19, 1949. *National Academy of Sciences Biographical Memoirs* 35: 333–350.

Bloor, David, 1976. *Knowledge and Social Imagery*. Chicago: University of Chicago Press.

Boltwood, Bertram B. 1904a. On the ratio of radium to uranium in some minerals. *American Journal of Science*, 168: 96–103.

Boltwood, Bertram B., 1904b. Relation between uranium and radium in some minerals. *Nature* 70, no. 1804(May 26): 80.

Boltwood, Bertram B., 1905a. On the ultimate disintegration products of the radio-active elements. *American Journal of Science* 170: 253–267.

Boltwood, Bertram B., 1905b. The origin of radium. *Philosophical Magazine*, 6th ser., 9: 599–613.

Boltwood, Bertram B., 1905c. The production of radium from uranium. *American Journal of Science*, 170: 239–244.

Boltwood, Bertram B., 1907a. Note on a new radio-active element. *American Journal of Science* 174: 370–72.

Boltwood, Bertram B., 1907b. The origin of radium. *Nature* 76, no. 1980(October 10): 589.

Bonatti, Enrico, 1994. The Earth's mantle below the oceans. *Scientific American* (March): 44–51.

Bork, Kennard Baker, 1994. *Cracking Rocks and Defending Democracy: Kirtley Fletcher Mather, 1888–1978*. San Francisco: American Association for the Advancement of Science.

Bouguer, M. Pierre, 1749. *La Figure de la Terre determinée par les observations faites au Pérou*. Paris: Guerin et de la Tour.

Bowen, N. L., 1956. *The Evolution of the Igneous Rocks*. With a new introduction by J. F. Schairer. New York: Dover.

Bowie, William, 1922. The Earth's crust and isostasy (with discussion following). *Geographical Review* 12: 613–627.

Bowie, William, 1924a. A Gravimetric Test of the "Roots of Mountains" Theory. U.S. Coast and Geodetic Survey Serial No. 291. Washington, D.C.: U.S. Government Printing Office, pp. 3–8.
Bowie, William, 1924b. Abnormal densities in the Earth's crust disclosed by analysis of geodetic data. Geographical Journal 13: 26–33.
Bowie, William, 1925a. Scientist to weigh the floating Earth's crust. New York Times, September 20, p. xx.
Bowie, William, 1925b. Scientists to test "drift" of continents. New York Times, September 6, p. 5.
Bowie, William, 1925c. Miscelleneous scientific intelligence. American Journal of Science 209: 180–185.
Bowie, William, 1926. Proposed theory, in harmony with isostasy, to account for major changes in the elevation of the Earth's surface. Gerlands Beiträge zur Geophysik 15: 103–115.
Bowie, William, 1927a. The effects of the shape of the geoid on values of gravity at sea. American Journal of Science 214: 222–227.
Bowie, William, 1927b. Geophysics: The part played by isostasy and geology. Journal of the Washington Academy of Sciences 17(5): 101–117.
Bowie, William, 1927c. Isostasy. New York: Dutton.
Bowie, William, 1928. Comments on the Wegener hypothesis. In The Theory of Continental Drift: A Symposium, edited by W. A. J. M. van Waterschoot van der Gracht. New York: American Association of Petroleum Geologists, pp. 178–186.
Bowie, William, 1929a. Possible origins of oceans and continents. Gerlands Beiträge zur Geophysik 21: 178–182.
Bowie, William, 1929b. Weighing the earth from a submarine. Scientific American: 218–221.
Bowie, William, 1929c. The figure of the Earth derived by triangulation methods. American Journal of Science 218: 53–64.
Bowie, William, 1929d. Isostasy and geological thought. Scientific Monthly 28: 385–392.
Bowie, William, 1929e. Annual Office Report of the Division of Geodesy, U.S. Coast and Geodetic Survey. Washington, D.C.: U.S. Government Printing Office.
Bowie, William, 1929f. Zones of weakness in the Earth's crust. Science 70: 589–592.
Bowie, William, 1930. International cooperation in geographical work. Science 71: 425–429.
Bowie, William, 1931. The Earth as an engineering structure. Scientific Monthly 32: 457–460.
Bowie, William, 1935. The origin of continents and oceans. Scientific Monthly 41: 444–449.
Bowie, William, 1936. The place of geodesy in geophysical research. Paper read at National Research Council. Transactions of the American Geophysical Union, Seventeenth Annual Meeting, 1936.
Bowie, William, 1938. Round-table discussion: Geophysical exploration of the ocean bottom. Proceedings of the American Philosophical Society 79: 127–144
Boyd, Richard, 1991. On the current status of scientific realism. In The Philosophy of Science, edited by Richard Boyd, Philip Gaspar, and J. D. Trout. Cambridge: MIT Press, pp. 195–222.
Braverman, Harry, 1974. Labor and Monopoly Capital: The Degradation of Work in the Twentieth Century. New York: Monthly Review Press.
Brennecke, E., 1927. Die Aufgaben und Arbeiten des Geodätischen Instituts in Potsdam in der Zeit nach dem Weltkriege. Zeitschrift für Vermessungswesen 743–766, 815–828.
Brown, Michael Chandos, 1989. Benjamin Silliman: A Life in the Young Republic. Princeton, N.J.: Princeton University Press.
Bruce, Robert V., 1987. The Launching of Modern American Science. Ithaca, N.Y.: Cornell University Press.

Bruck, M. T., 1995. Lady computers at Greenwich in the early 1890s. *Quarterly Journal of the Royal Astronomical Society* 36: 83–95.

Brush, Stephen G., 1978a. A geologist among astronomers: The rise and fall of the Chamberlin–Moulton cosmogony: Part 1. *Journal for the History of Astronomy* 9, part 1 (February): 1–41.

Brush, Stephen G., 1978b. A geologist among astronomers: The rise and fall of the Chamberlin–Moulton cosmogony: Part 2. *Journal for the History of Astronomy* 9, part 1 (June): 77–104.

Brush, Steven G., 1979. Scientific revolutionaries of 1905: Einstein, Rutherford, Chamberlin, Wilson, Stevens, Binet, Freud. In *Rutherford and Physics at the Turn of the Century*, edited by M. Bunge and W. R. Shea. New York: Dawson and Science History Publications, pp. 140–171.

Brush, Stephen G., 1996a. *Nebulous Earth: The Origin of the Solar System and the Core of the Earth from Laplace to Jeffreys*. Vol. 1. Cambridge: Cambridge University Press.

Brush, Stephen G., 1996b. *Transmuted Past: The Age of the Earth and the Evolution of the Elements from Lyell to Patterson*. Vol. 2. Cambridge: Cambridge University Press.

Brush, Stephen G., 1996c. *Fruitful Encounters: The Origin of the Solar System and of the Moon from Chamberlin to Apollo*. Vol. 3. Cambridge: Cambridge University Press.

Bucher, Walter H., 1933. *The Deformation of the Earth's Crust*. Princeton, N.J.: Princeton University Press.

Bucher, Walter H., 1952. Continental drift versus land bridges. *Bulletin of the American Museum of Natural History* 99(3): 93–103.

Buchwald, J. Z., 1991. Review of *Energy and Empire*. *British Journal for the History of Science* 24: 85–94.

Bull, A. J., 1921. A hypothesis of mountain building. *Geological Magazine* 58: 364–367.

Bull, A. J., 1927. Some aspects of the moutain building problem. *Proceedings of the Geological Association* 38: 145–156.

Bull, A. J., 1929. Further aspects of the moutain building problem. *Proceedings of the Geological Association* 40: 105–114.

Bull, A. J., 1931. The convection current hypothesis of mountain building. *Geological Magazine* 68: 495–498.

Bullard, Edward Crisp, 1964. Continental drift. *Quarterly Journal of the Geological Society of London* 120: 1–33.

Bullard, Edward Crisp, 1975. William Maurice Ewing, 1906–1974. *Biographical Memoirs of the Fellows of the Royal Society* 21: 268–311.

Bunge, Mario, and William R. Shea, eds., 1979. *Rutherford and Physics at the Turn of the Century*. New York: Dawson and Science History Publications.

Burchfield, Joe D., 1974. *Lord Kelvin and the Age of the Earth*. Chicago: Chicago University Press.

Burger, William H., 1931. Biographical memoir of John Fillmore Hayford 1868–1925. *National Academy of Sciences Biographical Memoirs* 16(5): 156–292.

Burrard, Sidney, 1921. On the origin of mountain ranges. *Geographical Journal*: 199–219.

Carey, Samuel Warren, ed., 1958. *Continental Drift: A Symposium*. Hobart: University of Tasmania, Department of Geology.

Carey, Samuel Warren, 1976. *The Expanding Earth*. New York: Elsevier.

Carey, Samuel Warren, ed., 1983. *The Expanding Earth: A Symposium*. Hobart: University of Tasmania, Department of Geology.

Carey, Samuel Warren, 1988. *Theories of the Earth and Universe: A History of Dogma in the Earth Sciences*. Stanford, Calif.: Stanford University Press.

Carozzi, A. V., 1970. A propos de l'origine de la théorie des dérives continentales: Francis Bacon (1620), François Placet (1668), A. von Humboldt (1801), et A. Snider (1858).

Comptes Rendu et Séances de la Société des Physiques et de la Histoire Naturelle n.s., 4: 171–179.
Carozzi, A. V., 1985. The reaction in continental Europe to Wegener's theory of continental drift. *Earth Sciences History* 4: 131.
Cartwright, Nancy, 1983. *How the Laws of Physics Lie*. Oxford: Clarendon Press.
Caster, Kenneth, 1952. Stratigraphic and paleontologic data relevant to the problem of Afro-American ligation during the Paleozoic and Mesozoic. In The Problem of Land Connections Across the South Atlantic with Special Reference to the Mesozoic. *Bulletin of the American Museum of Natural History* 99: 105–152.
Chamberlin, Rollin T., 1928. Some of the objections to Wegener's theory. In *The Theory of Continental Drift: A Symposium*, edited by W. A. J. M. van Waterschoot van der Gracht. London: Murray, pp. 83–87.
Chamberlin, T. C., 1890. The method of multiple working hypotheses. *Science* 15(366): 92–96.
Chamberlin, T. C., 1897a. The method of multiple working hypotheses. *Journal of Geology* 5: 837–848.
Chamberlin, T. C., 1897b. A group of hypotheses bearing on climatic changes. *Journal of Geology* 5: 653–683.
Chamberlin, T. C., 1898. The ulterior basis of time divisions and the classification of geologic history. *Journal of Geology* 6: 449–462.
Chamberlin, T. C., 1899a. An attempt to frame a working hypothesis of the cause of glacial periods on an atmospheric basis (I). *Journal of Geology* 7: 545–584.
Chamberlin, T. C., 1899b. An attempt to frame a working hypothesis of the cause of glacial periods on an atmospheric basis (II). *Journal of Geology* 7: 667–685.
Chamberlin, T. C., 1899c. An attempt to frame a working hypothesis of the cause of glacial periods on an atmospheric basis (III). *Journal of Geology* 7: 751–787.
Chamberlin, T. C., 1899d. Editorial. *Journal of Geology* 7: 295–297.
Chamberlin, T. C., 1899e. Lord Kelvin's address on the age of the Earth as an abode fitted for life. *Science* 9: 889–901.
Chamberlin, T. C., 1899f. Lord Kelvin's address on the age of the Earth as an abode fitted for life. *Science* 10: 11–18.
Chamberlin, T. C., 1905. The methods of the earth sciences. *Popular Science Monthly* 66: 66–75.
Chamberlin, T. C., 1906. Letter to George Darwin, June 30, 1906. In *Science in America: A Documentary History 1900–1939*, edited by Nathan Reingold and Ida H. Reingold. Chicago: University of Chicago Press.
Chamberlin, T. C., 1909. A geologic forecast of the future opportunities of our race. *Science* 30(783): 937–949.
Chamberlin, T. C., 1916a. The evolution of the Earth: I. Earth-Genesis. *Scientific Monthly* 2: 417–437.
Chamberlin, T. C., 1916b. The evolution of the Earth: II. Earth-Growth. *Scientific Monthly* 2: 536–556.
Chamberlin, T. C., 1916c. Isostasy in light of planetesimal theory. *American Journal of Science* 192: 371.
Chamberlin, T.C., 1916d. *The Origin of the Earth*. Chicago: University of Chicago Press.
Chamberlin, T. C., 1919. Investigation versus propagandism. *Journal of Geology* 27(5): 305–338.
Chamberlin, T. C., 1924. Seventy-five years of American geology. *Science* 59(1519): 127–135.
Chamberlin, T. C., 1965. The method of multiple working hypotheses. *Science* 148: 754–759.

Chamberlin, T. C., and Rollin Salisbury, 1909. *A College Text-book of Geology*. New York: H. Holt.
Chatterjee, Sankar, and Nicholas Hotton III, eds., 1992. *New Concepts in Global Tectonics*. Lubbock: Texas Tech University Press.
Chhibber, H. L., 1947. The origin of forces responsible for disruption of continents, mountain building, and continental drift: Origin and permanence of ocean basins and distribution of land and sea. *Bulletin of the National Geographical Society of India* 5(May): 1–11.
Cohen, I. Bernard, 1985. *Revolution in Science*. Cambridge: Harvard University Press, Belknap Press.
Collins, H. M., 1992. *Changing Order: Replication and Induction in Scientific Practice*. Chicago: University of Chicago Press.
Conant, James B., 1947. *On Understanding Science: An Historical Approach*. New Haven: Yale University Press.
Continental drift loses a dedicated opponent, 1989. *Geology Today*, July–August, p. 111.
Coonan, Stephen, 1991. John Joly's Irish Tricolor. *Technology Ireland* (October): 18–21.
Cox, Allan, ed., 1973. *Plate Tectonics and Geomagnetic Reversals*. San Francisco: W. H. Freeman.
Cox, A., R. R. Doell, and G. B. Dalrymple, 1963. Geomagnetic polarity epochs and Pleistocene geochronometry. *Nature* 198: 1049–1051.
Creath, Richard, ed., 1990. *Dear Carnap, Dear Van: The Quine–Carnap Correspondence and Related Work*. Berkeley: University of California Press.
Creer, K. M., E. Irving, and S. K. Runcorn, 1954. The direction of the geomagnetic field in remote epochs in Great Britain. *Journal of Geomagnetism and Geoelectricity* 6: 163–168.
Daly, Reginald A., 1912. *Geology of the North American Cordillera*. Memoir No. 38, *Canada Department of Mines Geological Survey*. Ottawa: Government Printing Bureau.
Daly, Reginald A., 1923. The earth's crust and its stability. *American Journal of Science* 205: 349–377.
Daly, Reginald A., 1926. *Our Mobile Earth*. New York: Charles Scribner's Sons.
Daly, Reginald A., 1927. Review of *The Geology of South Africa*, by Alexander L. du Toit. *Journal of Geology* 35(7): 671–672.
Daly, Reginald A., 1935. *The Changing World of the Ice Age*, 2d ed. New Haven: Yale University Press.
Daly, Reginald A., 1938. *Architecture of the Earth*. Edited by K. F. Mather. Century Earth Science Series. New York: Appleton-Century.
Daly, Reginald A., 1940. *Strength and Structure of the Earth*. New York: Prentice-Hall.
Daly, Reginald A., and G. A. F. Molengraaf, 1924. Structural relations of the Bushveld Igneous Complex, Transvaal. *Journal of Geology* 32(1): 1–35.
Dana, James Dwight, 1846. On the volcanoes of the Moon. *American Journal of Science* 52(2d ser., vol. 2): 335–355.
Dana, James Dwight, 1847a. Origin of the grand outline features of the Earth. *American Journal of Science* 53(2d ser., vol. 3): 381–398.
Dana, James Dwight, 1847b. On the origin of continents. *American Journal of Science* 53(2d ser., vol. 3): 94–100.
Dana, James Dwight, 1847c. Geological results of the Earth's contraction in consequence of cooling. *American Journal of Science* 53(2d ser., vol. 3): 176–188.
Dana, James Dwight, 1873a. On the origin of mountains. *American Journal of Science* 105(3d ser., vol. 5): 347–350.
Dana, James Dwight, 1873b. On some results of the Earth's contraction from cooling, including a discussion of the origin of mountains, and the nature of the Earth's interior. *American Journal of Science* 105(3d ser., vol. 5): 423–443.

Dana, James Dwight, 1875. *The Geological Story Briefly Told*. New York: Ivison, Blakeman, Taylor.
Daniels, George H., 1967. The process of professionalization in American science: The emergent period, 1820–1860. *Isis* 58: 151–166.
Daniels, George H., ed., 1972. *Nineteenth Century Science: A Reappraisal*. Evanston, Ill.: Northwestern University Press.
Darwin, G. H., 1879. The precession of a viscous spheroid and the remote history of the Earth. *Philosophical Transactions of the Royal Society of London* 170: 447–538.
Daston, Lorraine, 1992. Objectivity and the escape from perspective. *Social Studies of Science* 22: 597–618.
Daston, Lorraine, and Peter L. Galison, 1992. The image of objectivity. *Representations* 40: 81–128.
Davis, Geoffrey F., 1992. Plates and plumes: Dynamos of the Earth's mantle. *Science* 257: 493–494.
Dawson, J. William, 1894. Some recent discussions in geology: Annual address by the president. *Bulletin of the Geological Society of America* 5: 101–116.
De Geer, Gerard, 1888. *Om Skandinaviens nivaforandringar under qvartarperioden, Geologiska Foreningens, Stockholm, Forhandlingar*. Stockholm: Kongl. Boktryckeriet. P. A. Norstedt & Soneb.
De la Beche, Henry Thomas, 1832. *A Geographical Manual*. 3d ed. Philadelphia: Carey and Lea.
De la Beche, Henry Thomas, 1834. *Researches in Theoretical Geology*. London: C. Knight.
Dietz, Robert S., 1961. Continent and ocean basin evolution by spreading of the sea floor. *Nature* 190(June 3): 854–857.
Dietz, Robert S., 1968. Reply to comment by A. A. Meyerhoff. *Journal of Geophysical Research* 73: 6567.
Dixon, H. H., 1934. John Joly 1857–1933. *Obituary Notices of the Royal Society of London* 3: 259–286.
Dixon, K., 1957. John Joly, 1857–1933. *Trinity, An Annual Record Published by Trinity College, Dublin*, no. 9: 13–19.
Doel, R. E., 1996. *Solar System Astronomy in America: Communities, Patronage and Interdisciplinary Science, 1920–1960*. Cambridge: Cambridge University Press.
Doel, R. E., 1997. The earth sciences and geophysics. In *Science in the Twentieth Century*, edited by John Krige and Dominque Pestre. Paris: Harwood Academic Publishers.
Donovan, Arthur, Larry Laudan, and Rachel Laudan, 1992. Testing theories of scientific change. In *Scrutinizing Science: Empirical Studies of Scientific Change*, edited by A. Donovan, L. Laudan, and R. Laudan. Baltimore: Johns Hopkins University Press, pp. 3–46.
Dott, R. H., Jr., 1974. The geosynclinal concept. In *Modern and Ancient Geosynclinal Sedimentation*, edited by R. H. Dott Jr. and R. H. Shaver. Society of Economic Paleontologists and Mineralogists Special Publication 19. Tulsa, Okla., 1–13.
Dott, R. H., Jr., 1979. The geosyncline: First major geological concept "Made in America." In *Two Hundred Years of Geology in America*, edited by C. Schneer. Hanover, N.H.: University Press of New England, pp. 239–264.
Dott, R. H., Jr., 1985. James Hall's discovery of the craton. In *Geologists and Ideas: A History of North American Geology*. Geological Society of America, pp. 157–168.
Dott, R. H., Jr., 1997. James Dwight Dana's old tectonics—Global contraction under divine direction. *American Journal of Science* 297: 283–311.
Drifting apart, 1925. *New York Times*, September 6, sec. 2, p. 6.
Duchesneau, François, 1982. The role of hypotheses in Descartes's and Buffon's theories of the Earth. In *Problems of Cartesianism*, edited by T. Lennon, J. M. Nicholas,

and J. W. Davis. Kingston and Montreal: McGill–Queen's University Press, pp. 113–125.
Dunbar, Carl O., 1949 (1960). *Historical Geology.* New York: Wiley.
Dunham, K. C., 1966. Arthur Holmes. *Biographical Memoirs of Fellows of the Royal Society* 12: 291–310.
Dupree, A. H., 1986. *Science in the Federal Government.* Baltimore: Johns Hopkins University Press.
du Toit, Alexander L., 1927. A geological comparison of South America with South Africa. *Publication of the Carnegie Institution of Washington,* No. 381. Washington, D.C.: Carnegie Institution of Washington.
du Toit, Alexander L., 1929. The continental displacement hypothesis as viewed by du Toit (A reply to Charles Schuchert). *American Journal of Science* 217(5th ser., vol. 17): 179–183.
du Toit, Alexander L., 1937. *Our Wandering Continents.* Edinburgh: Oliver and Boyd.
du Toit, Alexander L., 1944. Tertiary mammals and continental drift: A rejoinder to George G. Simpson. *American Journal of Science* 242: 145–163.
du Toit, Alexander L., 1945. Further remarks on continental drift. *American Journal of Science* 243: 404–408.
Dutton, C. E., 1871–1872. The causes of regional elevations and subsidences. *Proceedings of the American Philosophical Society* 12: 70–72.
Dutton, C. E., 1874. A criticism upon the contractional hypothesis. *American Journal of Science* 108: 113–123.
Dutton, C. E., 1876a. Critical observations on theories of the Earth's physical evolution. *Penn Monthly* 7(May): 364–378.
Dutton, C. E., 1876b. Critical observations on theories of the Earth's physical evolution (A review). *American Journal of Science* 112: 142–145.
Dutton, C. E., 1880. *Report on the Geology of the High Plateaus of Utah; U.S. Geographical and Geological Survey of the Rocky Mountain Region.* Washington, D.C.: U.S. Government Printing Office.
Dutton, C. E., 1889. On some of the greater problems of physical geology. *Bulletin of the Philosophical Society of Washington* 11: 51–64.
Elsasser, Walter M., 1968. The mechanics of continental drift. *Proceedings of the American Philosophical Society* 112(5): 344–353.
Evans, J. W., 1921. The East Indian Archipelago: Discussion. *Geographical Journal* (February): 119–120.
Evans, J. W., 1923. The Wegener hypothesis of continental drift. *Nature* 111(March 24): 393.
Eve, Arthur Stewart, 1939. *Rutherford: Being the Life and Letters of the Rt. Hon. Lord Rutherford, OM.* New York: MacMillan.
Ewing, Maurice, 1937. Gravity measurements on the U.S.S. Barracuda. *Transactions of the American Geophysical Union* 18: 66–69.
Ewing, Maurice, 1938. Marine gravimetric methods and surveys. *Proceedings of the American Philosophical Society* 79: 47–70.
Ewing, Maurice, and J. Lamar Worzel, 1945. *Long Range Sound Transmission.* Woods Hole, Mass.: Woods Hole Oceanographic Institution, Contract NObs-2083, Report No. 9, declassified March 12, 1946.
Faye, H. A., 1880. Sur la réduction des observations du pendule au niveau de la mer. *Comptes rendus des séances de L'Académie des sciences, Comptes rendus hebdomadaires des séances* 90: 1443–1446.
Faye, H. A., 1883. Sur la réduction du baromètre et du pendule au niveau de la mer. *Comptes rendus des séances de L'Académie des sciences, Comptes rendus hebdomadaires des séances* 96: 1259–1262.

Faye, H. A., 1895. Réduction au niveau de la mer pesanteur observée à la surface de la terre (Coast and Geodetical Survey), par M. G.-R. Putnam. *Comptes rendus des séances de L'Académie des sciences, Comptes rendus hebdomadaires des séances* 120: 1081–1086.

Field, Richard M., 1936. Recent developments in the geophysical study of oceanic basins. Paper read at American Geophysical Union Seventeenth Annual Meeting.

Field, Richard M., 1937a. Structure of continents and ocean basins. *Journal of the Washington Academy of Sciences* 27: 181–195.

Field, Richard M., 1937b. *Report of Special Committee on Geophysical and Geological Study of Oceanic Basins*. Washington, D.C., American Geophysical Union.

Field, Richard M., ed. 1938. Geophysical exploration of the ocean bottom. *Proceedings of the American Philosophical Society* 79.

Field, Richard M., 1941. Geophysics and world affairs: A plea for geoscience. Paper read at American Geophysical Union Presidential Address and Third Award of the William Bowie Medal, 1941, at Elihu Root Hall, Carnegie Institution of Washington, April 30, 1941.

Field, Richard M., T. T. Brown, E. B. Collins, and H. H. Hess, 1933. *The Navy–Princeton Gravity Expedition to the West Indies in 1932*. Washington, D.C.: U.S. Government Printing Office.

Fisher, Osmond, 1873. On the inequalities of the Earth's surface viewed in connection with the secular cooling. *Transactions of the Cambridge Philosophical Society* 2: 324–331.

Fisher, Osmond, 1874. On the formation of mountains viewed in connexion with the secular cooling of the Earth. *Geological Magazine* 1: 60–63.

Fisher, Osmond, 1878. On the possibility of changes in the latitudes of places on the Earth's surface; Being an appeal to physicists. *Geological Magazine*, 2d ser., 2d decade, 5: 291–297.

Fisher, Osmond, 1879. On the inequalities of the earth's surface as produced by lateral pressure, upon the hypothesis of a liquid substratum. *Transactions of the Cambridge Philosophical Society* 12: 434–454.

Fisher, Osmond, 1881 (1889). *Physics of the Earth's Crust*. London: Macmillan.

Fisher, Osmond, 1882a. On the physical cause of the ocean basins. *Nature* 25 (January 12): 243–244.

Fisher, Osmond. 1882b. Physics of the Earth's crust. *Nature* 27 (November 23): 76–77.

Fisher, Osmond, 1882c. On the depression of ice-loaded lands. *Geological Magazine* 9: 526.

Fisher, Osmond, 1886. On the variations of gravity at certain stations of the Indian arc of the meridian in relation to their bearing upon the constitution of the Earth's crust. *Philosophical Magazine*, 5th ser., 22: 1–29.

Fisher, Osmond, 1887. A reply to the objections raised by Mr. Charles Davison, M.A., to the argument on the insufficiency of the theory of the contraction of a solid Earth to account for the inequalities or elevations of the surface. *Philosophical Magazine*, 5th ser., 24: 391–394.

Fisher, Osmond, 1888. On the mean height of the surface-elevations, and other quantitative results of the contraction of a solid globe through cooling; regard being paid to the existence of a level of no strain, as lately announced by Mr. T. Mellard Reade and by Mr. C. Davison. *Philosophical Magazine*, 5th ser., 25: 7–20.

Fisher, Osmond, 1895. On the age of the world as depending on the condition of the interior. *Geological Magazine*, 2d ser., 4th decade, 2: 244–46.

Fleming, James, 1992. T. C. Chamberlin and H_2O feedbacks: A voice from the past. *EOS* 73(47): 505–509.

Fleming, J. A., 1938. Terrestrial magnetism and oceanic structure. In *Geophysical Exploration of the Ocean Bottom*, edited by Richard M. Field. *Proceedings of the American Philosophical Society* 79: 109–125.

Fleming, J. A., 1951. William Bowie. *Biographical Memoirs of the National Academy of Sciences* 26: 61–97.

Fleming, J. A., and C. S. Piggott, 1956. Frederick Eugene Wright 1877–1953. *Biographical Memoirs of the National Academy of Sciences* 29: 317–344.

Forbush, Scott, 1946. Gravity determinations on the "Carnegie." *Scientific Results of Cruise VII of the "Carnegie," 1928–1929 under Command of Captain J. P. Ault.* Carnegie Institution of Washington Publication No. 571. Washington, D.C.: Carnegie Institution of Washington.

Forman, Paul, 1971. Weimar culture, causality, and quantum theory, 1918–1927: Adaptation of German physicists and mathematicians to a hostile intellectual environment. *Historical Studies in Physical Sciences* 3: 1–116.

4,000 fathom depth found in Atlantic Ocean, 1932. *New York Times*, March 21, p. 17.

Frangsmyr, Tore, J. L. Heilbron, and Robin Rider, eds., 1990. *The Quantifying Spirit in the 18th Century*. Berkeley: University of California Press.

Frankel, Henry, 1976. Wegener and the specialists. *Centaurus* 20: 305–324.

Frankel, Henry, 1978a. Arthur Holmes and continental drift. *British Journal of the History of Science* 11: 130–150.

Frankel, Henry, 1978b. The non-Kuhnian nature of the recent revolution in the earth sciences. In *Proceedings of the 1978 Biennial Meeting of the Philosophy of Science Association*, edited by P. J. Asquith and I. Hacking. East Lansing, Mich.: Philosophy of Science Association, pp. 197–214.

Frankel, Henry, 1979a. The career of continental drift theory: An application of Imre Lakatos' analysis of scientific growth to the rise of drift theory. *Studies in the History and Philosophy of Science* 10(1): 10–66.

Frankel, Henry, 1979b. Why continental drift theory was accepted by the geological community with the confirmation of Harry Hess' concept of sea-floor spreading. In *Two Hundred Years of Geology in America*, edited by C. J. Schneer. Hanover, N.H.: University of New England Press, pp. 337–353.

Frankel, Henry, 1982. The development, reception and acceptance of the Vine–Matthews–Morley hypothesis. *Historical Studies in the Physical and Biological Sciences* 13: 1–39.

Frankel, Henry, 1984. Biogeography before and after the rise of sea floor spreading. *Studies in the History and Philosophy of Science* 15: 141–168.

Frankel, Henry. 1985a. The biogeographical aspect of the debate over continental drift. *Earth Sciences History* 4(1): 160–81.

Frankel, Henry, 1985. The continental drift debate. In *Resolution of Scientific Controversies: Theoretical Perspectives on Closure*, edited by A. Caplan and H. T. Engleheart. Cambridge: Cambridge University Press, pp. 312–373.

Franklin, Allan, 1986. *The Neglect of Experiment*. Cambridge: Cambridge University Press.

Franklin, Allan, 1993a. Discovery, pursuit and justification. *Perspectives on Science* 1: 252–284.

Franklin, Allan, 1993b. Experimental questions. *Perspectives on Science* 1: 127–146.

Freeman, C. S., 1930. The gravity-measuring cruise of the S-21. Paper read at the Fifth Annual Meeting of the American Geophysical Union (1929), Washington, D.C. In *Transactions of the American Geophysical Union*, pp. 95–98.

Galison, Peter L., 1985. Bubble chambers and the experimental workplace. In *Observation, Experiment, and Hypothesis in Modern Physical Science*, edited by P. Achinstein and O. Hannaway. Cambridge: MIT Press, pp. 309–373.

Galison, Peter L., 1987. *How Experiments End*. Chicago: University of Chicago Press.

Galison, Peter L., 1988. History, philosophy, and the central metaphor. *Science in Context* 2: 196–212.

Galison, Peter L., 1997. *Image and Logic*. Chicago: University of Chicago Press.
Galison, Peter L., and A. Assmus, 1989. Artificial clouds, real particles. In *The Uses of Experiment*, edited by D. Gooding, T. Pinch, and S. Schaffer. Cambridge: Cambridge University Press, pp. 225–274.
Gass, I. G., Peter J. Smith, and R. C. L. Wilson, eds., 1972. *Understanding the Earth*. 2d ed. Sussex: Open University Press.
Geikie, Archibald, 1880. Review of *Report on the Geology of the High Plateaus of Utah*, by C. E. Dutton. *Nature* 22 (August 5): 324–327.
Geikie, Sir Archibald, 1897 (1905). *The Founders of Geology*. London: Macmillan.
Geschwind, Carl-Henry, 1996. Earthquakes and their interpretation: The campaign for seismic safety in California, 1906–1933. Ph.D. diss., Johns Hopkins University.
Gilbert, G. K., 1886. The inculcation of scientific method by example with an illustration drawn from the Quaternary geology of Utah. *American Journal of Science* 131: 284–299.
Gilbert, G. K., 1889. The strength of the Earth's crust. *Bulletin of the Geological Society of America* 1: 23–27.
Gilbert, G. K., 1890. Lake Bonneville. Washington: U.S. Geological Survey.
Gilbert, G. K., 1893. Continental problems. *Bulletin of the Geological Society of America* 4: 179–190.
Gilbert, G. K., 1895a. New light on isostasy. *Journal of Geology* 3: 331–334.
Gilbert, G. K., 1895b. Notes on the gravity determinations reported by Mr. G. R. Putnam. *Philosophical Society of Washington* 13: 61–75.
Gilbert, G. K. 1896. The origin of hypothesis, illustrated by the discussion of a topographic problem. *Science*, n.s., 3, no. 53: 1–13.
Gilbert, G. K., 1900. Rhythms and geologic time. *Science*, n.s. 11 (June 29, 1900): 1001–1012.
Gilbert, G. K., 1902. John Wesley Powell. *Science* 16: 561–567.
Gilbert, G. K., 1914. Interpretation of anomalies of gravity. *U.S. Geological Survey Professional Paper* 85–C: 29–37.
Gilman, Daniel C., 1899. *The Life of James Dwight Dana*. New York: Harper.
Glen, William, 1982. *The Road to Jaramillo*. Stanford, Calif.: Stanford University Press.
Glen, William, ed., 1994. *The Mass-Extinction Debates: How Science Works in a Crisis*. Stanford, Calif.: Stanford University Press.
Goetzmann, W. H., 1959. *Army Exploration in the American West*. New Haven, Conn.: Yale University Press.
Goetzmann, W. H., 1966. *Exploration and Empire*. New York: Knopf.
Good, Gregory, ed., 1994. *The Earth, the Heavens and the Carnegie Institution of Washington*. History of Geophysics 5. Washington, D.C.: American Geophysical Union.
Gooding, David, Trevor Pinch, and Simon Schaffer, eds., 1989. *The Uses of Experiment: Studies in the Natural Sciences*. Cambridge: Cambridge University Press.
Gossel, Patricia Peck, 1992. A need for standard methods: The case of American bacteriology. In *The Right Tools for the Job: At Work in Twentieth Century Life Sciences*, edited by A. E. Clarke and J. Fujimura. Princeton, N.J.: Princeton University Press, pp. 287–311.
Gould, Stephen Jay, 1965. Is uniformitarianism necessary? *American Journal of Science* 263: 223–228.
Gould, Stephen Jay, 1979. *Ever Since Darwin; Reflections in Natural History*. New York: Norton.
Gould, Stephen Jay. 1983. False premise, good science. *Natural History* 92(October): 20–26.
Greene, John C. 1979. Protestantism, science, and American enterprise: Benjamin Silliman's moral universe. In *Benjamin Silliman and His Circle*, edited by L. G. Wilson. New York: Science History Publications, pp. 11–27.
Greene, Mott T., 1982. *Geology in the Nineteenth Century: Changing Views of a Changing World*. Ithaca, N.Y.: Cornell University Press.

Greene, Mott T., forthcoming. *Alfred Wegener and the Origins of Modern Earth Science in the Theory of Continental Drift.*
Gregory, H. E., 1923. Memorial of Joseph Barrell. *Geological Society of America* 43: 18–28.
Gregory, J. W., 1925. Continental drift. *Nature* 115(February 21): 255–257.
Griggs, D. T., 1936. Deformation of rocks under high confining pressure: I. Experiments at room temperature. *Journal of Geology* 44: 541–577.
Griggs, D. T., 1938. Convection currents and mountain building (Abstract). *Geological Society of America Bulletin* 49: 1884.
Griggs, D. T., 1939a. Creep of rocks. *Journal of Geology* 47: 225–251.
Griggs, D. T., 1939b. A theory of mountain building. *American Journal of Science* 237(9): 611–650.
Griggs, D. T., 1974. Presentation of the Arthur L. Day award to David T. Griggs: Response by David Griggs. *Geological Society of America Bulletin* 85: 1342–1343.
Gross, Paul, and Norman Leavitt, 1994. *Higher Superstitions: The Academic Left and its Quarrels with Science.* Baltimore: Johns Hopkins University Press.
Gutenberg, Beno, 1939. Hypotheses on the development of the earth's crust and their implications. In *Physics of the Earth VII: Internal Constitution of the Earth*, edited by B. Gutenberg. New York: McGraw-Hill, pp. 177–217.
Gutenberg, Beno, and C. F. Richter, 1939. Evidence from deep-focus earthquakes. In *Internal Constitution of the Earth*, edited by B. Gutenberg. New York: McGraw-Hill, pp. 291–299.
Hacking, Ian, 1983. *Representing and Intervening: Introductory Topics in the Philosophy of Natural Science.* Cambridge: Cambridge University Press.
Hall, James, 1859. Paleontology: Containing descriptions and figures of the organic remains of the Lower Helderberg Group and Oriskany Sandstone, 1855–1859. In *Paleontology of New York.* Albany, N.Y., vol. 3, part 1.
Hallam, A., 1973. *A Revolution in the Earth Sciences.* Oxford: Oxford University Press.
Hallam, A., 1983. *Great Geological Controversies.* Oxford: Oxford University Press.
Hannaway, Owen, 1976. The German model of chemical education in America: Ira Remsen at Johns Hopkins (1876–1913). *Ambix:* 145–164.
Hanson, N. R., 1958. *Patterns of Discovery.* Cambridge: Cambridge University Press.
Harding, Sandra, 1986. *The Science Question in Feminism.* Ithaca: Cornell University Press.
Harding, Sandra, 1991. *Whose Science? Whose Knowledge?* Ithaca: Cornell University Press.
Hayford, John F., 1906. The geodetic evidence of isostasy, with a consideration of the depth and completeness of the isostatic compensation and of the bearing of the evidence upon some of the greater problems of geology. *Proceedings of the Washington Academy of Sciences* 8(May 18, 1906): 25–39, with discussion by Clarence E. Dutton, pp. 39–40.
Hayford, John F., 1907. The earth, a failing structure. *Bulletin of the Philosophical Society of Washington* 15: 57–74.
Hayford, John F., 1909. *The Figure of the Earth and Isostasy from Measurements in the United States.* Washington, D.C.: U. S. Government Printing Office.
Hayford, John F., 1911. The relations of isostasy to geodesy, geophysics and geology. *Science*, n.s. 3, 33(February 10): 199–208.
Hayford, J. F., and W. Bowie, 1912. The effect of topography and isostatic compensation upon the intensity of gravity. In U.S. Coast and Geodetic Survey Special Publication 10: Geodesy. Washington, D.C.: U. S. Government Printing Office.
Heck, N. H., 1937. The role of earthquakes and the seismic method in submarine geology. *Proceedings of the American Philosophical Society* 79(1): 104, 107.
Heck, N. H., 1938. The role of earthquakes and the seismic method in submarine geology. In *Geophysical Exploration of the Ocean Bottom*, edited by R. M. Field. *Proceedings of the American Philosophical Society* 79: 97–108.

Heezen, B. C., and Marie Tharp, 1965. Tectonic fabric of the Atlantic and Indian Oceans and continental drift. In *A Symposium on Continental Drift*, edited by P. M. S. Blackett, Edward Bullard, and S. K. Runcorn. *Philosophical Transactions of the Royal Society*, no. 1088: 90–106.
Heim, Albert, 1878. *Untersuchungen über den Mechanismus der Gebirgsbildung*. Basel: Benno Schwabe.
Heirtzler, J. R., 1968. Sea-floor spreading. *Scientific American* 219(6): 60–70.
Heiskanen, W. A., and F. A. Vening Meinesz, 1958. *The Earth and Its Gravity Field*. McGraw-Hill Series in the Geological Sciences, edited by R. R. Schrock. New York: McGraw-Hill.
Hempel, Carl, 1966. *Philosophy of Natural Science*. Englewood Cliffs, N.J.: Prentice-Hall.
Hendry, John, 1980. Weimar culture and quantum causality. In *Darwin to Einstein: Historical Studies on Science and Belief*, edited by C. Chant and J. Flauvel. Milton Keynes: Open University, pp. 303–326.
Hess, H. H., 1937. Geological interpretation of data collected on cruise of U.S.S. *Barracuda*, in the West Indies—Preliminary report. *Transactions of the American Geophysical Union* 18: 69–77.
Hess, H. H., 1938. Gravity anomalies and island arc structure with particular reference to the West Indies. *Proceedings of the American Philosophical Society* 79(1): 71–96.
Hess, H. H., 1946. Drowned ancient islands of the Pacific basin. *American Journal of Science* 244: 772–791.
Hess, H. H., 1962. History of the ocean basins. In *Petrologic Studies: A Volume to Honor A. F. Buddington*. Denver: Geologic Society of America.
Hess, H. H. 1968. Reply to comment by A. A. Meyerhoff. *Journal of Geophysical Research* 73: 6569.
Hess, H. H., n.d. Recent advances in interpretion of gravity anomalies and island-arc structures. *Advanced Report of the Commission on Continental and Oceanic Structure*: 46–48.
Hevly, Bruce, 1996. The heroic science of glacier motion. *Osiris* 11: 66–68.
Hofstadter, Richard, 1963. *Anti-Intellectualism in American Life*. New York: Vintage Books.
Holmes, Arthur, 1911a. The association of lead with uranium in rock-minerals, and its application to the measurement of geological time. *Proceedings of the Royal Society of London*, ser. A, 85: 248–256.
Holmes, Arthur, 1911b. The duration of geological time. *Nature* 87, no. 2175(July 6): 9–10.
Holmes, Arthur, 1913a. Radium and the evolution of the Earth's crust. *Nature* 91, no. 2277 (June 19): 398.
Holmes, Arthur, 1913b. The terrestrial distribution of the radio-elements. *Nature* 91, no. 2284 (August 7): 582–583.
Holmes, Arthur, 1913c. *The Age of the Earth*. London: Harper and Brothers.
Holmes, Arthur, 1914. Lead and the final product of thorium. *Nature* 93, no. 2318(April 2): 109.
Holmes, Arthur, 1925a. Radioactivity and the Earth's thermal history: Part IV. *Geological Magazine* 62: 504–528.
Holmes, Arthur, 1925b. Radioactivity and the Earth's thermal history: Part V. The control of geological history by radioactivity. *Geological Magazine* 62: 529–544.
Holmes, Arthur, 1926a. Estimates of geological time, with special reference to thorium minerals and uranium haloes. *Philosophical Magazine*, 7th ser., 1: 1055–1074.
Holmes, Arthur, 1926b. The geological age of the Earth. *Nature* 117, no. 2947(April 24): 592–594.
Holmes, Arthur, 1926c. Contributions to the theory of magmatic cycles. *Geological Magazine* 63: 306–329.
Holmes, Arthur, 1927. *The Age of the Earth, an Introduction to Geological Ideas*. Edited by W. Rose. London: Benn.

Holmes, Arthur, 1928. Radioactivity and continental drift. *Geological Magazine* 65: 236–238.
Holmes, Arthur, 1929a. A review of the continental drift hypothesis. *Mining Magazine* 40: 205–209, 286–288, 340–347.
Holmes, Arthur, 1929b. Radioactivity and earth movements. *Transactions of the Geological Society of Glasgow* 18: 559–606.
Holmes, Arthur, 1931. Radioactivity and geological time. *National Research Council Bulletin* 80, part 4: 125–459.
Holmes, Arthur, 1933. The thermal history of the Earth. *Journal of the Washington Academy of Science* 23: 169–195.
Holmes, Arthur, 1937. *The Age of the Earth*. New York: Nelson.
Holmes, Arthur, 1945. *Principles of Physical Geology*. New York: Ronald Press.
Holmes, Arthur, 1946. An estimate of the age of the Earth. *Nature* 157: 680–684.
Holmes, Frederic L., 1992. Manometers, tissue slices, and intermediary metabolism. In *The Right Tools for the Job: At Work in Twentieth Century Life Sciences*, edited by A. E. Clarke and J. Fujimura. Princeton, N.J.: Princeton University Press, pp. 151–171.
Holton, Gerald, 1973. *Thematic Origins of Scientific Thought: Kepler to Einstein*. Cambridge: Harvard University Press.
Holton, Gerald, 1993. *Science and Anti-Science*. Cambridge: Harvard University Press.
Hooykaas, Reijer, 1963. *Natural Law and Divine Miracle, a Historicocritical Study of the Principle of Uniformity in Geology, Biology, and Theology*. Leiden: Brill.
Hooykaas, Reijer, 1975. Catastrophism in geology, its scientific character in relation to actualism and uniformitarianism. In *Philosophy of Geohistory: 1785–1970*, edited by C. C. Albritton Jr. Stroudsberg, Pa.: Dowden, Hutchinson and Ross, pp. 310–356.
Houser, Nathan, and Christian Kloesel, eds., 1992. *The Essential Peirce: Selected Philosophical Writings: Volume 1 (1867–1893)*. Indianapolis: Indiana University Press.
Hovey, E. O., 1901. Geology and geography at the Denver meeting of the American Association for the Advancement of Science. *Scientific American Supplement* 52(1342): 21504.
Hubbert, M. King, 1937. Theory of scale models as applied to the study of geologic structures. *Bulletin of the Geological Society of America* 48: 1949–1520.
Hubbert, M. King, 1945. Strength of the Earth. *Bulletin of the American Association of Petroleum Geologists* 29(11): 1630–1653.
Hubbert, M. King, 1967. Critique of the principle of uniformity. In *Geological Society of America Special Paper* 89, edited by C. C. Albritton. Denver: Geological Society of America, pp. 3–33.
Hurlbut, Cornelius, Jr., and Cornelis Klein, 1977. *Manual of Mineralogy (after James D. Dana)*. 19th ed. New York: Wiley.
Hurley, Patrick M., 1968. The confirmation of continental drift. *Scientific American* 218(4): 52–64.
Irving, Edward, 1956. Paleomagnetic and palaeoclimatological aspects of polar wanderings. *Geofisica Pura e Applicata* 33: 23–41.
Isacks, Bryan, Jack Oliver, and Lynn R. Sykes, 1968. Seismology and the new global tectonics. *Journal of Geophysical Research* 73: 5855–5899.
J. B., 1906. Review of *Geology*, by T. C. Chamberlin and Rollin D. Salisbury. *American Journal of Science* 171: 400–401.
Jackson, Patrick Wyse, 1992. A man of invention: John Joly, engineer, physicist, and geologist. In *Treasures of the Mind: A Trinity College Quatercentenary Exhibition*, edited by D. Scott. London: Sotheby's, pp. 89–96.
James, William, 1907. What pragmatism means. In *The American Pragmatists*, edited by M. R. Konvitz and G. Kennedy. New York: World, pp. 28–43.
Jamieson, T. F., 1865. On the history of the last geological changes in Scotland. *Quarterly Journal of the Geological Society of London* 21: 161–203.

Jamieson, T. F., 1882. Oscillation of land in the Tertiary glacial period. *Geological Magazine* 9: 400–418.
Jeffreys, Harold, 1926. The Earth's thermal history and some related problems. *Geological Magazine* 63: 516–525.
Jeffreys, Harold, 1935. *Earthquakes and Mountains*. London: Methuen.
Johnson, J. G., 1990. Method of multiple working hypotheses: A chimera. *Geology* 18: 44–45.
Joly, John, 1889. On the steam calorimeter. *Proceedings of the Royal Society of London* 47: 67.
Joly, John, 1899. An estimate of the geological age of the Earth. *Scientific Transactions of the Royal Dublin Society*, 2d ser., 7: 23–66.
Joly, John, 1900. On the geological age of the Earth. *British Association Report* (Bradford): 369–379.
Joly, John, 1903. Radium and the geological age of the Earth. *Nature* 68, no. 1770(October 1): 526.
Joly, John, 1909. *Radioactivity and Geology: An Account of Radioactive Energy on Terrestrial History*. London: Constable.
Joly, John, 1910. Pleochroic haloes. *Philosophical Magazine*, 6th ser., 19: 326–30.
Joly, John, 1915. *The Birth-Time of the World and Other Scientific Essays*. New York: Dutton.
Joly, John, 1917. The genesis of pleochroic haloes. *Philosophical Transactions of the Royal Society of London* 217, section A (Jan. 29): 51–79.
Joly, John, 1923. The age of the Earth. *Scientific Monthly* 16(Jan.–June): 205–216.
Joly, John, 1924. *Radioactivity and the Surface-History of the Earth*. Oxford: Clarendon Press.
Joly, John, 1925. *The Surface-History of the Earth*. Oxford: Clarendon Press. 2d ed., 1930.
Joly, John. 1926a. The surface-history of the Earth. *Gerlands Beiträge zur Geophysik* 15: 189–200.
Joly, John, 1926b. The surface-history of the Earth. *Philosophical Magazine*, 7th ser., 1: 932–939.
Joly, John, 1927. Dr Jeffreys and the Earth's thermal history. *Philosophical Magazine*, 7th ser., 2: 338–348.
Joly, John, 1928a. Continental movement. In *Theory of Continental Drift: A Symposium*, edited by W. A. J. M. van Waterschoot van der Gracht. London: Murray, pp. 88–89.
Joly, John, 1928b. The Earth's thermal history. *Philosophical Magazine*, 7th ser., 5: 215–221.
Joly, John, 1928c. The theory of thermal cycles. *Gerlands Beiträge zur Geophysik* 19: 415–441.
Joly, John, and J. H. J. Poole, 1927. On the nature and origin of the Earth's surface structure. *Philosophical Magazine*, 7th ser., 3: 1233–1246.
Joly, J., and E. Rutherford, 1913. The age of pleochroic haloes. *Philosophical Magazine*, 7th ser., 25: 644–657.
Joseph Barrell (1869–1919), 1919. *American Journal of Science* 198, no. 286: 251–261.
Kaiser, David, 1992. More roots of complementarity. *Studies in the History and Philosophy of Science* 23: 213–239.
Kaiser, David, 1998. A Mannheim for all seasons: Bloor, Merton, and the roots of the sociology of scientific knowledge. *Science in Context* 11, no. 1.
Kant, Immanuel, 1755. *Allgemeine Naturgeschichte und Theorie des Himmels*. Königsberg, Leipzig: Petersen.
Kay, Lily E., 1997. Rethinking institutions: Philanthropy as an historiographic problem of knowledge and power. *Minerva* 35: 283–293.
Keith, M. L., 1993. Geodynamics and mantle flow: An alternative Earth model. *Earth-Science Reviews* 33: 153–337.
Keller, Evelyn Fox, 1985. *Reflections on Gender and Science*. New Haven, Conn.: Yale University Press.
Kelvin, Lord. *See* Thomson, William.

Kelvin, Lord, 1899. The age of the Earth as an abode fitted for life. *Science*, n.s. 9(May 12): 665–673; (May 19): 704–711.
Kevles, Daniel J., 1971. *The Physicists*. New York: Vintage Books.
King, Clarence, 1878. *Systematic Geology*. Washington, D.C.: U.S. Government Printing Office.
Kitts, David B., 1974. Continental drift and the scientific revolution. *American Association of Petroleum Geologists Bulletin* 58: 2490–2496.
Kitts, David B., 1977. *The Structure of Geology*. Dallas: Southern Methodist University Press.
Kitts, David, 1980. Analogies in G. K. Gilbert's philosophy of science. *Geological Society of America Special Paper* 183: 143–148.
Knopf, Adolf, 1952. *National Academy of Sciences Biographical Memoirs* 27: 363–389.
Kohler, Robert E., 1990. The Ph.D. machine: Building on the collegiate base. *Isis* 81: 638–662.
Kohler, Robert E., 1991. *Partners in Science: Foundations and Natural Scientists 1900–1945*. Chicago: University of Chicago Press.
Koppen, W., and Alfred Wegener, 1924. *Die Klimate der geologischen Vorzeit*. Berlin: Bornträger.
Kozenko, A. V., 1991. Relationship between astronomical and geophysical approaches to the study of the Earth as a planet by Harold Jeffreys. Unpublished ms. at the Earth Physics Institute, Moscow.
Kuenen, Ph. H., 1933. Turbidities in South Africa. *Transactions of the Geological Society of South Africa* (August): 1–5.
Kuenen, Ph. H., 1936. The negative isostatic anomalies in the East Indies (with experiements). *Overdrukuit Leidsche Geologische Mededeelingen* 8: 169–214.
Kuenen, Ph. H., 1937. Experiments in connection with Daly's hypothesis on the formation of submarine canyons. *Leidsche Geologische Mededeelingen* 8(2): 327–351.
Kuenen, Ph. H., 1938. Density currents in connection with the problem of submarine canyons. *Geological Magazine* 75: 241–249.
Kuenen, Ph. H., 1948. Slumping in the Carboniferous rocks of Pembrokeshire. *Quarterly Journal of the Geological Society of London* 104, part 3: 365–385.
Kuenen, Ph. H., 1951a. *Properties of Turbidity Currents of High Density*. Special Publication no. 2. Denver: Society of Economic Paleontologists and Mineralogists, pp. 14–33.
Kuenen, Ph. H., 1951b. Mechanics of varve formation and the action of turbidity currents. *Geologiska Föreningens Förhandlingar* 73(1): 69–84.
Kuenen, Ph. H., and Henry W. Menard, 1952. Turbidity currents, graded and non-graded deposits. *Journal of Sedimentary Petrology* 22(2): 83–96.
Kuenen, Ph. H., and C. L. Migliorini. 1950. Turbidity currents as a cause of graded bedding. *Journal of Geology* 58(2): 91–127.
Kuhn, Thomas S., 1957. *The Copernican Revolution*. Cambridge: Harvard University Press.
Kuhn, Thomas S., 1962. *The Structure of Scientific Revolutions*. Chicago: University of Chicago Press.
Kushner, David S., 1990. The emergence of geophysics in nineteenth century Britain. Ph.D. diss., Department of History of Science, Princeton University.
Kushner, David S., 1993. Sir George Darwin and a British school of geophysics. *Osiris* 18: 196–223.
Lakatos, Imre, and A. Musgrave, eds., 1970. *Criticism and the Growth of Knowledge*. Cambridge: Cambridge University Press.
Lake, Philip, 1922. Wegener's displacement theory. *Geological Magazine* 59: 338–346.
Lake, Philip, 1926. The origins of continents and oceans (Review). *Geological Magazine* 63: 181–182.
Lambert, W. D., 1921. Some mechanical curiosities connected with the Earth's field of force. *American Journal of Science* 202: 129–158.

Lambert, W. D., 1922. *An Investigation of the Latitude of Ukiah, California, and the Motion of the Pole*. U.S. Coast and Geodetic Survey Special Publication no. 80, Serial no. 183. Washington, D.C.: U. S. Government Printing Office.

Lambert, W. D., 1925. The variation of latitude. *Bulletin of the National Research Council* 10, part 3, no. 53: 43–45.

Lamson, E. A., 1930. Gravity expedition of the United States Navy in 1928. In *Geodetic Operations in the United States*, edited by William Bowie. Special Publication 166. Washington, D.C.: U.S. Coast and Geodetic Survey pp. 29–32.

Lankford, John, 1994. Women and women's work at Mt. Wilson observatory before World War II. *History of Geophysics* 5: 125–127.

Laplace, Pierre-Simon Marquis de, 1796. *Exposition du système du monde*. 2 vols. Paris: Imprimerie du Cercle-social.

Laporte, Léo F., 1985. *Wrong for the Right Reasons: G. G. Simpson and Continental Drift*. Centennial Special Paper No. 1. Denver: Geological Society of America, pp. 273–285.

Latour, Bruno, and Steven Woolgar, 1979. *Laboratory Life*. Beverly Hills: Sage.

Laudan, Larry, 1977. *Progress and Its Problems*. Los Angeles: University of California Press.

Laudan, Larry, 1984. Explaining the success of science: Beyond epistemic realism and relativism. In *Science and Reality: Recent Work in the Philosophy of Science*, edited by J. T. Cushing, C. F. Delaney, and G. M. Gutting. Notre Dame, Ind.: University of Notre Dame Press, pp. 83–105.

Laudan, Rachel, 1977. Ideas and organizations: The case of the Geological Society of London. *Isis* 68: 527–38.

Laudan, Rachel, 1978. The recent revolution in geology and Kuhn's theory of scientific change. In *Proceedings of the 1978 Biennial Meeting of the Philosophy of Science Association*, edited by P. J. Asquith and I. Hacking. East Lansing, Mich.: Philosophy of Science Association, pp. 227–239.

Laudan, Rachel, 1980a. Oceanography and geophysical theory in the first half of the twentieth century: The Dutch school. In *Oceanography: The Past*, edited by M. Sears and M. Merriman. New York: Springer-Verlag, pp. 656–666.

Laudan, Rachel, 1980b. The method of multiple working hypotheses and the discovery of plate tectonic theory. In *Scientific Discovery: Case Studies*, edited by T. Nickles. Boston Studies in the Philosophy of Science 60. Dordrecht: Reidel, pp. 331–343.

Laudan, Rachel, 1982. The role of methodology in Lyell's science. *Studies in the History and Philosophy of Science* 13: 215–249.

Laudan, Rachel, 1983. Redefinitions of a discipline: Histories of geology and geological history. In *Functions and Uses of Disciplinary History*, edited by L. Graham, W. Lepenies, and P. Weingart. Dordrecht: Reidel, pp. 79–104.

Laudan, Rachel, 1985. Frank Bursey Taylor's theory of continental drift. *Earth Science History* 5: 118–121.

Laudan, Rachel, 1987. *From Mineralogy to Geology*. Chicago: University of Chicago Press.

Laudan, Rachel, and Larry Laudan, 1989. Dominance and the disunity of method: Solving the problems of innovation and consensus. *Philosophy of Science* 56(2): 221–237.

Lawson, Andrew, 1924. The geological implications of the doctrine of isostasy. *Bulletin of the National Research Council* 8(4): 3–22.

Lawson, Andrew, 1938. The flotation of mountains. *Scientific Monthly* 47: 429–438.

Lawson, Andrew, 1939. Subsidence by thrusting: The discussion of a hypothetical fault. *Bulletin of the Geological Society of America* 50: 1381–1394.

Layton, Edwin T., Jr., 1981. European origins of the American engineering style of the nineteenth century. In *Scientific Colonialism: A Cross-cultural Comparison*, edited by N. Reingold and M. Rothenberg. Washington, D.C.: Smithsonian Institution Press, pp. 151–166.

LeConte, Joseph, 1872. The theory of the formation of the great features of the Earth's surface. *American Journal of Science* 104: 345–472.

LeConte, Joseph, 1893. Theories of the origin of mountain ranges. *Journal of Geology* 1: 543–573.

Le Grand, H. E., 1986. Specialties, problems and localism: The reception of continental drift in Australia, 1920–1940. *Earth Sciences History* 5: 84–95.

Le Grand, H. E., 1988. *Drifting Continents and Shifting Theories*. Cambridge: Cambridge University Press.

Le Grand, H. E., 1994. Chopping and changing at the DTM 1946–1958: M. A. Tuve, rock magnetism, and isotope dating. In *The Earth, the Heavens, and the Carnegie Institution of Washington*, edited by G. Good. History of Geophysics 5. Washington D.C.: American Geophysical Union, pp. 173–184.

Lenoir, Timothy, 1988. Practice, reason, context: The dialogue between theory and experiment. *Science in Context* 2(1): 3–22.

Le Pichon, Xavier, 1968. Sea-floor spreading and continental drift. *Journal of Geophysical Research* 73(12): 3661–3697.

Leviton, Alan E., and Michele L. Aldrich, 1991. The origin of the notion of Gondwanaland: The Indian Geological Survey, 1851–1889. *Geological Society of America Abstracts with Programs* 23(5): A47.

Lightman, Alan, and Owen Gingerich, 1991. When do anomalies begin? *Science* 255: 690–695.

Littell, F. B., and J. C. Hammond, 1928. World Longitude Operation—Results of observations at San Diego and Washington. *Astronomical Journal* 38(August 14): 185–188.

Littell, F. B., J. C. Hammond, C. B. Watts, and P. Sollenberger, 1929. World Longitude Operation of 1926—Results of observations at San Diego and Washington. In *Publications of the U.S. Naval Observatory*. Washington, D.C.: U.S. Government Printing Office, pp. 527–592.

Longino, Helen. 1990. *Science as Social Knowledge*. Princeton: Princeton University Press.

Longwell, Chester R., 1927. "Our Mobile Earth" (Review). *American Journal of Science* 213: 524–525.

Longwell, Chester R., 1928. Some physical tests of the displacement hypothesis. In *The Theory of Continental Drift: A Symposium*, edited by W. A. J. M. van Waterschoot van der Gracht. London: Murray, pp. 145–157.

Longwell, Chester R., 1938. Continents adrift. Review of *Our Wandering Continents: An Hypothesis of Continental Drifting*, by Alexander L. du Toit. *Geographical Review* 28: 704–705.

Longwell, Chester R., 1944a. The mobility of Greenland. *American Journal of Science* 242: 624.

Longwell, Chester R., 1944b. Some thoughts on the evidence for continental drift. *American Journal of Science* 242: 218–231.

Longwell, Chester R., 1944c. Further discussion of continental drift. *American Journal of Science* 242: 514–515.

Lyell, Charles, 1990. *Principles of Geology*. Facsimile edition, with introduction by Martin J. S. Rudwick, ed. Vol. 1. Chicago: University of Chicago Press. Original edition, 1830; 8th edition, 1850.

Mabey, Don R., 1980. Pioneering work of G. K. Gilbert on gravity and isostasy. In *The Scientific Ideas of G. K. Gilbert*, edited by E. L. Yochelson. Boulder, Colo.: Geological Society of America, pp. 61–68.

MacCarthy, Gerald, 1928. Experiments in underthrusting. *American Journal of Science*, 216: 51–67.

Bibliography

Mackin, J. Hoover, 1963. Rational and empirical methods of investigation in geology. In *The Fabric of Geology*, edited by C. C. Albritton. San Francisco: Freeman, Cooper, pp. 135–163.
Maienschein, Jane, ed., 1986. *Defining Biology: Lectures from the 1890s*. Cambridge: Harvard University Press.
Maienschein, Jane, 1991a. *Transforming Traditions in American Biology, 1880–1915*. Baltimore: Johns Hopkins University Press.
Maienschein, Jane, 1991b. Epistemic styles in German and American embryology. *Science in Context* 4(2): 407–427.
Manning, Thomas G., 1967. *Government in Science: The U.S. Geological Survey, 1867–1894*. Lexington: University of Kentucky Press.
Manning, T. G., 1988. *U.S. Coast Survey vs. Naval Hydrographic Office: A 19th Century Rivalry in Science and Politics*. Tuscaloosa: University of Alabama Press.
Markham, Clements R., 1878. *A Memoir on the Indian Surveys*. 2d ed. London: W.H. Allen. Reprinted Amsterdam: Meridian, 1969.
Marsden, George M., 1994. *The Soul of the American University: From Protestant Establishment to Established Non-Belief*. New York: Oxford University Press.
Martin, Schwarzbach, 1986. *Alfred Wegener, The Father of Continental Drift*. Translated by Carla Love. Scientific Revolutionaries: A Biographical Series. Madison, Wisc.: Science Tech.
Marvin, Ursula B., 1973. *Continental Drift: The Evolution of a Concept*. Washington, D.C.: Smithsonian Institution Press.
Marvin, Ursula B., 1985. The British reception of Alfred Wegener's continental drift hypothesis. *Earth Sciences History* 4: 138–159.
Mason, R. G., and A. D. Raff, 1961. A magnetic survey off the west coast of North America, 32°N to 42°N. *Geological Society of America Bulletin* 72: 1259–1265.
Mather, K. F., and S. L. Mason, eds., 1967. *A Source Book in Geology*. Cambridge: Harvard University Press. Original edition 1939.
Maxwell, John C., 1968. Continental drift and a dynamic Earth. *American Scientist* 56(1): 35–50.
Mayo, D. E., 1968. The development of the idea of the geosyncline. Ph.D. diss., Department of History of Science, University of Oklahoma.
Mayo, D. E., 1985. Mountain-building theory: The nineteenth century origins of isostasy and the geosyncline. In *Geologists and Ideas: A History of North American Geology*. Denver: Geological Society of America, pp. 1–18.
Mayr, Ernst, 1952. Introduction to the problem of land connections across the South Atlantic, with special refernce to the Mesozoic. *Bulletin of the American Museum of Natural History* 99(3): 85.
McGee, William J., 1896. James Hall, founder of American stratigraphy. *Science*, n.s., 4: 702–703.
McGrayne, Sharon Bertsch, 1993. *Nobel Prize Women in Science: Their Lives, Struggles, and Momentous Discoveries*. Secaucus, N.J.: Carol.
Menard, H. W., 1964. *Marine Geology of the Pacific*. New York: McGraw-Hill.
Menard, H. W., 1986. *The Ocean of Truth: A Personal History of Global Tectonics*. Princeton, N.J.: Princeton University Press.
Mendenhall, W. C., 1920. Memorial of Grove Karl Gilbert. *Bulletin of the Geological Society of America* 31: 26–64.
Merton, Robert, 1942. The normative structure of science. In *The Sociology of Science: Theoretical and Empirical Investigations*. Chicago: Chicago University Press, pp. 267–278.
Meyerhoff, A. A., 1968. Arthur Holmes: Originator of the spreading ocean floor hypothesis. *Journal of Geophysical Research* 73: 6563–6565.

Meyerhoff, A. A., and Howard A. Meyerhoff, 1972. The new global tectonics: Major inconsistencies. *American Association of Petroleum Geologists Bulletin* 56: 269–336.
Meyrick, Edward, 1925. Wegener's hypothesis and the distribution of Micro-Lepidoptera. *Nature* 115: 834–835.
Mill, John Stuart, 1989. *On Liberty.* Edited by S. Collini. Cambridge Texts in the History of Political Thought. Cambridge: Cambridge University Press. Originally published in 1859.
Miller, Howard S., 1970. *Dollars for Research: Science and Its Patrons in Nineteenth Century America.* Seattle: University of Washington Press.
Miller, Howard S., 1972. The political economy of science. In *Nineteenth Century American Science: A Reappraisal,* edited by G. H. Daniels. Evanston, Ill.: Northwestern University Press, pp. 95–112.
Miller, Ralph L., 1940. Revision of the concept of uniformitarianism. *Proceedings of the Pennsylvania Academy of Sciences* 14: 68–77.
Molengraaf, G. A. F., 1898. The glacial origins of the Dwyka Conglomerate. *Transactions of the Geological Society of South Africa* 4(5): 103–115.
Molengraaf, G. A. F., 1912. On recent crustal movements in the island of Timor and their bearing on the geological history of the East-Indian Archipelago. *Koninklijke Akademie van Wetenschappen te Amsterdam* 2(September): 224–235.
Molengraaf, G. A. F., 1915a. Folded mountain chains, overthrust sheets, and block-faulted mountains in the East Indian Archipelago. *Compte-Rendu du XIIe Congrès Géologique International, Toronto, Canada,* 1913: 689–702.
Molengraaf, G. A. F., 1915b. L'Expédition Néerlandaise à Timor en 1910–1912. *Archives des Néerlandaise des Sciences Exactes et Naturelles,* Serie 3 B (2): 395–404.
Molengraaf, G. A. F. 1916, The coral reef problem and isostasy. *Koninklijke Akademie van Wetenschappen te Amsterdam* 19(4): 610–627.
Molengraaf, G. A. F. 1921. Modern deep-sea research in the East Indian Archipelago. *Geographical Journal* (February): 95–118.
Molengraaf, G. A. F., 1928. Wegener's continental drift. In *Theory of Continental Drift: A Symposium,* edited by W. A. J. M. van Waterschoot van der Gracht. London: Murray, pp. 90–92.
Moore, Edward C., ed., 1993. *Charles S. Peirce and the Philosophy of Science.* Tuscaloosa: University of Alabama Press.
Morley, Lawrence W., 1986. Early work leading to the explanation of the banded geomagnetic imprinting of the ocean floor. *Eos* (September 9): 665–666.
Nansen, F., 1922. The strandflat and isostasy. *Videnskapsselskabets Skrifter, mat.-naturv. Kl.* 11(1921).
Nansen, F., 1928. The Earth's crust, its surface forms, and isostatic adjustment. *Avhand. Norske Videnskaps. Akad. Oslo, mat.-naturv. Kl.* 12(1927): 199.
Newcomb, Sally, 1990. Contributions of British experimentalists to the discipline of geology: 1780–1820. *Proceedings of the American Philosophical Society* 134(2): 161–225.
Newell, Julie, 1993. American geologists and their geology, 1780–1865. Ph.D. diss., Department of History of Science, University of Wisconsin.
Newman, Robert, 1995. American intransigence: The rejection of continental drift in the great debates of the 1920s. *Earth Sciences History* 14(1): 62–83.
Norris, J. A., and E. T. Witherspoon, 1904. *Telegraphic Determination of Longitudes.* Washington, D.C.: U.S. Hydrographic Office.
Novick, Peter, 1988. *That Noble Dream: The "Objectivity Question" and the American Historical Profession.* Cambridge: Cambridge University Press.
Nudds, John R., 1986. The life and work of John Joly (1857–1933). *Irish Journal of Earth Sciences* 8: 81–94.

Numbers, Ronald L., 1977. *Creation by Natural Law: Laplace's Nebular Hypotheses in American Thought.* Seattle: University of Washington Press.
Numbers, Ronald L., 1985. Science and religion. In *Historical Writing on American Sciences,* edited by S. G. Kohlstedt and M. W. Rossiter. Baltimore: History of Science Society, pp. 59–80.
Numbers, Ronald, 1992. *The Creationists.* New York: Knopf.
Nunan, Richard, 1988. The theory of an expanding earth and the acceptability of guiding assumptions. In *Scrutinizing Science: Empirical Studies of Scientific Change,* edited by A. Donovan, L. Laudan, and R. Laudan. Baltimore: Johns Hopkins University Press, pp. 289–314.
Nur, Amos, 1976. The Earth: Its Origin, History, and Physical Constitution (Review). *Geology* (December): 708.
Nye, Mary Jo, 1996. Testing big theories: Patrick Blackett and the Earth's magnetism. Paper read at History of Science Society, Atlanta, Ga.
Oldroyd, David, 1980. Sir Archibald Geikie (1835–1924), geologist, romantic aesthete, and historian of science: The problem of Whig historiography of science. *Annals of Science* 9(37): 441–462.
Oldroyd, David, 1987. Punctuated equilibrium theory and time: A case study in problems of coherence in the measurement of geological time (The KBS Tuff controversy and the dating of rocks in the Turkana Basin, Kenya). In *Realism and Objectivity: Essays on Measurements in the Social and Physical Sciences,* edited by J. Forge. Dordrecht: Reidel, pp. 89–152.
Oldroyd, David, 1990. *The Highlands Controversy.* Chicago: Chicago University Press.
Oldroyd, David, 1996. *Thinking about the Earth: A History of Ideas in Geology.* London: Athlone Press.
Olesko, Kathryn, 1991. *Physics as a Calling: Discipline and Practice in the Königsberg Seminar for Physics.* Ithaca, N.Y.: Cornell University Press.
Oliver, J., and B. Isacks, 1967. Deep earthquakes zones, anomalous structures in the upper mantle and lithosphere. *Journal of Geophysical Research* 72: 4259–4276.
Oreskes, Naomi, 1988. The rejection of continental drift. *Historical Studies in the Physical and Biological Sciences* 18(2): 311–348.
Oreskes, Naomi, 1994. Weighing the earth from a submarine: The gravity measuring cruise of the U.S.S. S-21. In *The Earth, the Heavens, and the Carnegie Institution of Washington,* edited by G. Good. History of Geophysics 5. Washington D.C.: American Geophysical Union, pp. 53–68.
Oreskes, Naomi, 1996. Objectivity or heroism? On the invisibility of women in science. *Osiris* 11: 87–113.
Oreskes, Naomi, 1998. La lente plongée vers le fond des océans. *Science et Vie* (March): 84–90.
Ospovat, Alexander M., 1959. Werner's influence on American geology. *Proceedings of the Oklahoma Academy of Sciences,* 1959: 98–103.
Ospovat, Alexander M., 1976. The distortion of Werner in Lyell's *Principles of Geology. British Journal for the History of Science* 9(32): 190–198.
Paper read at Second General Assembly, International Astronomical Union, July 14–22, 1925, at Cambridge. *Transactions of the International Astronomical Union, Second General Assembly.* Cambridge: 1925.
Paper read at Third General Assembly, International Astronomical Union, July 5–13, 1928, at Leiden. *Transactions of the International Astronomical Union, Third General Assembly.* Leiden: 1929.
Paul, Diane B., 1991. The Rockefeller Foundation and the origins of behavoir genetics. In

The Expansion of American Biology, edited by Keith R. Benson, Jane Maienschein, and Ronald Rainger. New Brunswick, N.J.: Rutgers University Press, pp. 262–283.

Pauly, Philip J., 1987. *Controlling Life: Jacques Loeb and the Engineering Ideal in Biology*. New York: Oxford University Press.

Pauly, Philip J., 1991. Summer resort and scientific discipline: Woods Hole and the structure of American biology, 1882–1925. In *The American Development of Biology*, edited by Ronald Rainger, Keith R. Benson, and Jane Maienschein. New Brunswick, N.J.: Rutgers University Press, pp. 121–150.

Penrose, R. A. F., Jr., 1929. The early days of the department of geology at the University of Chicago. *Journal of Geology* 37(May-June): 320–327.

Pettijohn, F. J., 1984. *Memoirs of an Unrepentant Field Geologist*. Chicago: University of Chicago Press.

Pickering, W. H., 1907. The place of origin of the moon—The volcanic problem. *Journal of Geology* 15: 23–38.

Pickering, W. H., 1924. The separation of continents by fission. *Journal of Geology* 61: 31–34.

Pitman, W. C., and J. R. Heirtzler, 1966. Magnetic anomalies over the Pacific-Antarctic Ridge. *Science* 154: 1164–1171.

Popper, Karl R., 1963. *Conjectures and Refutations*. London: Routledge.

Porter, Roy, 1976. Charles Lyell and the principles of the history of geology. *British Journal for the History of Science* 9: 91–103.

Porter, Theodore M., 1995a. Precision and trust: Early Victorian insurance and the politics of calculation. In *The Values of Precision*, edited by M. N. Wise. Princeton, N.J.: Princeton University Press, pp. 173–198.

Porter, Theodore M., 1995b. *Trust in Numbers: The Pursuit of Objectivity in Science and Public Life*. Princeton, N.J.: Princeton University Press.

Pratt, J. H., 1855. On the attraction of the Himalaya Mountains, and of the elevated regions beyond them, upon the plumb-line in India. *Philosophical Transactions of the Royal Society of London* 145: 53–100.

Pratt, J. H., 1859a. On the deflection of the plumb-line in India caused by the attraction of the Himalaya mountains and of the elevated regions beyond; and its modification by the compensating effect of a deficiency of matter below the mountain mass. *Philosophical Transactions of the Royal Society of London* 149: 745–778.

Pratt, J. H., 1859b. On the influence of the ocean on the plumb-line in India. *Philosophical Transactions of the Royal Society of London* 149: 779–96.

Pratt, J. H., 1871. On the constitution of the solid crust of the Earth. *Philosophical Transactions of the Royal Society of London* 161: 335–357.

Press, Frank, and Raymond Siever, 1974. *Earth*. San Francisco: Freeman.

Putnam, G. R., 1896. Results of recent pendulum observations. *American Journal of Science* 151: 186–193.

Putnam, G. R., 1922. Condition of the Earth's crust and the earlier American gravity observations. *Bulletin of the Geological Society of America* 33(June 30): 287–291.

Pyenson, Lewis, 1981. The limits of scientific condominium: Geophysics in Western Samoa, 1914–1940. In *Scientific Colonialism: A Cross-cultural Comparison*, edited by N. Reingold and M. Rothenberg. Washington, D.C.: Smithsonian Institution Press.

Pyenson, Lewis, 1990. Habits of mind: Geophysics at Shanghai and Algiers. *Historical Studies in Physical Sciences* 21: 161–196.

Pyne, Stephen J., 1978. Methodologies for geology: G. K. Gilbert and T. C. Chamberlin. *Isis* 69: 413–424.

Pyne, Stephen J., 1979. Certain allied problems in mechanics: Grove Karl Gilbert at the Henry Mountains. In *Two Hundred Years of Geology in America*, edited by C. J. Schneer. Hanover, N.H.: University of New England Press, pp. 225–238.

Pyne, Stephen J., 1980. *Grove Karl Gilbert: A Great Engine of Research*. Austin: University of Texas Press.
Quinn, Susan, 1995. *Marie Curie: A Life*. New York: Simon and Schuster.
Railsback, L. Bruce, 1990. Comment on "Method of multiple working hypotheses: A chimera." *Geology* (September): 917–918.
Rainger, Ronald, 1981. The continuation of the morphological tradition: American palentology, 1880–1910. *Journal of the History of Biology* 14(1): 129–158.
Rainger, Ronald, 1993. Biology or geology or neither or both? Vertebrate palentology at the University of Chicago, 1892–1950. *Perspectives on Science* 1: 478–519.
Rainger, Ronald, Keith R. Benson, and Jane Maienschein, eds., 1991. *The American Development of Biology*. New Brunswick, N.J.: Rutgers University Press.
Rankin, G. A., 1915. The Ternary system CaO-Al2O3-SiO2, with optical study by F. E. Wright. *American Journal of Science* 139: 1–79.
Rastall, R. H., 1928. A geological comparison of South America with South Africa (Review). *Geological Magazine* 65(3): 139–140.
Rayner, Dorothy H., [1967] 1976. *The Stratigraphy of the British Isles*. Cambridge: Cambridge University Press.
Read, H. H., 1957. *The Granite Controversy, or, Geological Addresses Illustrating the Evolution of a Disputant*. London: Murby.
Reade, T. Mellard, 1886. *The Origin of Mountain Ranges Considered Experimentally, Structurally, Dynamically, and in Relation to Their Geological History*. London: Francis and Taylor.
Reichenbach, Hans, 1938. *Experience and Prediction; An Analysis of the Foundations and the Structure of Knowledge*. Chicago: University of Chicago Press.
Reid, Harry Fielding, 1911. Isostasy and mountain ranges. *Proceedings of the American Philosophical Society*, pp. 444–451.
Reid, Harry Fielding, 1922. Drift of the Earth's crust and displacement of the pole. *Geographical Review* 22: 672–674.
Reingold, Nathan, 1964. *Science in Nineteenth Century America: A Documentary History*. Chicago: University of Chicago Press.
Reingold, Nathan, 1978. National style in the sciences: The United States case. In *Human Implications of Scientific Advance: Proceedings of the XVth International Congress of the History of Science*, edited by E. G. Forbes. Edinburgh: Edinburgh University Press, pp. 163–173.
Reingold, Nathan, 1985. The sciences, 1850–1900: A North Atlantic perspective. *Biological Bulletin* 168(supplement): 44–61.
Reingold, Nathan, 1991. *Science, American Style*. New Brunswick, N.J.: Rutgers University Press.
Reingold, Nathan, and Ida H. Reingold, eds., 1981. *Science in America: A Documentary History 1900–1939*. Chicago: University of Chicago Press.
Reingold, Nathan, and Marc Rothenberg, eds., 1981. *Scientific Colonialism: A Cross-cultural Comparison*. Washington, D.C.: Smithsonian Institution Press.
Reyer, E., 1892. On the causes of the deformation of the Earth's crust. *Nature* 46(1184): 224–227.
Richter, Charles F., 1958. *Elementary Seismology*. San Francisco: W. H. Freeman.
Roger, Jacques, 1973. La théorie de la terre au XVIIIème siècle. *Revue d'Histoire des Sciences* 26: 23–48.
Roger, Jacques, 1982. The Cartesian model and its role in eighteenth-century 'Theory of the Earth.' In *Problems of Cartesianism*, edited by T. M. Lennon, J. M. Nicholas, and J. W. Davis. Kingston and Montreal: McGill-Queen's University Press, pp. 95–125.

Rogers, W. B., and H. D. Rogers, 1843. On the physical structure of the Appalachian chain, as exemplifying the laws which have regulated the elevation of great mountain chains, generally. *Reports of the Meetings of the Association of American Geologists and Naturalists*, pp. 474–531.

Romer, A., ed., 1964. *The Discovery of Radioactivity and Transmutation.* New York: Dover.

Rosenberg, Charles E., 1972. Science, technology and economic growth: The case of the agricultural experiment station scientist, 1875–1914. In *Nineteenth Century American Science: A Reappraisal*, edited by G. H. Daniels. Evanston, Ill.: Northwestern University Press, pp. 181–209.

Rosenberg, Charles E., 1976. *No Other Gods: On Science and American Social Thought.* Baltimore: Johns Hopkins University Press.

Rosenberg, Charles E., 1988. Woods or trees? Ideas and actors in the history of science. *Isis* 79: 565–570.

Ross, Dorothy, 1991. *The Origins of American Social Science*, Cambridge: Cambridge University Press.

Rossiter, Margaret, 1982. *Women Scientists in America: Struggles and Strategies to 1940.* Baltimore: Johns Hopkins University Press.

Rudwick, Martin J. S., 1961. The principle of uniformity. *History of Science* 1: 82–86.

Rudwick, Martin J. S., 1963. The foundation of the Geological Society of London: Its scheme for co-operative research and its struggle for independence. *British Journal for the History of Science* 1: 117–135.

Rudwick, Martin J. S., 1970. The strategy of Lyell's *Principles of Geology*. *Isis* 61(206): 5–33.

Rudwick, Martin J. S., 1971. Uniformity and progression: Reflections on the structure of geological theory in the age of Lyell. In *Perspectives in the History of Science and Technology*, edited by D. H. D. Roller. Norman: University of Oklahoma Press, pp. 209–227.

Rudwick, Martin J. S., 1985. *The Great Devonian Controversy.* Chicago: University of Chicago Press.

Rudwick, Martin J. S., 1990. Introduction. In *Principles of Geology*, by Charles Lyell. Chicago: University of Chicago Press.

Runcorn, S. K., 1959. Rock magnetism. *Science* 129: 1002–1012.

Rupke, N. A., 1970. Continental drift before 1900. *Nature* 227: 349–350.

Rupke, N. A., 1983. *The Great Chain of History: William Buckland and the English School of Geology (1814–1849).* Oxford: Oxford University Press.

Ruse, Michael, 1978. What kind of revolution occurred in geology? In *Proceedings of the 1978 Biennial Meeting of the Philosophy of Science Association*, edited by P. J. Asquith and I. Hacking. East Lansing, Mich.: Philosophy of Science Association, pp. 240–273.

Rutherford, Sir Ernest, 1913. *Radioactive Substances and Their Radiations.* Cambridge: Cambridge University Press.

Sack, Dorothy, 1989. Reconstructing the chronology of Lake Bonneville: An historical review. In *History of Geomorphology*, edited by K. J. Tinkler et al. London: Unwin Hyman, pp. 223–256.

Sack, Dorothy, 1992. New wine in old bottles: The historiography of a paradigm change. *Geomorphology* 5: 251–263.

Sagan, Carl, 1996. *The Demon-Haunted World: Science as a Candle in the Dark.* New York: Random House.

Saull, Vincent A., 1986. Wanted: Alternatives to plate tectonics. *Geology* 14(June): 536.

Schairer, J. F., 1954. Memorial of Frederick Eugene Wright. *American Mineralogist* 39: 284–292.

Schuchert, Charles, 1918. A century of geology—The progress of historical geology in North America. In *A Century of Science in America, with Special Reference to the American Journal of Science, 1818–1918.* New Haven, Conn.: Yale University Press, pp. 60–121.

Schuchert, Charles, 1919. Joseph Barrell (1869–1919). *American Journal of Science* 198(4th ser., vol. 48): 251–280.
Schuchert, Charles, 1922. The Andean geosyncline. *Bulletin of the Geological Society of America* 33: 90–91.
Schuchert, Charles, 1923. Sites and nature of the North American geosynclines. *Bulletin of the Geological Society of America* 34: 151–230.
Schuchert, Charles, 1924. Review of *The New Geology: A Textbook for Colleges, Normal Schools, and Training Schools*. *Science* 59: 486–487.
Schuchert, Charles, 1928a. The hypothesis of continental displacement. In *The Theory of Continental Drift: A Symposium*, edited by W. A. J. M. van Waterschoot van der Gracht. London: Murray, pp. 104–144.
Schuchert, Charles, 1928b. Memorial of Frank Springer. *Bulletin of the Geological Society of America* 39: 65–80.
Schuchert, Charles, 1928c. The continental displacement hypothesis as viewed by du Toit. *American Journal of Science* 216: 266–274.
Schuchert, Charles, 1929a. Thomas Chrowder Chamberlin 1843–1928. *American Journal of Science* 217: 194–196.
Schuchert, Charles, 1929b. The geological history of the Antillean region. *Science* 69: 139–145.
Schuchert, Charles, 1929c. Chamberlin's philosophy of correlation. *Journal of Geology* 37(May-June): 328–340.
Schuchert, Charles, 1932a. Gondwana land bridges. *Geological Society of America Bulletin* 43: 875–915.
Schuchert, Charles, 1932b. The periodicity of oceanic spreading, mountain-making, and paleogeography. In *Physics of the Earth V: Oceanography*, edited by T. W. Vaughan. Washington, D.C.: National Research Council, pp. 537–557.
Schuchert, Charles, 1932c. Permian floral provinces and their interrelations. *American Journal of Science* 224: 405–413.
Schuchert, Charles, 1936. Some recent geologic philosophy. *American Journal of Science* 232: 147–150.
Schuchert, Charles, and Carl O. Dunbar, 1960. *Historical Geology*, 4th ed. New York: Wiley. Original edition 1915.
Schultz, Susan, F., 1976. Thomas C. Chamberlin: An intellectual biography of a geologist and educator. Ph.D. diss., Department of History of Science, University of Wisconsin.
Schwarzbach, Martin, 1986. *Alfred Wegener: The Father of Continental Drift*. Translated by Carla Love. Madison, Wisc.: Science Tech.
Schweber, S. S., 1986. The empiricist temper regnant: Theoretical physics in the United States, 1920–1950. *Historical Studies in the Physical and Biological Sciences* 17: 55–98.
Secord, James A., 1986. *Controversy in Victorian Geology: The Cambrian–Silurian Dispute*. Princeton, N.J.: Princeton University Press.
Servos, John W., 1984. To explore the borderland: The foundation of the Geophysical Laboratory of the Carnegie Institution of Washington. *Historical Studies in the Physical and Biological Sciences* 14: 147–186.
Servos, John W., 1986. Mathematics and the physical sciences in America, 1880–1930. *Isis* 77: 611–629.
Servos, John W., 1990. *Physical Chemistry from Ostwald to Pauling: The Making of a Science in America*. Princeton, N.J.: Princeton University Press.
Shaler, N. S., 1874. Notes on some of the phenomena of elevation and subsidence of the continents. *Proceedings of the Boston Society of Natural History*, 17(December 16): 288–295.
Shapin, Steven, 1994. *A Social History of Truth*. Chicago: University of Chicago Press.

Shapin, Steven, and Simon Schaffer, 1985. *Leviathan and the Air-Pump.* Princeton, N.J.: Princeton University Press.
Shea, James H., 1982. Twelve fallacies of uniformitarianism. *Geology* 10: 455-460.
Shea, James H., ed., 1985. *Continental Drift: Benchmark Papers in Geology.* New York: Van Nostrand Reinhold.
Shepard, E. S., and G. A. Rankin, 1911. Preliminary report on the Ternary system CaO-$Al_2O_3 0SiO_2$: A study of the constitution of Portland cement clinker, with optical study by F. E. Wright. *Journal of Industrial and Chemical Engineering* 3: 211-227.
Simpson, George Gaylord, 1943. Mammals and the nature of continents. *American Journal of Science* 241: 1-31.
Singer, Maxine, 1994. History and science at the Carnegie Institution of Washington. In *The Earth, the Heavens, and the Carnegie Institution of Washington,* edited by G. Good. History of Geophysics 5. Washington, D.C.: American Geophysical Union, pp. v-vii.
Singewald, Joseph T., Jr., 1928. Discussion of the Wegener theory. In *The Theory of Continental Drift: A Symposium,* edited by W. A. J. M. van Waterschoot van der Gracht. London: Murray, pp. 189-193.
Slotten, Richard Hugh, 1994. *Patronage, Practice, and the Culture of American Science: Alexander Dallas Bache and the U.S. Coast Survey.* New York: Cambridge University Press.
Smith, Crosbie, 1985. Geologists and mathematicians: The rise of physical geology. In *Wranglers and Physicists: Studies on Cambridge Mathematical Physics in the Nineteenth Century,* edited by P. M. Harman. Manchester: Manchester University Press, pp. 49-84.
Smith, Crosbie, and M. Norton Wise, 1989. *Energy and Empire: A Biographical Study of Lord Kelvin.* Cambridge: Cambridge University Press.
Smith, George Otis, 1918. A century of government geological surveys. In *A Century of Science in America, with Special Reference to the American Journal of Science, 1818-1918,* edited by E. S. Dana. New Haven: Yale University Press, pp. 193-216.
Smith, Robert W., 1982. *The Expanding Universe.* Cambridge: Cambridge University Press.
Snider-Pelligrini, Antonio, 1858. *La création et ses mystères dévoilés.* Paris: Franck et Dentu.
Sopka, Katherine R., 1988. *Quantum Physics in America: The Years through 1935,* History of Modern Physics and Astronomy. College Park, Md.: American Institute of Physics.
Spanagel, David, 1996. Chronicles of a land etched by God, water, fire, time and ice. Ph.D. diss., Department of History of Science, Harvard University.
Stewart, John A., 1986. Drifting continents and colliding interests: A quantitative application of the interests perspective. *Social Studies of Science* 16: 261-279.
Stewart, John A., 1990. *Drifting Continents and Colliding Paradigms: Perspectives on the Geoscience Revolution.* Bloomington: Indiana University Press.
Struik, Dirk J., 1957. *The Origins of American Science.* New York: Cameron.
Strutt, R. J., 1903. Radium and the sun's heat. *Nature* 68, no. 1772(Oct. 15): 572.
Strutt, R. J., 1905. On the radio-active minerals. *Proceedings of the Royal Society of London,* ser. A, 76: 88-101.
Suess, Eduard, 1904-1924. *The Face of the Earth.* Translated by Hertha B. C. Sollas. Oxford: Clarendon Press.
Sullivan, Walter, 1964. *Continents in Motion: The New Earth Debate.* New York: McGraw-Hill.
Suppe, Frederick, 1993. Credentialing scientific claims. *Perspectives on Science* 1(2): 153-203.
Switjink, Zeno, 1988. The objectification of observation: Measurement and statistical methods in the nineteenth century. In *The Probabilistic Revolution,* edited by L. Kruger, L. J. Daston, and M. Heidelberger. Cambridge, Mass.: MIT Press, pp. 261-284.

Sykes, Lynn R., 1967. Mechanism of earthquakes and nature of faulting on the mid-oceanic ridges. *Journal of Geophysical Research* 72, no. 8(April 15): 2131–2153.
Sykes, Lynn R., 1968. Seismological evidence for transform faults, sea floor spreading, and continental drift. In *The History of the Earth's Crust: A Symposium*, edited by R. A. Phinney. Princeton, N.J.: Princeton University Press, pp. 120–150.
Taylor, F. B., 1910. Bearing of the Tertiary mountain belt on the origin of the Earth's plan. *Geological Society of America Bulletin* 21: 179–226.
Taylor, F. B., 1925. Movement of continental masses under action of tidal forces. *Pan-American Geologist* 43: 15–50.
Taylor, F. B., 1926. Greater Asia and isostasy. *American Journal of Science* 212: 47–67.
Taylor, F. B., and Frank Leverett, 1915. The Pleistocene of Indiana and Michigan and the history of the Great Lakes. U.S. Geological Survey Monograph 53. Washington, D.C.: U.S. Geological Survey.
Taylor, Kenneth, 1992. The historical rehabilitation of theories of the earth. *Compass* 69(4): 334–345.
Termier, Pierre, 1924. The drifting of the continents. *Annual Report of the Smithsonian Institution*. Publication No. 2795 (1923–24): 219–236.
Thagard, Paul, 1992. *Conceptual Revolutions*. Princeton, N.J.: Princeton University Press.
Tharp, Marie, 1982. Mapping the ocean floor—1947 to 1977. In *The Ocean Floor*, edited by R. A. Scrutton and M. Talwani. New York: Wiley, pp. 19–31.
Tharp, Marie, and Henry Frankel, 1986. Mappers of the deep: How two geologists plotted the mid-Atlantic ridge and made a discovery that revolutionized the earth sciences. *Natural History* (October): 49–62.
Thomson, William, 1862a. On the rigidity of the Earth. *Proceedings of the Royal Society of London* 12: 103–108.
Thomson, William, 1862b. On the secular cooling of the Earth. *Transactions of the Royal Society of Edinburgh* 23(1): 157–169.
Thomson, William, 1863. On the rigidity of the Earth. *Philosophical Transactions of the Royal Society of London*: 573–582.
Thomson, William, 1876. Opening address in section A: Mathematical and physical. *Nature* 14(September 14): 426–431.
Thomson, William, and R. G. Tait, 1862. 'Energy.' *Good Words* 3: 601–607.
Toulmin, Stephen, 1975. The discovery of time. In *The Philosophy of Geohistory: 1785–1970*, edited by C. C. Albritton Jr. Stroudsburg, Pa.: Dowden, Hutchinson and Ross, pp. 11–23.
Toulmin, Stephen, and June Goodfield, 1965. *The Discovery of Time*. Chicago: University of Chicago Press.
Touret, Lydie, 1997. Le rayonnement de la minéralogie parisienne aux XVIII et XIXe Siècles. Paper read at the 20th International Congress of History of Science, Liège, Belgium, July 1997.
Traweek, Sharon, 1988. *Beamtimes and Lifetimes*. Cambridge: Harvard University Press.
Trümpy, Rudolf, 1991. The Glarus nappes: A controversy of a century ago. In *Controversies in Modern Geology*. London: Academic Press, pp. 385–404.
Uyeda, Seiya, 1978. *The New View of the Earth: Moving Continents and Moving Oceans*. Translated by Masako Ohnuki. San Francisco: Freeman.
Vacquier, Victor, 1959. Measurement of horizontal displacement along faults in the ocean floor. *Nature* 183(4659): 452–453.
van Andel, Tjeerd H., 1984. Plate tectonics at the threshold of middle age. *Geology end Mijnbouw* 64: 337–343.
van Andel, Tjeerd H., 1985. *New Views on an Old Planet: Continental Drift and the History of the Earth*. Cambridge: Cambridge University Press.

van der Gracht, W. A. J. M. van Waterschoot, 1928a. Remarks regarding the papers offered by the other contributors to the symposium. In *The Theory of Continental Drift: A Symposium*, edited by W. A. J. M. van Waterschoot van der Gracht. London: Murray, pp. 197–226.
van der Gracht, W. A. J. M. van Waterschoot, ed., 1928b. *The Theory of Continental Drift: A Symposium*. London: Murray.
van der Gracht, W. A. J. M. van Waterschoot, 1928c. The problem of continental drift. In *The Theory of Continental Drift: A Symposium*. Edited by W. A. J. M. van Waterschoot van der Gracht. London: Murray, pp. 1–75.
van Hise, Charles R., 1903. Powell as a geologist. *Proceedings of the Washington Academy of Sciences* 5: 113–118.
van Hise, Charles R., 1904. The problems of geology. *Journal of Geology* 12(7): 589–616.
Vening Meinesz, F. A., 1929. Gravity expedition of the U.S. Navy. *Nature* 123: 473–475.
Vening Meinesz, F. A., 1932. *Gravity Expeditions at Sea, 1923–30*. 3 vols. Vol. 1. Delft: Netherlands Geodetic Commission.
Vening Meinesz, F. A., 1941. *Gravity Expeditions at Sea, 1934–1939*. 3 vols. Vol. 3. Delft: Netherlands Geodetic Commission.
Vening Meinesz, F. A., 1952. Convection-currents in the Earth and the origin of the continents I. Paper read at Koninkl. Nederl. Akademie van Wetenschappen, Amsterdam.
Vening Meinesz, F. A., and Frederick E. Wright, 1930. The gravity measuring cruise of the U.S. submarine S-21, with an appendix on computational procedure by Miss Eleanor A. Lamson. In *Publications of the U.S. Naval Observatory*. Washington, D.C.: U.S. Government Printing Office.
Vening Meinesz, F. A., J. H. F. Umbrove, and Ph. H. Kuenen, 1934. *Gravity Expeditions at Sea, 1923–32*. 3 vols. Vol. 2. Delft: Netherlands Geodetic Commission.
Veysey, Laurence R., 1970. *Emergence of the American University*. Chicago: University of Chicago Press.
Vine, Frederick J., 1977. The continental drift debate. *Nature* 266: 19–22.
Vine, Frederick J., and Drummond Matthews, 1963. Magnetic anomalies over oceanic ridges. *Nature* 199: 947–949.
von Engelhardt, Wolf, and Jörg Zimmerman, 1988. *Theory of Earth Science*. Translated by Lenore Fischer. Cambridge: Cambridge University Press.
Walcott, Charles D., 1902. Outlook of the geologist in America. *Bulletin of the Geological Society of America* 13: 99–118.
Walcott, Charles D., 1925. Science and service. *Science* 61(1566): 1–5.
Wallace, Stewart R., 1992. Presidential perspective. *Society of Economic Geologists Newsletter* 10(July 1992): 3.
Warwick, Andrew, 1995. The laboratory of theory or what's exact about the exact sciences? In *The Values of Precision*, edited by M. N. Wise. Princeton, N.J.: Princeton University Press, pp. 311–351.
Watts, C. B., P. Sollenberger, and J. E. Willis, 1938. World Longitude Operation of 1933 at San Diego and Washington. In *Publications of the U.S. Naval Observatory*, vol. 14, part 5. Washington: U.S. Government Printing Office, pp. 243–283.
Wegener, Alfred L., 1911. *Thermodynamik der Atmosphäre*. Leipzig: Barth.
Wegener, Alfred L., 1912. Die Entstehung der Kontinente. *Geologische Rundschau* 3: 276–292.
Wegener, Alfred L., 1915. *Die Entstehung der Kontinente und Ozeane*. Braunschweig. Friedr. Viewig.
Wegener, Alfred L., 1924. *The Origin of Continents and Oceans*. 3d ed. Translated by J. G. A. Skerl. London: Methuen.

Wegener, Alfred L., [1929] 1966. *The Origin of Continents and Oceans.* 4th ed. Translated by John Biram, (1929). Reprinted, New York: Dover, 1966.

Weir, Gary, 1993. *Forged in War: The Naval–Industrial Complex and American Submarine Construction, 1940–1961.* Washington, D.C.: U.S. Government Printing Office.

Weiss, Malcolm P., 1992. Geological Society of America election of 1921: Attack on the candidacy of Charles Schuchert for the presidency. *Earth Sciences History* 11(2): 90–102.

Westman, Robert S., 1994. Two cultures or one? A second look at Kuhn's *The Copernican Revolution. Isis* 85(1): 79–115.

Westoll, T. S., 1965. Geological evidence bearing on continental shift. In *A Symposium on Continental Drift,* edited by P. M. S. Blackett, S. E. Bullard, and S. K. Runcorn. *Philosophical Transactions of the Royal Society of London,* ser. A, 258: 12–26.

Whewell, William, 1858. *History of the Inductive Sciences from the Earliest to the Present Time.* Vol. 2. New York: Appleton.

Whitten, C., 1992. William Bowie: Engineer, administrator, diplomat. *Eos* 73: 113–125.

Wilde, Garner L., 1990. Memorial to John Wesley Skinner, 1907–1988. *Geological Society of America Memorials* 20: 33–35.

Willis, Bailey, 1893. Mechanics of Appalachian structure. *Annual Report of the U.S. Geological Survey 1891–1892,* 13: 211–281.

Willis, Bailey, 1907. Systematic geology. In *Research in China, in Three Volumes and Atlas.* Washington, D.C.: Carnegie Institution of Washington.

Willis, Bailey, 1910. Principles of paleogeography. *Science* 31: 241–60.

Willis, Bailey, 1922. Role of isostatic stress. *Bulletin of the Geological Society of America* 33: 371–374.

Willis, Bailey, 1924. American geology, 1850–1900. *Science* 96: 167–172.

Willis, Bailey, 1927. Folding or shearing, which? *Bulletin of the American Association of Petroleum Geologists* 11: 31–47.

Willis, Bailey, 1928. Continental drift. In *The Theory of Continental Drift: A Symposium,* edited by W. A. J. M. van Waterschoot van der Gracht. London: Murray, pp. 76–82.

Willis, Bailey, 1929. "Dynamics is the soul of the problem." *Journal of Geology* 37(May–June): 357–367.

Willis, Bailey. 1932. Isthmian links. *Geological Society of America Bulletin* 43: 917–952.

Willis, Bailey, 1942a. Address to Sigma Xi. Stanford University Archives, Bailey Willis Papers, Biographical File.

Willis, Bailey, 1942b. American geology, 1850–1900. *Science,* n.s., 96, no. 2486(August 21): 167–172.

Willis, Bailey, 1944. Continental drift, ein Märchen. *American Journal of Science* 242: 509–513.

Willis, Bailey, and Robin Willis, 1934. *Geologic Structures.* 3d ed. New York: McGraw-Hill.

Wilson, Daniel J., 1990. *Science, Community, and the Transformation of American Philosophy 1860–1930.* Chicago: Chicago University Press.

Wilson, J. Tuzo, 1963. A possible origin of the Hawaiian Islands. *Canadian Journal of Physics* 41(863): 863–870.

Wilson, J. Tuzo, 1965. A new class of faults and their bearing on continental drift. *Nature* 207 (July 24): 343–347.

Wilson, J. Tuzo, 1968. Static or mobile earth: The current scientific revolution. *Proceedings of the American Philosophical Society* 112(5): 309–320.

Wilson, J. Tuzo, ed., 1971. *Continents Adrift: Readings from* Scientific American. San Francisco: Freeman. 2d ed. 1976.

Wilson, J. Tuzo, ed., 1976. *Continents Adrift and Continents Aground: Readings from* Scientific American. 2d ed. San Francisco: Freeman.

Wilson, Leonard, 1967. The emergence of geology as a science in the United States. *Journal of World History* 10: 416–437.
Wilson, Leonard G., ed., 1979. *Benjamin Silliman and His Circle: Studies on the Influence of Benjamin Silliman on Science and America—Prepared in Honor of Elizabeth H. Thompson.* New York: Science History Publications.
Wilson, Leonard G., 1980. Geology on the eve of Lyell's first visit to America, 1841. *Proceedings of the American Philosophical Society* 124: 168–202.
Winnik, Herbert C., 1970. Science and morality in Thomas C. Chamberlin. *Journal of the History of Ideas* 31: 441–456.
Wise, M. Norton, 1986. Measurement, work and industry in Lord Kelvin's Britain. *Historical Studies in the Physical and Biological Sciences* 17: 147–173.
Wise, M. Norton, ed., 1995. *The Values of Precision.* Princeton, N.J.: Princeton University Press.
Wise, M. Norton, and Crosbie Smith, 1987. The practical imperative: Kelvin challenges the Maxwellians. In *Kelvin's Baltimore Lectures and Modern Theoretical Physics*, edited by P. Achinstein and R. Kargon. Cambridge: MIT Press, pp. 323–348.
Wood, Harry O., 1915. On a possible causal mechanism for heave-fault slipping in the California Coast Range region. *Seismological Society of America Bulletin* 5: 214–229.
Wood, R. M., 1985. *The Dark Side of the Earth.* London: Allen and Unwin.
Woodward, Horace, 1907. *The History of the Geological Society of London.* London: Geological Society.
Woodward, R. S., 1888. *On the Form and Position of Sea Level.* Bulletin 48. Washington, D.C.: U.S. Geological Survey.
Woodward, R. S., 1889. The mathematical theories of the Earth. *American Journal of Science*, 3d ser., 38(227): 337–355.
Worzel, J. L., 1985. *Pendulum Gravity Measurements at Sea, 1936–1959.* New York: Wiley.
Wright, F. E., 1926. *Apparatus for the Measurement of the Variations in the Force of Gravity.* Washington, D.C.: U.S. Patent Office.
Wright, F. E., 1927. Polarization of light reflected from rough surfaces with special reference to light reflected by the moon. *Proceedings of the National Academy of Sciences* 13: 535–540.
Wright, F. E., 1929. *Gravity Measuring Cruise of the Submarine U.S.S. S-21.* Washington, D.C.: Carnegie Insititution of Washington.
Wright, F. E., 1937. *The Public Relations Problem of the Carnegie Institution of Washington.* Washington, D.C.: Carnegie Institution of Washington.
Wright, F. E., 1941. Gravity measurements in Guatemala. *Transactions of the American Geophysical Union* 22: 512–515.
Wright, F. E., and J. L. England, 1938. An improved torsion gravity meter. *American Journal of Science* 235A: 373–383.
Wright, W. B., 1923. The Wegener hypothesis: Discussion at the British Association, Hull. *Nature* 111(Jan. 6): 30–31.
Yochelson, Ellis L., ed., 1980. The scientific ideas of G. K. Gilbert. *Geological Society of America Special Papers* 183. Boulder, Colo.: Geological Society of America.
Yochelson, Ellis L., 1994. Andrew Carnegie and Charles Doolittle Walcott: The origin and early years of the Carnegie Institution of Washington. In *The Earth, the Heavens, and the Carnegie Institution of Washington*, edited by G. Good. History of Geophysics 5. Washington, D.C.: American Geophysical Union, pp. 1–20.
Yoder, H. S., Jr. 1989. Scientific highlights of the Geophysical Laboratory, 1905–1989. *Carnegie Institution of Washington Annual Report of the Director of the Geophysical Laboratory*, pp. 143–203.

Yoder, H. S., Jr., 1994. Development and promotion of the initial scientific program for the Geophysical Laboratory. In *The Earth, the Heavens, and the Carnegie Institution of Washington*, edited by G. Good. History of Geophysics 5. Washington, D.C: American Geophysical Union, pp. 21–28.

Zittel, Karl Alfred Ritter von, 1901. *History of Geology and Paleontology to the End of the Nineteenth Century*. Translated by Maria Matilda Ogilvie-Gordon. Edited by H. Ellis. The Contemporary Science Series. London: Scribner's.

INDEX

Adams, John Quincy, 128
Adjustment and readjustment, isostatic, 32, 169–170, 254
Africa-India ridge, 214, 216
Agassiz, Louis, 58, 279
Age of the Earth, The (Holmes), 64, 114
Airy, George Biddell
 fluid substrate and, 67
 isostasy and, 36, 77, 172, 175–177, 197, 241–242, 259–260, 276–277: compensation, isostatic, 24–25, 27
 mountain-building and, 24–25, 28, 36
Allegre, Claude, 65
Allison Commission Hearings (1886), 37
Alpine overthrusts and thrusts, 64, 84, 196–197
Alps (Swiss), 21–23, 32, 51, 82, 103–104, 254
American earth science. *See also* Geology
 Chamberlin (T. C.) and, 129–133, 140–141
 continental drift and, 130–133: rejection of, 5–6, 125–129, 153–156
 empiricism and, 144
 gradualist approach and, 150–152, 201
 independence of developing, 138–140
 interpretive framework of, 142–144
 neptunist-plutonist debate and, 134–136
 nonsectarianism and, 152
 patrons of, 133–134
 physics and chemistry in, 281–282, 285, 302
 plate tectonics and, 127
 scientific theory and method in, 127–134, 157
 theory-driven science and, 284–285
 theory and practice in, 127–134
 theory versus hypothesis debate and, 148–150
 U.S. Navy studies and, 231, 233, 262–263, 288–289, 302–303
 Willis and, 129
American Geophysical Union, 167, 254
American Philosophical Society, 254
Ampferer, Otto, 119
Andes Mountains, 83
Antiauthoritarian logic of discovery, 134–136
Antlitz der Erde, Das (The Face of the Earth) (Seuss), 10, 14, 21, 75, 113
Apatite, 50
Appalachian Mountains, 17–18, 23, 30–32, 103, 119–120, 129, 148, 209, 224, 254
Archimedes' principle, 65
Argand, Emile, 84, 115, 119, 276
Astatic magnetometer, 263–264
Asthenosphere, 94, 182–183, 189, 194, 272
Asymmetric patterns of deformation and overthrusting, 17–18
Atlantic land bridge, 211, 214, 216, 218, 277

404

Index

Atomic physics, 52
Authoritarianism, scientific, 135–136
Autocracy, rejection of European theoretical, 135–136

Babbage, Charles, 31, 73, 104
Bache, Alexander Dallas, 37, 128
Bacon, Francis, 156, 221, 315
Bahamas, crustal structure of, 244
Baker, Howard, 54, 129, 198, 219, 293
Baltic Sea, 75
Barracuda submarine expedition, 244, 253, 264
Barrell, Joseph
 biographical background of, 181–182
 Chamberlin (T. C.) and, 181–182
 continental drift and, 141, 188–191
 continental fragmentation and, 188–191
 Daly and, 94
 geosyncline theory and, 73–74
 Gilbert and, 185
 independent science and, 138
 isostasy and, 181–188, 192, 236, 286
 Longwell and, 192
 neptunist-plutonist debate and, 135
 ruling theory and, 139
 Schuchert and, 188, 210
 scientific theory and method and, 144
 terminology of, 182–183, 189, 191, 194, 272
 uniformitarianism and, 206
Bartlett Deep, 240–241, 244
Barus, Carl, 283–284
Basalts, 105, 117, 188–191, 194, 211–212, 271
Becker, George F., 71–72, 149, 284
Becquerel, Henri, 48
Belief
 formulation of, 6
 representation as reified, 175–177
Berkey, Charles, 206
Berry, Edward W., 206, 218, 278
Bertrand, Marcel, 12, 21–22
Biblical accounts of life, 205
Billings, Marlin, 226–229
Biotite mica, 50–51
Birch, Francis, 227
Blackett, P. M. S., 112, 263–267
Blake expedition, 33
Block-faulting, 189
Boltwood, Bertram, 49–50, 102, 113–114

Bouguer correction, 39, 46–47
Boulder beds, 125
Bowen, Norman Levi, 88
Bowie, William
 acclaim of, 167
 advisory committee on oceanography, assembled by, 262
 continental drift and, 149, 167–177, 192, 194–195, 230, 276–277, 279; evidence of, 170–179, 231–236, 238–239
 continental fragmentation and, 192
 death of, 262
 Dutton and, 169
 earthquakes and, 254
 Field and, 288–289
 geophysical constraints and, 277
 Hayford and, 168–169, 173–175, 183
 Holmes and, 171
 isostasy and, 37, 45–47, 129, 149, 167–171, 175–178, 182–184, 239, 241–242, 254, 259–261; compensation, isostatic, 276
 paleontological evidence and, 276
 prominence of, 73
 quantification and instrumentation and, 299–301
 Schuchert and, 123, 167, 171–172, 178–179, 191, 197, 212–213
Box-of-rocks geology, 290
Branner, J. C., 133
Brazil-Guinea ridge, 214
Bridgman, Percy W., 89, 94, 152, 288
British Association in Hull, 124–126, 158
British earth science, 124–125, 138–139, 156. *See also* Geology
Bruce, Robert, 128
Bryan, William Jennings, 205
Bucher, Walter, 285–287, 298–299, 305
Buddington, A. F., 267
Buffon, George-Louis Leclerc de, 14, 278
Bull, A. J., 119
Bullard, Edward, 263–264, 266–267, 304–306
"Bullard fit" of continents, 304–306
Burchfield, Joe D., 126, 128
Burke, Kevin, 306–307
Bushveld Igneous Complex, 91, 94, 97

Camus, Albert, 121
Cape Province, 165
Caribbean crustal structure, 244–245

Caribbean Deeps, 240
Carnap, Rudolf, 303
Carnegie, Andrew, 133, 283
Caspian Sea, 189
Caster, Kenneth, 309–310
Catastrophism, 12, 204–207
Causal mechanism, 6, 62–65
Causal theory, 153. *See also specific types*
Causation, occurrence versus, 63
Challenger expedition, 33
Chamberlin, Rollin T., 98, 155, 304–305, 313
Chamberlin, T. C.
  acclaim of, 69, 130
  American earth science and, 129–133, 140–141
  Barrell and, 181–182
  biographical background of, 130
  contraction theory and, 153
  discovery and, 138–140
  fluid substrate and, 69
  glacial deposits and landforms and, 58, 136–138
  gradualist approach and, 150–152
  interior of earth and, 69–70, 72
  isostatic creep and, 69–70
  plantesimal theory and, 132
  prominence of, 73
  scientific theory and method and, 136–142, 144–145, 148, 281–283
  solid earth accretion theory and, 49, 98
  "strong crust" school and, 182
  Taylor (Frank) and, 81
  uniformitarianism and, 206
  Willis and, 70, 136, 150–151
Chhibber, H. L., 293
CIW (Carnegie Institution of Washington), 132–133, 238, 283, 288
Climate zones, 199, 207
Coast Survey. *See* U.S. Coast and Geodetic Survey
Coleman, A. P., 87
College Text-book of Geology, A (Salisbury and Chamberlin), 69
Colorado Plateau, 29, 36, 185, 190
Columbia River Plateau, 190
Compensation, isostatic, 23–25, 27, 44–46, 70, 177, 182, 184–187, 241, 277
Comte, Auguste, 303
Conceivable mechanism, 65
Continental creep. *See* Continental drift
Continental crust, 77–80

Continental deformation, 97, 104–105
Continental drift. *See also* Drift mechanisms
  acceptance of, 312
  American earth science and, 130–133
  American geology support and, 133–134
  American response to, 125–129, 153–156, 308–311
  Barrell and, 141, 188–191
  Billings and, 226–229
  Bowie and, 149, 167–177, 192, 194–195, 230, 276–277, 279
  British response to, 124–125, 156
  comparison of South Africa and South America, 157–167
  consequences of, 59–62
  Daly and, 157–158, 202, 227, 230, 279
  du Toit and, 158–159, 161–167, 279
  evidence of: asymmetries and, 276–278; Bowie and, 170–179, 231–236, 238–239; certainty through measurement and, 280–284; defining, 227–228; direct and indirect, 227–236; du Toit and, 264, 291–298, 308–309; epistemic traditions and, 278–279; faunal and flora similarities, 57; geological, 117, 171; jigsaw-puzzle fit, 57–58; Joly and, 230 paleontological, 56–57; qualitative, 296–298; radio telegraphy and, 230–236; tectogenes and, 243–249; types of, 223–226; underdetermination and, 276–278; USS *S-21* expedition and, 236–240; Wegener and, 56–59, 124–125, 227–228, 230, 264, 273, 275–276, 309
  Fennoscandian uplift and, 74–76, 84
  Gondwana and, 148, 157–167
  gravity and, 236–240
  Gutenberg and, 223–226
  Holmes and, 6, 124, 171–172, 192–194, 196–200, 207, 218
  instrumentation and, 311–312
  interior of earth and, 66–69
  isostasy and, 76–79, 129–130; undertow, isostatic, 69–70
  isthmian links and, 214–219, 226
  Joly and, 110, 124, 230
  kinematic account of, plausible, 79–80, 94, 119
  Longwell and, 64, 125, 155, 280

mechanism issues and: causal, 6, 62–65; conceivable, 65; Daly and, 90, 93–99; lack of, 292; physical, 65–66; plausible, 118–119
Menard and, 223–228
mobile substrate and, 71–73, 79–80
mountain-building and, 59, 85, 224, 226
permanence theory and, 254
plastic substrate and, 69–70, 72
plate tectonics and, 64–65, 120
rejection of: by Americans, 5–6, 125–129, 153–156; causal mechanism and, lack of, 62–65; du Toit and, 291–292; historical context of, 123; Simpson, 295; Wegener and, 123–124; Willis, 296
Schuchert and, 157, 178–179, 196–202, 234, 308
test of, 157–167
truth and, scientific, 5–6
uniformitarianism and, 83, 199–200, 207
as violation of geology standards, 153–156
Wegener and, 5–6, 54–62, 77, 80, 93, 124, 153–155, 167, 224, 260–261, 267
westward, 242
Willis and, 86–87, 126, 143–144, 296, 308
Wright and (Frederick), 157–161
Continental fragmentation, 188–192, 194
Continental reconstructions, 59–62
Contraction theory
Chamberlin (T. C.) and, 153
collapse of, 51–52, 55
contradiction of, 154
Dana and, 15–17, 23, 29, 48, 52, 153
Darwin (George) and, 13
De la Beche and, 12–13
Dutton and, 29–33, 50
field mapping and, 51–52
Fisher and, 13, 25–29, 50
Gilbert and, 33–35
isostasy and, 35, 48, 51–52, 55–56
mountain-building and, 19, 21, 251
orogeny and, 30
permanence theory and, 17
Pratt and, 48
problems with: Fisher, 25–29; isostasy, 23–25, 35, 48, 51–52; mountain-building, 19–21; radioactivity, 48–52; Suess and, 11–12, 15, 17, 21, 23, 48
Swiss Alps and, 21–23
Willis and, 35

"Contributions to the Theory of Magmatic Cycles" (Holmes), 115
Convection theory, 103–104, 112–119, 171–172, 199, 213, 227–228, 251, 259, 265–271, 292, 296
Conybeare, William, 12
Cooling
  progressive, 115, 203
  secular, 12–14, 19, 30, 130
Cordillera of the sea, 214
Cox, Allan, 9, 127, 271
Creep, isostatic, 69–70
Creer, Kenneth, 265
Crustal compression, 17–18, 207–208, 226
Crustal deformation, 31
Crustal motion. *See* Continental drift
Crustal root, 259
Crustal warping, 96
Crystalline materials, 93–94
Crystallography, 13
Curie, Marie, 48, 50–51, 102
Curie, Pierre, 48, 102, 263, 311
Curie point, 263
Cuvier, Georges, 14, 278

Dalrymple, Brent, 271
Daly, Reginald A.
  Barrell and, 94
  basaltic elevation and, 211
  biographical background of, 87–89
  Bridgman and, 288
  continental drift and, 157–158, 202, 227, 230, 279
  drift mechanisms and, 89–91, 93–99, 108, 112–119, 126, 129
  du Toit and, 291, 294, 297
  employment of, geological, 133–134
  geophysical constraints and, 277
  geosyncline theory and, 96–98
  gravity sliding and, 87–89, 93–99, 119
  Griggs and, 251
  Holmes and, 193–194
  Longwell and, 89, 99
  mountain-building and, 123
  notebooks of, 145–147
  observation versus interpretation and, 145–147, 152
  orogeny and, 270
  subduction and, 119
  theory versus hypothesis and, 149

Dana, James Dwight
  contraction theory and, 15–17, 23, 29, 48, 52, 153
  Hall's work and, 63
  mountain-building and, 18–19, 59, 63
  permanence theory and, 17, 55
  plate tectonics and, 10, 14–20, 129
  scientific theory and method and, 281
  sediment accumulation and, 18–19
Danish Geodetic Institute, 61
Darwin, Charles, 11, 16, 28, 52, 56, 201, 276
Darwin-Fisher fissiparturition hypothesis, 81, 168–170, 195
Darwin, George H., 13, 28, 131, 168–170
Daston, Lorraine, 300
Data collection, 298–300
Davis, William Morris, 88, 182
Dawson, J. William, 23
Day, Arthur, 283–284
Death Valley, 189
Deccan lava flows, 266
Deccan Traps, 105, 108–109, 190, 266, 293
Deduction, 135, 304–305
Deflections of the vertical, 39–44
Deformation of the Earth's Crust, The (Bucher), 285
De Geer, Gerard, 75
De la Beche, Henry, 12–13, 17, 203, 279, 299
Density deficiencies, 44–45
Depersonalization of geology
  deduction and, 303–304
  discountenance of du Toit and, 291–296
  distance and detachment and, 298–303
  induction and, 303–304
  move from field work to laboratory and, 288–291
  objectivity and, 298–300
  quality and, quantifying, 296–298
  self-validating power of success and, 304–308
Depth of isostatic compensation, 44–46, 70, 177
Descartes, René, 14
Devonian controversy, 204
De Vries, Hugo, 277
Diastrophism, 143, 249
Dietz, Robert, 270–271
Discovery
  antiauthoritarian logic of, 134–136
  Chamberlin (T. C.) and, 138–140
  context of, 4

Displacement hypothesis, 9–10, 54, 64, 123, 163
Distance and detachment from work, 298–303
Dixon, H. H., 112
Dixon, Kendal, 112
Doell, Richard, 271
Double-fold theory, 22–23
Down-warpings, 245, 247, 250
Drakensburg scarp, 147–148
Drift mechanisms. *See also* Mechanism issues
  Alpine thrust and, 84–86, 89
  Daly and, 89–91, 93–99, 108, 112–119, 126, 129
  du Toit and, 292
  Eötvös force and, 83–84
  flexure and, upward and downward, 112
  geological activity and, 15, 105–107
  gravity sliding and, 87–89, 94–95, 97, 119
  Holmes and, 112–117, 119–120
  Jeffreys and, 86
  Joly and, 99–112, 119
  overthrusting and, 115–116
  periodic fusion and, 99–102, 119, 126
  pole-flight force and, 84–86, 89
  seismic waves and, 86, 108
  sub-crustal convection currents and, 103–104, 112–117
  Taylor and, 81–82
  tidal forces and, 110–111
  Willis and, 86–87, 264–265
  Wood and, 83–84
Drift theory. *See* Continental drift
Duhem, Pierre, 138
Duluth gabbro, 190
du Toit, Alexander Logie
  comparison of South Africa with South America and, 157–167, 240, 276
  continental drift and, 158–159, 161–167, 279: evidence of, 264, 291–298, 308–309; rejection of, 291–292
  Daly and, 291, 294, 297
  discountenance of, 291–296
  drift mechanisms and, 292
  Gondwana and, 292
  Holmes's convection theory and, 292, 296
  Longwell and, 294–296
  recognition of work of, lack of, 278

Schuchert and, 294
Simpson and, 295
in South Africa, 92–93, 160–161
in South America, 161–163, 238
uniformitarianism and, 292–293
Vening Meinesz and, 296
Willis and, 293–294, 298
Wright (Frederick) and, 294
Dutton, Clarence Edward
acclaim of, 128
Bowie and, 169
contraction theory and, 29–33, 50
geosyncline theory and, 148
Hayford and, 70
isostasy and, 25, 29–33, 35–36, 67–68, 73–74, 129
Kelvin and, 30, 33
Rocky Mountain survey and, 29
scientific theory and method and, 136, 142, 281
"weak crust" school and, 182
Dynamics, kinematics versus, 65–66

Earth. See also Continental drift; Earth science; Isostasy; Plate tectonics
age of, 48–50, 102, 114, 118, 126, 130, 203, 278–279
crust of, 77–80
evolution of, 19–20, 28
fluidity of, 26–27, 65–67, 69–72, 76, 108–112, 118–119, 279, 307–308
gravity of shape of, 90
interior of, 66–70, 72, 94
oscillations of, 66–67, 131, 149
rigidity of, 33–34, 66, 86, 93, 224
secular cooling of, 19
solidity of, 26–27, 29–30, 66–68, 93
"Earth as Engineering Structure, The" (Bowie), 170
Earth (Press and Siever), 62
Earthquakes, 105, 254
Earth science. *See also* American earth science; Depersonalization of geology; Geology
alternative, 205
authoritarianism versus egalitarianism and, 135–136
British, 124–125, 138–139, 156
concept of, 280
controversies, 204
epistemic tradition and, 278–279

evidentiary tradition and, 278–279
Geophysical Laboratory of the Carnegie Institution of Washington and, 283–284
practice of, 204
professionalization of, 204–205
term of, 280
uniformitarianism and, 204–207
Earth, The (Jeffreys), 86
Earthworms, lumbricidae, 57
East African Rift, 189, 191, 235
East Indian ridge, 214
East Indies, 244
Eclogites, 117
Eddington, Arthur, 86
Egalitarianism, scientific, 135–137, 151–152
Elementary Seismology (Richter), 277
Elevation, average, 39
Elie de Beaumont, Léonce, 12–14, 17, 139, 157, 203, 278
Eliot, Charles William, 152
Empiricism, 128, 144
Entstehung der Kontinents und Ozeane, Die (Wegener), 9–10, 54, 59, 65, 76–79, 81–84, 87, 91, 115, 153, 157, 226, 230
Eötvös force, 83–84
Eötvös, Roland (Lorand) von, 82–84
Epistemic tradition, 278–279
Epistemological affinity, 51–53
Epistemological chauvinism, 52–53
Equanimity, scientific, 294
Equilibrium, isostatic, 169, 236
"Escape from perspective," 300
Euler's theorem, 305
Europe, North America comparison with, 57
Evans, John W., 54, 92, 115, 124, 230
Everest, George, 23, 40
Evidentiary tradition, 278–279
Evolution theory, 52, 56, 204–206
Ewing, Maurice, 244, 253–254, 264

Face of the Earth, The (Suess), 10, 14, 21, 75, 113
Falkland Islands, 165
False theories, 315–316
Faults, 69, 189, 272
Faunal similarities, 56–59, 207, 218, 276
Faye reduction, 39
Fear, War, and the Bomb (Blackett), 263

Fennoscandian uplift, 74–76, 84, 182
Field mapping
  accuracy and, 307
  contraction theory and, 51–52
  isostasy and, 51–53
  laboratory work versus, 288–291
  limitations of, 307
  radioactivity and, 51–53
Field, Richard M., 243–244, 246–247, 259, 262, 264, 272, 288–289
Figure of the Earth, The (Hayford), 48, 173–174, 235
Fisher, Osmond
  contraction theory and, 13, 25–29, 50
  fissiparturition theory of, 81, 168–170
  fluid substrate and, 27, 67, 76, 118–119
  isostasy and, 25–29, 35–36, 67, 74, 241
  tidal forces and, 26–27
  yielding substrate and, 67
Fissiparturition hypothesis, 81, 168–170, 195
Fleming, John, 254, 257, 263, 271
Flexure, upward and downward, 112
Flood basalts, 108
Floral similarities, 56–59, 218
Fluid substrate, 26–27, 65–67, 69–72, 76, 108–112, 118–119, 279, 307–308
Forbes, John D., 30, 279
Fossil assemblages, 11, 16–17, 58, 189, 199
Founders of Geology (Geikie), 139
Fourier, Joseph, 14
Fourier's theorem, 30
Fragmenting the continents, 188–191
Frankel, Henry, 52, 275
Free-air correction, 39, 46
Freeman, C. S., 238

Galison, Peter, 311
Gass, I. G., 9
Geikie, Archibald, 49, 139
Geochronology, 114, 120, 275–276
Geodesy, 52, 70, 84, 90–91, 94, 183–184, 233–235, 280–281, 316
Geognosy, 63–64
Geological activity, 15, 105–107, 115. *See also specific types*
Geological Comparison of South America with South Africa, A (du Toit), 163–167, 240, 294
Geological Society of America, 73, 128

Geology. *See also* Depersonalization of geology; Earth science; Scientific theory and method
  alternative, 205
  box-of-rocks, 290
  Bucher and, 285–287
  controversy and, 204
  geodesy and, 94
  geognosy and, 63–64
  historical, 56, 167, 192, 204, 207
  imagination and, 89
  of moon, 190
  outcrop, 289–290
  physics and chemistry in, 281–282, 285, 302
  process-oriented, 282
  professionalization of, 204–205
  theory-driven science and, 284–285
  time scale and, 114, 120, 276–277
  tradition, 279
  uniformitarianism, 204–207
Geology of the High Plateaus of Utah (Dutton), 136, 142
Geophysical Exploration of the Ocean Bottom (Field), 254
Geophysical Laboratory of the Carnegie Institution of Washington, 283–284, 288
Geophysics, 86, 277–278, 280
Geosyncline theory
  Barrell and, 73–74
  Daly and, 96–98
  down-warping in, 96
  Dutton and, 148
  Hall and, 63, 129
  isostasy and, 73–74
  mountain-building and, 17–19
  passing of, 254
  plate tectonics and, 17–19
Gevers, T. W., 292
Giant's Causeway, 105
Gilbert, Grove Karl
  acclaim of, 128
  Barrell and, 185
  contraction theory and, 33–35
  De Geer and, 75
  glacial deposits and landforms and, 58, 168
  isostasy and, 33–39, 47, 68, 149, 182, 241
  Lake Bonneville and, 145
  mobile substrate and, 71

multiple working hypotheses and, 136, 140–141
plate tectonics and, 168
prominence of, 73
Rocky Mountain survey and, 29
scientific theory and method and, 140–142, 144–145, 156, 281–282
Taylor and, 81
uniformitarianism and, 206
Gilbert Survey, 33
Gilman, Daniel Coit, 283
Glacial deposits, 58, 125, 136–138, 165.
See also Marine deposits
Glacier motion, 279
Glass formation, 94
Glissement, 119
Gondwana
Baker and, 219
breakup of, 293, 295
continental drift and, 148, 157–167
du Toit and, 292
evidence of, 11
land bridges and, 190
Longwell and, 191–193
passing of, 213–219
Schuchert and, 179–181, 183, 191–196, 213–214, 218
Suess and, 55, 93, 157–158
Willis and, 208–209, 213–214, 218
"Gondwana Land Bridges" (Schuchert), 213–214
Gould, Stephen Jay, 65, 200, 218
Gradualist approach, 150–152, 201
Graham, John, 265
Grand Canyon, 29
Gravimeter, 238
"Gravimetric Test of the Roots of Mountain Theory, A" (Bowie), 241
Gravity
anomalies, 111–112, 184–187, 239, 242–245, 248–249, 257, 259
continental drift and, 236–240
contraction of, 130–131
of earth's shape, 90
Himalayas and, 36
instability of, 90
isostasy and, 36–48
laws of, formulation of, 87
magnetism and, 309–310
marine, 309–310
measuring, 37, 236, 238–244, 301, 303

mountain-building and, 36
oscillations of earth and, 131
sliding, 87–89, 93–99, 119
stress of, 90
Great Lakes, 81–82
Great Plains of America, 37–38
Great Trigonometrical Survey of India, 23
Greene, Mott, 129
Greenland expedition, 59, 61–62, 230
Gregory, J. W., 124–126, 198
Griggs, David, 119, 225, 249–253, 259, 264–265, 268
Gulf of California, 189–190
Gulf of Mexico, 241
Gunnison Peak, 38
Gutenberg, Beno, 223–226, 253, 257, 270, 279, 293
Guyot, Arnold, 262

Haarman, Erich, 198–199
Hacking, Ian, 301
Hallam, Anthony, 63
Hall, James, Jr. 17–19, 31, 63, 73, 104, 129, 133, 179, 281
Hanson, Gilbert, 304
Haüy, René Just, 13
Hayford, John
  Bowie and, 168–169, 173–175, 183
  continental deformation and, 104
  Dutton and, 70
  horizontal displacement and, 77
  isostasy and, 37, 39–47, 70, 129, 149, 172, 174–177, 182–184, 238–239: compensation, isostatic, 70, 182, 276
  quantity of data and, 234
  Wright (Frederick) and, 236
Heck, N. H., 254–257
Heezen, Bruce, 267, 272
Heim, Albert, 21–23
Heirtzler, J. R., 271
Helmert, F. R., 44–45, 78
Hempel, Carl, 304
Henry, Joseph, 33, 151
Henry Mountains, 33
Herodotus, 64
Herschel, John, 18, 31, 73, 104, 169
Hess, Harry H., 244–245, 248–251, 253–257, 259, 262, 264, 267–270, 295, 311
Highlands controversy, 204
High-water marks, 75
Himalayas, 23–24, 27, 36, 82, 103, 235

Historical Geology of North America (Schuchert), 180
Historicity, 317–318
"History of Ocean Basins" (Hess), 267
Hitchcock, Edward, 135–136
Holmes, Arthur
  acclaim of, 112–113
  biographical background of, 112–114
  Bowie and, 171
  continental drift and, 6, 124, 171–172, 192–194, 196–200, 207, 218
  convection theory and, 112–117, 119, 171–172, 227, 259, 265, 268, 270–271, 292, 296
  Daly and, 193–194
  as disciplinary outsider, 123
  drift mechanisms and, 112–117, 119–120
  evidence and, geological, 298
  Griggs and, 251
  Longwell and, 7, 120, 193
  radioactivity and, 207
  Read and, 264–265
  Schuchert and, 148, 183, 192–193, 196, 198–199, 202
  temporal relations and, 276
  time available and, 64
  uranium-lead technique and, 50
  Wegener and, 270
  Willis and, 212–213
Holton, Gerald, 304
Hopkins, William, 66–67, 86, 118, 278, 280
Horizontal compression, 17–18, 207–208
Horizontal displacement, 77
Hot springs, 66
Hubbert, M. King, 224–226
Hull meeting, 124–126, 158
Humboldt, Alexander von, 118
Hurley, Patrick M., 310
Hutton, James, 134
Hypothetico-deductive method of science, 141, 157

IAU (International Astronomic Union), 231–232, 235
Iceberg analogy, 24–25, 28
Igneous intrusions, 11, 28, 31, 91–92, 119–120
Igneous rock compositions, 88–89, 91–92, 94–95
Induction, 142–144, 303–304

Instrumentation, 300–301, 307, 311–312
Interpretation, observation versus, 145–147, 149, 152
"Investigation vs. Propagandism" (Chamberlin, T. C.), 136
Irving, Edward (Ted), 265–268
Isacks, Brian, 272
Isostasy
  adjustment and readjustment and, 32, 169–170, 254
  Airy model of, 36, 77, 172, 175–177, 197, 241–242, 259–260, 276–277
  Barrell and, 181–188, 192, 236, 286
  Bowie and, 37, 45–47, 129, 149, 167–171, 175–178, 182–184, 239, 241–242, 254, 259–261
  Bucher and, 286
  compensation and, 23–25, 27, 44–46, 70, 177, 182, 184–187, 241, 277
  concept of, 23–25
  confirmation of, 71
  continental drift and, 76–79, 129–130
  contraction theory and, 35, 48, 51–52, 55–56
  creep and, 69–70
  deflections of the vertical and, 39–44
  density deficiencies and, 44–45
  Dutton and, 25, 29–33, 35–36, 67–68, 73–74, 129
  in early twentieth century, 129–130
  equilibrium and, 169, 236
  field mapping and, 51–53
  Fisher and, 25–29, 35–36, 67, 74, 241
  geodetic data of, 70
  geological theory and, implications for, 47–48, 56
  geosyncline theory and, 73–74
  Gilbert and, 33–39, 47, 68, 149, 182, 241
  gravity and, 36–48
  Gutenberg and, 223
  Hall and, 73
  Hayford and, 37, 39–47, 70, 129, 149, 172, 174–177, 182–184, 238–239
  historical geology and, 56, 167, 192
  hypothesis or fact and, 148–150
  Joly and, 29, 47, 105, 107–108, 149
  limitations of, 34, 39
  mobile substrate and, 68, 71–74, 76–77
  oceanic crust and, 79
  Pratt-Hayford model of, 172, 174–177

Pratt model of, 36, 77, 172–177, 182, 197, 241, 259–260, 276–277, 313–314, 316
  radioactivity and, 48–52
  reduction and, 47
  reexamination of, 181–188
  representations of, 175–177
  restoration and, 34
  seismic studies and, 254–255
  sunken continents and, 11, 55
  terminology of, 149
  terrestrial heat and, 71–72
  triangulation and, 37–39
  truth and, scientific, 35–39, 45–47
  underdetermination and, 276–278
  undertow and, 69–70, 96
  Wegener and, 76–79
  Willis and, 35, 48, 208–211
  Woodward and, 35, 40
  yielding substrate and, 67–68
Isthmian links, 214–219, 226, 277–278, 308
"Isthmian Links" (Willis), 213–214
IUGG (International Union of Geodesy and Geophysics), 45, 231, 233, 235

Jamieson, T. F., 74
Java Trench, 238–239, 244, 248
Jeans, J. H., 86, 111
Jefferson, Thomas, 37, 128
Jeffreys, Harold, 47, 86, 90, 93–94, 112, 212, 245, 248, 251, 264, 266, 279, 305
Jigsaw-puzzle fit of continents, 57–58, 60, 154, 178, 303–304
Joly, John
  biographical background of, 99–102
  causal mechanism and, 64
  continental drift and, 110, 124, 230;
    evidence of, 230
  convection and, 103–104
  drift mechanisms and, 99–112, 119
  fluid substrate and, 108–112, 119
  geological tradition and, 279
  isostasy and, 29, 47, 105, 107–108, 149
  Longwell and, 7, 120
  periodic fusion and, 99–102, 119
  plate tectonics and, 103–104, 107–108
  radioactivity and, 49–52, 102–104
  uniformitarianism and, 108
  van der Gracht and, 126
Joly process, 101
Justification, inductive logic of, 142–144

Kant-Laplace nebular hypothesis, 14
Karoo Basin, 92, 147, 157–167, 192
Kay, Marshall, 63–64, 243
Kelvin, Lord
  age of earth and, 48–50, 102, 118, 126, 130, 203, 207, 278
  Dutton and, 30, 33
  fluid substrate and, 307–308
  followers of, 108
  geophysics and, 86, 280
  instrumentation and, 301, 307
  radioactivity and, 103
  solidity of earth and, 26–27, 29–30, 66–68
  thermodynamic view of earth and, 132, 203
Kevles, Daniel, 128, 152, 278
Keweenawan floods and basalts, 108, 190
K-II submarine expedition, 238
Kinematics, dynamics versus, 65–66
King, Clarence, 128, 281, 284
King, Lester C., 296
Kirsch, G., 119
Knopf, Eleanora Bliss, 134
Knowledge, scientific. See Truth, scientific
Koch, J. P., 61, 230
Kuenen, Philip H., 249–251, 259, 296
Kuhn, Thomas, 304, 316
K-XIII submarine expedition, 238

Laboratory work, field mapping versus, 288–291
Lake Bonneville, 145, 182, 185
Lake Bonneville (Gilbert), 142
Lake, Philip, 123, 125
Lake Superior, 190
Lambert, Walter, 61–62, 64, 84–87, 89–91, 95–96, 123
Lamson, Eleanor, 239
Land bridges, 11, 55, 190, 211, 213–219, 264–265, 277–278, 307
Langevin, Paul, 102
Laplace, Pierre-Simon Marquis de, 14, 136, 203
Laudan, Rachel, 65, 145
Laurentia, 292, 295
Lawson, Andrew, 73, 119, 172, 191
Lead, 49–50
LeConte, Joseph, 66, 73, 118, 281, 284
Le Grand, Homer, 52, 126, 305
Leibniz, Gottfried Wilhelm, 14

Lemuria (sunken continent), 56
Lemurs, Madagascarian, 56
LePichon, Xavier, 272, 274
LeVene, Clara M., 180
Lithosphere, 182, 189, 191, 194, 272, 274
Localism, 52
Longitude measurement, 61, 231–236
Longwell, Chester
  Barrell and, 192
  continental drift and, 64, 125, 155, 280
  Daly and, 89, 99
  displacement theory and, 123
  du Toit and, 294–296
  Gondwana and, 191–193
  Holmes and, 7, 120, 193
  Joly and, 7, 120
  Schuchert and, 120
  speculation and, 149, 211
  Willis and, 211
Lord Kelvin and the Age of the Earth (Burchfield), 126
Lyell, Charles, 12–13, 74, 103, 139, 200–201, 278, 298

MacKenzie, Dan, 272
Madagascar (island), 56
Magmatic streaming, system of, 292
Magnetism, 263–267, 309–311
Magnetometers, 263–264
Maienschein, Jane, 152
Manning, Thomas, 240
Manual of Geology (De la Beche), 12
Mapping. See Field mapping
Maria, lunar, 190
Marine deposits, 11, 16–17, 58, 189, 199
Marvin, Ursula, 64
Matthews, Drummond, 271
Maxwell, John, 267
Mayr, Ernst, 309
Mechanics of Appalachian Structure (Willis), 70
Mechanism issues. See also Drift mechanisms
  causal, 6, 62–65
  conceivable, 65
  Daly and, 90, 93–99
  lack of, 292
  physical, 65–66
  plausible, 118–119
Memoirs of an Unrepentant Field Geologist (Pettijohn), 289

Menard, Henry William, 223–229, 267, 270, 272, 277, 299, 306–308
Mendenhall, W. C., 37
Merriam, John, 132–133, 148–149, 158–161, 239, 284, 294
Method. See Scientific theory and method
Meyerhoff, Arthur, 270
Mica, 50–51
Michelson, A. A., 283
Microseisms, 223
Mill, John Stuart, 313, 315
Mississippi Delta, 241
Mobile substrate, 68, 70–74, 76–77, 79–80
Mohorovicic discontinuity, 259
Molengraaf, G. A. F., 92, 126, 201, 238, 240, 276
Moon, 48, 190
Morgan, Jason, 272
Morley, Lawrence, 271
Moulton, F. R., 86, 130
Mountain belts, 13–14, 17, 28. See also specific mountains
Mountain-building. See also specific mountains
  Airy's theory of, 24–25, 28, 36
  Appalachians and, 17–18, 23
  continental drift and, 59, 84, 224–227
  contraction theory and, 19, 21, 251
  Daly and, 123
  Dana's theory of, 18–19, 59, 63
  geosyncline theory and, 17–19
  gravity and, 36
  Hall's theory of, 18, 63
  Menard and, 225
  mountain belts and, 13–14, 17, 28
  multiple working hypotheses of, 229
  nappe formation and, 22–23
  pentagonal network and, 13–14
  Pratt's theory of, 24–25, 28, 36
  radioactivity and, 103
  Rogers brothers' theory of, 17–18
  roots of mountains hypothesis and, 24–25, 172–173
  sediment accumulation and, 18
  Suess and, 59
  Swiss Alps and, 21–23
  uniform depth of compensation hypothesis and, 24–25
Multiple working hypotheses, 136–138, 140–141, 147, 151, 227, 229, 282–283
Murchison, Roderick, 299

Nansen, Fritjof, 76, 239
Nappe formation, 22–23
Nares Deep, 240–241
Natural selection, 11, 52, 276
Nebular hypothesis for origin of solar system, 130
Negative gravity anomalies, 248
Neptunism, 134–137, 139
Neptunist-plutonist debate, 134–137, 204
Newton, Isaac, 87, 136
Nile Delta, 239
Noncrystalline materials, 93–94
Nonsectarianism, 152
North American Datum, 39
North American, Europe comparison with, 57
NSF (National Science Foundation), 290–291
Numbers, Ronald, 205

Objectivity, 298–300
Observation, interpretation versus, 145–147, 149, 152
Occurrence, causation versus, 63
Ocean basins and deeps, 117, 188, 206, 208–209, 278
Oceanic crust, 77–80, 94, 97
Oceanic sinking, 16
Oceanography, 289
  advisory committee on, 262
Office of the Naval Research (ONR), 263
Oldroyd, David, 290, 316
Oliver, Jack, 272
"On the Form and Position of Sea Level" (Woodward), 68
ONR (Office of Naval Research), 263
Opdyke, Neil, 271
Operationalism, 152
Origin of Continents and Oceans, The (Wegener), 9–10, 54, 59, 65, 76–79, 81–84, 87, 91, 115, 153, 157, 226, 230
"Origin of Forces Responsible for Disruption of Continents, Mountain Building, and Continental Drift, The" (Chhibber), 293
Origins of American Science, The (Struik), 127
Orogeny, 30–31, 104, 107, 229, 270
Oscillations of earth, 66–67, 131, 149
Our Mobile Earth (Daly), 95–96, 126, 149, 193, 211

Our Wandering Continents (du Toit), 292, 294
Outcrop geology, 289–290
Overthrusting, 17–18, 64, 84, 90, 115–116, 196–197

Paleomagnetism, 265, 271, 310–311
Paleontological evidence, 55–57, 276
Paleozoic period, 216
Panama, isthmus of, 211, 214
Park, Charles, 290
Patagonia, 165
Pauly, Philip, 300
Peirce, Benjamin, 37
Peirce, Charles Saunders, 37
Pentagonal network, 13–14
Periodic fusion, 99–102, 107, 119, 126
Permanence theory, 17–19, 55–56, 130–131, 195, 208, 254
Permian period, 211, 213–214, 264
Permo-Carboniferous glaciation, 207
Petrogenesis, 137
Pettijohn, Francis J., 289–290
Physical mechanism, 65–66
Physical upheaval, periodic, 14. *See also* Continental drift; Contraction theory
Physics of the Earth's Crust (Fisher), 26
Pickering, W. H., 54, 129

Piggot, Charles, 254, 257
Pike's Peak, 38–39
Pirsson, L. V., 179, 182–183
Pitman, Walter, 271
Plantesimals, 130
Plantesimal theory, 98, 130–132
Plastic flow in rocks, 72
Plastic substrate, 69–70, 72, 77, 223, 241
Plate tectonics
  American earth science and, 127
  continental crust and, 79–80
  continental drift and, 64–65, 120, 275
  convection theory and, 103–104
  Dana and, 10, 14–20, 129
  development of, 126–127
  displacement hypothesis and, 9–10, 54
  evidence of, 273
  geological standards and, 288
  geosyncline theory and, 17–19
  Gilbert and, 168
  instrumentation and, 311–312
  Joly and, 103–104, 107–108

Plate tectonics (*continued*)
  juxtaposing views of, 19–20
  oceanic crust and, 79–80
  radioactivity and, 102–104
  rejection of, 9–10
  secular cooling and, 12–14, 19
  Suess and, 10–12, 14, 19–20
  theory of, 9
  Wegener and, 9–10, 275
  Wood and, 127
Playfair, John, 134
Pleochroic haloes, 50–51, 102
Pluralism, theoretical, 136–138, 153, 155–156, 206–207
Poland, Joseph, 218
Pole-flight force, 84–86, 89
Popper, Karl, 304
Porter, Louise Haskell, 89
Porter, Theodore, 300
Positive gravity anomalies, 239, 241–242
Positive Philosophy, 303
"Possible Origin of Oceans and Continents" (Bowie), 170–171
Postglacial rebound, 74, 76
Powell, John Wesley, 29, 129, 136, 284
Pratt, John
  contraction theory and, 48
  Himalayas analysis and, 23, 27
  isostasy and, 36, 77, 172–177, 182, 197, 241, 259–260, 276–277, 313–314, 316; compensation, isostatic, 24–25, 27, 44–45
  mountain-building and, 24–25, 28, 36
Precipitate explanation, 138
Press, Frank, 62
Price, George McCready, 205
Principles of Geology (Lyell), 74, 139
Principles of Physical Geology (Holmes), 112–113, 264, 268
"Problems of Geology, The" (van Hise), 281
Proceedings of the American Philosophical Society (Berkey), 206
Processes and Their Results (Chamberlin, T. C.), 282
Process-oriented geology, 282
Progressionism, 203
Progressive cooling, 115, 203
Provincialism, 52
Putnam, G. R., 38–39, 69
Pyne, Stephen, 140

Qualitative data and methods, 296–298
Quantitative data and methods, 280–284, 296, 296–298
Quine, W. V. O., 303

Radioactivity
  creation of, 104
  discovery of, 48–49
  field mapping and, 51–53
  Holmes and, 207
  isostasy and, 48–52
  Joly and, 49–52, 102–104
  Kelvin and, 102, 103
  mountain-building and, 103
  plate tectonics and, 102–104
  Willis and, 51, 213
"Radioactivity and Earth Movements" (Holmes), 198, 270
Radioactivity and Geology (Joly), 103
Radiogenic heat. *See* Radioactivity
Radio telegraphy, 61, 230–236, 275
Radium, 50
Rafting (dispersal of species), 295
Rambaut, Arthur, 101
Read, H. H., 264–265, 298
"Recent Trends in Geophysical Research" (Bowie), 259
Red Sea, 189–191, 235
Reduction, isostatic, 47
Reichenbach, Hans, 4
Reid, Harry Fielding, 157
Reingold, Nathan, 128, 278, 283
Remanent magnetism, 263–267
Remson, Ira, 129, 152
Report on the Geology of the High Plateaus of Utah (Dutton), 29
Representation as reified belief, 175–177
Research in China (Willis), 143, 213
Researches in Theoretical Geology (De la Beche), 12
Réseau pentagonal (pentagonal network), 13–14
Residual gravity anomalies, 184–187
Restoration, isostatic, 34
Revelle, Roger, 267
Richter, Charles F., 253, 277
Rift valleys, 189
Rigidity of earth, 33–34, 66, 86, 93, 224
Rocky Mountains, 29, 30, 38, 83, 103
Rogers, Henry Darwin, 17–18, 128–129
Rogers, William B., 17–18, 128–129

Roots of mountains hypothesis, 24–25, 172–173
Rothpletz, August, 22
Rudwick, Martin, 63, 200–201, 290, 299
Ruling theory, 138–139, 142
Runcorn, Stanley Keith, 112, 264–268, 309
Rutherford, Ernest, 48, 50, 102–103, 113

Salisbury, R., 69
San Andreas Fault, 233
Sarton, George, 303
Saull, Vincent, 65
Schuchert, Charles
  Baconian overtones of, 156
  Barrell and, 188, 210
  biographical background of, 179–181
  Bowie and, 123, 167, 171–172, 178–179, 191, 197, 212–213
  continental drift and, 157, 178–179, 196–202, 234, 308
  death of, 262
  du Toit and, 294
  faunal and floral similarities and, 56–59
  geological data and, 277–278
  Gondwana and, 179–181, 183, 191–196, 213–214, 218
  gradualist approach and, 151–152
  Haarman and, 198–199
  Hall and, 133
  Holmes and, 148, 183, 192–193, 196, 198–199, 202
  jigsaw-puzzle fit of continents and, 305
  land bridges and, 264–265
  Longwell and, 120
  Pirsson and, 179–180
  scientific theory and method and, 156
  Tennessee trials and, 205
  uniformitarianism and, 199–202, 207–217, 287
  van der Gracht and, 196–197
  Vaughan and, 226
Schwinner, R., 119
Scientific theory and method. *See also* Earth science
  in American earth science, 127–134, 157
  antiauthoritarian logic of discovery and, 134–136
  Barrell and, 144
  in British earth science, 124–125
  Chamberlin (T. C.) and, 136–142, 144–145, 148, 281–283
  conceptual framework and, 144–150
  Dana and, 281
  data collection and, 298–300
  deduction and, 135, 304–305
  distance and, 298–303
  diversity of, 5
  Dutton and, 136, 142, 281
  egalitarianism in, 135–137, 151–152
  epistemic status of, 5
  false, 315–316
  Gilbert and, 140–142, 144–145, 156, 281–282
  gradualist approach and, 150–152
  hypothetico-deductive method of science and, 141, 157
  induction and, 303–304
  inductive logic of justification and, 142–144
  interpretation versus observation and, 145–147
  interpretive framework and, 142–144
  multiple working hypotheses and, 136–138, 140–141, 147, 282–283
  neptunist-plutonist debate and, 134–137
  nonsectarianism and, 152
  objectivity and, 298–300
  pluralism in, 136–138, 153, 155–156, 206–207
  qualitative, 296–298
  quantitative, 234, 280–284, 296–298
  ruling theory and, 138–139, 142
  Schuchert and, 156
  self-validating power and, 304–308
  singular, myth of, 5
  social context of, 4–5
  underdetermination and, 138
  Wegener and, 87, 123, 126, 153–156
  Willis and, 136, 142–144
  Wright (Frederick) and, 284
Sclater, Philip, 56
Scopes, John Thomas, 204–206
Seafloor spreading, 267–271, 310–311
  confirmation of, 271–273
Sealing wax analogy, 225
Secord, James, 290
Secular cooling, 12–14, 19, 30, 130
Sedgwick, Adam, 12, 25
Sediment accumulation, 18, 73, 96, 104
Seismic waves and studies, 86, 108, 254–255
Self-ratification, 317–318

Self-validating power of success, 304–308
Sensitivity analysis, 185
Servos, John, 302
Shaler Memorial Expedition, 91–92
Sierra Nevada, 30
Siever, Raymond, 62
Sigsbee, Charles, 240
Sigsbee Deep, 240
Silliman, Benjamin, 14–15, 135
Simpson, George Gaylord, 295
Singewald, Joseph, 126
Skerl, J. G. A., 154
"Slab-pull" model of subduction, 87–89, 94–95, 97
Sliding gravity, 87–89, 94–95, 97, 119
Smithsonian Institution, 133
Snake River Plain, 271
Snellius expedition, 249
Social context of science, 4–5
Society for the Promotion of Engineering Education, 302
Soddy, Frederick, 48, 102
Solar system, origin of, 130
Solid earth accretion theory, 49, 98
Solidification of subcrust, 107
Solidity of earth, 26–27, 29–30, 66–68, 93
Sollas, Hertha, 10
Sollas, W. J., 110–111
South Africa, South America comparison with, 157–167, 240, 276
South America, South Africa comparison with, 157–167, 240, 276
Static earth concept, 251
Steinman, G., 196
Stewart, John, 127, 305–307
"Strength of the Earth's Crust, The" (Barrell), 182
"Strong crust" school, 182, 188, 242
"Structure of Continents and Ocean Basins" (Field), 259
Struik, Dirk J., 127
Strutt, R. J., 49–50, 103–104, 113–114
Sub-crustal convection currents. *See* Convection theory
Subduction, 119
Submarine ridge, 211
Sudbury intrusion of Ontario, 190
Suess, Edward
    contraction theory and, 11–12, 15, 17, 21, 23, 48
    Fennoscandian uplift and, 75
    Gondwana and, 55, 93, 157–158
    mountain-building and, 59
    plate tectonics and, 10–12, 14, 19–20
    ruling theory and, 139
    uniformitarianism and, 19
    Wegener and, 153
Surface-History of the Earth, The (Joly), 109, 111–112
Swiss Alps, 21–23, 32, 51, 82, 103–104, 254
Sykes, Lynn, 272
System of Mineralogy (Dana), 14

Taconic controversy, 204
Tangential movement and pressure, 17, 208
Taylor, Frank Bursey, 54, 81–82, 87, 124, 129, 202
Taylor-Wegener hypothesis, 82. *See also* Continental drift
Tectogenes, 243–249, 254–256
    experimental support for, 249–253
Tennessee trials, 204–206
Termier, Pierre, 22, 136, 280
Terrestrial degradation, 34
Terrestrial heat, 71–72
Tertiary basalts, 105, 271
Textbook of Geology, A (Pirsson and Schuchert), 179–180
Textbook of Geology (Dana), 15
Tharp, Marie, 267, 272
Theory-driven science, 284–285
Theory. *See* Scientific theory and method; *specific types*
Theory versus hypothesis, terminology, 148–150
Thermal contraction. *See* Contraction theory
Thermal cycles, 107
Thermal equilibrium, 115
Thermodynamics, 132, 203
Thermodynamik der Atmosphäre (Wegener), 58
Thomson, William. *See* Kelvin, Lord
Tidal forces, 26–27, 66, 82–83, 90, 109–111
Tidal retardation, 90
Time scale, geologic, 114, 120, 275–276
Topographic corrections, 39–44, 173
Torridonian sandstones, 265
Transform faults, 272
Triangulations, transcontinental, 37–39
Trigonometrical Survey, 52

Truth, scientific
  certainty through measurement and, 280–284
  continental drift and, 5–6
  discovery and, context of, 4
  epistemological affinity and, 51–53
  expertise and, 317–318
  historicity and, 317
  human factors and, 4
  isostasy and, 35–39, 45–47
  search for, 3–6
  self-ratification and, 317–318
  social context of, 4–5
  unmaking of, 316
  utility and, 313–318
"Two Antagonist Doctrines of Geology, The" (Whewell), 13
Tyndall, John, 279

Underdetermination, 138, 276–278
Underthrusting, 250, 259
Undertow, isostatic, 69–70, 96
Uniform depth of compensation hypothesis, 24–25
Uniformitarianism
  Barrell and, 206
  catastrophism and, 203
  Chamberlin (T.C.) and, 206
  continental drift and, 83, 199–200, 207
  crustal motion and, 83
  du Toit and, 292–293
  earth science and, 204–207
  geology and, 204–207
  Gilbert and, 206
  Joly and, 108
  laws and rules of, 202–207
  Lyell and, 12–13
  preserving practice of, 200–207
  Schuchert and, 199–202, 207–217, 287
  solution of, 207–213
  Suess and, 19
  uniformity and, 202–203
  van Hise and, 285
  Willis and, 208–217
Uniformity, 202–203
United States Standard Datum, 39
Untersuchungen über den Mechanismus der Gebirgsbildung (Heim), 23
Upthrusting, 216
Ural Basin, 195
Uranium, 104–105

Uranium decay, 50
Uranium-lead technique, 49–50
U.S. Coast and Geodetic Survey, 37–40, 127–128, 133, 233, 236, 238, 243, 280, 302
USGS (U.S. Geological Survey), 133
U.S. Navy studies, 231, 233, 262–263, 288–289, 302–303
USS *Cape Johnson* expedition, 262
USS *S-21* expedition, 236–244, 253, 276, 303
USS *S-48* expedition, 244, 253
Utility, truth and, 313–318
Uyeda, Seiya, 62–63

Vacquier, Victor, 272
van der Gracht, W. J. A. M. van Waterschoot, 84, 125–126, 196, 299
van Hise, Charles, 141, 281–285, 287, 289
Vaughan, T. Wayland, 226
Vening Meinesz, Felix A.
  convection theory and, 251, 265, 268
  crustal root versus underthrusting and, 259
  du Toit and, 296
  gravity anomalies and, 111–112, 239, 242–245, 248–249, 257
  gravity measurements and, 236, 238–244, 301, 303
  Griggs and, 251
  papers of work and, 253
  USS *S-21* expedition and, 236, 238–244, 303
  in World War II, 295
Vertical tectonics, 110
Vine, Frederick, 271
Volcanoes, 11, 14, 28–29, 66, 94, 106
von Buch, Leopold, 139
von Ihering, Herbert, 56

Walcott, Charles Doolittle, 132, 283–284
Watts, W. W., 113
"Weak crust" school, 182, 188, 242
Wegener, Alfred
  continental drift and, 5–6, 54–62, 77, 80, 83, 93, 124, 153–155, 167, 224–225, 260–261, 267: evidence of, 56–59, 124–125, 227–230, 264, 273, 275–276, 309; rejection of, 123–124
  death of, 123
  Fennoscandian uplift and, 74, 76

Wegener, Alfred (*continued*)
  fluid substrate and, 65–66, 76
  glacial deposits and, 125
  Holmes and, 270
  horizontal displacement and, 77
  isostasy and, 76–79
  jigsaw-puzzle fit of continents and, 57–58, 60, 154, 304–305
  land bridges and, 307
  plate tectonics and, 9–10
  scientific theory and method and, 87, 123, 126, 153–156
  sealing wax analogy and, 224
  Suess and, 153
  westward drift and, 242
Wegener, Kurt, 59
Weltall theories, 139
Werner, Abraham Gottlob, 63, 134–135, 139
West Indies, 248
Westward drift, 242
Wheeler, G. M., 33
Wheeler Survey, 33
Whewell, William, 13, 203
Whitehead, Alfred North, 287
Whitten, Charles, 236
Wilkes expedition, 14–15, 17, 281
William Bowie Medal, 167
Willis, Bailey
  American earth science and, 129
  Appalachians and, structure of, 224
  Chamberlin (T. C.) and, 70, 136, 150–151
  continental drift and, 86–87, 126, 143–144, 296, 308
  contraction theory and, 35
  drift mechanisms and, 86–87, 264–265
  du Toit and, 293–294, 298
  employment of, geological, 134
  Gondwana and, 208–209, 213–214, 218
  Holmes and, 212–213
  isostasy and, 35, 48, 208–211
  land bridges and, 264–265
  Longwell and, 211
  ocean basins and, 278
  oceanic crust and, 97
  permanence theory and, 17, 55, 208
  plantesimals and, 130
  plantesimal theory and, 98
  radioactivity and, 51, 213
  scientific theory and method and, 136, 142–144
  uniformitarianism and, 208–217
Wilson, J. Tuzo, 127, 272–273
Witting, R., 75
Wood, Harry O., 83, 172
Wood, Robert Muir, 127
Woodward, Robert, 35, 40, 68–69, 128, 135–136, 283–284
Worzel, J. Lamar, 244
Wright, Frederick E., 91–92, 132–133
  continental drift and, 157–161
  du Toit and, 294
  Hayford and, 236
  scientific theory and method and, 284
  USS S-21 expedition and, 203, 238–243
Wright, W. B., 124

Yielding substrate, 67–68

Zircon, 50
Zittel, Karl, 11–12, 21
Zone of flow, 23
Zone of fracture, 23

Printed in the United Kingdom
by Lightning Source UK Ltd.
103781UKS00001B/292